OT 41
Operator Theory: Advances and Applications
Vol. 41

Birkhäuser Verlag
Basel · Boston · Berlin

The Gohberg Anniversary Collection

Volume II: Topics in Analysis and Operator Theory

Edited by

H. Dym
S. Goldberg
M. A. Kaashoek
P. Lancaster

1989

Birkhäuser Verlag
Basel · Boston · Berlin

Volume Editorial Office:

Department of Mathematics
and Computer Science
Vrije Universiteit
Amsterdam, The Netherlands

CIP-Titelaufnahme der Deutschen Bibliothek

The **Gohberg anniversary collection** / [vol. ed. office: Dep. of Mathematics and Computer science, Vrije Univ., Amsterdam, The Netherlands]. Ed. by H. Dym ... – Basel ; Boston ; Berlin : Birkhäuser
(Operator theory ; ...)
ISBN 3-7643-2283-7 (Basel ...)
ISBN 0-8176-2283-7 (Boston)
NE: Dym, Harry [Hrsg.]; Vrije Universiteit <Amsterdam> / Faculteit der Wiskunde en Informatica; Gochberg, Izrail': Festschrift
Vol. 2. Topics in analysis and operator theory. – 1989

Topics in analysis and operator theory / [vol. ed. office: Dep. of Mathematics and Computer Science, Vrije Univ., Amsterdam, The Netherlands]. Ed. by H. Dym ... – Basel ; Boston ; Berlin : Birkhäuser, 1989
(The Gohberg anniversary collection ; Vol. 2)
(Operator theory ; Vol. 41)
ISBN 3-7643-2308-6 (Basel ...) Pb.
ISBN 0-8176-2308-6 (Boston) Pb.
NE: Dym, Harry [Hrsg.]; 2. GT

Printed in Germany on acid-free paper
ISBN 3-7643-2283-7 (complete set)
ISBN 3-7643-2307-8 (vol. 1)
ISBN 3-7643-2308-6 (vol. 2)
ISBN 0-8176-2283-7 (complete set)
ISBN 0-8176-2307-8 (vol. 1)
ISBN 0-8176-2308-6 (vol. 2)

Table of contents of Volume II

Israel Gohberg

Editorial Preface

The Gohberg Anniversary Collection is dedicated to Israel Gohberg. It contains the proceedings of the international conference on "Operator Theory: Advances and Applications", which was held in Calgary, August 22 - 26, 1988, on the occasion of his 60th birthday. The two volumes present an uptodate and attractive account of recent advances in operator theory, matrix theory and their applications. They include contributions by a number of the world's leading specialists in these areas and cover a broad spectrum of topics, many of which are on the frontiers of reseach. The two volumes are a tribute to Israel Gohberg; they reflect the wide range of his mathematical initiatives and personal interests.

This second volume consists of original research papers on linear operator theory, on nonlinear Toeplitz operators and nonlinear lifting theorems, on complex function theory, on numerical analysis, and on applications of operator theory to problems in analysis, in the theory of differential equations, in control theory, and in mathematical physics. Recent theoretical developments are presented as well as new advances in the study of specific classes of operators.

The Editors are grateful to the Department of Mathematics and Computer Science of the Vrije Unversiteit for providing secretarial assistance. In particular, they thank Marijke Titawano for her work.

The Editors

Operator Theory:
Advances and Applications, Vol. 41

DUALITY METHODS FOR THE BOUNDARY CONTROL OF SOME EVOLUTION EQUATIONS.

Frederic Abergel and Roger Temam

Dedicated to Israel Gohberg on the occasion of his 60th birthday

We present some new and simple proofs, using convex analysis, for some results related to the boundary control of parabolic and hyperbolic evolution equations.

INTRODUCTION

The quadratic cost problem for the boundary control of evolution equations has been extensively studied, and, thanks to recent regularity results [L-2] - [L-T] - [L-L-T], its study has made important progress. The main results concerning these problems are to be found in [L-1] - [L-2] - [L-T], in which the optimality conditions are derived, some regulartiy results for the optimal control are proven, and the existence of a pointwise feedback operator is established.

It has appeared to us that, for the first part of the results evoked above, i.e. the existence and regularity of the optimal control as well as the system of optimality conditions, the use of classical duality methods in Convex Analysis provides us with straightforward and simpler proofs of these results; we therefore find it interesting to expose how these mehtods apply for such problems.

We shall deal with the boundary control problem for parabolic and hyperbolic dynamics. In the parabolic case, we recover the existence and regularity results of [L-2], and our proofs are essentially self contained. In the hyperbolic case, we have to restrain ourselves to the finite time interval problem, recovering only partly the results of [L-T]; moreover, our proofs rely crucially on the recent regularity results in [L-L-T].

In the first two sections, we are interested in the parabolic dynamics:

$$(1.1)\qquad \begin{cases} (\frac{\partial}{\partial t} - \Delta)\ y = o \text{ in } Q_T = \Omega \times (0,T), \\ y \qquad = u \text{ on } \sum_T = \Gamma \times (0,T), \\ y(o) \qquad = y_0 \text{ in } \Omega, \end{cases}$$

where Ω is a bounded open set of $\mathbb{R}^n, N \geq 2$, with a smooth boundary Γ, y_o is given in $L^2(\Omega)$, and the boundary control u is in $L^2(\Sigma_T)$. T is an (extended) real number, strictly larger than zero. We are interested in the quadratic cost problem:

$$(P_T)\ (1.2)\qquad \begin{cases} \text{To find } U \text{ in } L^2\ (\Sigma_t), \text{minimizing the cost function} \\ y^T(u) = \frac{1}{2}\ \|y(u)\|^2_{L^2(Q_T)} + \frac{1}{2}\ \|u\|^2_{L^2(\Sigma_T)} \\ \text{with } y(u) \text{ solution of (1.1)} \end{cases}$$

Our results will be obtained by using duality methods [E-T] for (P_T), and are summed up in the

THEOREM A *Let* u *be in* $L^2(\Sigma_T)$, y(u) *be the solution of* (1.1);

(a) (u,y(u)) *is an optimal pair if and only if there exists* f *in* $L^2(Q_T)$ *such that*

$$y(u) = -f \quad \text{in } Q_T \qquad (A.1)$$

$$u = \frac{\partial}{\partial \nu}\ (v(f)) \quad \text{on } \Sigma_T \qquad (A.2),$$

where v(f) *is the solution of the adjoint evolution problem*:

$$(1.3)\qquad \begin{cases} (\frac{\partial}{\partial t} + \Delta)\ v = f & \text{in } Q_T, \\ v = o & \text{on } \Sigma_T, \\ v(T) = o. \end{cases}$$

(b) *the optimal control* $\bar{u}$ *is in* $L^2(o,T;H^{\frac{1}{2}}(\Gamma))$.

Due to technical differences, the cases $T < +\infty$ and $T = +\infty$ will be dealt with separately.

In the third section, we consider the hyperbolic dynamics

$$(1.4)\qquad \begin{cases} (\frac{\partial^2}{\partial t^2} - \Delta)\ z = 0 \quad \text{in } Q_T, \\ (z(0), \frac{\partial z}{\partial t}(0)) \text{ given in } L^2(\Omega) \times H^{-1}(\Omega), \\ z = u \text{ on } \Sigma_T. \end{cases}$$

Here, we restrict ourselves to $T < +\infty$; we study the associated quadratic cost problem.

$$(1.5)\qquad (Q_T)\qquad \begin{cases} \text{To find } u \text{ in } L^2(\Sigma_T) \text{ minimizing the cost function} \\ K^T(u) = \frac{1}{2}\|z(u)\|^2_{L^2(Q_T)} + \frac{1}{2}\|u\|^2_{L^2(\Sigma_T)} \\ \text{with } z(u) \text{ solution of (1.4).} \end{cases}$$

We proceed to prove the following result:

THEOREM B: *Let* u *be in* $L^2(\Sigma_T)$, *and* $z(u)$ *be the solution of* (1.4); $(v,z(v))$ *is the optimal pair for* (Q_T) *if and only if there exists* f *in* $L^2(Q_T)$ *such that:*

$$u = -\frac{\partial}{\partial \nu}(p(f)) \quad \text{on } \Sigma_T \qquad (B.1)$$

$$y(u) = f \quad \text{in } Q_T \qquad (B.2)$$

with $p(f)$ *solution of the adjoint evolution problem:*

$$(1.6)\qquad \begin{cases} \left(\dfrac{\partial^2}{\partial t^2} - \Delta\right) p = f & \text{in } Q_T \\ \qquad p = 0 & \text{on } \Sigma_T \\ \left(p(T), \dfrac{\partial p}{\partial t}(T)\right) = (0,0) & \text{in } \Omega \end{cases}$$

The plan of the article is as follows:

I THE PARABOLIC BOUNDARY CONTROL PROBLEM ON A FINITE TIME INTERVAL

1. Duality and Green's formula

We follow the presentation of [E-T]: let Z be $L^2(\Sigma_T)$, Y be $L^2(\Sigma_T) \times L^2(Q_T)$ and Λ be the operator

$$\Lambda: \quad Z \to Y$$
$$u \mapsto (u, y(u)),$$

where $y(u)$ is solution of (1.1).

We set: $F \equiv 0,\ G(p_1, p_2) = G_1(p_1) + G_2(p_2)$

with:

$$G_1(p_1) = \frac{1}{2}\|p_1\|^2_{L^2(\Sigma_T)} \tag{I.1.1}$$

$$G_2(p_2) = \frac{1}{2}\|p_2\|^2_{L^2(Q_T)}. \tag{I.1.2}$$

Problem (P_T) is obviously equivalent to:

$$(\mathcal{P}_T) \qquad \underset{u \in Z}{\operatorname{Inf}} \left\{ F(u) + G(\Lambda u) \right\}, \tag{I.1.3}$$

and its dual problem $(\mathcal{P}_T{}^*)$ has the following form:

$$(\mathcal{P}_T{}^*) \qquad \underset{P \in Y}{\operatorname{Sup}} \left\{ - G^*(-p) - F^*(\Lambda^* p) \right\},$$

F^* (resp. G^*) being the conjugate function of F (resp.G), and Λ^*, the transposed operator of Λ. In (I.1.4), the supremum is taken on Y, for the latter is identified with its dual space Y^*, thanks to its Hilbertian structure. Let us now determine the expression of F^* and G^*; computing G^* is easy, and we find:

$$G^*(p_1,p_2) = G(p_1,p_2) = \frac{1}{2}\|p_1\|^2_{L^2(\Sigma_T)} + \frac{1}{2}\|p_2\|^2_{L^2(Q_T)}. \qquad (I.1.5)$$

For F^*, we have:

PROPOSITION I.1.1: *Let* y_h *be the solution of* (1.1) *with* $u \equiv 0$; *for* $p = (p_1,p_2)$ *in* Y, *and* $v(p_2)$ *being the solution of* (1.3) *associated to* p_2, *we have*:

$$F^*(\Lambda^*p) = \begin{cases} \int_{Q_T} p_2 y_h \, dxdt & \text{if } p_1 = -\frac{\partial}{\partial\nu}(v(p_2)) \text{ on } \Sigma_T, \\ +\infty & \text{otherwise.} \end{cases}$$

PROOF: We write:

$$F^*(\Lambda^*p) = \operatorname*{Sup}_{u \in Z}\left(\int_{Q_T} p_2 y(u)\, dxdt + \int_{\Sigma_T} u.p_1 \, d\sigma dt\right)$$

$$= \operatorname*{Sup}_{u \in Z}\left(\int_{Q_T} (\frac{\partial}{\partial t} + \Delta)(v(p_2)).y(u).dxdt + \int_{\Sigma_T} u.p_1 d\sigma dt\right).$$

Proposition I.1.1 will then follow directly from the

LEMMA I.1.2: *Let* y(u) (*resp.*v(f)) *be the solution of* (1.1)(*resp.* (1.3)) *related to* u (*resp.* f); *we have the Green's formula*:

$$\text{(I.1.6)} \quad \int_{Q_T} (\frac{\partial}{\partial t} + \Delta)\, v(f).y(u).dxdt = \int_{\Sigma_T} \frac{\partial}{\partial\nu}(v(f)).\, u\, d\sigma dt + \int_{Q_T} (\frac{\partial}{\partial t} + \Delta)(v(f)).y_h dxdt.$$

PROOF: We set $y(v) = y_h + \tilde{y}(v)$ where $\tilde{y}$ (u) satisfies:

$$\text{(I.1.7)}\qquad \begin{cases} (\frac{\partial}{\partial t} - \Delta)\tilde{y}(u) = 0 & \text{in } Q_T, \\ \tilde{y}(u) = u & \text{on } \Sigma_T, \\ \tilde{y}(u)(0) = 0 & \text{in } \Omega. \end{cases}$$

We then approximate u (resp. p_2) by smooth, compactly supported functions u_n in $\mathcal{D}(\Sigma_T)$(resp.p_2^n in $\mathcal{D}(Q_T)$), and consider the solutions $\tilde{y}(u_n)$ and $v(p_2^n)$. Both are in $C^\infty(\overline{Q}_T)$, for the assumptions on the support of u_n and P_2^n ensures that all the necessary compatibility conditions are satisfied; hence, we have the Green's formula:

$$\int_{Q_T} (\frac{\partial}{\partial t} + \Delta)v(p_2^n).\tilde{y}(u_n)dxdt = \int_{\Sigma_T} \frac{\partial}{\partial \nu} (v(p_2^n)).u_n \, d\,\sigma\, dt \qquad \text{(I.1.8)}$$

(all the other terms being zero), and Lemma I.1.2 is then obtained through a straight forward passage to the limit in (I.1.8). One just has to notice that, according to [L-M], $v(p^n)$ converges to $v(p\ )$ in $L^2(0,T;H^2(\Omega))$, and therefore, $\frac{\partial}{\partial \nu}(v(p_2^n))$ converges to $\frac{\partial}{\partial \nu}(v(p_2))$ in $L^2(o,T;\ H^{\frac{1}{2}}(\Gamma))$.

We can now give the expression of $(\mathcal{P}_T{}^*)$:

$$(\mathcal{P}_T{}^*)\qquad \sup_{\begin{cases} p \in L^2(Q_T) \\ v \equiv v(p) \text{ solution of (1.3).} \end{cases}} \left\{ -\int_{Q_T} p\bullet y_h dxdt - \frac{1}{2}\|p\|^2_{L^2(Q_T)} - \frac{1}{2}\|\frac{\partial v}{\partial \nu}\|^2_{L^2(\Sigma_T)} \right\}. \qquad \text{(I.1.9)}$$

2. System of optimality conditions

We first state a simple result about $(\mathcal{P}_T)$ and $(\mathcal{P}_T{}^*)$

PROPOSITION I.2.1 (*i*) *There exists a unique optimal pair* $(\overline{y},\overline{u})$ *for* $(\mathcal{P}_T)$.

(*ii*) *There exists a unique optimal pair* $(\overline{v},\overline{p})$ *for* $(\mathcal{P}_T{}^*)$.

(*iii*) *We have the equality*: $\text{Inf}(\mathcal{P}_T) = \text{Sup}\,(\mathcal{P}_T{}^*)$.

PROOF: (*i*) and (*ii*) are the results for the minimization of a coercive strictly convex continuous functional on a Hilbert space; as for (*iii*), one can easily show that problem $(\mathcal{P}_T)$ is normal in the sense of [E-T]. We can now state and prove the main result of this section:

THEOREM I.2.1: (*i*) *Let* $(y(u),u)$ [*resp.*$(v(p),p)$] *be admissible for* $(\mathcal{P}_T)$ [*resp.* $\mathcal{P}_T{}^*)$]; *the necessary and sufficient conditions for* $(y(v),u)$ [*resp.* $(v(p),p)$] *to be an optimal couple for* $(\mathcal{P}_T)$ [*resp.*$(\mathcal{P}_T{}^*)$] *are*:

(I.2.*i*) $$u = \frac{\partial v}{\partial \nu} \quad \text{on} \quad \Sigma_T$$

(I.2.*ii*) $$y(u) = - p \quad \text{in} \quad Q_T.$$

(*ii*) *The optimal control* $\bar{u}$ *is in* $L^2(0,T;\ H^{\frac{1}{2}}(\Gamma))$

PROOF: for (I.2.*i*) and (I.2.*ii*), one can use [E-T], p.61; (*ii*) follows from (I.2.*i*), and the fact that $v(p)$ belongs to $L^2(0,T;\ H^2(\Omega))$, see [L-M].

II THE CASE "T = + ∞"

1. An existence result for a non well posed parabolic problem

In order to give a sense to the condition "$v(T) = o$", when $T = +\infty$, we prove the

LEMMA II 1.1 *Let* f *be given in* $L^2(Q_\infty)$; *there exists a unique* $v(f)$ *satisfying the following conditions*:

(*i*) $v(f) \in L^2(o, +\infty;\ H^2(\Omega)\cap H^1_o(\Omega)\Big)\cap H^1(o, +\infty;\ L^2(\Omega))$

(*ii*) $(\frac{\partial}{\partial t} + \Delta)v(f) = f$ in $L^2(Q_\infty)$.

(*iii*) $\lim_{t\to+\infty} \|v(f)\|_{L^2(\Omega)} = 0.$

PROOF: We first proceed to show the existence of $v(f)$; let f_n be the truncation of f

$$f_n \begin{cases} = f & \text{if } \quad o<t<n \\ = o & \text{if } \quad t \geq n, \end{cases} \qquad \text{(II.1.1)}$$

and let v_n be the solution of

$$\begin{cases} (\frac{\partial}{\partial t} + \Delta)\, v_n = f_n & \text{in } \; Q_n, \\ v_n = 0 \quad \text{on } \; \Sigma_n, \\ v_n(n) = 0 \quad \text{in } \Omega. \end{cases} \qquad \text{(II.1.2)}$$

We denote by $\tilde{v}_n$ the extension of v_n

$$\tilde{v}_n \begin{cases} = v_n & \text{if } \quad 0<t<n \\ = o & \text{if } \quad t \geq n \end{cases} \quad . \qquad \text{(II.1.3)}$$

Using [L-M], we have:

$$\tilde{v}_n \in X = L^2(o, +\infty; H^2 \cap H^1{}_o(\Omega)) \cap H^1(o, +\infty; L^2(\Omega)) \qquad \text{(II.1.4)}$$

$$|\tilde{v}_n|_X \leq c.\ |f|_{L^2(Q_\infty)}. \qquad \text{(II.1.5)}$$

The crucial point is that the constant c in (II.1.5) does not depend on n, for the operator $(-\Delta)$ is time-independent. Therefore, we can extract a subsequence weakly converging to some v in X. Moreover, the operator $(\frac{\partial}{\partial t} + \Delta)$, being a linear continuous operator from X into $L^2(Q_\infty)$, is weakly continuous, and we then have:

$$(\frac{\partial}{\partial t} + \Delta)\ v = f \text{ in } L^2(Q_\infty). \qquad \text{(II.1.6)}$$

We now prove the uniqueness of v : let v_1, v_2 be two elements of X, satisfying (II.1.6), and set $w = v_1 - v_2$.

From (II.1.6), with $f \equiv o$, we derive the energy equality

$$\frac{1}{2}\frac{d}{dt}\|w\|^2_{L^2(\Omega)} - \|\nabla w\|^2_{(L^2(\Omega))^3} = o. \qquad \text{(II.1.7)}$$

Integrating (II.1.7) with respect to time, and using the fact that w is in X, we have:

$$\lim_{(t_1,t_2)\to+\infty} \left(\|w(t_1)\|^2_{L^2(\Omega)} - \|w(t_2)\|^2_{L^2(\Omega)}\right) = 0. \qquad \text{(II.1.8)}$$

Therefore, there exists $\ell = \lim_{t\to+\infty} \|w(\epsilon)\|^2_{L^2(\Omega)}$. We now use

Poincarré's inequality to obtain from (II.1.7)

$$\frac{1}{2}\frac{d}{dt}\|w\|^2_{L^2(\Omega)} - c\|w\|^2_{L^2(\Omega)} \geq 0. \qquad \text{(II.1.8)}$$

Integrating between t and $+\infty$, which is legitimate, yields:

$$0 \leq \|w(t)\|^2_{L^2(\Omega)} \cdot e^{-2ct} \leq 0, \qquad \text{(II.1.9)}$$

which obviously implies that $w \equiv 0$.

Let us prove assertion (*iii*): first, we remark that $\|v(f)\|_{L^2(\Omega)}$ is a bounded function of time. As a matter of fact, $v(f)$ belongs to $H^1(0, +\infty; L^2(\Omega))$, which is continuously imbedded in $L^\infty(0, +\infty; L^2(\Omega))$. Keeping this fact handy, we go back to the equation (II.1.6), and derive the a priori inequality:

$$\frac{1}{2}\frac{d}{dt}\|v(f)\|^2_{L^2(\Omega)} - c\|v(f)\|^2_{L^2(\Omega)} \geq \int_\Omega f.v(f)\,dx$$

Using Young's inequality, we obtain:

$$\frac{d}{dt}\|v(f)\|^2_{L^2(\Omega)} - c\|v(f)\|^2_{L^2(\Omega)} \geq -\frac{1}{c}\|f\|^2_{L^2(\Omega)}$$

Integrating the inequality above between t_1 and t_2 yields (with $y(t) = \|v(f)(t)\|^2_{L^2(\Omega)}$):

$$y(t_1)e^{-ct_1} - y(t_2)\,e^{-ct_2} \leq \frac{e^{-ct_1}}{c}\int_{t_1}^{t_2} \|f\|^2_{L^2(\Omega)}\,dt.$$

Thanks to the boundedness of $y(t)$, we can let t_2 go to infinity, which implies:

$$y(t_1) \le \frac{1}{C}\int_{t_1}^{+\infty} \|f\|^2_{L^2(\Omega)} dt.$$

As f belongs to $L^2(0,+\infty;\ L^2(\Omega))$, we can conclude that $y(t)$ tends to zero as t tends to infinity, and Lemma II.1.1 is proven.

REMARK: II.1.2 One can use the Fourier expansion of $v(f)$ to prove that one actually has: $\lim \|v(f)\|_{H^1_o(\Omega)} = 0$.

2. Duality and Optimality System

Before applying the method of Section I, we need to extend the Green's formula (I.1.6) to the case "$T = +\infty$"; this is

PROPOSITION II.2.1: *Let* $y(u)$ *(resp.* $v(f)$*) be the solution of* (1.1) *(resp.* (1.3)*) with* $T = +\infty$; *the following Green's formula holds:*

$$\int_{Q_\infty} (\frac{\partial}{\partial t} + \Delta)v(f).y(u).dxdt = \int_{\Sigma_\infty} \frac{\partial}{\partial \nu}(v(f)).u\ d\sigma dt + \int_{Q_\infty} (\frac{\partial}{\partial t} + \Delta)v(f).y_h dxdt$$

(we recall that y_h was defined in (I.1.7)).

PROOF: We use Lemma I.1.2, applied to $y(v)$ and the approximate solutions $\tilde{v}_n$ of (1.3), and let n tend to ∞.

One can easily check that Green's formula (II.2.1) is the key tool for applying the duality method to (P_∞); there is no major change in the process of establishing the system of optimality conditions, and we shall only state without proof the

THEOREM II.2.1: *(i) The necessary and sufficient conditions for* $(u,y(u))$ *[resp.*$(p,v(p)]$ *to be an optimal couple for problem* $(\mathcal{P}_\infty)$ *[resp.* $(\mathcal{P}^*_\infty)$] are:

(II.2.i) $$u = \frac{\partial v}{\partial \nu} \text{ on } \Sigma_\infty$$

(II.2.ii) $$y(u) = -p \text{ in } Q$$

(ii) The optimal control u *belongs to* $L^2(0,\ +\infty;\ H^{\frac{1}{2}}(\Gamma))$.

III THE HYPERBOLIC BOUNDARY CONTROL PROBLEM

In this section, we just sketch the study of the boundary control problem for the hyperbolic dynamics (1.4), which is definitely similar to the one we made in section I ($T< +\infty$). As before, the critical result is the Green's formula below, allowing us to use the duality method:

PROPOSITION III.1.1: *Let* $z(v)$ [resp. $q(f)$] *be the solution of* (1.4)[*resp.*(1.6)]; *the following Green's formula holds*:

$$\int_{Q_T} \left(\frac{\partial^2}{\partial t^2} - \Delta\right)q(f).z(u) = -\int_{\Sigma_T} \frac{\partial}{\partial \nu}(q(f)).u + \int_{Q_T} \left(\frac{\partial^2}{\partial t^2} - \Delta\right)q(f).z_h(u), \quad \text{(III.1.1)}$$

where $z_h(u)$ *is the solution of* (1.4) *with* $u \equiv o$.

PROOF: First of all, one can notice that all the terms in (III.1.1) are well defined, see [L.L.T]; as a matter of fact, we have

$$z_h(u), z(u) \in C([0,T];L^2(\Omega)), \quad \text{(III.1.2)}$$

$$q(f) \in C([0,T];L^2(\Omega)), \quad \text{(III.1.3)}$$

$$\frac{\partial}{\partial \nu}(q(f)) \in L^2(\Sigma_T). \quad \text{(III.1.4)}$$

Now, Green's formula (III.1.1) is then proven by regularizing the data and passing to the limit.

Using the same tools as in section I, we can now state the

THEOREM III.1.1: *Let* u *be in* $L^2(\Sigma_T)$, *and* $z(u)$ *be the solution of* (1.4);

then $(u,z(u))$ *is an optimal pair for* (Q_T) *if and only if there exists* f *in* $L^2(Q_T)$ *such that*:

$$u = -\frac{\partial}{\partial \nu}(q(f)) \quad \text{on } \Sigma_T, \quad \text{(III.1.}i\text{)}$$

$$z(u) = f \quad \text{in } Q_T, \quad \text{(III.1.}ii\text{)}$$

where $q(f)$ *is the solution of* (1.6).

PROOF: Using Green's formula (III.1.1), one easily determines the dual problem of (Q_T), the optimization problem associated to (Q_T):

$$(Q_T^*) \quad \sup_{f \in L^2(Q_T)} \left(- \int_{Q_T} y_H \cdot f - \frac{1}{2} \|f\|^2_{L^2(Q_T)} - \frac{1}{2} \|\frac{\partial p(f)}{\partial \nu}\|^2_{L^2(\Sigma_T)} \right) \quad \text{(III.1.5)}$$

where y_H is the solution of (1.4) with $u \equiv o$, and $p(f)$ is the solution of the adjoint dynamics (1.6). Equalities (III.1.*i*), (III.1.*ii*) are now the optimality system associated to the problems (Q_T) and (Q_T^*).

REFERENCES:

[E-T] I. Ekeland - R. Temam: Convex Analysis and Variational problems, North Holland, 1979.

[L.1] I. Lasiecka: App. math. Optimiz., 4, p. 301-327, 1978.

[L.2] I. Lasiecka: App. Math. Optimiz., 6, p. 287-333, 1980.

[L-M] J.L. Lions- E. Magenes: Problems Aux Limites Non Homogenes et Applications, *vol*. 2, Dunod, 1968.

[L-L-T] I. Lasiecka-J.L. Lions-R. Triggiani: J. Math. Pures-Appl., 65, p. 149-192, 1986.

[L-T] I. Lasiecka - R. Triggiani: S.I.A.M. J. Control Opt., 24 (5), P. 884-925, 1986.

F. Abergel
Department of Mathematics
The Pennsylvania State University
University Park, PA 16802
U.S.A.

and

R. Temam
Laboratoire d'Analyse Numerique
C.N.R.S et Universite Paris Sud
91405-Orsay Cedex
France

Operator Theory:
Advances and Applications, Vol. 41

UNITARY EXTENSIONS OF ISOMETRIES AND CONTRACTIVE INTERTWINING DILATIONS

Rodrigo Arocena

To Professor Israel Gohberg, with admiration for his mathematical work as a researcher and as a teacher.

It is shown that unitary extensions of a suitable isometry give a simple proof of the Nagy-Foias lifting theorem and lead directly to a Schur type analysis of contractive intertwining dilations.

INTRODUCTION

To each commutant à la Nagy-Foias an isometry acting in a Hilbert space can be associated, in such a way that each unitary extension of it gives in a natural way a contractive intertwining dilation of the commutant. This will be done in section I, where a salient lifting will be identified.

In section II, the elements of a Schur-Szegö type analysis of the unitary extensions of an isometry are sketched and the connection with choice sequences is established.

Applications of that material to the Arsene-Ceausescu-Foias intertwining dilation theory are presented in section III.

This note simplifis and develops the approach of [A.1] which was extended to the bidimensional case in [A.4], where some conditions for the existence of the lifting of a commutant of two pairs of contractions are given.

I - THE CORRESPONDENCE BETWEEN CID AND UNITARY EXTENSIONS OF AN ISOMETRY

Let $T_j \in \mathbf{L}(E_j)$, $j = 1,2$, be contractions in Hilbert spaces and $X \in \mathbf{L}(E_1,E_2)$ an operator that intertwines them, i.e., $X T_1 = T_2 X$. If $W_j \in \mathbf{L}(G_j)$ denotes the minimal isometric dilation of T_j to $G_j = \mathbf{V}\{W_j^n E_j : n \geq 0\}$ and P_j the orthogonal projection of G_j onto E_j, $j = 1,2$, the set of all the Contractive Intertwining Dilations of X is defined as

$$CID(X) = \{Y \in \mathbf{L}(G_1,G_2): Y W_1 = W_2 Y,\ P_2 Y = X P_1,\ \|Y\| = \|X\|\}$$

The Nagy-Foias theorem says that $CID(X) \neq \phi$. [N-F] We now give an alternative proof of this result.

Leaving aside the trivial case $X = 0$, we assume that $\|X\| = 1$. Let $W_- \in \mathbf{L}(G_-)$ be the minimal isometric dilation of $T_- := T_2^*$ and P_- the orthogonal projection of G_- onto E_2. Set $G_+ \equiv G_1$, $W_+ \equiv W_1$, $P_+ \equiv P_1$. In $G_+ \times G_-$ we define a scalar product setting, for every $(h_1,h_2), (h'_1,h'_2) \in G_+ \times G_-$,
$\langle (h_1,h_2),(h'_1,h'_2)\rangle = \langle h_1,h'_1\rangle_{G_+} + \langle X P_1 h_1,h'_2\rangle_{G_-} + \langle h_2, X P_1 h'_1\rangle_{G_-} + \langle h_2,h'_2\rangle_{G_-}$
which is positive semidefinite since $\|X\| = 1$. Thus, we get a Hilbert space H such that $H = G_+ \vee G_-$. Next we define an isometry V with domain $D = W_- G_- \vee G_+$ by $V(W_- v + u) = v + W_+ u$, $v \in G_-$, $u \in G_+$; in fact, $\|W_- v + u\|_H^2 = \|W_- v\|_{G_-}^2 + 2\,\mathrm{Re}\langle X P_+ u, W_- v\rangle_{G_-} + \|u\|_{G_+}^2 = \|v\|_{G_-}^2 + 2\,\mathrm{Re}\langle X P_+ W_+ u,v\rangle_{G_-} + \|W_+ u\|_{G_+}^2 = \|v + W_+ u\|_H^2$, since $\langle W_-^* X P_+ u,v\rangle_{G_-} = \langle T_2 X P_+ u,v\rangle_{G_-} = \langle X T_1 P_1 u,v\rangle_{G_-} = \langle X P_+ W_+ u,v\rangle_{G_-}$. Now, let $U \in \mathbf{L}(F)$ be a unitary operator such that $F \supset H$ and $U_{|D} = V$. If $v,u \in E_2$ and $n \geq 0$ then $\langle U^n v,u\rangle_F = \langle v,V^{-n}u\rangle_{G_-} = \langle v,P_- W_-^n u\rangle_{E_2} = \langle T_2^n v,u\rangle_{E_2}$, so setting $G_2 = \vee\{U^n E_2 : n \geq 0\}$ and $W_2 = U_{|G_2}$ we have the minimal isometric dilation of T_2 to $G_2 \subset F$. We shall now prove that the restriction $Y = P^F_{G_2|G_1}$ to G_1 of the orthogonal projection of F onto G_2 belongs to CID(X).

Let $u \in E_1$, $v \in E_2$ and $n \geq 0$. Since $\langle (P_2 Y) W_1^n u,v\rangle_{E_2} = \langle W_1^n u,v\rangle_H = \langle (X P_1) W_1^n u,v\rangle_{E_2}$, we see that $P_2 Y = X P_1$. Consequently, $1 = \|X\| \leq \|Y\| \leq 1$, i.e., $\|Y\| = \|X\|$. If $m > 0$ then $\langle (Y W_1) W_1^n u, W_2^m v\rangle_{G_2} = \langle U^{n+1}u,U^m v\rangle_F = \langle Y U^n u,U^{m-1}v\rangle_F = \langle (W_2 Y) W_1^n u,W_2^m v\rangle_{G_2}$. Also, since $T_2^* = W_2^*{}_{|E_2}$, $\langle (Y W_1) W_1^n u,v\rangle_{G_2} = \langle W_1^{n+1}u,v\rangle_H = \langle X P_1 W_1^{n+1} u,v\rangle_{E_2} = \langle T_2 X T_1^n u,v\rangle_{E_2} = \langle X P_1 W_1^n u,T_2^* v\rangle_{E_2} = \langle X P_1 W_1^n u,W_2^* v\rangle_{E_2} = \langle P_2 Y W_1^n u,W_2^* v\rangle_{E_2} = \langle (W_2 Y) W_1^n u,v\rangle_{G_2}$. Thus, $(Y W_1) W_1^n u = (W_2 Y) W_1^n u$ and so $Y W_1 = W_2 Y$. The proof is over.

Remark. As it was said in the Introduction, the above proof of the lifting theorem can be seen as a simplification of the one given in [A.1]. Nevertheless, as the referee of a first version of this paper pointed out, the present proof can also be seen as an extension of the scattering theoretic proof of Nehari's theorem, due to Adamjan, Arov and Krein [A-A-K].

Now, when U is a unitary operator on a Hilbert space F such that $H \subset F \subset \vee\{U^nH: n\in\mathbf{Z}\}$ and $U_{|D} = V$ we say that (U,F) belongs to the family $\mathcal{U}$, where we identify (U,F) and (U',F') if there exists a unitary operator S from F to F' such that $SU = U'S$ and $S_{|H} = I_H$. As we just saw, $G_1, G_2 \subset F$ and $U_{|G_1} = W_1$, $U_{|G_2} = W_2$. Moreover

THEOREM I.1 *There is a bijection from* $\mathcal{U}$ *to CID(X) given by*

$$(U,F) \to Y = P^F_{G_2|G_1}.$$

PROOF. If $(U,F) \in \mathcal{U}$ then F is generated by $\vee\{U^nE_1: n\in\mathbf{Z}\}$ and $\vee\{U^nE_2: n\in\mathbf{Z}\}$. Given $n,m \in \mathbf{Z}$, set $t = |m| + |n|$; if $u,v \in E_j$, $j = 1,2$, $\langle U^nv,U^mu\rangle_F = \langle U^{t+n}v,U^{t+m}u\rangle_F = \langle W_j^{t+n}v,W_j^{t+m}u\rangle_{J_j}$; if $v \in E_1$, $u \in E_2$, $\langle U^nv,U^mu\rangle_F = \langle P^F_{G_2} W_1^{t+n}v,W_2^{t+m}u\rangle_{G_2}$. Thus, with obvious notation, if $Y = Y'$, $S(U^nv) \equiv U'^nv$ defines a unitary isomorphism that shows that $(U,F) \approx (U',F')$, so the correspondence is injective. Now let $U_j \in \mathbf{L}(F_j)$ be the minimal unitary dilation of T_j, so $G_j \subset F_j$, $W_j = U_{j|F_j}$, $j = 1,2$ and $G_- = \vee\{U_2^nE_2: n\leq 0\}$, $W_- = U_2^{-1}{}_{|G_-}$. Given Y in CID(X) an operator $\tau \in \mathbf{L}(F_1,F_2)$ is well defined by $\langle \tau U_1^nv_1,U_2^mu_2\rangle_{F_2} = \langle YW_1^{t+n}v_1,W_2^{t+m}v_2\rangle_{G_2}$, for any t such that $t+n, t+m \geq 0$, and $\tau U_1 = U_2\tau$ holds. In $F_1\times F_2$ a scalar product is given by $\langle(h_1,h_2),(h'_1,h'_2)\rangle = \langle h_1,h'_1\rangle_{F_1} + \langle \tau h_1,h'_2\rangle_{F_2} + \langle h_2,\tau h'_1\rangle_{F_2} + \langle h_2,h'_2\rangle_{F_2}$ and thus we get a Hilbert space $F = F_1 \vee F_2 \supset H$; if $f_1 \in F_1$, $f_2 \in F_2$, $\langle U_1f_1,U_2f_2\rangle_F = \langle \tau U_1f_1,U_2f_2\rangle_{F_2} = \langle U_2\tau f_1,U_2f_2\rangle_{F_2} = \langle \tau f_1,f_2\rangle_{F_2} = \langle f_1,f_2\rangle_F$, so $U(f_1+f_2) = U_1f_1 + U_2f_2$ defines a unitary operator in F. Clearly, $(U,F) \in \mathcal{U}$. Moreover, $\langle Y W_1^nv_1,W_2^mv_2\rangle_{G_2} = \langle U^nv_1,U^mv_2\rangle_F = \langle P^F_{G_2} W_1^nv_1,W_2^mv_2\rangle_{G_2}$, so $Y = P^F_{G_2|G_1}$. The proof is over.

Remark. Each $Y \in$ CID(X) is determined by the sequence $\{Y(k): k>0\}$ of operators belonging to $\mathbf{L}(E_1,E_2)$ defined as $Y(k) = P_2 W_2^{*k} Y_{|E_1}$, because $\langle Y W_1^n v_1, W_2^m v_2\rangle_{G_2}$ equals $\langle XT_1^{n-m} v_1, v_2\rangle_{E_2}$ if $n \geq m$ and $\langle W_2^n Y v_1, W_2^m v_2\rangle_{G_2} = \langle Y(m-n)v_1, v_2\rangle_{E_2}$ if $n < m$. We have $\langle Y(k)v_1, v_2\rangle_{E_2} \equiv \langle v_1, U^k v_2\rangle_F$. Call $L = E_1 \vee E_2$, which is a closed subspace of $H \subset F$. Thus the characterization of the sequence $\{Y(k): k>0\}$, i.e., of CID(X), is related to the characterization of the sequence $g(k) := P_L U^k_{|L}$ of the moments of the unitary extensions of V with scale subspace L. Moreover, a salient element $\tilde{Y}$ is naturally distingüished in CID(X); it is the one corresponding to $(\tilde{U},\tilde{F}) \in \mathcal{U}$ such that $\tilde{U}$ is the minimal unitary dilation of the contraction $V P_D \in \mathbf{L}(H)$; in that case we have $\tilde{Y}(k) = P_{E_2} (V P_D)^{*k}{}_{|E_1}$, $\forall k > 0$.

II - SCHUR-SZEGÖ TYPE ANALYSIS OF THE UNITARY EXTENSIONS OF AN ISOMETRY

Let V be an isometry such that its domain D and its range R are closed subspaces of the Hilbert space H, $\mathcal{U}$ as before the family of the essentially different minimal unitary extensions of V to spaces that contain H, N and M the defect subspaces of V, i.e., the orthogonal complements in H of D and R, respectively.

Given $(U,F) \in \mathcal{U}$ set $F_- = \vee\{U^n H: n\leq 0\}$; the Szegö operator of (U,F) - its prediction from the past-error operator - is $\Delta_+[U] = P^F{}_H(I - P^F{}_{F_-})_{|H}$. It is a positive operator such that $\langle\Delta_+[U]v,v\rangle = \mathrm{dist}^2(v,F_-)$ for every v in H and

$$0 \leq \Delta_+[U] = P^H{}_N \Delta_+[U] = \Delta_+[U] P^H{}_N \leq P^H{}_N$$

The prediction from the future-error operator $\Delta_-[U]$ can be defined in the same way and then $\Delta_-[U] = \Delta_+[U^*]$, if U^* is considered as an extension of V^{-1}. It is not difficult to prove (details concerning the material in this section will be presented in [A-2]) that:

PROPOSITION II.1 *Given $(U,F) \in \mathcal{U}$ the following conditions are equivalent: (a) $U^k N$ is orthogonal to H for every $k > 0$; (b) $U^{-j} M$ is orthogonal to H for every $j > 0$; (c) $\Delta_+[U] = P^H{}_N$; (d) $\Delta_-[U] = P^H{}_M$. There exists only one $(\tilde{U},\tilde{F}) \in \mathcal{U}$ that verifies them, given by the minimal unitary dilation $\tilde{U}$ of $V P^H{}_D \in \mathbf{L}(H)$ and $\tilde{F} = (\oplus_{j>0} U^{-j} M) \oplus H \oplus (\oplus_{k>0} U^k N)$.*

We shall say that $(\tilde{U},\tilde{F})$ is the most innovative unitary extension of V.

Let us remark that every $(U,F) \in \mathcal{U}$ defines a generalized resolvent R^U

of V given by $R^U(z) = P^F{}_H(I - zU)^{-1}|_H$, $|z| \neq 1$. According to a theorem of Chumakin[Ch], if $\mathbf{B}(N,M)$ denotes the family of all the contractive analytic functions defined in the unit disk and with values in $\mathbf{L}(N,M)$, then the association of $\Phi \in \mathbf{B}(N,M)$ to R defined by

$$R(z) = [I - z[V P^H{}_D \oplus \Phi(z)]]^{-1}, \quad |z| \neq 1,$$

gives a bijection between $\mathbf{B}(N,M)$ and the set of all the generalized resolvents of V. Thus, the one corresponding to the most innovative extension is obtained by setting $\Phi \equiv 0$ in Chumakin's formula.

Each $(U,F) \in \mathcal{U}$ can be obtained by means of an iterative process where the m-step is the extension of an isometry V_m acting in a Hilbert space H_m and with domain H_{m-1} to an isometry V_{m+1} with domain H_m and acting in $H_{m+1} = H_m \vee (V_{m+1}H_m)$; the process starts with $V_0 = V$, $H_0 = H$, $H_{-1} = D$. Thus, $H_m = \vee\{U^nH: 0 \leq n \leq m\}$. Let N_m and M_m be the orthogonal complements in H_m of H_{m-1} and $U H_{m-1}$, respectively, if $m > 0$, $N_0 = N$, $M_0 = M$; the "degrees of freedom" in each step are given by the contraction $\gamma_{m+1} := P_{M_m}U|N_m$.

Now, in the classical Caratheodory-Fejer problem, we can define an isometry V such that there exists a bijection between $\mathcal{U}$ and the solutions of the problem while the γ_m are identified with the Schur parameters. (See [A-3]). So in general we shall say that $\{\gamma_m: m>0\}$ are the Schur parameters of (U,F) as an element of $\mathcal{U}$. Let us remark that $(\tilde{U},\tilde{F})$ is determined by the condition $\gamma_m \equiv 0$.

Since $N_{m+1} = \{(U-\gamma_{m+1})h: h \in N_m\}^-$ and $M_{m+1} = \{(I-U\gamma_{m+1}{}^*)h: h \in M_m\}^-$ for every $m > 0$, we see that $N_{m+1} \approx \mathbf{D}_{\gamma_{m+1}}$, $M_{m+1} \approx \mathbf{D}_{\gamma_{m+1}{}^*}$, so an (N,M)-choice sequence can be associated naturally to each $(U,F) \in \mathcal{U}$. The following holds.

THEOREM II.2 *A bijection between the family of all the (N,M)-choice sequences and* $\mathcal{U}$ *may be obtained as follows. Given a choice sequence* $\{\Gamma_k: k>0\}$, $\Gamma_1 \in \mathbf{L}(N,M)$, *let* $U \in \mathbf{L}(F)$ *be the minimal unitary extension of the isometry in* $F_+ = H \oplus (\overset{\infty}{\underset{1}{\oplus}} \mathbf{D}_{\Gamma_j})$ *given by the iterative process*

$$U|_D = V,\ U|_N = \Gamma_1 \oplus \mathbf{D}_{\Gamma_1},\ U|_{\mathbf{D}_{\Gamma_k}} = \eta_k{}^* \Gamma_{k+1} \oplus \mathbf{D}_{\Gamma_{k+1}},\ \forall k > 0,$$

where η_k *is a unitary operator from the orthogonal complement* M_k *of* UH_{k-1} *in*

H_k, $H_N = H \oplus (\overset{N}{\underset{1}{\oplus}} \mathcal{D}_{\Gamma_j})$, onto $\mathcal{D}_{\Gamma_k^*}$ given, for every $h \in M_{k-1}$ and $k \geq 1$, by

$\eta_k [I - U(P_{M_{k-1}} U|\mathcal{D}_{\Gamma_{k-1}})^*] h = \mathcal{D}_{\Gamma_k^*} \eta_{k-1} h$, with $M_0 = M$ and $\eta_0 = I$. The Schur

parameters of such (U,F) are $\gamma_{k+1} = \eta_k^* \Gamma_{k+1}$, $\forall k \geq 0$.

The connection between the Szegö operator of $(U,F) \in \mathcal{U}$ and its choice sequence is given by the formula

$$\Delta_-[U]_{|M} = \text{strong limit}_{k \to \infty} D_{\Gamma_1^*} \cdots D_{\Gamma_{k-1}^*} D_{\Gamma_k^*}^2 D_{\Gamma_{k-1}^*} \cdots D_{\Gamma_1^*},$$

analogous to the one in [C], which in the scalar case reduces to Verblunsky's formula, $\Delta_-[U]_{|M} = \Pi\{(1 - |\Gamma_k|^2): k=1,2 \ldots\}$.

Let $L \subset H$ be a closed subspace such that $H = D \vee L = R \vee L$. Setting $g_n = P_L U^n{}_{|L}$ we shall say that $\{g_n: n>0\} \subset \mathbf{L}(L)$ is the sequence of the moments of (U,F) with scale subspace L. If $\{\Gamma_k: k>0\}$ is the corresponding choice sequence then

$$\text{(II.3)} \quad g_{k+1} = P_L(U P_{H_{k-1}} \oplus \eta_k^* \Gamma_{k+1} P_{\mathcal{D}_{\Gamma_k}}) U^k{}_{|L}, \quad \forall k \geq 0,$$

so $\{\Gamma_1, \ldots \Gamma_{k+1}\}$ determines $\{g_1, \ldots g_{k+1}\}$. The converse statemente stems from the equality $\langle \Gamma_{k+1} P_{\mathcal{D}_{\Gamma_k}} U^k l, \eta_k P_{M_k} l' \rangle = \langle g_{k+1} l, l' \rangle - \langle U P_{H_{k-1}} U^k l, l' \rangle$, $k \geq 0$, $l, l' \in L$.

Schur type analysis of some classical problems motivates the consideration of the following moment problems for the isometry V with scale subspace L.

($\mathcal{M}_k$) Given $\{g_j: 1 \leq j \leq k\} \subset \mathbf{L}(L)$, $\exists\, (U,F) \in \mathcal{U}$ such that $g_j = P_L U^j{}_{|L}$, $1 \leq j \leq k$?

($\mathcal{M}$) Given $\{g_j: j=1,2 \ldots\} \subset \mathbf{L}(L)$, $\exists\, (U,F) \in \mathcal{U}$ such that $g_j = P_L U^j{}_{|L}$, $j=1,2 \ldots$?

For $\mathcal{M}_1$ to have solution a necessary and sufficient condition is

$g_1 = A_0 + X_0 \Gamma Y_0$, $\Gamma \in \mathbf{L}(N,M)$ an arbitrary contraction and $A_0 = P_L V P_{D|L}$, $X_0 = P_{L|M}$, $Y_0 = P_{N|L}$. In general, if $\mathcal{M}_k$ has a solution then two subspaces $\bar{N}_k$ and $\bar{M}_k$ of H and three operators $A_k \in \mathbf{L}(H)$, $X_k \in \mathbf{L}(\bar{M}_k, L)$, $Y_k \in \mathbf{L}(L, \bar{N}_k)$ are well defined and are such that $\mathcal{M}_{k+1}$ has a solution iff g_{k+1} belongs to the "disk"

$$\{A_k + X_k \Gamma Y_k : \Gamma \in \mathbf{L}(\bar{N}_k, \bar{M}_k), \|\Gamma\| \leq 1\}$$

The solution of $\mathcal{M}$ is $(\tilde{U}, \tilde{F})$ iff g_j is always given by the centre of the disk, i.e., if the moments are given by $\tilde{g}_n = P_L(V P_D)^n{}_{|L}$. The above posed problems may be solved by means of the following extension of Schur's algorithm.

- In step (k+1), $k \geq 0$, it is assumed that we have isometries V_j acting in Hilbert spaces H_j, $0 \leq j \leq k$, $V_0 = V$, $H_0 = H$, with non-trivial defect subspaces N_j, M_j, while, if $k > 0$, V_{j+1} is defined in H_j, extends V_j and satisfies $H_{j+1} = H_j \mathbf{V}(V_{j+1}H_j) = H_j \mathbf{V}(V_{j+1}{}^{j+1}L)$, $0 \leq j \leq k$. If the following conditions hold

(a_{k+1}) $[g_{k+1} - P_L V_k P_{H_{k-1}} V_k{}^k]l = 0$, $\forall\, l \in L$ such that $P_{N_k} V_k{}^k l = 0$,

$[g_{k+1} - P_L V_k P_{H_{k-1}} V_k{}^k]L \perp l'$, $\forall\, l' \in L$ such that $P_{M_k} l' = 0$,

then a form $\varepsilon_{k+1}: (P_{N_k} V_k{}^k L) \times (P_{M_k} L) \to C$ is defined by

(b_{k+1}) $\varepsilon_{k+1}(P_{N_k} V_k{}^k l, P_{M_k} l') = \langle g_{k+1} l, l' \rangle - \langle V_k P_{H_{k-1}} V^k l, l' \rangle$

Iff $\|\varepsilon_{k+1}\| \leq 1$ then there exists a contraction $\gamma_{k+1} \in \mathbf{L}(N_k, M_k)$ determined by

(c_{k+1}) $g_{k+1} = P_L (V_k P_{H_{k-1}} + \gamma_{k+1} P_{N_k}) V_k{}^k{}_{|L}$;

then there exists an isometry V_{k+1} acting in $H_{k+1} = H_k \mathbf{V}(V_{k+1} H_k) = H_k \mathbf{V}(V_{k+1}{}^{k+1} L)$, such that it is defined in H_k, extends V_k and satisfies

$$\gamma_{k+1} = P_{M_k} V_{k+1|N_k} , \quad \langle V_{k+1}{}^{k+1} l, l' \rangle \equiv \langle g_{k+1} l, l' \rangle .$$

Considering $(U,F) \in \mathcal{U}$ such that $U_{|H_k} = V_{k+1}$ it follows that

(d_{k+1}) $\mathcal{M}_{k+1}$ can be solved iff (a_{k+1}) holds and (b_{k+1}) defines in $(P_{N_k} V_k{}^k L) \times (P_{M_k} L)$ a bounded form ε_{k+1} such that $\|\varepsilon_{k+1}\| \leq 1$.

In that case, the solutions of $\mathcal{M}_{k+1}$ are given by the $(U,F) \in \mathcal{U}$ with the (k+1)-first Schur parameters equal to $\gamma_1 \ldots \gamma_{k+1}$. If one of the defect subspaces of γ_{k+1} is trivial, then the solution is unique; if not, there is an infinite number of solutions.

- If the process never ends, i.e., if we never have either $\gamma_k{}^*\gamma_k = I$ or $\gamma_k \gamma_k{}^* = I$, every $\mathcal{M}_k$ has an infinite number of solutions and $\mathcal{M}$ has only one solution (U,F) determined in $\mathcal{U}$ by $\{\gamma_k: k>0\}$, which is the one with choice sequence $\{\Gamma_k: k>0\}$ determined by $\{g_k: k>0\}$ by means of (II.3).

III - APPLICATIONS TO THE STUDY OF CID(X)

In order to show that this approach leads to some of the fundamental results of Arsene, Ceausescu and Foias concerning CID(X) - see [A-C-F] and references therein - we use the notation introduced in section I and relate N and M with T_1, T_2 and X.

Identification of the defect subspaces (for another method see [C])

The definiton of the scalar product in the space H shows that $P^H_{G_-|G_+}$ $= X P_1$, so we see that $G_- \perp (G_+ \ominus E_1)$, $(G_- \ominus E_2) \perp G_+$ and

$$H = (G_- \ominus E_2) \oplus (E_2 \vee E_1) \oplus (G_+ \ominus E_1).$$

If $\mathfrak{D}'_X$ denotes the closure in H of $\{(I - X)v: v \in E_1\}$, then $\mathfrak{D}'_X \approx \mathfrak{D}_X$ under the correspondence $(I - X)v \to D_X v$ and

$$E_2 \vee E_1 = E_2 \oplus \mathfrak{D}'_X .$$

In fact, from the definition we see that $E_2 \vee E_1 = E_2 \vee \mathfrak{D}'_X$; since $X = P^H_{G_-|E_1}$, $E_2 \perp \mathfrak{D}'_X$; also, if $v \in E_1$, then $\|(I-X)v\|_H^2 = \|v\|_{E_1}^2 - 2 \operatorname{Re} \langle P_{G_-}v, Xv\rangle_{E_2} + \|Xv\|_{E_2}^2 = \|D_X v\|_{E_1}^2$. Consequently

$$H = G_- \oplus \mathfrak{D}'_X \oplus (G_+ \ominus E_1).$$

Note that $R = G_- \vee W_+G_+ = (G_- \ominus E_2) \oplus (E_2 \vee W_1E_1) \oplus W_1(G_+ \ominus E_1)$; as before we see that $E_2 \vee W_1E_1 = E_2 \oplus \{(I-XP_+)W_1v: v \in E_1\}^-$. Thus:

$$R = G_- \oplus \{(I-XP_+)W_1v: v \in E_1\}^- \oplus W_1(G_+ \ominus E_1).$$

As it is well known (see [N-F]), if $W \in \mathbf{L}(G)$ is the minimal isometric dilation of the contrction $T \in \mathbf{L}(E)$ then $G = E \oplus [\oplus_{n \geq 0} W^n \Lambda(T)] = W(G) \oplus \Lambda_*(T)$, with $\Lambda(T) = \{(W - T)E\}^-$, $\Lambda_*(T) = \{(I - W T^*)E\}^-$.

Comparing the expressions we got for H and R it follows that

$$M = [\mathfrak{D}'_X \oplus \Lambda(T_1)] \ominus \{(I-XP_+)W_1v: v \in E_1\}^-.$$

Now, the correspondence $(I-X)v'_1 + (W_1-T_1)v''_1 \to D_Xv'_1 \oplus D_{T_1}v''_1$, $v'_1, v''_1 \in E_1$, takes $(I-XP_+)W_1v = (I-X)T_1v + (W_1-T_1)v_1$ to $D_X T_1 v \oplus D_{T_1}v$, so

(1) $$M \approx M' := [\mathfrak{D}_X \oplus \mathfrak{D}_{T_1}] \ominus \{D_X T_1 v \oplus D_{T_1} v: v \in E_1\}^- .$$

From $D = V^{-1}R = V^{-1}[G_- \oplus \{(W_1-X T_1)v: v \in E_1\}^- \oplus W_1(G_+ \ominus E_1)]$ it follows that $D = W_-G_- \oplus \{(I-W_-XT_1)v: v \in E_1\}^- \oplus (G_+ \ominus E_1)$ and consequently

$N = [\Lambda_*(T_2^*) \oplus \mathfrak{D}'_X] \ominus \{(I-W_-XT_1)v: v \in E_1\}^- =$

$[\{(I-W_-T_2)v_2: v_2 \in E_2\}^- \oplus \{(I-X)v_1: v_1 \in E_1\}^-] \ominus \{(I-W_-XT_1)v: v \in E_1\}^-$.

The correspondence $(I-W_-T_2)v_2 + (I-X)v_1 \to D_{T_2}v_2 \oplus D_Xv_1$ takes $(I-W_-XT_1)v = (I-W_-T_2)Xv + (I-X)v$ to $D_{T_2}Xv \oplus D_Xv$; thus:

(2) $N \approx N' := [\mathfrak{D}_{T_2} \oplus \mathfrak{D}_X] \ominus \{D_{T_2} Xv \oplus D_X v: v \in E_1\}^-$.

<u>CID(X) and choice sequences</u>

From the above calculations and theorem (II.2) it follows that

THEOREM [A-C-F] *There exists a bijection between CID(X) and the set of all the (N',M') - choice sequences.*

Given an (N,M)-choice sequence $\{\Gamma_k: k>0\}$ we obtain an $(U,F) \in \mathfrak{U}$ in the way described in theorem (II.2) and from (U,F) we get an $Y \in CID(X)$ such that $Y = P^F{}_{G_2|G_1}$; the operator Y is determined by the sequence $\{Y(k): k>0\} \subset \mathbf{L}(E_1,E_2)$ given by $Y(k) = P_2 W_2^{*k} Y_{|E_1}$, i.e., such that $\langle Y(k)v_1,v_2\rangle_{E_2} \equiv \langle v_1,U^k v_2\rangle_F$.

Now, for any $h \in H$, we get an explicit expression for $U^k h$. In fact, with the notation of the above mentioned theorem and, as before,
$N_j = V\{U^n H: 0 \le n \le j\} \ominus V\{U^n H: 0 \le n \le j-1\}$ if $j > 0$ and $N_0 = N$, we have $U^k h = UU^{k-1}h = U(P_D + P_N + P_{N_1} + \ldots + P_{N_{k-1}})U^{k-1}h =$
$(VP_D + \gamma_1 P_N + D_{\gamma_1} P_N + \ldots + \gamma_k P_{N_{k-1}} + D_{\gamma_k} P_{N_{k-1}})U^{k-1}h$, so an iteration gives
$U^k h = [VP_D + \Sigma\{(\gamma_j + D_{\gamma_j})P_{N_{j-1}}: 1 \le j \le k\}]\,[VP_D + \Sigma\{(\gamma_j + D_{\gamma_j})P_{N_{j-1}}: 1 \le j \le k-1\}] \ldots$
$[VP_D + (\gamma_1 + D_{\gamma_1})P_N]h$.

<u>On Schur's algorithm</u>

Given a sequence $\{Y(k): k>0\} \subset \mathbf{L}(E_1,E_2)$ the problem of determining if there exists an $Y \in CID(X)$ such that $Y(k) = P_2 W_2^{*k} Y_{|E_1}$ for $1 \le k \le n$ where $n \le \infty$ is fixed can be handled by means of the Schur type algorithm described above, with $L = E_1 \vee E_2 \subset H$ ($\Rightarrow D \vee L = R \vee L = H$). In fact, assume that there exists a solution; then, for the corresponding $(U,F) \in \mathfrak{U}$ and $g_k = P_L U^k{}_{|L}$, we have
$\langle v_1 + v_2, g_k(v'_1 + v'_2)\rangle_L =$
$\langle v_1, W_1{}^k v'_1\rangle_{G_1} + \langle v_1, U^k v'_2\rangle_F + \langle v_2, U^k v'_1\rangle_F + \langle v_2, W_2{}^k v'_2\rangle_{G_2} =$
$\langle v_1, T_1{}^k v'_1\rangle_{E_1} + \langle Y(k)v_1, v'_2\rangle_{E_2} + \langle v_2, XT_1{}^k v'_1\rangle_{E_2} + \langle v_2, T_2{}^k v'_2\rangle_{E_2}$.

Thus, an explicit bijection is established between the set of sequences

$\{(P_2 W_2^{*k} Y_{|E_1}: k > 0) , Y \in CID(X)\}$

and the solutions of the moment problem for the isometry V with scale subspace

L. So, given $\{Y(k): k>0\} \subset \mathbf{L}(E_1,E_2)$, the problem we are considering leads to:

i) analyze if the correspondence $(v_1+v_2, v'_1+v'_2) \to$

$\langle v_1,T_1^k v'_1\rangle_{E_1} + \langle Y(k)v_1,v'_2\rangle_{E_2} + \langle v_2,XT_1^k v'_1\rangle_{E_2} + \langle v_2,T_2^k v'_2\rangle_{E_2}$,

with $v_1,v'_1 \in E_1$, $v_2,v'_2 \in E_2$ defines a bounded form in L, i.e., a form given by an operator $g_k \in \mathbf{L}(L)$;

ii) if that is so, apply the Schur's algorithm to the sequence $\{g_k\}$.

On a parameterization formula

The bijection between $\mathfrak{U}$ and CID(X) gives an alternative proof of the bijection between CID(X) and $\mathfrak{B}(N,M)$. In fact, basic results concerning characteristic functions of operators ([B-S], [N-F]) establish a bijection associating to each $(U,F) \in \mathfrak{U}$ a function $\Phi \in \mathfrak{B}(N,M)$ given by

$$\Phi(z) = P_M(I - z\,U\,P_{F\ominus H})^{-1}|N$$

Now, those results also show that each $(U,F) \in \mathfrak{U}$ can be recovered from its so-called characteristic function with scale subspace L, $S^{U,L} \in \mathfrak{B}(L,L)$, which satisfies $S^{U,L}(z) = [P_L(I - zU)^{-1}|L]^{-1}[P_L(I - zU)^{-1}U|L]$, $z \in \mathbb{D}$. If $\{g_n\}$ are the moments of (U,F) with scale subspace L and we set $\psi(z) = \Sigma\{z^n g_{n+1}: n\geq 0\}$, we have

$$\psi(z) = S^{U,L}(z)[I - zS^{U,L}(z)]^{-1}$$

Each $Y \in$ CID(X) is determined by the function $\hat{Y}: \mathbb{D} \to \mathbf{L}(E_1,E_2)$ given by

$$\hat{Y}(z) = \Sigma\{z^k Y(k): k>0\}, \quad |z| < 1.$$

The connection between $\{Y(k)\}$ and $\{g_k\}$ shows that

$$\langle \hat{Y}(\bar{z})v_1,v_2\rangle_{E_2} = \langle v_1, z\psi(z)v_2\rangle_L$$

holds for every $v_1 \in E_1$, $v_2 \in E_2$, $z \in \mathbb{D}$ so

$$\hat{Y}(\bar{z})^* = zP^L{}_{E_1}\psi(z)|E_2 = zP^L{}_{E_1}\{S^{U,L}(z)[I - zS^{U,L}(z)]^{-1}\}|E_2.$$

Thus, the bijection that to each $\Phi \in \mathfrak{B}(N,M)$ assigns an operator $Y \in$ CID(X) can be obtained by means of a theorem of Arov and Grossman [A-G], according to which the formula

$$S^{U,L}(z) = S_{12}(z) + S_{11}(z)\Phi(z)[I_N - S_{21}(z)\Phi(z)]^{-1}S_{22}(z)$$

with:

$$S_{11}(z) = P_L[I - zVP_D P_{H\ominus L}]^{-1}|M \quad , S_{12}(z) = P_L[I - zVP_D P_{H\ominus L}]^{-1}VP_D|L$$

$$S_{21}(z) = zP_N[I - zP_{H\ominus L}VP_D]^{-1}P_{H\ominus L}|M, \; S_{22}(z) = P_N[I - zP_{H\ominus L}VP_D]^{-1}|L$$

establishes a bijection between $\{\Phi: \Phi \in \mathfrak{B}(N,M)\}$ and $\{S^{U,L}: (U,F) \in \mathfrak{U}\}$.

For a related approach to this subject, see [M].

On maximum entropy

Dym and Gohberg have shown [D-G] that each element of sequence that solves the Nehari problem belongs to a well defined "disk" and that the maximum entropy solution is obtained by taking in each case the "centre" of the disk. Now, the first assertion is a general property of moment problems for an isometry with scale subspace, while the sequence of such "centres" is the sequence of moments given by $(\tilde{U},\tilde{F})$ [A,2]. Thus, as the approach in [C] suggests, the most innovative extension corresponds to the maximum entropy solution.

REFERENCES

[A-A-K] Adamjan, V. M.; Arov, D.Z.; Krein, M. G.: Infinite Hankel matrices and Carathéodory-Féjer and Schur problems, *Funct. Anal. and Appl.* **2** (1968), 1-17.

[A-G] Arov, D. Z. and L. Z. Grossman: Scattering matrices in the theory of dilation of isometric operators, *Soviet Math. Doklady* **27** (1983), 518-522.

[A-C-F] Arsene, Gr.; Ceausescu, Z.; Foias, C.: On intertwining dilations. VIII, *J. Operator Theory* **4** (1980), 55-91.

[A.1] Arocena, R.: Generalized Toeplitz kernels and dilations of intertwining operators, *Integral Equations and Operator Theory* **6** (1983), 759-778.

[A.2] Arocena, R.: Schur analysis of a class of translation invariant forms, to appear in a volume in honor of M. Cotlar.

[A.3] Arocena, R.: On a geometric interpretation of Schur parameters, *preprint.*

[A.4] Arocena, R.: On the extension problem for a class of translation invariant forms, to appear in *J. Operator Theory* .

[B-S] Brodskii, V. M. and Ja. S. Shvartsman: On invariant subspaces of contractions, *Soviet Math. Doklady* **12** (1971), 1659-1663.

[C] Constantinescu, T.: A maximum entropy principle for contractive intertwining dilations, *Operator Theory: Advances and Applications* , **24** (1987), 69-85.

[Ch] Chumakin, M. E.: Generalized Resolvent of Isometric Operators, translated from *Sibirskii Mat. Zhurnal,* No.4 (1967), 876-892.

[D-G] Dym, H. and I. Gohberg: A maximum entropy principle for contractive interpolants, *J. Funct. Anal.* **65** (1986), 83-125.

[M] Morán, M. D.: On intertwining operators, to appear in *J. of Math. Anal. and Appl.*

[N-F] Nagy, B. Sz.- and C. Foias: *Harmonic analysis of operators on Hilbert spaces* , Amsterdam-Budapest, 1970.

Rodrigo Arocena - Postal address:
José M. Montero 3006, ap. 503 - Montevideo, URUGUAY

Operator Theory:
Advances and Applications, Vol. 41

FACTORIZATION AND GENERAL PROPERTIES OF NONLINEAR TOEPLITZ OPERATORS[1]

Joseph A. Ball and J. William Helton

Dedicated to Israel Gohberg on the occasion of his sixtieth birthday

We discuss basic properties of nonlinear smooth norm-preserving maps on a Hilbert space and nonlinear analogues of various types of operators (Toeplitz, analytic Toeplitz, inner, outer) associated with a shift on a Hilbert space.

1. Introduction. In the last couple of decades the theory of analytic operator and matrix valued functions and its applications in operator and systems theory has achieved a sophisticated and intricate level of development (see e.g. the books [NF, BGK, GLR1, GLR2, F, RR, He, BRG1, BRG2]). In many aspects of this theory and in the applications, a central role is played by various types of factorization problems (e.g. Wiener-Hopf, spectral and inner-outer) for matrix or operator valued functions on the unit circle. Factorization problems for such operator functions can be reformulated as purely operator-theoretic factorization problems concerning Toeplitz operators associated with a shift operator S on an abstract Hilbert space; this is one of the main themes in [RR]. Recently, motivated by problems in systems theory and examples in circuits theory, the authors [BH2-5] have been pursuing a nonlinear analogue of this theory.

Here we separate out some of the basic elements of this new nonlinear shift analysis. One goal of this paper is to provide a gentle introduction to the area. A key ingredient is the nonlinear analogue of an isometry.

Literally an isometry is a map which preserves the metric, i.e. which preserves the distance between two points. As a model for the nonlinear analogue we consider a nonlinear lossless circuit;

[1] Supported in part by the Air Force Office of Scientific Research and the National Science Foundation.

here "losslessness" means simply preservation of the norm (i.e. the distance of a given point to the origin), or more generally, of an energy (possibly indefinite) quadratic form. Thus the nonlinear analogue of an isometry we take to be a (possibly nonlinear) norm preserving map (or more generally, a map preserving some quadratic form). It turns out that the only such maps which are complex analytic are linear, so we restrict the nonlinear theory to real Hilbert spaces.

In Section 2 we introduce and develop some elementary properties of nonlinear norm preserving maps on an abstract Hilbert space. In particular we present a nonlinear analogue of the characterization of a linear isometry U through the identity $U^*U = I$, despite the fact that the adjoint U^* makes no sense for a nonlinear map U. In Section 3 we introduce a shift operator S and define nonlinear S-Toeplitz, S-analytic, S-inner, and S-outer. We show that any S-Toeplitz map dilates to a map commuting with the minimal unitary extension U of S and discuss inner-outer and spectral factorization for the nonlinear case.

Related work goes under the heading of Volterra series. These are power series expansions of analytic shift invariant operators. For work in this direction, see [BFHT2, FT, A1, A2]. We do not assume that our maps are analytic so we take a different "softer" approach.

We thank C. Foias for very helpful discussions in connection with this work.

2. Norm preserving maps. A linear transformation A on a Hilbert space H is an isometry if either

$$||Ax||^2 = ||x||^2 \text{ for all } x \in H \tag{2.1}$$

or equivalently

$$A^*A = I_H. \tag{2.2}$$

By linearity, A also preserves the metric, i.e.

$$||Ax - Ay|| = ||x - y|| \text{ for all } x, y \in H \tag{2.3}$$

(and hence the term *isometry*). For a nonlinear map $A : H \to H$, the equation defining the adjoint

$$< A(x), y >_H = < x, A^*(y) >_H \tag{2.4}$$

is rarely satisfied by some $A^*(y) \in H$ for a given $y \in H$, since the right side of (2.4) is linear in x, while the left side in general is nonlinear in x; we conclude that condition (2.2) does not even make sense for a nonlinear map A. On the other hand, while condition (2.3) certainly makes sense for a general (possibly nonlinear) mapping A, if A is continuously Frechet differentiable then (2.3) in fact forces A to be affine (i.e. translation of a linear map by a fixed vector).

For A a mapping on a Hilbert space H, we say that A is (Frechet) *differentiable at the point* $x \in H$ if there is a bounded linear transformation $DA(x) : H \to H$ (called the *derivative* of A at x)

$$DA(x) : h \to DA(x)[h] \in H \text{ for } h \in H$$

such that

$$\lim_{t \to 0} \|\frac{1}{t}[A(x+th) - A(x)] - DA(x)[h]\| = 0$$

for each $h \in H$. If A is differentiable at each point $x \in H$, we say simply that A is differentiable. In the case where A is linear, then A is differentiable and $DA(x) = A$ is independent of $x \in H$.

The following result is probably well known in the theory of monotone operators and nonlinear evolution equations [Br]; for the sake of completeness we include the elementary proof.

PROPOSITION 2.1 *A (possibly nonlinear) continuously differentiable isometry A (i.e. A satisfies (2.3)) on a Hilbert space H is affine.*

PROOF: By hypothesis

$$\|\frac{1}{t}[A(x+th) - A(x)]\| = \|h\|$$

for all x and $h \in H$, so

$$\begin{aligned}\|DA(x)[h]\| &= \lim_{t \to 0} \|\frac{1}{t}[A(x+th) - A(x)]\| \\ &= \|h\|\end{aligned}$$

i.e. $DA(x)$ is a (linear) isometry for each x in H. Next observe by the fundamental theorem of calculus,

$$A(x) - A(0) = \int_0^1 DA(tx)[x]dt$$

and hence

$$\begin{aligned}\|x\|^2 = \|A(x) - A(0)\|^2 &= < A(x) - A(0),\ A(x) - A(0) > \\ &= \int_0^1 < DA(tx)[x],\ A(x) - A(0) > dt \\ &\le \int_0^1 \|DA(tx)[x]\|\ \|A(x) - A(0)\| dt \\ &= \|x\|\ \|A(x) - A(0)\| = \|x\|^2.\end{aligned}$$

Thus the inequality appearing in this chain must actually be an equality

$$\int_0^1 < DA(tx)[x],\ A(x) - A(0) > dt = \int_0^1 \|DA(tx)[x]\|\ \|A(x) - A(0)\|.$$

As the integrand on the left is pointwise less than the integrand on the right, by continuity equality of the integrals forces the equality to hold pointwise

$$< DA(tx)[x],\ A(x) - A(0) >= \|DA(tx)[x]\|\ \|A(x) - A(0)\|.$$

Thus the equality holds in the Schwarz inequality so $DA(tx)[x]$ must be a positive multiple of $A(x) - A(0)$. As $\|DA(tx)[x]\| = \|x\| = \|A(x) - A(0)\|$, we conclude that

$$DA(tx)[x] = A(x) - A(0)$$

for $0 \le t \le 1$. In particular, setting $t = 0$ gives

$$A(x) = A(0) + DA(0)[x].$$

As $x \in H$ is arbitrary, this shows that A is affine as claimed. ■

A similar result holds for condition (2.1) if the Hilbert space H is complex and A is (complex) analytic.

PROPOSITION 2.2: *A (possibly nonlinear) complex analytic map A on a complex Hilbert space which preserves norms (2.1) is linear.*

PROOF: See the introduction of [BH1].

As we do not wish to exclude analytic mappings on H, to get a worthwhile nonlinear analogue of isometric or unitary operators, we assume that the Hilbert space H is real and consider *norm preserving* (possibly nonlinear) maps A, i.e. maps A satisfying (2.1). If $dim H = 1$ and A is continuous, it is still the case that A necessarily is linear. If $dim H \geq 2$, it is not difficult to write down (real) analytic norm preserving maps which are not linear. For example, let $H = \mathbb{R}^2$ and let $(x, y) \to \theta(x, y)$ be any real valued real entire function of two real variables and define $A : \mathbb{R}^2 \to \mathbb{R}^2$ by

$$A : (x, y) \to (x cos\theta(x, y) - y sin\theta(x, y),\ x sin\theta(x, y) + y cos\theta(x, y)).$$

Henceforth we assume that all Hilbert spaces are real. In general, for L a bounded linear transformation on the real Hilbert space H, we denote by L^T (rather than L^*) the transpose (or real adjoint) of L:

$$< L[x], y > = < x, L^T[y] >$$

for all $x, y \in H$. Then of course condition (2.2) should be written as

$$A^T A = I_H. \tag{2.2'}$$

As we have seen, condition (2.2) or (2.2') on the surface does not appear to make any sense for nonlinear maps. However, if A is differentiable, the condition

$$DA(x)^T[A(x)] = x \text{ for } x \in H,\ A(0) = 0 \tag{2.5}$$

makes sense for the nonlinear case and collapses to (2.2') if A is linear. This turns out to be the nonlinear version of (2.2') equivalent to (2.1), as the following theorem shows.

THEOREM 2.3: *Suppose A is a continuously differentiable (possibly nonlinear) map on the real Hilbert space H. Then A is norm preserving if and only if A satisfies (2.5).*

PROOF: Suppose A is norm preserving and $x \in H$. Then differentiation of the identity

$$\|A(tx)\|^2 = t^2\|x\|^2$$

with respect to the real parameter t and evaluation at $t = 1$ gives

$$2 < A(x),\ DA(x)[x] > = 2 < x, x >$$

which is clearly equivalent to

$$DA(x)^T[A(x)] = x$$

for all x in H.

Trivially $A(0) = 0$ since $\|A(0)\| = \|0\|$, and thus (2.5) holds.

Conversely suppose that (2.5) holds. Note that

$$\frac{d}{dt}\|A(tx)\|^2 = 2 < A(tx),\ DA(tx)[x] >$$
$$\frac{d}{dt}\|tx\|^2 = 2 < tx, x > .$$

Then (2.5) (with tx in place of x) gives us that $\|A(tx)\|^2 - \|tx\|^2$ is independent of t. Evaluation at $t = 0$ and using $A(0) = 0$ gives that the constant value is 0 and hence in particular $\|A(x)\|^2 = \|x\|^2$. As $x \in H$ was arbitrary, we conclude that A is norm preserving. ∎

An elementary fact about bounded linear operators A and B on H is that $A^TA = B^TB$ if and only if $A = UB$ where U is a uniquely determined isometry from the image ImB of B onto ImA, or equivalently, if and only if $\|Ax\| = \|Bx\|$ for all $x \in H$. The following is a nonlinear analogue of this fact.

THEOREM 2.4: *Suppose A and B are differentiable (possibly nonlinear) maps on H and suppose that $A(0) = 0 = B(0)$. Then the following are equivalent.*

(i) $DA(x)^T[A(x)] = DB(x)^T[B(x)]$ for all $x \in H$.

(ii) $\|A(x)\|^2 = \|B(x)\|^2$ for all $x \in H$.

If B is injective then either of (i) and (ii) is also equivalent to

(iii) There exists a unique norm preserving map $U : ImB \to ImA$ for which $A = UB$.

PROOF: The equivalence of (i) and (ii) follows much as in the proof of Theorem 2.3. Specifically, note that

$$\frac{d}{dt}\{\|A(tx)\|^2\} = 2 < DA(tx)^T[A(tx)],\ x >$$
$$\frac{d}{dt}\{\|B(tx)\|^2\} = 2 < DB(tx)^T[B(tx)],\ x > .$$

Thus (i) is equivalent to $\|A(tx)\|^2 - \|B(tx)\|^2$ being independent of t; evaluation at $t = 0$ gives that the constant value is 0. This verifies (i)$\Leftrightarrow$(ii). If B is injective we may define a mapping U on the image of B by

$$U(B(x)) = A(x). \tag{2.6}$$

The injectivity of B guarantees that U is well-defined, and clearly U is norm preserving if and only if (ii) holds.

REMARK. In the linear case (2.6) gives a well-defined mapping U if (ii) holds even if B is not injective. Indeed, if $Bx = By$ and B is linear, then $B(x-y) = 0$ so by (ii) $\|A(x-y)\| = \|B(x-y)\| = 0$ so $0 = A(x-y) = Ax - Ay$ as required.

3. Nonlinear Toeplitz operators. We introduce here a nonlinear analogue of the abstract shift analysis as presented e.g. in [NF, RR]. We first review the elements of the linear theory on which we wish to focus.

Let H be a real Hilbert space. A linear operator S on H is said to be a *shift* if S is an isometry and

$$\bigcap_{n \geq 0} S^n H = (0).$$

Associated with a shift operator are collections of linear operators on H called S-Toeplitz, S-analytic, S-inner and S-outer, defined as follows.

Definition 3.1 Let A be a bounded linear operator on H and S a shift on H.

i) A is *S-Toeplitz* if $S^T A S = A$

ii) A is *S-analytic* if $SA = AS$

iii) A is *S-inner* if A is an *S-analytic* isometry

iv) A is *S-outer* if A^{-1} exists and both A and A^{-1} are S-analytic.

More generally these notions could be defined for operators A mapping one Hilbert space H_1 on which the shift S_1 acts to another Hilbert space H_2 on which the shift S_2 acts; for our nonlinear generalizations we restrict ourselves to $S_1 = S_2$ for the sake of simplicity. Also the definition of outer may be weakened somewhat; one may allow A^{-1} to be defined only on a dense subspace; this is what is done in [NF] and [RR] and corresponds more precisely to the classical scalar case (see [Ho]).

Any shift operator S on H has a minimal unitary extension U on a Hilbert space K

containing H; by a *minimal unitary extension* U on $K \supset H$ we mean that the operator U on K has the properties

i) U is unitary ($U^TU = UU^T = I_K$)

ii) $U|H = S$

and

iii) $\bigcup_{n\geq 0} U^{Tn}H$ is dense in K.

We obtain a model for a shift S and its minimal unitary extension U as follows. Let $\mathcal{C}$ be an auxiliary Hilbert space and take $H = l^{2+}_{\mathcal{C}} = \{x = \{x_n\}_{n=0}^{\infty} : x_n \in \mathcal{C}$ and $||x||^2 = \sum_{n=0}^{\infty} ||x_n||^2_{\mathcal{C}} < \infty\}$ and $K = l^2_{\mathcal{C}} = \{x = \{x_n\}_{n=-\infty}^{\infty} : x_n \in \mathcal{C},\ ||x||^2 = \sum_{n=-\infty}^{\infty} ||x_n||^2_{\mathcal{C}} < \infty\}$ with $S : \{x_0, x_1, x_2, \ldots\} \to \{0, x_0, x_1, x_2, \ldots\}$ the unilateral shift on $l^{2+}_{\mathcal{C}}$ and $U = \{\ldots, x_{-1}; x_0, x_1, \ldots\} \to \{\ldots, x_{-2}; x_{-1}, x_0, \ldots\}$ the bilateral shift on $l^2_{\mathcal{C}}$; here we consider $l^{2+}_{\mathcal{C}}$ embedded in $l^2_{\mathcal{C}}$ in the natural way. The abstract situation can be mapped to the model situation by taking $\mathcal{C} = KerS^T \subset H$ and defining $\mathcal{U} : H \to l^{2+}_{\mathcal{C}}$ and $\mathcal{V} : K \to l^2_{\mathcal{C}}$ by

$$\mathcal{U} : h \to \{(I - SS^T)S^{Tn}h\}_{n=0}^{\infty}$$

$$\mathcal{V} : h \to \{(I - SS^T)P_H U^{Tn}h\}_{n=-\infty}^{\infty}$$

(where P_H is the orthogonal projection of K onto H).

Operators on $l^{2+}_{\mathcal{C}}$ and $l^2_{\mathcal{C}}$ can be represented as matrices $[A_{ij}]_{i,j\geq 0}$ and $[A_{ij}]_{-\infty<i,j<\infty}$ respectively with matrix entries A_{ij} equal to operators on $\mathcal{C}$. In terms of these matrix representations, an operator $A = [A_{ij}]_{i,j\geq 0}$ is Toeplitz (with respect to the unilateral shift) if and only if $[A_{ij}]$ has the Toeplitz form $A_{ij} = \tilde{A}_{i-j}$, and is S-analytic if and only if in addition $[\tilde{A}_{i-j}]$ is lower triangular ($\tilde{A}_k = 0$ for $k < 0$). An operator A on $l^2_{\mathcal{C}}$ commutes with the bilateral shift if and only if the associated matrix $[A_{ij}]_{-\infty<i,j<\infty}$ has the Laurent form $A_{ij} = \tilde{A}_{i-j}$. In the context of the model, it is then clear (at least formally and turns out to be correct in general) that any Toeplitz operator A has the form $A = P_H W|H$ for a unique W on K which commutes with the unitary extension U of S. Our first result on nonlinear Toeplitz operators is an extension of this result.

For concreteness, we now assume that S is the unilateral shift on $H = l_C^{2+}$ and that U is the bilateral shift on $K = l_C^2$ and we consider l_C^{2+} as embedded in l_C^2 in the natural way; here C is an auxiliary real Hilbert space. For A a (possibly nonlinear) mapping on l_C^{2+} we say that A is *Toeplitz* if $S^T \cdot A \cdot S = A$ on l_C^{2+}. As in the linear case it is easily verified that $A = P_H W|H$ is S-Toeplitz whenever W is a (possibly nonlinear) mapping on K which commutes with U. We show that essentially any Toeplitz map is of this form. Recall that a mapping $T : H \to H$ is said to be *Lipschitz* if there is a constant $r < \infty$ for which

$$\|T(x) - T(y)\| \leq r\|x - y\|$$

for all $x, y \in H$.

THEOREM 3.1: *Suppose the mapping A on $H = l_C^{2+}$ is Toeplitz and is Lipschitz on bounded subsets of H. Then there is a mapping W on $K = l_C^2$ which is also Lipschitz on bounded sets such that*

$$(i) \quad W \cdot U = U \cdot W$$

and

$$(ii) \quad A = P_H W|H.$$

We remark that the articles [A1, A2] study analytic nonlinear maps which satisfy (i) and prove a representation theorem for them.

PROOF: Let A be as in the Theorem and define W first on $\bigcup_{k \geq 0} l_C^2[-k, \infty)$ by

$$W(x) = \lim_{n \to \infty} U^{Tn} \cdot A \cdot U^n(x). \tag{3.1}$$

Note that $U^n(x) \in l_C^{2+}$ if $x \in l_C^2[-k, \infty)$ as soon as $n \geq k$ so the expression on the right side of (3.1) makes sense for n large. To show that the limit in (3.1) exists we show that

$$P_{l_C^2[-k,\infty)} U^{Tn} A U^n(x)$$

is independent of K for $n \geq K \geq k$ if $x \in l_C^2[-k, \infty)$. Indeed, use that $P_{l_C^2[-k,\infty)} = U^{Tk} P_H U^k$ and the Toeplitz property of A to get

$$P_{l_C^2[-k,\infty)} U^{Tn} A U^n(x)$$

$$= U^{Tk} P_H U^{k-n} A U^n(x)$$
$$= U^{Tk} S^{T(n-k)} A S^{n-k} (U^k x)$$
$$= U^{Tk} A U^k(x).$$

From this is clear that the sequence $y_n = U^{Tn} \cdot A \cdot U^n(x)$ converges entrywise to a sequence $y^{(\infty)} = \{y_j^{(\infty)}\}_{j=-\infty}^{\infty}$. To show $y^{(\infty)} \in l^2_{\mathcal{C}}$, we need only check that the sequence $\{y_n\}$ of elements of $l^2_{\mathcal{C}}$ is uniformly bounded in $l^2_{\mathcal{C}}$-norm. But this follows easily from the boundedness of A on bounded subsets of $l^{2+}_{\mathcal{C}}$.

For x a general element of $l^2_{\mathcal{C}}$, we may approximate x by a sequence $\{x_k\}$ of elements $x_k \in l^2_{\mathcal{C}}[-k, \infty)$:

$$x = \lim_{k \to \infty} x_k, \quad x_k \in l^2_{\mathcal{C}}[-k, \infty).$$

From the Lipschitz property of A on bounded sets, it is easily seen that the sequence $\{y_k\}$ given by

$$y_k = \lim_{n \to \infty} U^{Tn} \cdot A \cdot U^n(x_k)$$

is Cauchy in $l^2_{\mathcal{C}}$. We may then define

$$W(x) = \lim_{k \to \infty} W(x_k)$$
$$= \lim_{k \to \infty} U^{Tk} \cdot A \cdot U^k(x_k). \tag{3.2}$$

In this way we now have W defined on all of $l^2_{\mathcal{C}}$. From the definition it is easily seen that $W = U^T \cdot W \cdot U$, i.e. $U \cdot W = W \cdot U$. Moreover, for $x \in H$,

$$P_H W(x) = P_H[\lim_{k \to \infty} U^{Tk} \cdot A \cdot U^k(x)]$$
$$= \lim_{k \to \infty} S^{Tk} \cdot A \cdot S^k(x)$$
$$= A(x)$$

again by the Toeplitz property of A. From the construction it is clear that W is Lipschitz on bounded sets of K with Lipschitz constant α_M for a given bound M the same as for A. This completes the proof of Theorem 3.1. ∎

To define nonlinear analogues of S-analytic, S-inner and S-outer, we first need the notion of *causality.* A mapping A on $l_{\mathcal{C}}^{2+}$ is said to be *causal* if

$$P_n \cdot A = P_n \cdot A \cdot P_n, \tag{3.3}$$

where $P_n : l_{\mathcal{C}}^{2+} \to l_{\mathcal{C}}^2[0,n]$ is the natural projection of $l_{\mathcal{C}}^{2+}$ to the space of sequences supported on $[0,n]$:

$$P_n : (x_0, x_1, \ldots, x_n, x_{n+1}, \ldots) \to (x_0, \ldots, x_n, 0, 0, \ldots).$$

When the variable n is thought of as a (discrete) time variable and A is thought of as an input-output map, then the causality condition (3.3) has the interpretation that "future inputs have no influence on past outputs." A related notion is that of *stability*; the map A on $l_{\mathcal{C}}^{2+}$ (or more generally on the space $l_{\mathcal{C}}^{+}$ of all (possibly not square-summable) $\mathcal{C}$-valued sequences) is said to be *stable* if there is a constant $\gamma < \infty$ for which

$$\|P_n \cdot A(x)\| \leq \gamma \|P_n x\| \tag{3.4}$$

for $n = 0, 1, 2, \ldots$. We are now ready to give the complete nonlinear analogue of Definition 3.1; note that "S-Toeplitz" has already been defined and is repeated here for completeness.

Definition 3.2 Let A be a (possibly nonlinear) mapping on $H = l_{\mathcal{C}}^{2+}$, and let S be the unilateral shift on H.

i) A is *S-Toeplitz* if $S^T \cdot A \cdot S = A$.

ii) A is *S-analytic* if A is causal and stable and if $S \cdot A = A \cdot S$

iii) A is *S-inner* if A is S-analytic and is norm preserving

iv) A is *S-outer* if A^{-1} exists and both A and A^{-1} are causal and stable on H.

We remark that a bounded linear operator A on $H = l_{\mathcal{C}}^{2+}$ is automatically causal and stable; thus Definition 3.2 when restricted to linear maps gives notions exactly equivalent to those in Definition 3.1.

A linear S-analytic operator A is determined by its "symbol" Ω; the symbol is a bounded linear mapping Ω from $\mathcal{C}$ (identified with the image of $I - SS^T$) into $l_{\mathcal{C}}^{2+}$, and determines A via

$$A(x) = \sum_{n=0}^{\infty} S^n \Omega (I - SS^T) S^{Tn} x,$$

where the infinite series converges in the strong topology. A nonlinear analogue of this representation for a nonlinear S-analytic map is given in [BFHT1].

The following is an immediate consequence of Theorem 2.3 and the definitions.

THEOREM 3.2: *A continuously differentiable S-analytic mapping A on l_C^{2+} is S-inner if and only if*

$$DA(x)^T[A(x)] = x$$

for all $x \in l_C^{2+}$ and

$$A(0) = 0.$$

In the linear case it is well known how inner-outer factorization can be reduced to spectral factorization. By this we mean that an S-analytic F has inner-outer factorization $F = \Theta \cdot A$ if and only if the outer factor A provides the outer factor in a spectral factorization of $F^T F$:

$$F^T F = A^T A.$$

With the aid of Theorem 2.4 we provide a nonlinear analogue of this connection.

THEOREM 3.3: *Suppose that F is S-analytic, that A is an S-outer mapping on l_C^{2+}, and that both F and A are continuously differentiable. Then F has an inner-outer factorization of the form*

$$F = \Theta \cdot A \tag{3.4}$$

for some S-inner Θ if and only if the "nonlinear positive definite Toeplitz" operator $x \to DF(x)^T[F(x)]$ has the "nonlinear spectral factorization"

$$DF(x)^T[F(x)] = DA(x)^T[A(x)]. \tag{3.5}$$

for all $x \in l_C^{2+}$.

PROOF: If F has the inner-outer factorization (3.4), then it is immediate from Theorem 2.4 that the identity (3.5) holds. Conversely, suppose (3.5) and define Θ by

$$\Theta = F \cdot A^{-1}.$$

Then from the proof of Theorem 2.4, we see that Θ is norm preserving. Since A is outer, A^{-1} is causal and stable. Therefore Θ is the composition of stable, causal maps and hence is itself stable and causal. We conclude that Θ is S-inner as desired. ∎

Theorem 3.3 does not solve the problem of inner-outer factorization, since it only reduces the problem to another one of equal difficulty. This problem is analyzed in [BH2] from the point of view of nonlinear Beurling-Lax theorems. Here we state the main result of [BH2] more in the framework of Toeplitz operators.

THEOREM 3.4: (see Theorem 6.1 of [BH2]) *Suppose F is a twice continuously differentiable S-analytic mapping on $l_{\mathcal{C}}^{2+}$. If F has an inner-outer factorization*

$$F = \Theta \cdot A$$

with Θ S-inner and with A an S-outer diffeomorphism of $l_{\mathcal{C}}^{2+}$, then necessarily

(i) For each $a \in l_{\mathcal{C}}^2[0,n]$ there is a unique $\nu(a) \in l_{\mathcal{C}}^2[n+1,\infty)$ for which

$$P_{l_{\mathcal{C}}^2[n+1,\infty)} DF(a,\nu(a))^T [F(a,\nu(a))] = 0. \tag{3.6}$$

and

(ii) The quadratic form

$$h \to < D^2F(a,\nu(a))[(0,h),(0,h)], F(a,\nu(a)) > + \|DF(a,\nu(a))[0,h]\|^2$$

is positive definite on $l_{\mathcal{C}}^2[n+1,\infty)$.

Conversely, conditions (i) and (ii) together with

(iii) [For each $a \in l_N^2[0,n]$, let $b_a \in l_N^2[n+1,n+p]$ be chosen so that $\nu(a) = (b_a, \nu(a,b_a))$ (where

$\nu(a)$ is as in (i))]. Then for each $a \in l_{\mathcal{C}}^2[0,n]$, there is a $p > 0$, a bounded neighborhood U of b_a in $l_{\mathcal{C}}^2[n+1,n+p]$, and a $\delta > 0$ for which

$$\|[0, I, D_2\nu(a,b)^T] DF(a,b,\nu(a,b))^T \cdot F(a,b,\nu(a,b))\| \geq \delta$$

for all $b \in l_{\mathcal{C}}^2[n+1,n+p]$ not in U

guarantee that F has an "asymptotic" inner-outer factorization $F = \Theta \cdot A$ defined on $\bigcup_{n\geq 0} l^2_{\mathcal{C}}[0,n]$.

For the precise definition of "asymptotic" inner-outer factorization, we refer the reader to [BH2].

Condition (i) is equivalent to the map γ_a on $l^2_{\mathcal{C}}[n+1,\infty)$ defined by

$$\gamma_a(x) = \|F(a,x)\|^2_{l^{2+}_{\mathcal{C}}}$$

having a unique critical point $x = \nu(a) \in l^2_{\mathcal{C}}[n+1,\infty)$ for each $a \in l^2_{\mathcal{C}}[0,n]$. Condition (ii) then requires that the Hessian of γ_a at this unique critical point be positive definite.

Computation of the inner-outer factorization for a given S-analytic map F then requires solving the critical point equation (3.6) as a first step. The linear analogue is to compute the "wandering subspace" $\mathcal{L} = Fl^{2+}_{\mathcal{C}} \ominus S \cdot F \cdot l^{2+}_{\mathcal{C}}$ in the Halmos proof of the Beurling-Lax theorem [Ha]. The second step requires construction of a nonlinear congruence, analogous to constructing an orthonormal basis for $\mathcal{L}$ in the linear case.

In practice (even in the linear case) it is often more convenient to express F in terms of difference equation recursion relations (or in engineering parlance, "state space equations"). Then the key critical point equation (3.6) can be expressed in a more condensed form. For progress in this direction and applications to nonlinear control theory, see [BH2, BH3, BH4]. For the state space approach in the linear case, see [BGR1, BR, AG, F].

In applications to interpolation and approximation problems, what is usually required is J-inner-outer factorization rather than simply inner-outer factorization (see [BH1] for the linear case and [BH2] for the nonlinear). Here J is a signature operator on $\mathcal{C}(J = J^T = J^{-1})$ and a causal, stable map Θ on $l^{2+}_{\mathcal{C}}$ is said to be J-inner if Θ preserves the quadratic form on $l^{2+}_{\mathcal{C}}$ induced by J:

$$\rho^+_J \cdot \Theta(x) = \rho^+_J \cdot x \text{ for } x \in l^{2+}_{\mathcal{C}}$$

where

$$\rho^+_J(x) = \sum_{n=0}^{\infty} < Jx_n, x_n >_{\mathcal{C}} \text{ if } x = \{x_n\}_{n=0}^{\infty}.$$

More generally, when $F : l_C^{2+} \to l_C^2$ is a noncausal map of l_C^{2+} into l_C^2 one may want a J-phase-factorization $F = \Theta \cdot A$ where Θ is a J-phase and A is outer on l_C^{2+}. By a J-phase we mean a map $\Theta : l_C^{2+} \to l_C^2$ which intertwines S with U and for which

$$\rho_J \cdot \Theta(x) = \rho_J^+(x) \text{ for all } x \in l_C^{2+}$$

where ρ_J is the quadratic form on l_C^2 induced by J:

$$\rho_J(x) = \sum_{n=-\infty}^{\infty} < Jx_n, x_n >_C \ \text{ if } x = \{x_n\}_{-\infty}^{\infty}.$$

With the exception of Theorem 3.4, all the ideas presented above extend routinely to the setting where a nondegenerate Hermitian form (or more precisely a Krein space inner product) replaces the Hilbert space inner product, so these extensions are not done explicitly here. Theorem 3.4 also extends if $dim C \leq 2$ and in [BH2] it is conjectured that this restriction is removable.

4. Conclusion. This paper concerns the extension of shift analysis ideas, which have played a prominent role in operator theory in the past two decades, to nonlinear settings. Applications to factorization, interpolation and approximation as required in nonlinear engineering applications are in the process of development.

REFERENCES

[AG] D. Alpay and I. Gohberg, Unitary rational matrix functions, in: *Topics in Interpolation Theory of Rational Matrix-valued Functions* (ed. I. Gohberg), Operator Theory Advances and Applications 33 (1988), Birkhauser (Basel), pp. 175-222.

[A1] W. Arveson, Spectral theory for nonlinear random processes, Symposia Mathematica Vol. XX, Academic Press (London), 1976, pp. 531-537.

[A2] W. Arveson, A spectral theorem for nonlinear operators, Bull. Amer. Math. Soc. 82 (1976), 511-513.

[BFHT1] J.A. Ball, C. Foias, J.W. Helton and A. Tannenbaum, On a local nonlinear commutant lifting theorem, Indiana Univ. Math. J. 36 (1987), 693-709.

[BFHT2] J.A. Ball, C. Foias, J.W. Helton and A. Tannenbaum, A Poincaré-Dulac approach to a nonlinear Beurling-Lax-Halmos theorem, J. Math. Anal. and Appl., to appear.

[BGR1] J.A. Ball, I. Gohberg and L. Rodman, Realization and interpolation of rational matrix functions, in: *Topics in Interpolation Theory of Rational Matrix-valued Functions* (ed. I. Gohberg) 33 (1988), Birkhauser (Basel), pp. 1-72.

[BRG2] J.A. Ball, I. Gohberg and L. Rodman, Interpolation problems for rational matrix functions, monograph in preparation.

[BH1] J.A. Ball and J.W. Helton, A Beurling-Lax theorem for the Lie group $U(m,n)$ which contains most classical interpolation theory, J. Operator Theory 9 (1983), 107-142.

[BH2] J.A. Ball and J.W. Helton, Shift invariant manifolds and nonlinear analytic function theory; Integral Equations and Operator Theory, 11 (1988), 615-725.

[BH3] J.A. Ball and J.W. Helton, Sensitivity bandwidth optimization for nonlinear feedback systems, Proceedings MTNS (Math. Theory of Networks and Systems) Conference, Phoenix, 1987.

[BH4] J.A. Ball and J.W. Helton, Factorization of nonlinear systems: toward a theory for nonlinear H^∞ control, in Proceeding of IEEE Conference on Decision and Control, Austin, 1988.

[BH5] J.A. Ball and J.W. Helton, Interpolation problems for null and pole structure of nonlinear systems, in Proceedings of IEEE Conference on Decision and Control, Austin, 1988.

[BR] J.A. Ball and A.C.M. Ran, Global inverse spectral problems for rational matrix functions, Linear Algebra and Applications 86 (1987), 237-282.

[BGK] H. Bart, I. Gohberg and M.A. Kaashoek, *Minimal Factorization of Matrix and Operator Functions*, Birkhauser, (Basel), 1979.

[Br] H. Brezis, *Operateurs maximaux monotone et semigroups de contractions dans les espaces de Hilbert*, Math. Studies 5, North Holland, 1973.

[FT] C. Foias and A. Tannenbaum, Weighted optimization theory for nonlinear systems, preprint.

[F] B. Francis, *A Course in H^∞ Control Theory*, Lecture Notes in Control and Information Sciences No. 88, Springer-Verlag (New York), 1987.

[GLR1] I. Gohberg, P. Lancaster and L. Rodman, *Matrix Polynomials*, Academic Press (New York), 1982.

[GLR2] I. Gohberg, P. Lancaster and L. Rodman, *Invariant Subspaces of Matrices with Applications*, J. Wiley (New York), 1986.

[Ha] P.R. Halmos, Shifts on Hilbert space, J. Reine Angew. Math. 208 (1961), 102-112.

[He] J.W. Helton et al, Operator Theory, Analytic Functions, Matrices and Electrical Engineering, CBMS No. 68, American Math. Soc. (Providence), 1987.

[Ho] K. Hoffman, *Banach Spaces of Analytic Functions*, Prentice-Hall (Englewood Cliffs), 1965.

[NF] B. Sz.-Nagy and C. Foias, *Harmonic Analysis of Operators on Hilbert Space*, North Holland, 1970.

[RR] M. Rosenblum and J.R. Rovnyak, *Hardy Classes and Operator Theory*, Oxford Univ. Press, 1985.

J.A. Ball
Department of Mathematics
Virginia Tech
Blacksburg, VA 24061

J.W. Helton
Department of Mathematics
University of California at San Diego
La Jolla, CA 92093

Operator Theory:
Advances and Applications, Vol. 41

QUASILOCAL ALGEBRAS OVER INDEX SETS WITH A MINIMAL CONDITION

H. Baumgärtel

Dedicated to I. Gohberg on the occasion of his 60th birthday.

The paper discusses quasilocal algebras over index sets with a minimal condition. In particular the "invariance of the centers of the local algebras" and the so-called "Wightman inequality" are considered under the assumption of "global primitive causality".

INTRODUCTION

In this paper we consider quasilocal algebras over certain index sets I. I is assumed to be partially ordered, directed and equipped with a separability condition, i.e. it is assumed that there is a sequence of indices $\alpha_1 < \alpha_2 < \alpha_3 < \ldots$ such that to each index $\beta \in I$ there corresponds a natural $n = n(\beta)$ with $\beta < \alpha_n$. It is further assumed that finite index sets $\{\beta_1, \beta_2, \ldots, \beta_m\}$ have a supremum $\beta = \sup\{\beta_1, \beta_2, \ldots, \beta_m\}$. For an arbitrary index set $M \subset I$ in general the supremum does not exist. However, there are also infinite index sets that have a supremum.

A local set of C^*-algebras $\{A_\alpha\}_{\alpha \in I}$ with respect to an index set I is a set $\{A_\alpha\}$ of C^*-algebras, indexed by I and equipped with the following properties:

$$\text{(I)} \quad \alpha < \beta \text{ implies } A_\alpha \subset A_\beta, \text{ (isotony)},$$

i.e. one can define the so-called quasilocal algebra A by

$$A = \mathrm{clo}_{||\cdot||} \cup_{\alpha \in I} A_\alpha = \mathrm{clo}_{||\cdot||} \bigcup_{n=1}^{\infty} A_{\alpha_n},$$ according to the separability condition.

$$\text{(II) If } \beta = \sup M \text{ then } A_\beta = \bigvee_{\alpha \in M} A_\alpha, \text{ (additivity)}.$$

Here $\vee$ denotes the C^* algebra, generated (in the C^*-sense) by all algebras A_α, $\alpha \in M$. In particular, (II) is true if M is finite.

The theory of local nets of C^*-algebras and their quasilocal algebras was developed by many authors. At present it is developed to a large extent. The first collected presentations of the material were given by Emch [1] and Bratteli/Robinson [2].

In particular, such nets are of interest if the index set I is equipped with a relation, $\alpha \perp \beta$, usually called causal disjointness and if the net is causal with respect to this relation.

The causal disjointness relation is characterized by the following properties

(1) $\alpha \perp \beta$ implies $\beta \perp \alpha$, (symmetry),

(2) $\alpha \perp \alpha$ is impossible for any index $\alpha \in I$,

(3) $\alpha < \beta$ and $\beta \perp \gamma$ implies $\alpha \perp \gamma$, (heredity to smaller indices),

(4) If $\alpha \perp \gamma$ for all $\alpha \in M \subset I$ and if $\beta := \sup M$ exists then $\beta \perp \gamma$ follows, (heredity to the supremum),

(5) To each $\alpha \in I$ there is an index $\beta \in I$ such that $\alpha \perp \beta$.

DEFINITION 1. A local net $\{A_\alpha\}_{\alpha\in I}$ of C^*-algebras A_α with respect to I is said to be causal with respect to the causal disjointness relation $\perp$ if

(6) $\alpha \perp \beta$ implies: A_α commutes with A_β by elements, (causality).

A famous result of the abstract theory of causal nets of C^*-algebras is the theorem of Lanford/Ruelle, where the concept of the "algebra at infinity" plays the decisive role (see, for example, Bratteli/Robinson [2, p. 156] for details).

The theory of quasilocal algebras with causal disjointness relation was strongly influenced from Mathematical Physics by the famous paper of Haag/Kastler [3], which turned out to be a cornerstone in algebraic Quantum Field Theory. In this paper Haag/Kastler emphasize the quasilocal nature of the C^*-algebras (of observables) in Quantum Field Theory (see also Haag [4]). In the context of Quantum Field Theory the index set I is the set of all open and bounded sets in the Minkowski space R^4 and $\perp$ is the usual causal disjointness (also called spacelike separation in this case) which is defined by the Minkowski scalar product $x_0y_0 - \sum_{j=1}^{3} x_jy_j$ in R^4.

This index set has the property that to each $G \in I$ there is a sequence $G_n \in I$ with $G \supset G_1 \supset G_2 \supset G_3 \supset \ldots$

In the present paper the theory is developed (to some extent) for an index set that behaves opposite to those just mentioned: we consider index sets with a minimal condition.

(7) To each $\alpha \in I$ there is an index $\beta \leq \alpha$ and β is minimal, i.e. there does not exist any index γ with $\gamma < \beta$.

The set of all minimal (or atomic) indices is denoted by $I_{min} \subset I$. Note that it is not required that the set I_{min} is denumerable, i.e. card (I_{min}) may be larger than $\aleph_0$. For some purposes a second condition is imposed.

(8) Every index $\alpha \in I$ is a supremum of a finite set of minimal indices $\beta_j \in I_{min}$, $j = 1, 2, \ldots, n$, $\alpha = \sup\{\beta_1, \ldots, \beta_n\}$.

The motivation for introducing the minimal condition is, on the one hand, that of discretization: one claims to consider, for example in the context of a Minkowski space theory, discrete variants of the theory. On the other hand, the concepts to begin with should be sufficiently general to include the often used index set of all finite-dimensional subspaces of a given infinite-dimensional separable Hilbert space.

EXAMPLE 1. Let S be a complex infinite-dimensional separable Hilbert space. Let $R \subset S$ be a subspace, $1 \leq \dim R < \infty$. Then $I = \{R\}$ is an index set equipped with all conditions mentioned above, where $R_1 < R_2$ means $R_1 \subset R_2$ (inclusion). One has $I_{min} = \{R: \dim R = 1\}$. Here card $I_{min} = 2^{\aleph_0}$. The causal disjointness relation $\perp$ is the usual orthogonality $R_1 \perp R_2$.

EXAMPLE 2. Consider the lattice $\mathbb{Z}^n \subset \mathbb{R}^n$. Let $A \subset \mathbb{Z}^n$ be a finite subset. Then $I = \{A\}$ is an index set where $A_1 < A_2$ means $A_1 \subset A_2$ (inclusion). I_{min} is the subset of all one-point subsets $\{a\}$ of $\mathbb{Z}^n$, $a \in \mathbb{Z}^n$. One has card $I_{min} = \aleph_0$. The causal disjointness relation $\perp$ in $\mathbb{Z}^n$ may be introduced by a system of subsets $M_x \subset \mathbb{Z}^n$ indexed by $x \in \mathbb{Z}_n$ equipped with the properties $x \in M_y$ implies $y \in M_x$ and $x \in M_x$. Then $N \perp M$ can be defined by $N \perp M$:= "for all $n \in N$, $m \in N$ one has $n \in M_m$". A trivial example is M_x: $= \mathbb{Z}^n \setminus \{x\}$ which leads to $\perp$ as the simple set-theoretical disjointness; another example is given by a non-degenerate indefinite 2-form $\langle x,y \rangle$ of signature +, -, -, ..., where $M_0 := \{x: \langle x,x \rangle < 0\}$. This example leads for n=4 to $\mathbb{Z}^4$ as a sublattice of the Minhowski space with real time axis. This example reflects the desire to work with discretized variants of the Minkowski space.

2. PROBLEMS

In the present paper there will be discussed the following questions. Firstly let the four-tuple $\{I, \{A_\alpha\}_{\alpha \in I}, A, \perp\}$ be given.

Further let ω be a state on A. We aim to discuss simple conditions on ω such that, at least for a sufficiently large class of indices α, the center

$$Z_\omega(\alpha) = \pi_\omega(A_\alpha)'' \cap \pi_\omega(A)'$$

is constant, i.e. independent of α (the canonical GNS-dates of ω with respect to A are denoted by H_ω, π_ω, Ω_ω).

This problem arises, for example, in the context of Algebraic Quantum Field Theory, where one claims to classify the possible types of the local algebras $\pi_\omega(A_\alpha)''$ and where it is important to consider, without essential loss of generality, the case where the local algebras $\pi_\omega(A_\alpha)''$ are factors. See, for example, in this connection the paper Buchholz/D'Antoni/Fredenhagen [5]. As "simple conditions", mentioned above, we discuss (I) "relative bicommutant property" and (II) "global primitive causality". Both properties will be defined in the following.

The second question discussed here is known in Algebraic Quantum Field Theory as the "Wightman inequality" which says that under simple additional conditions $\alpha < \beta$ implies that the corresponding local algebras $\pi_\omega(A_\alpha)''$, $\pi_\omega(A_\beta)''$ satisfy $\pi_\omega(A_\alpha)'' \subset \pi_\omega(A_\beta)''$, i.e. $\pi_\omega(A_\alpha)''$ is <u>properly</u> included in $\pi_\omega(A_\beta)''$. If there is no minimal condition, this inequality can be used to prove that the algebras $\pi_\omega(A_\alpha)''$ are not finite-dimensional.

The conditions (I) and (II) play a role also in the discussion of this second question.

Of course, the Wightman inequality can also be discussed abstractly under different conditions on the index set (for example no minimal condition) and under certain transitivity conditions with respect to an invariance group of the causal disjointness relation. Some aspects in this direction are pointed out in a paper of Wollenberg [6].

3. INVARIANCE OF THE CENTER OF LOCAL ALGEBRAS

3.1 <u>The relative bicommutant property</u>

Firstly we discuss the relative bicommutant property. Let H be a Hilbert space, M and R von Neumann algebras over H and $M \subset R$. The relative commutant M^c of M with respect to R is defined by

$$M^c := M' \cap R,$$

where M' denotes, as usual, the commutant of M over H. Then

$$M^{cc} = (M' \cap R)' \cap R$$

DEFINITION 2. M is said to have the <u>relative bicommutant property</u> with respect to R if $M^{cc} = M$.

Obviously, each von Neumann algebra M over H has the relative bicommutant property with respect to $L(H)$. In the literature the relative bicommutant property sometimes is called "inner duality" (see for example Horuzii [7, p. 62] where also further references can be found). Horuzii motivates this terminology by the remark that "duality" implies "inner duality". Note that "duality" depends on the concept $\perp$ (causal disjointness) and means that $M_{\alpha^\perp} = M_\alpha{}'$ (where M_α are von Neumann algebras over H and $M_{\alpha^\perp} = \bigvee_{\beta\perp\alpha} M_\beta$). However the "relative bicommutant property" is completely independent of the introduction of any $\perp$. So it seems that the term "inner duality" is not an adequate one.

If the four-tuple $\{I, \{A_\alpha\}\ \alpha \in I, A, \omega\}$ is given, where ω denotes a state on A, the relative bicommutant property can be defined as a property of this four-tuple.

DEFINITION 3. The four-tuple $\{I, \{A_\alpha\}_{\alpha\in I}, A, \omega\}$ has the relative bicommutant property, if each von Neumann algebra $\pi_\omega(A_\alpha)''$, $\alpha \in I$, has this property with respect to $\pi_\omega(A)''$, according to Def. 2.

Now we connect the relative bicommutant property of a von Neumann algebra inclusion $M \subset R$ with properties of their centers.

PROPOSITION 1. <u>Let $M \subset R$ be an inclusion of von Neumann algebras over H. Then $M^{cc} = M$ implies $Z_R \subset Z_M$ where $Z_M := M \cap M'$, $Z_R := R \cap R'$ denote the centers of the corresponding algebras.</u>

Proof. Let $M^{cc} = M$. Firstly we show $Z_R \subset M$. $M^{cc} = M$ means $(M' \cap R)' \cap R = M$ or $(M \vee R') \cap R = M$, so we have $R' \cap R \subset (M \vee R') \cap R = M$, i.e. $Z_R \subset M$. Secondly we show $Z_R \subset M'$. Now $M \subset R$ implies $M \subset R \vee R' = (R \cap R')'$ which implies $M \subset Z_R'$ or $Z_R \subset M'$. ■

The question arises whether the converse of Proposition 1 is also true. We show this converse under a certain additional assumption:

We denote by $(M \vee R')_{fin}$ the algebra of all finite expressions $\sum_{j=1}^n M_j R_j'$, $M_j \in M$, $R_j' \in R'$. Then obviously $((M \vee R')_{fin})'' = M \vee R'$ and $(M \vee R')_{fin} \cap R \subset (M \vee R') \cap R$, hence $((M \vee R')_{fin} \cap R)'' \subset (M \vee R') \cap R$. The additional assumption refers to the last inclusion and requires

(9) $((M \vee R')_{fin} \cap R)'' = (M \vee R')_{fin}'' \cap R$

PROPOSITION 2. <u>Let</u> $M \subset R$ <u>be an inclusion of von Neumann algebras over H. Assume Condition (9) to be satisfied. Then</u> $Z_R \subset Z_M$ <u>implies</u> $M^{cc}=M$.

Proof. One has to prove $(M \vee R') \cap R = M$. However, because of $M \subset M \vee R'$ and $M \subset R$ the inclusion $M \subset (M \vee R') \cap R$ is trivial, so it remains to prove

(10) $(M \vee R') \cap R \subset M$

According to Condition (9) it is sufficient to prove this for an element $\hat{R} = \sum_{j=1}^{n} M_j R_j' \in (M \vee R')_{fin}$, i.e. one assumes $\hat{R} \in R$ and one has to show $\hat{R} \in M$. After that by taking weak closure one obtains (10). So let

$$\hat{R} = \sum_{j=1}^{n} M_j R_j' \in R, \qquad M_j \in M, \qquad R_j' \in R'$$

or

(11) $\sum_{j=0}^{n} M_j R_j' = 0, \qquad M_0 := -\hat{R} \in R, \qquad R_0' = 1$

The equation (11) is considered in $R \vee R'$. According to Kadison/Ringrose [8, p. 333, Theorem 5.5.4] there are operators $C_{jk} \in Z_R$, j, k =0, 1, ..., n, such that

$$\sum_{j=0}^{n} M_j C_{jk} = 0, \ k = 0, 1, \ldots, n; \quad \sum_{k=0}^{n} C_{jk} R_k' = R_j', \ j = 0, 1, \ldots n$$

These equations we can write in the following form:

(12) $$\hat{R} \cdot C_{ok} = \sum_{j=1}^{n} M_j C_{jk}; \ k = 0, 1, \ldots, n, \qquad C_{jk} \in Z_R,$$

(13) $$1 = C_{oo} + \sum_{k=1}^{n} C_{ok} R_k' ; \quad R_j' = C_{jo} + \sum_{k=1}^{n} C_{jk} R_k', \quad C_{jk} \in Z_R$$

Multiplying the first equation of (13) from the left by $\hat{R}$ we obtain

(14) $$\hat{R} \cdot 1 = \hat{R} \cdot C_{oo} + \sum_{k=1}^{n} (\hat{R} \cdot C_{ok}) R_k'$$

Since $Z_R \subset Z_M$ and $R' \subset M'$, from (12) we obtain $\hat{R} \cdot C_{ok} \in M$, k = 0, 1, 2, ..., n and $R_k' \in M'$. So one can look at the right hand side of (14) as an element of $M \vee M'$. According to the mentioned theorem one has an algebraic isomorphy between $(R \vee R')_{fin}$ and $R \odot R'$ as Z_R-modules, i.e. the equation

$$\hat{R} \otimes 1 = (\hat{RC}_{oo}) \otimes 1 + \sum_{k=1}^{u} (\hat{RC}_{ok}) \otimes R'_k$$

holds in $R \hat{\otimes} R'$. But the left hand side is also an element of $M \cdot M'$ as a Z_M-module. Hence $\hat{R} \in M$ follows. ■

3.2 <u>The global primitive causality (GPC)</u>

This property is intimately connected with the causal disjointness. To introduce this property we consider <u>spacelike</u> subsets of I_{min}. Let $M \subset I_{min}$.

DEFINITION 3. The subset $M \subset I_{min}$ is said to be spacelike if α, $\beta \in M$, $\alpha \neq \beta$ implies $\alpha \perp \beta$.

That is, in spacelike index sets of minimal indices two different indices are always causally disjoint.

LEMMA 1. <u>There are spacelike subsets</u> $M \subset I_{min}$ <u>with at least two different elements.</u>

Proof. According to property (5) to $\alpha \in I_{min}$ there is $\beta \in I$ with $\alpha \perp \beta$. Hence $\alpha \neq \beta$ follows. Now β "contains" a minimal index $\beta_{min} < \beta$. According to property (3) one obtains $\alpha \perp \beta_{min}$ and $\alpha \neq \beta_{min}$, so $\{\alpha, \beta_{min}\}$ is spacelike and a subset of I_{min}. ■

LEMMA 2. <u>There are maximal spacelike sets</u> $M \subset I_{min}$.

Proof. Use Zorn's lemma: Let $K = \{K_\alpha\}$ be a chain formed by spacelike sets of minimal indices. Then $K := \cup_\alpha K_\alpha \subset I_{min}$ and K is spacelike, because γ, $\delta \in K$, $\gamma \neq \delta$ implies $\gamma \in K_{\alpha'}$ for some α', $\delta \in K_{\alpha''}$ for some α'' and $K_{\alpha'} \subset K_{\alpha''}$ or $K_{\alpha''} \subset K_{\alpha'}$ is true. Hence $\gamma \perp \delta$, i.e. K is an upper bound for the chain. ■

It is easy to construct maximal spacelike sets in the examples mentioned before:

Example 1: $\perp$ is the usual orthogonality, the maximal spacelike sets of atomic (minimal) indices are exactly the systems $\{\mathbb{C}e_1, \mathbb{C}e_2, \mathbb{C}e_3, \ldots\}$ where $\{e_1, e_2, e_3, \ldots\}$ runs through all orthonormal bases of S.

Example 2: Let Z^{n+1} be equipped with $\langle x,y\rangle := x_0y_0 - \sum_{p=1}^{n} x_py_p$.

We denote $x \in \mathbb{Z}^{n+1}$ by $x = \{x_0, \phi\}$, $x_0 \in Z$ time, $\phi \in \mathbb{Z}^n$ space. A maximal spacelike set of atomic indices is given by $M_{max} := \{(0,\phi):\phi\in\mathbb{Z}^n\}$. Here $\mathbb{Z}^{n+1}$

is equipped with the causal disjointness resulting from $M_0 := \{x: < x,x > < 0\}$.

If Z^{n+1} is equipped with $\perp$ as the set-theoretical complement, then only Z^{n+1} itself is a maximal spacelike set.

Now we are ready to introduce GPC. Let the four-tuple $\{I, \{A_\alpha\}_{\alpha \in I}, A, \perp\}$ be given. We introduce GPC in two versions. The first version is the strong version, where GPC is related to the four-tuple $\{I, \{A_\alpha\}_{\alpha \in I}, A, \perp\}$. The second version is a weaker version. It is additionally related to a state ω, i.e. it is related to the five-tuple $\{I, \{A_\alpha\}_{\alpha \in I}, A, \perp, \omega\}$.

DEFINITION 4. The four-tuple $\{I, \{A_\alpha\}_{\alpha \in I}, A, \perp\}$ is said to have the GPC-property if for each maximal spacelike set $M \subset I_{min}$ already the algebras A_α, $\alpha \in M$, generate the quasilocal algebra A, i.e.

$$A = \text{clo}_{||\cdot||} \cup_{\alpha \in M} A_\alpha .$$

The second version formulates the GPC-property with respect to a given state ω.

DEFINITION 5. The five-tuple $\{I, \{A_\alpha\}_{\alpha \in I}, A, \perp, \omega\}$ is said to have the GPC-property with respect to ω if for each maximal spacelike set $M \subset I_{min}$ the von Neumann algebra $\pi_\omega(A)''$ is generated already by the local von Neumann algebras $\pi_\omega(A_\alpha)''$ for $\alpha \in M$, where M is an arbitrary (but fixed) maximal spacelike set.

Obviously, if the four-tuple $\{I, \{A_\alpha\}_{\alpha \in I}, A, \perp\}$ is GPC according to Definition 4 then $\{I, \{A_\alpha\}_{\alpha \in I}, a, \perp, \omega\}$ is GPC according to Definition 5 for each state ω on A.

In the context of the Algebraic Quantum Field Theory this property (axiom) was introduced very early into the investigations by Haag/Schroer (see Haag/Schroer [9]). Loosely speaking in that context this axiom reflects the "existence of a field equation". For detailed motivation see the mentioned paper [9] (see also Horuzii [7]). Also some stronger requirements are discussed there.

In order to obtain results, using GPC, with respect to the problem of invariant centers for large index sets, we have to discuss the relevant index property. Some preparations follow.

Let $N \subset I$ be an arbitrary index set. We introduce

(15) $N^\perp := \{\beta \in I: \beta \perp \alpha \text{ for all } \alpha \in N\}$

$N^{\perp}$ is called the <u>causal complement</u> of N. According to the properties (1)-(5) of the relation $\alpha \perp \beta$ we obtain the following simple properties of causal complements:

(16) $N \subseteq N^{\perp\perp}$

(17) $N \cap N^{\perp} = \emptyset$ for all N, hence also $N^{\perp} \cap N^{\perp\perp} = \emptyset$,

(18) if sup N exists then $\{\sup N\}^{\perp} = N^{\perp}$,

(19) $(\cup_j N_j)^{\perp} = \cap_j N_j^{\perp}$ for arbitrary collections $\{N_j\}_j$.

These properties imply further properties:

(20) For each $\alpha \in I$ one has $\{\alpha\}^{\perp} \neq \emptyset$, and, moreover, for each finite index set $\{\alpha_1, \alpha_2, \ldots \alpha_n\}$ one has $\{\alpha_1, \alpha_2, \ldots \alpha_n\}^{\perp} \neq \emptyset$,

(21) $N_1 \subseteq N_2$ implies $N_2^{\perp} \subset N_I^{\perp}$,

(22) $N^{\perp\perp\perp} = N^{\perp}$.

Note that if sup N exists then $\sup N \in N^{\perp\perp}$. Namely, $\beta \in N^{\perp}$ means $\beta \perp \alpha$ for all $\alpha \in N$, hence according to (4) this implies $\beta \perp \sup N$. But this is valid for all $\beta \in N^{\perp}$, hence $\sup N \in N^{\perp\perp}$. In general, if sup N exists, $\sup N^{\perp\perp}$ could be non-existent. This possibility we exclude by the following (additional) requirement.

(23) If sup N exists then also $\sup N^{\perp\perp}$ exists.

Obviously in this case (i.e. if (23) is satisfied) always

(24) $\sup N \leq \sup N^{\perp\perp}$

is true. Note that, according to the requirement (23), in particular $N^{\perp\perp}$ has a supremum if N is finite. Now we introduce the concept of a <u>causal closed</u> index.

DEFINITION 6. The index $\alpha \in I$ is called <u>causal closed</u> if $\alpha = \sup\{\alpha\}^{\perp\perp}$.

Note that $\sup\{\alpha\}^{\perp\perp}$ exists according to the requirement (23), further $\alpha \leq \sup\{\alpha\}^{\perp\perp}$ is obvious according to (24), hence α is causal closed if $\sup\{\alpha\}^{\perp\perp} \leq \alpha$. In particular, we consider causal closed indices that can be generated by spacelike sets.

Let $M \subset I_{min}$ and M spacelike and sup M exist. Then also $\sup M^{\perp\perp}$ exists.

DEFINITION 7. Let $\alpha \in I$. α is called <u>spacelike generated</u> if there is a spacelike set $M \subset I_{min}$ such that $\alpha = \sup M^{\perp\perp}$.

LEMMA 3. <u>If</u> $\alpha \in I$ <u>is spacelike generated then necessarily</u> α <u>is causal closed.</u>

Proof. Let $\alpha = \sup M^{\perp\perp}$, $M \subset I_{min}$, M spacelike, sup M exists. According to (18) and (22) we have

$$(\sup M^{\perp\perp})^{\perp} = (M^{\perp\perp})^{\perp} = M^{\perp\perp\perp} = M^{\perp},$$

hence

$$(\sup M^{\perp\perp})^{\perp\perp} = M^{\perp\perp}$$

follows. Thus we obtain

$$\sup\{(\sup M^{\perp\perp})^{\perp\perp}\} = \sup M^{\perp\perp},$$

i.e.

$$\sup\{\alpha\}^{\perp\perp} = \alpha \quad ■$$

The special set of spacelike generated indices we use in the next proposition where implications from GPC with respect to centers of local algebras are formulated. But firstly we check the property "spacelike generated" with respect to our examples:

Example 1: each index is spacelike generated because $R = \mathrm{spa}\{Ce_1, \ldots Ce_k\}$ where $\{e_1, e_2, \ldots, e_k\}$ denotes an ONB of R.

Example 2: if $^{\perp}$ means set-theoretical disjointness, obviously each index is spacelike generated. In the "Minkowskian case" of Example 2 we can form spacelike generated indices taking the maximal spacelike set $M_0 := \{(0, \phi), \phi \in Z_n\}$ and choosing a finite subset $M_{fin} \subset M_0$. Then $\sup M_{fin}^{\perp\perp}$ exists (i.e. the requirement (23) is fulfilled) and $\sup M_{fin}^{\perp\perp}$ is spacelike generated. Now we prove:

PROPOSITION 3. <u>Let the four-tuple</u> $\{I, \{A_\alpha\}_{\alpha\in I}, A, ^{\perp}\}$ <u>be given and assume (23) to be satisfied. Let</u> ω <u>be a state of</u> A <u>and assume that the GPC-property with respect to</u> ω <u>is satisfied. Then: for all spacelike generated indices</u> $\alpha \in I$ <u>one has</u> $Z_\omega(\alpha) \subseteq Z_\omega$, <u>where</u> Z_ω <u>denotes the center of</u> $\pi_\omega(A)''$ <u>and</u> $Z_\omega(\alpha)$ <u>the center of</u> $\pi_\omega(A_\alpha)''$.

Proof. Let $\alpha = \sup M^{\perp\perp}$, M spacelike. We consider $\{\alpha\}^{\perp} = \{\beta: \beta^{\perp}\alpha\}$. Now we choose a set $N \subset I_{min}$ which is maximal spacelike as a subset of $\{\alpha\}^{\perp}$ (i.e. there are maximal spacelike subsets relative with respect to $\{\alpha\}^{\perp}$). Then $M \cup N$ is obviously maximal spacelike with respect to the whole index set I. We put $M_\gamma := \pi_\omega(A_\gamma)''$, $\gamma \in I$. Then we have

$$M_\alpha \vee \bigvee_{\beta\in\{\alpha\}^{\perp}} M_\beta = R := \pi_\omega(A)'' \tag{25}$$

because the algebras M_m, M_n with $m \in M$, $n \in N$ generate R according to the GPC-assumption. Now we have

$$M_\beta \subseteq M_\alpha^c , \qquad \beta \in \alpha^\perp ,$$

because $M_\beta \subseteq R$ and $M_\beta \subseteq M'_\alpha$, i.e. $M_\beta \subseteq R \cap M'_\alpha$. Hence

$$\bigvee_{\beta \in \{\alpha\}^\perp} M_\beta \subseteq M_\alpha^c$$

follows. According to (25) this implies

$$M_\alpha \vee M_\alpha^c = R.$$

Hence we have

$$M'_\alpha \cap (M_\alpha^c)' = R',$$

but $(M_\alpha^c)' = (M'_\alpha \cap R)' = M_\alpha \vee R'$, i.e. we obtain

$$M'_\alpha \cap (M_\alpha \vee R') = R',$$

hence

(26) $\qquad M'_\alpha \cap M_\alpha \subset M'_\alpha \cap (M_\alpha \vee R') = R'$

follows. But from $M_\alpha \subseteq R$ we obtain $M_\alpha \cap M'_\alpha \subseteq M_\alpha \subseteq R$ and together with (26)

(27) $\qquad M'_\alpha \cap M_\alpha \subseteq R \cap R'$

is obtained. But (27) means

$$Z_\omega(\alpha) \subseteq Z_\omega. \quad \blacksquare$$

Remark. For Example 1 it is easy to construct nets that satisfy GPC with respect to some state. Take an ONB in S, $e_1, e_2, e_3, \ldots$, then the set $\{Ce_1, Ce_2, Ce_3, \ldots\}$ is maximal spacelike. Now we choose the CCR-algebra $A := CCR(S)$ as the quasilocal algebra. Then the localization we define by $A_R := CCR(R)$, $R \in I$, i.e. $1 \leq \dim R < \infty$. Further we choose the Fock state ω, defined by $\omega(W(f)) := e^{-\frac{1}{4}||f||^2}$, where $W(f)$ denote the generating "Weyl" elements of A. Then one has

(28) $$\pi_\omega(A)'' = \left(\bigcup_{\rho=1}^{\infty} \pi_\omega(A_{Ce_\rho})\right)'' = \left(\bigcup_{\rho=1}^{\infty} \pi_\omega(A_{Ce_\rho})''\right)''$$

and (28) expresses the GPC-property.

Combining Proposition 1 and Proposition 3 we obtain the following result.

THEOREM. <u>Let the five-tuple</u> $\{I, \{A_\alpha\}_{\alpha \in I}, A, \perp, \omega\}$ <u>be given. Assume the "relative bicommutant property" (Definition 3) and GPC</u>

(Definition 5) to be satisfied. Then for all spacelike generated indices $\alpha \in I$ the centers $Z_\omega(\alpha) = (\pi_\omega(A_\alpha))'' \cap (\pi_\omega(A_\alpha))'$ coincide with the center $Z_\omega = (\pi_\omega(A))'' \cap (\pi_\omega(A))'$.

Proof. Combine Proposition 1 and Proposition 3. ∎

Remark. In the case that ω is a factor state, i.e. $Z_\omega = \mathbb{C}1$, the assumption "relative bicommutant property" can be dropped in the Theorem.

Example 1: The Fock state ω considered above is irreducible, hence a factor state. Further: each index is spacelike generated and the GPC-property is satisfied. Hence the theorem yields $Z_\omega(\alpha) = \mathbb{C}1$ for all indices α.

4. WIGHTMAN'S INEQUALITY

In the general theory of quasilocal algebras Wightman's inequality plays a decisive role because it gives information on the behavior of the local algebras A_α if one "enlarges" the index, i.e. if one considers A_β with $\alpha < \beta$. In the present context Wightman's inequality is considered under the assumption of GPC. As already mentioned in Section 2 one can study this inequality also under quite different assumptions. To get smooth results we introduce additionally a homogeneity assumption concerning the local algebras for atomic indices.

(29) All algebras A_α, $\alpha \in I_{min}$, are mutually algebraically isomorphic.

Moreover, we consider only non-abelian quasilocal algebras, i.e. firstly we require

(30) A is non-abelian

and secondly we consider only states ω such that

(31) $\pi_\omega(A)''$ is non-abelian, i.e. $\pi_\omega(A)'' \supset Z_\omega(A)$

Under these additional assumptions the non-abelianess is implied also for all local algebras $\pi_\omega(A_\alpha)''$, if GPC is assumed.

LEMMA 4. Let the five-tuple $\{I, \{A_\alpha\}_{\alpha\in I}, A, \perp, \omega\}$ be given. Let (29), (31) be satisfied and assume GPC. Then $\pi_\omega(A_\alpha)''$ is non-abelian, i.e. $\pi_\omega(A_\alpha)'' \supset Z_\omega(\alpha)$ for all $\alpha \in I$.

Proof. Assume that there is an index α_0 with $\pi_\omega(A_{\alpha_0})'' = Z_\omega(\alpha_0)$.

Now take an atomic index $\beta_0 < \alpha_0$. Then

$$\pi_\omega(A_{\beta_0})'' \subset \pi_\omega(A_{\alpha_0})'' = Z_\omega(\alpha_0),$$

i.e. $\pi_\omega(A_\beta)''$ is abelian. According to (29) for each atomic index β one has $\pi_\omega(A_\beta)''$ is algebraically isomorphic to $\pi_\omega(A_{\beta_0})''$, in particular $\pi_\omega(A_\beta)''$ is abelian. However, if β_1 and β_2 are atomic and $\beta_1 \perp \beta_2$ then $\pi_\omega(A_{\beta_1})'' \vee \pi_\omega(A_{\beta_2})''$ is also abelian, because

$$\pi_\omega(A_{\beta_1})'' \subseteq \pi_\omega(A_{\beta_1})',$$

$$\pi_\omega(A_{\beta_2})'' \subseteq \pi_\omega(A_{\beta_1})',$$

hence

$$\pi_\omega(A_{\beta_1})'' \vee \pi_\omega(A_{\beta_2})'' \subseteq \pi_\omega(A_{\beta_1})'$$

and the same inclusion is valid for β_2, so one has

$$\pi_\omega(A_{\beta_1})'' \vee \pi_\omega(A_{\beta_2})'' \subseteq \pi_\omega(A_{\beta_1})' \cap \pi_\omega(A_{\beta_2})' = (\pi_\omega(A_{\beta_1}) \vee \pi_\omega(A_{\beta_2}))'$$

Obviously, this remains true also for an arbitrary set of mutually causally disjoint atomic indices. Hence, according to GPC, one obtains

$$\pi_\omega(A)'' = \bigvee_{\beta \in M} \pi_\omega(A_\beta)'', \qquad M \text{ maximal spacelike},$$

is abelian which contradicts (31). ∎

LEMMA 5. <u>Under the assumptions of Lemma 4 one has</u> $\pi_\omega(A_\alpha)'' \subset \pi_\omega(A)''$ <u>for all</u> $\alpha \in I$

Proof. Let $\alpha \in I$. Then there is $\beta \in I$ with $\beta \perp \alpha$, i.e. A_β commutes with A_α by elements. Assuming $\pi_\omega(A_\alpha)'' = \pi_\omega(A)''$ we would obtain

$$\pi_\omega(A_\beta)'' \subseteq Z_\omega(A)$$

i.e. $\pi_\omega(A_\beta)''$ is abelian which contradicts Lemma 4. ∎

Putting together the propositions of Lemma 4 and Lemma 5 we obtain that under the mentioned assumptions (of Lemma 3, 4) always

$$Z_\omega(A_\alpha) \subset \pi_\omega(A_\alpha)'' \subset \pi_\omega(A)'', \qquad \alpha \in I,$$

is true. Now we can prove, under the same assumptions, the desired inequality.

PROPOSITION 4 (Wightman's inequality). <u>Let the five-tuple</u> $\{I, \{A_\alpha\}_{\alpha\in I}, A, \perp, \omega\}$ <u>be given. Let (29), (31) be satisfied and assume GPC. Further let</u> $\alpha_2 < \alpha_1$ <u>be indices such that there exists an index</u> β <u>with</u> $\beta < \alpha_1$ <u>and</u> $\beta \perp \alpha_2$. <u>Then</u>

$$\pi_\omega(A_{\alpha_2})'' \subset \pi_\omega(A_{\alpha_1})'',$$

i.e. the inclusion is proper.

Proof. $\pi_\omega(A_\beta)''$ commutes with $\pi_\omega(A_{\alpha_2})''$ by elements according to the assumption $\beta \perp \alpha_2$. Furthermore, $\pi_\omega(A_\beta)'' \subseteq \pi_\omega(A_{\alpha_1})''$. If one assumes $\pi_\omega(A_{\alpha_2})'' = \pi_\omega(A_{\alpha_1})''$ then one obtains that $\pi_\omega(A_\beta)''$ commutes with $\pi_\omega(A_{\alpha_1})''$ by elements or

$$\pi_\omega(A_\beta)'' \subset \pi_\omega(A_{\alpha_1})'$$

hence

$$\pi_\omega(A_\beta)'' \subset Z_\omega(\alpha_1)$$

follows, i.e. $\pi_\omega(A_\beta)''$ is abelian in contradition to Lemma 4. ■

REFERENCES

[1] G. Emch: Algebraic Methods in Statistical Mechanics and Quantum Field Theory, Mir, Moscow 1976 (Russian translation).

[2] O. Bratteli, D.W. Robinson: Operator Algebras and Quantum Statistical Mechanics I, Springer-Verlag, New York 1979.

[3] R. Haag, D. Kastler: An Algebraic Approach to Quantum Field Theory, J. Math. Phys. 5, Nr. 7, 848-861 (1964).

[4] R. Haag: Local relativistic quantum physics, Physica, 124 A, 357-364 (1984).

[5] D. Buchholz, C. D'Antoni, K. Fredenhagen: The universal structure of local algebras, Commun. Math. Phys. 111, 113-135 (1987).

[6] M. Wollenberg: On causal nets of algebras, Proceedings Internat. Conference on Operator Theory, Romania 1988, to appear.

[7] S.S. Horuzii: Introduction to algebraic quantum field theory, Nauka, Moscow 1986 (in Russian).

[8] R.V. Kadison, J.R. Ringrose: Fundamentals of the Theory of Operator Algebras I, Academic Press, New York 1983.

[9] R. Haag, B. Schroer: Postulates of quantum field theory, J. Math. Phys. 3, Nr. 2, 248-256 (1962).

Prof. Hellmut Baumgärtel
Karl-Weierstraß Institut fur Mathematik
Akademie der Wissenschaften der DDR
Mohrenstr. 39
Berlin
DDR - 1086

Operator Theory:
Advances and Applications, Vol. 41

THE ANALOGUE OF KURODA'S THEOREM FOR n-TUPLES

Hari Bercovici and Dan Voiculescu

Dedicated to Professor Israel Gohberg on the occasion of his sixtieth birthday.

Let H denote a complex, separable, infinite-dimensional Hilbert space, and denote by CSA(n,H) the set of commuting n-tuples of selfadjoint operators on H. A normed ideal $(J, |\ |_J)$ of operators on H will be called a **diagonalization ideal** if for every t in CSA(n,H) there is a diagonalizable t' in CSA(n,H) such that $t-t' \in J^n$ (where the exponent indicates the Cartesian product of n copies of J). An ideal which is not a diagonalization ideal will be called an **obstruction ideal.** We shall only consider symmetrically normed ideals (or s.n. ideals); cf. [1].

When n=1 it was shown by Weyl and von Neumann that the ideals of compact and Hilbert-Schmidt operators, respectively, are diagonalization ideals. Later Kuroda proved that any normed ideal different from the trace class C_1 is a diagonalization ideal; see Theorem 2.3 in Chapter X of [2]. On the other hand, Kato and Rosenbloom proved that C_1 is indeed an obstruction ideal.

If $n\geq 2$ these problems were considered in [3]. It was shown there that the Schatten ideal C_n is a diagonalization ideal, and an ideal denoted $C_n^- \subset C_n$ (to be described shortly) is an obstruction ideal. In this paper we prove that in fact C_n^- is the largest obstruction ideal, thus extending Kuroda's result.

We record for further use some notation and a result from [3]. If $t = (T_1, T_2, \dots, T_n) \in CSA(n,H)$ and A is an operator on H then the commutator [A,t] is

$$[A,t] = ([A,T_1],[A,T_2],\dots,[A,T_n]),$$

and

$$[A,t]_J = \max\left\{|[A,T_j]|_J:\ 1\leq j\leq n\right\}.$$

The set R_1^+ consists of all operators A of finite rank such that $0 \leq A \leq I$; R_1^+ is a directed set in the natural order. Thus one can define the number

$$k_J(t) = \limsup_{A \in R_1^+} |[A,t]|_J$$

for every t in CSA(n,H).

1. PROPOSITION. (2.6 of [3]) If J is a separable s.n. ideal and $t \in$ CSA(n,H) then the following are equivalent:

a. there is a diagonalizable $t' \in$ CSA(n,H) such that $t-t' \in J^n$;

b. for every $\varepsilon > 0$ there exists a diagonalizable t' such that $t-t' \in J^n$ and $|t-t'|_J < \varepsilon$;

c. $k_J(t) = 0$.

We will use a special kind of ideals which we presently describe. Let $P = (p_j)_{j \in N}$ be a sequence of positive numbers decreasing to zero. The s.n. ideal $(J_P, |\ |_P)$ (cf. Chapter III in [1]) is defined as the set of those compact operators T such that

$$|T|_P = \sum_j p_j s_j(T) < \infty ,$$

where $s_j(T)$ denote the singular values of T, i.e., the eigenvalues of $(T^*T)^{\frac{1}{2}}$. If $p_j = j^{-1+1/n}$ then J_P will be denoted by C_n^-.

2. LEMMA. If J_P is not contained in C_n^- then

$$\liminf_{n \to \infty} \frac{p_1+p_2+\ldots+p_{m^n}}{m} = 0 .$$

Proof. Denote $P(k) = \sum_{j=1}^k p_j$ and $N(k) = \sum_{j=1}^k j^{-1+1/n}$. It is easily seen that there is a constant C>1, independent of m, such that

$$C^{-1}\frac{P(m^n)}{m} \leq \frac{P(k)}{N(k)} \leq C\frac{P((m+1)^n)}{m+1}$$

for all $k \in [m^n,(m+1)^n]$. Hence the assertion to be established is equivalent to $\liminf_k P(k)/N(k) = 0$. Assume that this is not true, so that $P(k) \geq MN(k)$ for some $M > 0$. Choose $T \in J_P \setminus C_n^-$ and note that

$$\begin{aligned}\sum_j s_j(T) j^{-1+1/n} &= \sum_j (s_j(T)-s_{j+1}(T))N(j) \\ &\leq M^{-1}\sum_j (s_j(T)-s_{j+1}(T))P(j) \\ &= M^{-1}|T|_P < \infty,\end{aligned}$$

which is a contradiction. QED

We are now able to prove the analogue of Kuroda's theorem for ideals of the form J_P.

3. PROPOSITION. If J_P is not contained in C_n^- then J_P is a diagonalization ideal.

Proof. We must prove that $k_p(t) = 0$ for every $t \in CSA(n,H)$, and by the results of [3] (see Proposition 1.4) we may restrict ourselves to the case in which t is a cyclic n-tuple. Assume therefore that t has a cyclic vector x. With the notation in the proof of Lemma 2, there are integers $m_j \to \infty$ such that $\lim_j P(m_j^n)/m_j = 0$. For each j we can partition the cube $[-\|t\|,\|t\|]^n$ into m_j^n Borel sets of diameter $\leq 2n^{\frac{1}{2}}\|t\|m_j^{-1}$. Denote by $Q_s^{(j)}$, $1 \leq s \leq m_j^n$, the spectral projections of t corresponding with these Borel subsets, and let $R_s^{(j)}$ denote the orthogonal projection onto the one-dimensional space generated by $Q_s^{(j)}x$ (set $R_s^{(j)} = 0$ if $Q_s^{(j)}x = 0$). Finally, let $P_j = \sum_{s=1}^{m_j^n} R_s^{(j)}$. We have

$$\begin{aligned}|[P_j,t]|_P &\leq 2P(m_j^n)\|[P_j,t]\| \\ &\leq 8P(m_j^n)n^{\frac{1}{2}}\|t\|m_j^{-1} \to 0\end{aligned}$$

as $j \to \infty$, so that $k_P(t) = 0$ by the results of [3]. Note that in the first inequality above we used the fact that the rank of $[P_j,t]$ is at most $2m_j^n$. QED

Finally, we extend the analogue of Kuroda's theorem to arbitrary s.n. ideals.

4. THEOREM. A s.n. ideal J is an obstruction ideal if and only if $J \subset C_n^-$.

Proof. We already know from [3] that C_n^-, and hence any smaller ideal, is an obstruction ideal. Conversely, assume that J is not contained in C_n^-. Denote by J_0 the closure in J of the ideal of finite-rank operators. We claim that J_0 is not contained in C_n^-. Indeed, if $J_0 \subset C_n^-$ we would have

$$\begin{aligned}J &\subset \{X: \sup_{A \in R_1^+} |XA|_J < \infty\} \\ &\subset \{X: \sup_{A \in R_1^+} |XA|_n^- < \infty\} = C_n^- ,\end{aligned}$$

(We recall that $|XA|_n^-$ denotes the norm of XA in C_n^-.) It suffices to show that J_0 is a diagonalization ideal, hence we may as well assume that J is separable. Assume now, to get a contradiction, that J is an obstruction ideal. By Proposition 1 there exists $t \in CSA(n,H)$ such that $k_J(t) > 0$. Now, (2.6) in [4] implies the existence of an ideal $J_P \supset J$ such that $k_P(t) > 0$. But then J_P is not contained in C_n^-, and Proposition 3 implies that J_P is a diagonalization ideal. Hence $k_P(t) = 0$, a contradiction. QED

Both authors were supported in part by grants from the National Science Foundation.

REFERENCES

1. I.T. Gohberg and M.G. Krein, Introduction to the theory of non-selfadjoint operators, Nauka, Moscow, 1965.

2. T. Kato, Perturbation theorey for linear operators, Springer, New York, 1966.

3. D. Voiculescu, Some results on norm-ideal perturbations of Hilbert space operators, J. Operator Theory 2(1979), 3-37.

4. D. Voiculescu, On the existence of quasicentral approximate units relative to normed ideals. I, preprint.

H. Bercovici,
Department of Mathematics
Indiana University
Bloomington, IN 47405
U.S.A.

D. Voiculescu,
Department of Mathematics
University of California
Berkeley, CA 94720
U.S.A.

Operator Theory:
Advances and Applications, Vol. 41

THE GEOMETRY OF REPRESENTING MEASURES AND THEIR CRITICAL VALUES

Kevin F. Clancey

On the occasion of the 60th birthday of Israel Gohberg

The author [1] has recently described a smooth parametrization of the convex set of representing measures M_0 for evaluation of analytic functions at a point q_0 in a g holed planar domain with analytic boundary. Here this parametrization is used to provide answers to three questions from the literature concerning the geometry of M_0 and the nature of the critical values of elements in M_0. First, in answer to a question of Nash [5] it is shown that for domains with $g > 2$ holes the set M_0 is not strictly convex. Further the details of a class of symmetric examples are completed. In these examples it is shown that M_0 is the span of 2^g isolated extreme points answering a question of Nash [5]. Finally, in answer to a question of Sarason [6] it is shown that elements in M_0 can have "double" critical values on the boundary of the domain.

PRELIMINARIES

Let D be a g-holed planar domain with positively oriented analytic boundary $\partial D = b_0 \cup b_1 \cup \cdots \cup b_g$ where it is assumed that the simple closed disjoint analytic curves $b_1, \cdots, b_g$ bound the holes of D and that b_0 is the boundary of the unbounded component of the complement of the closure of D. For q_0 in D, let M_0 denote the convex set of representing measures for evaluation at q_0. Thus the elements m in M_0 are non-negative measures in the space $M_{\mathbf{R}}(\partial D)$ of real measure on ∂D such that $f(q_0) = \int_{\partial D} f \, dm$ for every f in the uniform closure $R = R(\overline{D})$ in $C(\partial D)$ of the algebra of rational functions with poles off the closure of D.

The analytic nature of representing measures is made clear on the double X of the finite Riemann surface

$\overline{D}$. Recall that the double $X = D \cup \partial D \cup D'$, where D' is a second copy of D glued to D along ∂D. The conformal structure on D' is the conjugate of the usual conformal structure on D. Thus the involution $J: X \to X$ which fixes ∂D and interchanges points in D with their twins in D' is anticonformal. We mark the double by completing the cycles $b_1, \cdots, b_g$ to a canonical homology basis as follows. Fix p_0 in b_0 and let α_j be a crosscut in D from p_0 to b_j, $j = 1, \cdots, g$. Set $a_j = \alpha_j \cup -J\alpha_j$, $j = 1, \cdots, g$. Then $a_1, \cdots a_g$; $b_1, \cdots, b_g$ is a canonical homology basis for X having the requisite intersection properties. It will henceforth be assumed that X is the double marked with this homology basis.

Let $G(p) = G(p, q_0)$ denote the Green's function for D with pole at q_0. The meromorphic differential $dw_0 = \frac{-1}{\pi i} \partial G dz$ on $\overline{D}$ can be reflected to D' by setting

$dw_0 = \overline{J^* dw_0}$ on D'. The restriction dm_0 of dw_0 to ∂D is harmonic measure on ∂D based at q_0. That is

$$dm_0 = -\frac{1}{2\pi} \frac{\partial G}{\partial \eta} ds = -\frac{1}{\pi i} \partial G dz|_{\partial D}$$

where $\frac{\partial}{\partial \eta}$ is the outward normal derivative and ds is arclength measure on ∂D. Thus the representing measure dm_0 is the restriction to ∂D of an element dw_0 in the space of meromorphic differentials $\mathcal{M}^{(1)}(X)$ which is symmetric $(J^* dw_0 = \overline{dw_0})$ and non-negative $(\frac{dw_0}{ds} \geq 0)$ on ∂D. We next observe this is true of all representing measures.

Let $\omega_j = \omega_j(z)$ be the harmonic measure of b_j based at z in D and dw_j be the reflection of the holomorphic differentials $\partial \omega_j dz$ to holomorphic differentials on X. The measures $dm_j = i dw_j|_{\partial D}$ $j = 1, \cdots, g$ are real and form a basis for $R^{\perp}$ in $M_{\mathbf{R}}(\partial D)$.

Any element m in M_0 has the form

$m = m_0 + \sum_{j=1}^{g} c_j m_j$ where $c_m = (c_1,\cdots,c_g)^t$ is in $\mathbf{R}^g$. Thus dm is the restriction to ∂D of the element $dw = dw_0 + i\sum_{j=1}^{g} c_j dw_j$ in $\mathcal{M}^{(1)}(X)$. This observation identifies M_0 with the collection of elements dw in $\mathcal{M}^{(1)}(X)$ which are symmetric $(J^*dw = \overline{dw})$, non-negative $(\frac{dw}{ds} \geq 0)$, having only simple poles at q_0, Jq_0 with $2\pi i \operatorname{Res}_{p=q_0} dw = 1$.

For the remainder of this paper it will be assumed the basis $m_1,\ldots,m_g$ of $R^{\perp}$ is fixed as above. Using this basis the coefficient map $W_0: M_0 \to \mathbf{R}^g$, defined by $W_0(m) = c_m$, provides an affine linear bijection between M_0 and the compact convex body $C_0 = W_0(M_0)$ in $\mathbf{R}^g$. The fact that $\frac{\partial G}{\partial \eta} > 0$ on ∂D insures that C_0 contains a neighborhood of the origin in $\mathbf{R}^g$ and, consequently, C_0 is g dimensional.

The notion of critical divisor of a representing measure will be important in the sequel. The following conventions will be used on the divisor group $\mathrm{Div}(X)$ of X. Divisors will be written additively, so that the typical element $\mathcal{D}$ in $\mathrm{Div}(X)$ is a finite sum $\mathcal{D} = \sum_{p\in X} n_p p$, $n_p \in \mathbf{Z}$. Addition and comparison of divisors is done pointwise and $\deg \mathcal{D} = \sum n_p$ will be used to denote the degree of the divisor $\mathcal{D} = \sum n_p p$. The collection of non-negative divisors $\mathcal{D} = p_1 + \cdots + p_g$ of degree g will be identified with the g fold symmetric product $X^{(g)}$ of X. Recall $X^{(g)} = X^g/S_g$, where S_g is the symmetric group on g letters, is a compact g dimensional complex space.

Any meromorphic differential dw has a pole-zero divisor (dw) of degree $2g-2$. If dw in $\mathcal{M}^{(1)}(X)$ restricted to ∂D belongs to M_0, then

$$(dw) = \mathcal{D} - q_0 + J(\mathcal{D} - q_0) \ , \qquad (1.1)$$

where $\mathcal{D} = p_1 + \cdots + p_g$ is in $X^{(g)}$. It is crucial to note that the representation (1.1) is not unique. In fact if the divisor of dw restricted to D has the form $n_1p_1 + \cdots + n_sp_s$ where $p_1, \cdots, p_s$ are distinct, then (dw) can be written in the form (1.1) in $(n_1 + 1)(n_2 + 1)\cdots(n_s + 1)$ distinct ways. Usually, (dw) has g distinct points in D and then the decomposition (1.1) can be done in 2^g ways.

There is only one way of writing the divisors (dw) in the form (1.1) with $\mathcal{D}$ in $\overline{D}^{(g)}$. This divisor $\mathcal{D}_m$ will be called the critical divisor of the representing measure $dm = dw|_{\partial D}$. The divisor $\mathcal{D}_m$ is supported on the points in $\overline{D}$ where $\frac{dw}{dz} = 0$. Those zeros of $\frac{dw}{dz}$ in D are counted in $\mathcal{D}_m$ with exact multiplicity, whereas, the critical values on ∂D are only counted in $\mathcal{D}_m$ with one-half multiplicity. For example, a double critical value in $\mathcal{D}_m$ at p in ∂D means $\frac{dw}{dz}$ has a fourth order zero at p.

The collection of critical divisors in $\overline{D}^g$ of elements in M_0 will be denoted by B_0. The natural bijection which associates with each m in M_0 its critical divisor $\mathcal{D}_m$ in B_0 will be denoted by $U_0 : M_0 \rightarrow B_0$. Sitting over B_0 in $X^{(g)}$ is the collection V_0 of divisors $\mathcal{D}$ providing the decomposition (1.1) for elements dw in $\mathcal{M}^{(1)}(X)$ whose restriction to ∂D belong to M_0. The usual retraction $r : X^{(g)} \rightarrow \overline{D}^{(g)}$ induces a "covering" map $r_0 : V_0 \rightarrow B_0$ with $\mathcal{D}_m|_D = n_1p_1 + \cdots + n_sp_s$, where $p_1, \cdots, p_s$ are the distinct critical values of m in D.

The Abel Jacobi map allows one to identify V_0 with the real torus $\mathbf{R}^g/\mathbf{Z}^g$. We first recall the essentials of the Abel-Jacobi map. The holomorphic differentials $dw_1, \cdots, dw_g$ introduced above form a basis for the space $\Omega(X)$ of holomorphic one forms on X which is dual to our canonical homology basis. This means the following. Set

$\vec{dw} = (dw_1, \cdots, dw_g)^t$. The $g \times 2g$ Riemann period matrix has the form

$$\left[\int_{a_1} \vec{dw} \cdots \int_{a_g} \vec{dw} : \int_{b_1} \vec{dw} \cdots \int_{b_g} \vec{dw}\right] = [I : \tau]$$

where I is the $g \times g$ identity matrix. It follows from Riemann's bilinear relations that the $g \times g$ symmetric matrix τ has the form $\tau = iP$ with $P^t = P > 0$. The complex torus $Jac(X) = \mathbf{C}^g/\mathbf{Z}^g + \tau\mathbf{Z}^g$ is called the Jacobian variety of the marked Riemann surface X . Note that because we are working with a double of a planar domain the anticonformal map $J[z] = -[\bar{z}]$ is well defined on $Jac(X)$, where $[z]$ denotes the class of z in $\mathbf{C}^g$ modulo the period lattice $\mathbf{Z}^g + \tau\mathbf{Z}^g$.

The Abel Jacobi map $\xi_0 : X \to Jac(X)$ based at p_0 is defined by

$$\xi_0(p) = \int_{p_0}^{p} \vec{dw} \bmod(\mathbf{Z}^g + \tau\mathbf{Z}^g) .$$

This mapping can be extended linearly to $Div(X)$. Jacobi's theorem states that the holomorphic mapping $\xi_0 : X^{(g)} \to Jac(X)$ is surjective and Abel's theorem establishes that $\xi_0(\mathcal{D}_1) = \xi_0(\mathcal{D}_2)$ for divisors $\mathcal{D}_1, \mathcal{D}_2$ of the same degree if and only if $\mathcal{D}_1 = \mathcal{D}_2 + (f)$ for some element f in the space of meromorphic functions on X . Note that here we will always assume the base point p_0 is in b_0 . This leads to the symmetry $\xi_0 \circ J = J \circ \xi_0$.

It is slightly more convenient to work with a translate of the Abel Jacobi map. Let Δ_0 be the classical Riemann constant in $Jac(X)$ computed for the marked Riemann surface X with base point p_0 in b_0 . The explicit form of Δ_0 is

$$\Delta_0 = \left[- \sum_{k=1}^{g} \left\{ \int_{a_k} \vec{w}_0(p)\,dw_k - \frac{1}{2}\,\tau_{kk} e_k \right\} \right] ,$$

where $\vec{w}_0(p) = \int_{p_0}^{p} d\vec{w}$ and $e_1, \cdots, e_g$ is the standard basis in $\mathbf{C}^g$. The constant Δ_0 appears in Riemann's fundamental work on theta functions. One property of Δ_0 that is worth noting here is that $-2\Delta_0 = K_X^0$, where $K_X^0 = \xi_0((dw))$, for any dw in $\mathcal{U}^{(1)}(X)$. The constant Δ_0 also satisfies $J\Delta_0 = \Delta_0$.

Define $\Phi_0 : X^{(g)} \to \mathrm{Jac}(X)$ to be the holomorphic mapping

$$\Phi_0(\mathcal{D}) = \xi_0(\mathcal{D}) - \xi_0(q_0) + \Delta_0 . \tag{1.2}$$

We remark that Φ_0 is independent of the base point p_0 chosen for the Abel Jacobi map.

The following result formulated in [1] is a simple consequence of work of Fay [3] characterizing divisors of symmetric definite meromorphic differentials.

PROPOSITION 1.1. *Let* V_0 *be the collection of divisors* $\mathcal{D}$ *in* $X^{(g)}$ *providing the decomposition* (1.1) *for some* dw *in* $\mathcal{U}^{(1)}(X)$ *with* $dw|_{\partial D}$ *in* M_0. *A divisor* $\mathcal{D}$ *in* $X^{(g)}$ *belongs to* V_0 *if and only if* $\Phi_0(\mathcal{D})$ *is in the real g-dimensional torus* $T_0 = \mathbf{R}^g/\mathbf{Z}^g$ *in* $\mathrm{Jac}(X)$. *Indeed,* Φ_0 *is biholomorphic from a neighborhood of* V_0 *to a neighborhood of* T_0 *in* $\mathrm{Jac}(X)$.

At this stage of the discussion we have the following diagram of maps and spaces

$$\begin{array}{ccccc}
X^{(g)} \supset V_0 & & \xrightarrow[\cong]{\Phi_0} & & T_0 \subset \mathrm{Jac}(X) \\
\Big\downarrow r_0 & & & & \Big\downarrow \\
\overline{D}^{(g)} \supset B_0 & \xleftarrow[\cong]{U_0} & M_0 & \xrightarrow[\cong]{W_0} & C_0 \subset \mathbf{R}^g ,
\end{array} \tag{1.3}$$

where " ≅ " denotes "bijection".

The nice result is that Riemann's theta function provides a mapping $\pi_0 : T_0 \to C_0$ so that the diagram (1.3) commutes. Recall that given a symmetric $g \times g$ matrix τ with positive definite imaginary part $(\frac{1}{2i}(\tau + \tau^*) > 0)$ one defines on $\mathbf{C}^g$ the Riemann theta function

$$\theta(z) = \theta(z,\tau) = \sum_{n \in \mathbf{Z}^g} \exp\{2\pi i(\tfrac{1}{2}n^t\tau n + n^t z)\} .$$

The theta function is entire and quasi-periodic in the sense that

$$\theta(z + m + \tau n) = \exp\{-2\pi i(\tfrac{1}{2}n^t\tau n + n^t z)\}\theta(z)$$

for $m + \tau n$ in the lattice $\mathbf{Z}^g + \tau\mathbf{Z}^g$. The function θ is even.

From now on we will speak only of theta functions associated with period matrices $\tau = iP$ of the double of a planar domain. Such a theta function possesses the symmetry $\theta(Jz) = \overline{\theta(z)}$ and, consequently, is real on $\mathbf{R}^g \to \mathbf{C}^g$. It turns out that θ is actually positive on $\mathbf{R}^g$ (see, Fay [3] and [1]). The $\mathbf{Z}^g$ periodicity of θ implies that $\theta([x]) = \theta(x)$ is a well defined positive function on the torus $T_0 = \mathbf{R}^g/\mathbf{Z}^g$.

For q_0 in D let $\omega_0 = (\omega_1(q_0), \cdots, \omega_g(q_0))^t$, where as above $\omega_j(q_0)$ denotes the harmonic measure of b_j $(j = 1, \cdots, g)$ based at q_0. The following result was established in [1].

THEOREM 1.1. *The mapping* $\pi_0 : T_0 \to \mathbf{R}^g$ *defined by*

$$\pi_0([x]) = \frac{1}{2\pi}\vec{\nabla}\ln\frac{\theta(x)}{\theta(x + \omega_0)}$$

has range the convex set C_0. *Moreover, one has the commutative diagram*

$$\begin{array}{ccccc} V_0 & & \xrightarrow[\cong]{\Phi_0} & & T_0 \\ \big\downarrow r_0 & & & & \big\downarrow \pi_0 \\ B_0 & \xleftarrow[\cong]{U_0} & M_0 & \xrightarrow[\cong]{W_0} & C_0 \end{array} \tag{1.4}$$

where $\cong$ *denotes homeomorphism.*

The result in Theorem 1.1 provides an unexpected smooth parametrization of the convex set C_0. A priori there is no reason to expect the range of π_0 to be convex. The remainder of this paper will deal with applications of the commutative diagram to the geometry of M_0 and critical divisors. The explicit form of π_0 involving theta functions will not be important.

We conclude this section with some general remarks related to Theorem 1.1.

(i) It is quite easy to see that a measure m in M_0 belongs to the boundary of M_0 if and only if the critical divisor $\mathcal{D}_m$ has a point on ∂D . See, Sarason [6].

(ii) Let $\psi_0 : B_0 \to C_0$ be the homeomorphism $\psi_0 = W_0 \circ U_0^{-1}$. Then $\psi_0 \circ r_0 = \pi_0 \circ \phi_0$ so that the mappings $r_0 : V_0 \to B_0$ and $\pi_0 : T_0 \to C_0$ are topologically equivalent. These mappings are "branched" over points corresponding to ∂M_0 on points in M_0 having multiple critical values in D .

(iii) The involution $J : X^{(g)} \to X^{(g)}$ leaves V_0 invariant. It is easily verified that $\Phi_0 \circ J = R_0 \circ \Phi_0$, where $R_0 : T_0 \to T_0$ is the reflection $R_0(t) = -t - [\omega_0]$ in the point $-1/2[\omega_0]$. This reflection has the 2^g fixed points $-1/2[\omega_0 + \nu]$, $\nu \in \mathbf{Z}^g/2\mathbf{Z}^g$. The following two remarks are based on results of Sarason [6] and Nash [5].

(iv) The annihilator of $R(\overline{D})$ in $M_{\mathbf{R}}(\partial D)$ has been identified with the set of restrictions to ∂D of the real g dimensional vector space $\Omega_s(X)$ of symmetric holomorphic differentials on X . Thus given a measure ν in $R^{\perp}$ we let $\mathcal{D}_\nu$ be the divisor in $X^{(g-1)}$ supported on

$\overline{D}$ which provides the representation $(dw) = \mathcal{D}_\nu + J\mathcal{D}_\nu$, where dw in $\Omega_s(X)$ restricts to ν on ∂D . Sarason [6] has characterized the extreme points of M_0 in the following manner. A measure m in M_0 fails to be an extreme point if and only if $\mathcal{D}_\nu|_{\partial D} \geq \mathcal{D}_m|_{\partial D}$ for some ν in $R^\perp$.

(v) A non-trivial result of Nash [5] shows that if m is an isolated extreme point of M_0 , then $\mathcal{D}_m = p_1 + \cdots + p_g$ for g distinct points on ∂D . It will be seen that the converse of this last statement is not true. Note that if m in M_0 has critical divisor $\mathcal{D}_m$ supported on ∂D , then $\Phi_0(\mathcal{D}_m)$ is a fixed point of reflection in $-1/2[\omega_0]$. Consequently, there can be at most 2^g isolated extreme points of M_0.

EXAMPLES

The geometric theory of representing measures and their critical divisors as developed above can be nicely illustrated using a class of symmetric examples which were introduced by Sarason [6] and Nash [5]. Moreover, these examples can be combined with the above theory to answer questions from [5] and [6].

Every g holed planar domain with analytic boundary is conformally equivalent to a circular domain consisting of the open unit disc **D** from which g disjoint closed discs $\overline{D}_1, \cdots, \overline{D}_g$ have been punched. See, e.g., Goluzin [4] or Tsuji [7]. The examples considered by Nash [5] are such that the discs $D_1, \cdots, D_g$ are centered at $d_1, \cdots, d_g$ on the real axis. Consequently, the doubles $X = D \cup \partial D \cup D'$ of these domains possess the extra anticonformal symmetry $Q : X \to X$ of reflection of $\overline{D}$ and D' in the real axis. The composition $Q \circ J$ is a conformal involution with quotient the unit sphere. Thus these doubles are hyperelliptic with $Q \circ J$ the hyperelliptic involution.

Let D be such a symmetric domain. For definiteness it is assumed that the centers of the removed discs satisfy $d_1 < \cdots < d_g$. The intersection of D with the real axis is the union of the $g + 1$ open subintervals

$$I_0 = (s_0, s_1) \ , \ I_2 = (s_2, s_3), \ \cdots \ , \ I_g = (s_{2g}, s_{2g+1})$$

where $-1 = s_0 < s_1 < \cdots < s_{2g} < s_{2g+1} = 1$. The subsets $T_j = \bar{I}_j \cup J\bar{I}_j$, $j = 0,\cdots,g,$ form "circles" in the double. A convenient base point for all considerations is $p_0 = s_0 = -1$.

Natural crosscuts α_j from $p_0 = -1$ in b_0 to s_{2j-1} in $b_j = \partial D_j$, $j = 1,\cdots,g$, can be made by running along the axis and the top halves γ_k of the boundaries b_k of D_k , $k = 1,\cdots,g-1$. See, figure below for $g = 2$.

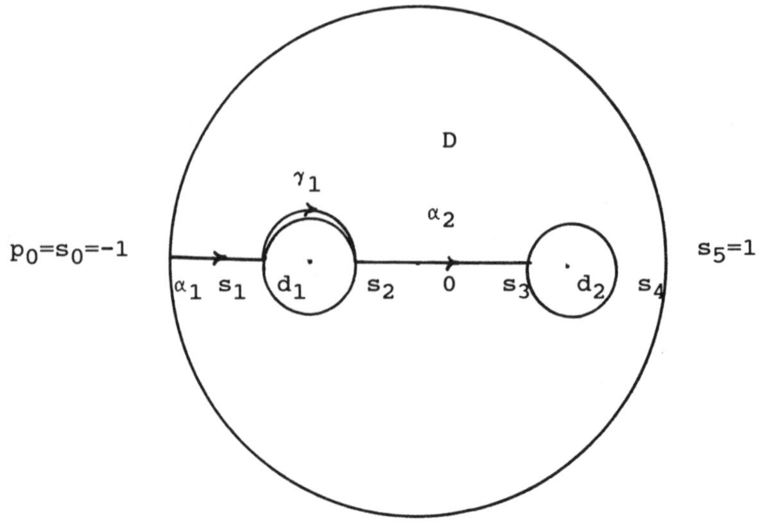

FIGURE

It follows from symmetry that

$$\int_{\gamma_k} d\vec{w} = \frac{1}{2}\, \tau e_k \ , \ k = 1,\cdots,g \ , \tag{2.1}$$

where τ is our usual period matrix. Note that in addition to the symmetry $J^*dw = -d\overline{w}$ our basis of $\Omega(X)$ satisfies $Q^*dw = d\overline{w}$.

The Riemann constant Δ_0 (based at $p_0 = -1$) can be computed to be $\Delta_0 = [\frac{1}{2}\vec{1} + \frac{1}{2}\tau\vec{g}]$ where $\vec{1} = (1,\cdots,1)^t$ and $\vec{g} = (g,g-1,\cdots,2,1)^t$.

Suppose q_0 in D is real. If dw_0 in $\mathcal{M}^{(1)}(X)$ restricts on ∂D to harmonic measure based at q_0 , then $Q^*dw_0 = d\overline{w}_0$. Consequently, both $Q^*dw = \overline{dw}$ and $J^*dw = \overline{dw}$ for any element dw in $\mathcal{M}^{(1)}(X)$ which restricted to ∂D is in M_0 . This implies the critical divisor $\mathcal{D}_m$ of any m in M_0 satisfies $Q\mathcal{D}_m = \mathcal{D}_m$.

If q_0 is real then it will be in one of the subintervals $I_0, I_1, \cdots, I_g$. Let I_g' be the intervals among $I_0, I_1, \cdots, I_g$ not containing q_0 and $\mathbf{T}_j' = \overline{I}_j' \cup J\overline{I}_j'$, $j = 1,\cdots,g$, the corresponding circles in X . Given m in M_0 the symmetry $Q\mathcal{D}_m = \mathcal{D}_m$ and the explicit nature of Δ_0 imply $\mathcal{D}_m$ consists of one point from each of the closed intervals $\overline{I}_1', \cdots, \overline{I}_g'$. Nash [5] gives a simpler proof of this fact. Further, the inclusion $\Phi_0(\mathbf{T}_1' \times \cdots \times \mathbf{T}_g') \subset T_0$ can be seen directly using (2.1). These remarks establish the following:

PROPOSITION 2.1. *Let* D *be a domain obtained from the unit disc by removing* g *disjoint closed discs centered at points on the real axis. Let* $I_0, I_1, \cdots, I_g$ *be the open intervals whose union forms the intersection of* D *with the real axis and* $\mathbf{T}_j = \overline{I}_j \cup J\overline{I}_j$ *the "circles" obtained by reflecting* $\overline{I}_j$ *into the double* X . *For* q_0 *real in* D *the collection* B_0 *of critical divisors of representing measures in* M_0 *is the product* $B_0 = \overline{I}_1' \times \cdots \times \overline{I}_g'$ *sitting in* $X^{(g)}$, *where* $\overline{I}_1', \cdots, \overline{I}_g'$ *are those intervals* $\overline{I}_0, \cdots, \overline{I}_g$ *not containing* q_0 . *The torus* V_0 *which is the preimage of* $T_0 = \mathbf{R}^g/\mathbf{Z}^g$ *under* $\Phi_0 : X^{(g)} \to \mathrm{Jac}(X)$ *is the product* $V_0 = \mathbf{T}_1'\times\cdots\times\mathbf{T}_g'$ *sitting in* $X^{(g)}$, *where* $\mathbf{T}_1',\cdots,\mathbf{T}_g'$ *are those circles* $\mathbf{T}_0,\cdots,\mathbf{T}_g$ *not containing* q_0 .

Remarks: (i) Nash [5] has shown (using Sarason's test for extreme points) that for D as in the above proposition and q_0 real, M_0 has at most 2^g extreme points. The preceding proposition implies that if p_j is one of the ends of $\overline{I}_j'$, $j = 1,\cdots,g$, then $\mathcal{D} = p_1+\cdots+p_g$ is a critical divisor of a measure in ∂M_0. Sarason's test (see, remark (iv) in the preceding section) implies that m is an extreme point of M_0. Thus M_0 is the convex hull of 2^g extreme points. This answers a question of Nash [5, p. 134].

(ii) In the situation covered by Proposition 2.1 the mapping $r_0 : V_0\to B_0$ is a 2^g to one covering map over $U_0(\mathrm{int}(M_0))$, where $\mathrm{int}(M_0)$ denotes the interior of M_0. The "branched" covering $r_0 : V_0\to B_0$ is normal with the group of deck transformations generated by $J_k(p_1\times\cdots\times p_g) = p_1\times\cdots\times Jp_k\times\cdots\times p_g$, $k = 1,\cdots,g$. The transplants of the group of deck transformations to the covering $\pi_0 : T_0 \to C_0$ is not easy to identify. The visual action of $r_0 : V_0\to B_0$ as the collapsing of $\mathbf{T}_j'$ onto I_j', $j = 1,\cdots,g$, means that $\pi_0 : T_0 \to C_0$ corresponds to the successive collapse of g circles onto arcs.

(iii) This remark answers question (2) of Sarason [6, p. 376]. (Questions (1) and (3) of Sarason [6, p. 376] are answered in Nash [5].) It will be shown in this remark that critical divisors of representing measures can have multiplicity greater than one on ∂D . Recall that if m is a representing measure which is the restriction of dw in $\mathcal{U}^{(1)}(X)$ to ∂D , then a multiple point p in $\mathcal{D}_m$ on ∂D means that $\frac{dw}{dz}$ has a zero at least of order four on ∂D .

Let D be the domain obtained from the unit disc the two closed discs of radius $\frac{1}{4}$ centered at $d_1 = -\frac{1}{2}$, $d_2 = \frac{1}{2}$, We are interested in the solutions in $X^{(2)}$ of

$$\xi_0(\mathcal{D}) - \xi_0(q) + \Delta_0 = -\frac{1}{2}[\omega(q)] \tag{2.2}$$

as q varies over D. We know that for any q in D the equation (2.2) has a unique solution $\mathcal{D}(q)$ such that $r_0(\mathcal{D}(q))$ is the critical divisor of an element in the space M_q of representing measures for the point q. A simple compactness argument shows that $q \to \mathcal{D}(q)$ is continuous on D. Further, $J\mathcal{D}(q) = \mathcal{D}(q)$ so that $\mathcal{D}(q)$ consists of two points on ∂D or $\mathcal{D}(q) = p + Jp$ for some p in D. Consider a path in D which joins a point q_0 in $I_1 = [-\frac{1}{4}, \frac{1}{4}]$ to a point q_1 in $I_0 = [-1, -\frac{3}{4}]$. It is easy to check that $\mathcal{D}(q_0) = (-1)+1$ and $\mathcal{D}(q_1) = \frac{1}{4} + \frac{3}{4}$. As q varies along our path from q_0 to q_1 the value $\mathcal{D}(q)$ must move continuously from $\mathcal{D}(q_0)$ to $\mathcal{D}(q_1)$. Thus $\mathcal{D}(q)$ always consists of two points on the boundary b_0 of the unit circle, $p + Jp$ for p in D or two points on the boundary b_2 of the circle centered at $d_2 = \frac{1}{2}$. Necessarily, for some q on the path $\mathcal{D}(q) = 2p$ for some point p in b_0 (similarly, at some q, $\mathcal{D}(q) = 2p$ for some p in b_2). This shows critical divisors of representing measures can have multiplicity on ∂D.

Note also in the above example there will be points q on the path where $\mathcal{D}(q) = p + Jp$, for p in D. Thus $r_0(\mathcal{D}(q)) = 2p$ is the critical divisor of a representing measure having multiplicity in D. this multiplicity was noted by Sarason [6, p. 376].

The above example generalizes to any domain covered by Proposition 2.1. Given such a domain there will always be points q in D such that M_q has elements with critical divisors having multiplicity on ∂D.

A CONJECTURE OF NASH

In this brief section we will sketch a proof of the following:

PROPOSITION 3.1. *If* D *is a* g *holed planar domain with analytic boundary and* q_0 *is a point in* D *, then when* $g > 2$ *the set of representing measures* M_0 *for evaluation at* q_0 *is not strictly convex.*

The statement that M_0 is not strictly convex is equivalent to the statement that ∂M_0 contains line segments. The result in the above proposition for odd $g > 2$ was established by Nash [5] who conjectured the result for all $g > 2$. Nash [5] also shows that when $g = 2$ the set M_0 is strictly convex only when q_0 lives on a finite number of distinguished subarcs of D .

PROOF. Suppose M_0 is strictly convex. Then given p in ∂D the point p can be in the support of at most one critical divisior $\mathcal{D}_m$ of some m in M_0. Otherwise, if p belongs to the support of the critical divisor m_1, m_2, then the line $tm_1 + (1 - t)m_2$, $0 \leq t \leq 1$, would be in ∂M_0 .

Consider the real analytic space $\partial D \times V_0$ and the subset $A = \{(p,\mathcal{D}):\ p$ is in the support of $\mathcal{D}\}$ of $\partial D \times V_0$. Note that $(p,\mathcal{D})$ is in A if and only if $\mathcal{D} - p$ is a non-negative divisor of degree $g - 1$. This means precisely that $\theta(\xi_0(\mathcal{D}) - \xi_0(p) + \Delta_0) = 0$. See, e.g., Farkas and Kra [2, p. 291]. In other words A is a real analytic subvariety of $\partial D \times V_0$.

Let F be the closed subset of ∂D obtained by projecting A onto ∂D . The set F consists of at most a finite union of intervals and isolated points. For every p in F there is exactly one critical divisor $\mathcal{D}(p)$ in B_0 with p in the support of $\mathcal{D}(p)$.

The mapping $p \to \mathcal{D}(p)$, p in F , is a piecewise real analytic mapping into V_0. Further, the function $\pi_0(\Phi_0(\mathcal{D}(p)))$, $p \in F$, must parametrize the boundary of C_0. This would only be possible when $g = 2$. The proof of the proposition is complete.

It is hoped that the remarks in this paper illustrate the use of theta functions and Abel-Jacobi theory in the investigation of rational functions on multiply connected domains.

REFERENCES

1. Clancey, K. F.: Applications of the theory of theta functions to Hardy spaces of representing measures on multiply connected domains, submitted for publication.

2. Farkas, H. M. and Kra, I.: Riemann Surfaces, Springer-Verlag, New York, 1980.

3. Fay, J. D.: Theta Functions on Riemann Surfaces, Lecture Notes in Mathematics No. 352, Springer-Verlag, New York, 1973.

4. Goluzin, G. M.: Geometric Theory of Functions of a Complex Variable, Moscow, 1952. English transl: American Mathematical Society, Providence, Rhode Island, 1974.

5. Nash, D.: Representing measures and topological type of finite bordered Riemann surfaces, Trans. Amer. Math. Soc. 192(1974), 129-138.

6. Sarason, D.: Representing measures for R(X) and their Green's functions, J. Functional Analysis 7(1971), 359-385.

7. Tsuji, M.: Potential Theory in Modern Function Theory, Maruzen, Tokyo, 1959.

Kevin F Clancey
Department of Mathematics
University of Georgia
Athens, Georgia 30602
U.S.A.

Operator Theory:
Advances and Applications, Vol. 41

BOUNDARY ELEMENT ANALYSIS OF A DIRECT METHOD FOR THE BIHARMONIC DIRICHLET PROBLEM

M. Costabel and J. Saranen

Dedicated to Prof. I. Gohberg on the occasion of his sixtieth birthday

Based on a direct method proposed by Christiansen and Hougaard we consider spline approximation solution of the clamped plate problem. Our discretization methods cover the Galerkin and collocation solutions. The system of boundary equations is not strongly elliptic. However we are able to derive optimal order error estimates for unknown boundary densities in some Sobolev spaces. The corresponding asymptotic convergence for approximation of the biharmonic function itself has been observed also in numerical calculations.

1. Introduction

Christiansen and Hougaard proposed in [3] a direct integral equation method for solution of the clamped plate equation

$$\begin{cases} \Delta^2\Phi = 0 & \text{in } \Omega \subset \mathbf{R}^2 \\ \Phi = f_1 & \text{on } \Gamma \\ \partial_n\Phi = f_2 & \text{on } \Gamma. \end{cases} \tag{1.1}$$

Here $\partial_n = \partial/\partial_n$ is the outward normal derivative on the smooth boundary Γ. Their approach uses the representation

$$\Phi = \int_\Gamma (G\partial_n\Delta\Phi - \partial_{n_y}G\Delta\Phi + \Delta_y G\partial_n\Phi - \partial_{n_y}\Delta G\Phi)ds_y \tag{1.2}$$

of the solution Φ in Ω. We have needed above the fundamental solution

$$G(x,y) := -\frac{1}{8\pi}|x-y|^2\ln|x-y| \tag{1.3}$$

of the biharmonic equation. Introducing the boundary values

$$u := -\partial_n \Delta\Phi|_\Gamma, \quad v := \Delta\Phi|_\Gamma \tag{1.4}$$

one obtains from the representation formula (1.2) by passing to the boundary

$$Au + Bv = f_1^* := (\frac{1}{2}I - D)f_1 - Sf_2 + \frac{1}{2\pi}\int_\Gamma f_2 ds. \tag{1.5}$$

Here we have used the boundary integral operators

$$\begin{aligned} Su(x) &:= \int_\Gamma g(x,y)u(y)ds_y, \quad Au(x) := -\int_\Gamma G(x,y)u(y)ds_y \\ Bu(x) &:= -\int_\Gamma \partial_{n_y} G(x,y)u(y)ds_y, \quad Du(x) := -\int_\Gamma \partial_{n_y} g(x,y)u(y)ds_y, \end{aligned} \tag{1.6}$$

where

$$g(x,y) = -\frac{1}{2\pi}\ln|x-y| \tag{1.7}$$

is the fundamental solution of the Laplace equation. In addition to (1.5) one obtains from the representation formula for the harmonic function $\Delta\Phi$ the equation

$$Su + (\frac{1}{2}I - D)v = 0. \tag{1.8}$$

The method of Christiansen and Hougaard consists of using the representation formula (1.2) such that the unknown boundary densities u and v are determined by the system

$$\begin{cases} Au + Bv = f_1^*, \\ Su + (\frac{1}{2}I - D)v = 0 \end{cases} \tag{1.9}$$

Although the original problem (1.1) is uniquely solvable for bounded domains the system (1.9) doesn't always posses a unique solution. Solvability properties of (1.9) were completely studied by Fuglede [9] who showed the conjecture of Christiansen and Hougaard that (1.9) is uniquely solvable if the logarithmic capacity (exterior mapping radius, transfinite capacity) cap Γ differs from 1 and from e^{-1}. In these exceptional cases where eigensolutions exists one can fix a unique solution requiring that two additional side conditions are satisfied [9].

From the above reasons it seems to obvious that the method of Christiansen and Hougaard could be adapted also for numerical solution of the biharmonic problem (1.1).

In fact Christiansen and Hougaard [3] carried out some numerical experiments based on the system (1.9). However a serious numerical analysis including estimation of the error is not available.

Here we analyze spline Galerkin and collocation approximation schemes for the solution of the system (1.9). This system is not strongly elliptic and therefore the standard argumentation for such systems [1], [19] cannot be utilized. The crucial point in the error analysis is that we consider the system (1.9) as a compact perturbation of simpler system having a lower triangular structure. For the analysis we use the powerful tool of pseudodifferential operators defined on smooth compact manifolds and acting between Sobolev spaces [7]. In the numerical approximation it turns out that the approximation schemes own a better behavior with respect to the first component u than with respect to the second unknown boundary density v. This phenomenon is clearly seen in calculating the potential function. Namely, it suffices (in the Galerkin method) to use lower degree trial functions for the first component than for the second one and still retain the same convergence rate. Besides of the conventional use of classical splines of degree $d \geq 0$ as trial functions we have included the use of Dirac's distributions as trial functions. For single equations this was proposed and analyzed by Ruotsalainen and Saranen in [13]. For midpoint collocation with splines of even degree d as trial functions we obtain the same rate of convergence for the potential as by using splines of the odd degree $d+1$ and nodal point collocation. Such a phenomenon was recently discovered by Saranen [16] for strongly elliptic equations. It is also worth pointing out that although the system (1.9) is not strongly elliptic we obtain for boundary densities optimal order convergence results with respect to suitable Sobolev norms.

Finally we remark that there are also other boundary element approaches to solve the biharmonic problem. A comparison of the method of Christiansen and Hougaard with an other direct method taken from Costabel and Wendland [6] and with the method of Hsiao and MacCamy is presented in [4].

2. The Boundary Integral Equations

We consider the mapping properties of the operator

(2.1) $$\mathcal{A} := \begin{pmatrix} A & B \\ S & \frac{1}{2}I - D \end{pmatrix}.$$

Theorem 2.1. *The operator $\mathcal{A}$ is a matrix of pseudodifferential operators of orders $\begin{pmatrix} -3 & -2 \\ -1 & 0 \end{pmatrix}$. Therefore it is a continuous mapping*

(2.2) $$\mathcal{A} : H^{s-1} \times H^{s} \longrightarrow H^{s+2} \times H^{s} \quad \textit{for any } s \in \mathbf{R}.$$

The operator A is strongly elliptic. The operator B has a vanishing principal symbol of order -2, hence it is actually a pseudodifferential operator of order -3. Thus there is a decomposition $\mathcal{A} = \mathcal{A}_0 + \mathcal{K}$ with

$$\mathcal{A}_0 = \begin{pmatrix} A_0 & 0 \\ S & \frac{1}{2}I \end{pmatrix}, \quad \mathcal{K} = \begin{pmatrix} K & B \\ 0 & -D \end{pmatrix}, \tag{2.3}$$

where A_0 is positive definite, hence invertible, and K has order $-\infty$. Hence

$$\mathcal{A}_0 : H^{s-1} \times H^s \longrightarrow H^{s+2} \times H^s$$

is bijective, and $\mathcal{K}$ maps $H^t \times H^s$ into $H^{s+3} \times H^{s+1}$ for any $s, t \in \mathbf{R}$.

Proof. The operators S and D from potential theory are well-studied. It is known [18] that S is a pseudodifferential operator of order -1 and D has a vanishing principal symbol of order zero, hence D is of order -1, too. For the operators A and B, we use a result from [7]. According to [7, 23.48.14.5], an operator P_{jk} defined by

$$P_{jk}g(x) := \partial^j_{n(x)} \int_\Gamma \partial^k_{n(y)} G(x,y)g(y)\,ds(y)$$

has a principal symbol $\sigma_{jk}(\xi')$ of order $-3+j+k$ given by

$$\sigma_{jk}(\xi') = c_{jk} \int_{\Gamma_+(\xi')} \tau^0(\xi', \zeta_n)\zeta_n^{j+k}\,d\zeta_n \quad (\xi' \in \mathbf{R}^{n-1}),$$

where

$$c_{jk} = (2\pi i)^{j+k}(-1)^k, \quad \tau^0(\xi', \zeta_n) = (\zeta_n^2 + |\xi'|^2)^{-2},$$

and $\Gamma_+(\xi')$ for $\xi' \in \mathbf{R}^{n-1}$ is a contour in $\mathbf{C}$, positively encircling the poles of $\tau_0(\xi', \cdot)$ in the upper half-plane $\mathbf{C}_+$. Thus

$$\begin{aligned}\int_{\Gamma_+(\xi')} \tau^0(\xi', \zeta_n)\zeta_n^{j+k}\,d\zeta_n &= \int_{\Gamma_+(\xi')} \zeta_n^{j+k}(\zeta_n + i|\xi'|)^{-2}(\zeta_n - i|\xi'|)^{-2}\,d\zeta_n \\ &= 2\pi i \frac{d}{d\zeta_n}\left(\zeta_n^{j+k}(\zeta_n + i|\xi'|)^{-2}\right)\Big|_{\zeta_n = i|\xi'|} \\ &= 2\pi i \cdot \frac{1}{4}(i|\xi'|)^{j+k-3} \cdot (j+k-1).\end{aligned}$$

This gives

$$\sigma_{jk}(\xi') = \frac{i\pi}{2} c_{jk}(j+k-1)(i|\xi'|)^{j+k-3}.$$

In particular $\sigma_{00}(\xi') = c\,|\xi'|^{-3}$, with $c > 0$. This shows that A is a strongly elliptic pseudodifferential operator of order -3. Thus there exists a pseudodifferential operator K of order -4 such that

$$A_0 := A - K$$

is positive definite. K can be chosen as a smoothing operator of order $-\infty$, or even as a finite dimensional operator. Furthermore, $\sigma_{01}(\xi') \equiv 0$, hence the principal symbol of B of order -2 vanishes, and B is actually of order -3. □

Note that $\mathcal{A}$ is elliptic in the Douglis-Nirenberg sense, but not strongly elliptic.

Fuglede [9] showed that under the hypothesis

$$\operatorname{cap}\Gamma \notin \{1, e^{-1}\}, \tag{2.4}$$

the system (1.9) is uniquely solvable in the classical sense of Hölder continuous solutions. Now, $\mathcal{A}$ being an elliptic pseudodifferential operator, its kernel and cokernel consists of smooth functions. It follows that under the hypothesis (2.4), the operator $\mathcal{A}$ in (2.2) is an isomorphism for any $s \in \mathbf{R}$. We shall assume the validity of (2.4) from now on.

3. Collocation and Galerkin Methods

We consider collocation and Galerkin approximation methods for the completely nonhomogeneous system

$$\begin{cases} Au + Bv = f_1^*, \\ Su + (\frac{1}{2}I - D)v = f_2^* \end{cases} \tag{3.1}$$

or equivalently

$$\mathcal{A}U = F^* \tag{3.2}$$

where $U = (u, v)$ and $F^* = (f_1^*, f_2^*)$. We discuss these approximation methods assuming that the solution U belongs to the Sobolev space $H^{s_1} \times H^{s_2}$. Conditions for the indices s_1 and s_2 are given later. Note that for the real system appearing in solving the clamped plate problem we have $f_2^* = 0$. Then it follows from the second equation in (3.1) that always $s_2 = s_1 + 1$. Therefore in such a case it would be completely sufficient for the real problem to consider the mapping $\mathcal{A} : H^{s-1} \times H^s \to H^{s+2} \times H^s$.

In the numerical solution of the boundary integral equations (3.1) we approximate the boundary densities by using smoothest splines or distributions as trial functions. Let $S^d = S^d(\Delta)$ be the space of smoothest splines of degree $d \geq 0$ with respect to the mesh $\Delta = \{x_k\}_1^N$ on the boundary Γ. Here we include also the case $d = -1$ where $S^{-1}(\Delta)$ denotes the space of Dirac's distributions

$$S^{-1}(\Delta) = \{\delta | \delta = \sum_{k=1}^{N} \alpha_k \delta_k, \quad \alpha_k \in \mathbf{C}\}.$$

The trial function δ_k corresponds to the node x_k of the mesh Δ. For the practical computation it is advantageous to use splines of different degrees for the approximation of the unknown components u and v. Therefore we use as trial subspace the general product space $S^{d_1} \times S^{d_2}$. The mesh parameter is $h = 1/N$.

The Galerkin approximation $U_h \in S^{d_1} \times S^{d_2}$ is defined by the requirement

$$(\mathcal{A}U_h|\Psi) = (F^*|\Psi), \quad \Psi \in S^{d_1} \times S^{d_2}. \tag{3.3}$$

Writing $U_h = (u_h, v_h)$ and $\Psi = (\xi, \eta)$ we have equivalently: $u_h \in S^{d_1}, \quad v_h \in S^{d_2}$ such that

$$(Au_h|\xi) + (Bv_h|\xi) = (f_1^*|\xi), \quad \xi \in S^{d_1}, \tag{3.3a}$$

$$(Su_h|\eta) + ((\frac{1}{2}I - D)v_h|\eta) = (f_2^*|\eta), \quad \eta \in S^{d_2}. \tag{3.3b}$$

We shall allow the values $d_1 \geq -1, \ d_2 \geq 0$ for degrees of the trial subspaces.

In addition to the Galerkin method we analyze the collocation method: find $U_h = (u_h, v_h) \in S^{d_1} \times S^{d_2}$ such that

$$(Au_h)(t_k^1) + (Bv_h)(t_k^1) = f_1^*(t_k^1), \quad k = 1, \ldots, N, \tag{3.4a}$$

$$(Su_h)(t_k^2) + ((\frac{1}{2}I - D)v_h)(t_k^2) = f_2^*(t_k^2), \quad k = 1, \ldots, N. \tag{3.4b}$$

We point out that the collocation points t_k^1 and t_k^2 are different if one of the degrees d_1 and d_2 is odd and the other is even. More precisely, if the degree d_l is odd, then the points t_k^l are nodal points of the mesh Δ and if the degree d_l is even, then the points t_k^l are midpoints of the mesh Δ. In the case of the collocation we shall allow the degrees $d_1 \geq 0$, $d_2 \geq 0$ for trial functions. If Δ^l is the mesh $\Delta^l = \{t_k^l\}_1^N$ we can write the collocation equations (3.4) equivalently as

$$(\mathcal{A}U_h|\Psi) = (F^*|\Psi), \quad \Psi \in S^{-1}(\Delta^1) \times S^{-1}(\Delta^2). \tag{3.5}$$

For the Galerkin method we only assume that the mesh is quasiuniform. In the case of collocation we partly use some additional assumptions concerning the mesh Δ.

Since the system (3.1) is not strongly elliptic, standard techniques are not applicable for error analysis. Our method of proof uses a perturbation type argumentation where we consider the operator $\mathcal{A}$ as a compact perturbation of the operator $\mathcal{A}_0$. This also enables for us to reduce estimation of the error to discussion of single scalar equations. The results which we shall need concerning single equations are somewhat scattered in the literature. Therefore we recall here the corresponding convergence results for a single, strongly elliptic pseudodifferential operator A of the order 2α. For the Galerkin approximation, $u_h \in S^d, \quad d \geq -1$ given by

$$(Au_h|\phi) = (Au|\phi), \quad \phi \in S^d, \tag{3.6}$$

the following result is true.

Theorem 3.1. *Assume that A is a strongly elliptic pseudodifferential operator of order 2α such that $A : H^\alpha \to H^{-\alpha}$ is an isomorphism. If $d \geq -1, \quad \alpha < d + \frac{1}{2}$ and if the solution u satisfies $u \in H^s$ where $2\alpha - d - \frac{1}{2} < s \leq d + 1$, then the Galerkin problem (3.6) is uniquely solvable for sufficiently small h, and for a quasiuniform mesh we have the error estimate*

$$||u - u_h||_t \leq ch^{s-t}||u||_s \tag{3.7}$$

for $2\alpha - d - 1 \leq t < d + \frac{1}{2}, \quad t \leq s$.

This result is known by [11] under the assumption that the solution u belongs to the energy space H^α. A proof covering also the values $2\alpha - d - \frac{1}{2} < s \leq \alpha$ is proved in [15]. We point out that in these articles only the case $d \geq 0$ of the classical spline spaces was considered. But since the conventional approximation and inverse properties are valid also for the space $S^{-1}(\Delta)$ of Dirac's distributions by Ruotsalainen and Saranen [13], all the arguments in [15] are applicable also with $d = -1$.

For the collocation method, $u_h \in S^d, \quad d \geq 0$ given by

$$(Au_h)(t_k) = (Au)(t_k) \qquad k = 1, \ldots, N \tag{3.8}$$

the existing convergence results are different depending on whether the degree of trial functions is odd or even. If the degree d is odd we use nodal collocation in (3.8) and if d is even then the collocation points t_k are midpoints of the mesh. Furthermore for error estimates the mesh is quasiuniform if d is odd and smoothly graded ([2]) if d is even. We have the following optimal order convergence result.

Theorem 3.2. *Assume that A is a strongly elliptic pseudodifferential operator of order 2α such that $A : H^{\alpha} \to H^{-\alpha}$ is an isomorphism. Furthermore we assume that $d > 2\alpha$ if d is odd and $d > 2\alpha - \frac{1}{2}$ if d is even. If the solution u satisfies $u \in H^s$ where $2\alpha + \frac{1}{2} < s \leq d+1$, then the collocation problem (3.8) is uniquely solvable for sufficiently small h and we have the error estimate*

$$||u - u_h||_t \leq ch^{s-t}||u||_s \tag{3.9}$$

for $2\alpha \leq t < d + \frac{1}{2}$, $\quad t \leq s$.

This result has been proved for odd integers with nodal collocation by Arnold and Wendland in [1] with values $\frac{1}{2}(d+1) + \alpha \leq s \leq d+1$. For the remaining values $2\alpha + \frac{1}{2} < s \leq \frac{1}{2}(d+1) + \alpha$ the result was proved by Ruotsalainen and Saranen in [14]. In the case of the even degree splines we refer to [17] and [2]. For even degree splines the optimal order convergence result (3.9) is not sufficient in order to conclude the true order of the convergence for the biharmonic potential. To cover this case we need the following suboptimal convergence result proved by Saranen in [16].

Theorem 3.3. *Let d be an even integer and let the assumptions of Theorem 3.2 be valid such that the principal symbol of A is an even function with respect to the Fourier transformed variable. If the solution u has the regularity $u \in H^{d+1+\tau}$, where $0 \leq \tau \leq 1$, then the collocation approximation u_h fulfils for a uniform mesh the estimate*

$$||u - u_h||_{2\alpha-\tau} \leq ch^{d+1+\tau-2\alpha}||u||_{d+1+\tau}. \tag{3.10}$$

In our further analysis we shall make use of the approximation and inverse properties mentioned above. These results are the following

Lemma 3.1. *(Approximation property) Let the mesh Δ be quasiuniform and $d \geq -1$. Assume in addition that $t_0 < d + \frac{1}{2}$ and $t_0 \leq s \leq d+1$. Then for any $u \in H^s$ there exists $u_\Delta \in S^d(\Delta)$ such that*

$$||u - u_\Delta||_t \leq c(t)h^{s-t}||u||_s \tag{3.11}$$

for all $t \leq t_0$. The constant $c(t)$ is independent of u and Δ.

Lemma 3.2. *(Inverse property) For any $t, s \in \mathbf{R}$ with $t \leq s < d + \frac{1}{2}$ and for any quasiuniform mesh Δ there exists a constant $c > 0$ such that*

$$||\phi||_s \leq ch^{t-s}||\phi||_t \tag{3.12}$$

for all $\phi \in S^d(\Delta)$, $d \geq -1$.

Proofs of these results can be found in Elschner and Schmidt [8] for smoothest splines, extended in Ruotsalainen and Saranen [13] to cover also Dirac's distributions.

4. Error Analysis

We shall analyze the discretization methods by considering the operator $\mathcal{A} = \mathcal{A}_0 + \mathcal{K}$ as a compact perturbation of the operator $\mathcal{A}_0$. Therefore we shall begin discussing first the case $\mathcal{A} = \mathcal{A}_0$. In transfering the error estimates from the case $\mathcal{A} = \mathcal{A}_0$ to the general case $\mathcal{A} = \mathcal{A}_0 + \mathcal{K}$ we shall need the following

Lemma 4.1. *The operator* $\mathcal{A}_0^{-1}\mathcal{K} : H^{t_1} \times H^{t_2} \to H^{t_2} \times H^{t_2+1}$ *is continuous for every* $t_1, t_2 \in \mathbf{R}$.

Proof. Since K is smoothing, B is of the order -3 and $D : H^s \to H^{s+1}$ is continuous, then the operator

$$\mathcal{K} = \begin{pmatrix} K & B \\ 0 & -D \end{pmatrix}$$

is continuous $\mathcal{K} : H^{t_1} \times H^{t_2} \to H^{t_2+3} \times H^{t_2+1}$. The inverse $\mathcal{A}_0^{-1}$ of $\mathcal{A}_0$ is given by

$$\mathcal{A}_0^{-1} = \begin{pmatrix} A_0^{-1} & 0 \\ -2SA_0^{-1} & 2I \end{pmatrix}$$

and it defines a continuous mapping $\mathcal{A}_0^{-1} : H^{s_1} \times H^{s_2} \to H^{s_1-3} \times H^{min\{s_1-2,s_2\}}$. Consequently $\mathcal{A}_0^{-1}\mathcal{K} : H^{t_1} \times H^{t_2} \to H^{t_2} \times H^{t_2+1}$ is continuous. □

4.1. Galerkin method

We consider the Galerkin method (3.3) by taking $\mathcal{A} = \mathcal{A}_0$. The equations now read

$$(\mathcal{A}_0 u_h|\xi) = (\mathcal{A}_0 u|\xi), \quad \xi \in S^{d_1}, \tag{4.1a}$$

$$(Su_h|\eta) + \frac{1}{2}(v_h|\eta) = (Su|\eta) + \frac{1}{2}(v|\eta), \quad \eta \in S^{d_2}. \tag{4.1b}$$

Here we require $d_1 \geq -1,\ d_2 \geq 0$. The solution $U = (u, v)$ is assumed to have the regularity $U \in H^{s_1} \times H^{s_2}$. We shall measure the error in the approximation of the components u and v independently of each other by using Sobolev norms with indices t_1 and t_2. Consequently we shall need various restrictions concerning the indices s_k, t_k and d_k. For this purpose we introduce the abbreviations

$$\underline{t}_1 = -d_1 - 4,\ \underline{s}_1 = \underline{t}_1 + \frac{1}{2},\ \overline{s}_1 = d_1 + 1,$$

$$\underline{t}_2 = \max\{-d_1 - 3, -d_2 - 1\},\ \underline{s}_2 = \underline{t}_2 + \frac{1}{2},\ \overline{s}_2 = \min\{d_1 + 2, d_2 + 1\}.$$

For the problem (4.1) we have

Theorem 4.1. *We assume that $d_1 \geq -1$, $d_2 \geq 0$ and that the indices s_1, s_2 satisfy $\underline{s}_k < s_k \leq \overline{s}_k$, $k = 1,2$. If $U \in H^{\sigma_1} \times H^{s_2}$ with $\sigma_1 = \max\{s_1, s_2 - 1\}$, then the Galerkin problem (4.1) has for sufficiently small h a unique solution $U_h \in S^{d_1} \times S^{d_2}$. Moreover for all $t_1 \leq s_1$, $t_2 \leq s_2$ satisfying $\underline{t}_k \leq t_k < d_k + \frac{1}{2}$ we have the error estimates*

$$(4.2) \qquad ||u - u_h||_{t_1} \leq ch^{s_1 - t_1}||u||_{s_1},$$

$$(4.3) \qquad ||v - v_h||_{t_2} \leq ch^{s_2 - t_2}(||u||_{s_2 - 1} + ||v||_{s_2}).$$

Proof. It is enough to derive the estimates (4.2) and (4.3) assuming that U_h is a Galerkin solution. The unique solvability follows immediately after having these estimates. Inequality (4.2) follows from Theorem 3.1 by choosing $\alpha = o$. For (4.3) we observe that by (4.1b) the function v_h can be considered as Galerkin approximation (with $A = I$) of the function $v + 2S(u - u_h)$. Therefore we decompose $v_h = v_h^1 + v_h^2$, where v_h^1, $v_h^2 \in S^{d_2}$ are solutions of the problems

$$(4.4) \qquad (v_h^1|\eta) = (v|\eta), \quad \eta \in S^{d_2},$$

$$(4.5) \qquad (v_h^2|\eta) = (2S(u - u_h)|\eta), \quad \eta \in S^{d_2}.$$

From (4.4) follows by Theorem 3.1

$$(4.6) \qquad ||v^1 - v_h^1||_{t_2} \leq ch^{s_2 - t_2}||v||_{s_2}.$$

Since s_2 satisfies $\underline{s}_2 < s_2 \leq \overline{s}_2$ we can choose a real number τ_2 such that $\underline{s}_2 < \tau_2 \leq \overline{s}_2$ with $\tau_2 < d_1 + \frac{3}{2}$. Writing $v^2 = 2S(u - u_h)$ one concludes by Theorem 3.1 and by estimate (4.2)

$$(4.7) \quad ||v^2 - v_h^2||_{\underline{t}_2} \leq ch^{\tau_2 - \underline{t}_2}||S(u - u_h)||_{\tau_2} \leq ch^{\tau_2 - \underline{t}_2}||u - u_h||_{\tau_2 - 1} \leq ch^{s_2 - \underline{t}_2}||u||_{s_2 - 1}.$$

From (4.6), (4.7) follows

$$(4.8) \qquad ||v - v_h + 2S(u - u_h)||_{\underline{t}_2} \leq ch^{s_2 - \underline{t}_2}(||u||_{s_2 - 1} + ||v||_{s_2}).$$

Furthermore we have by (4.2)

$$(4.9) \qquad ||S(u - u_h)||_{\underline{t}_2} \leq c||u - u_h||_{\underline{t}_2 - 1} \leq ch^{s_2 - \underline{t}_2}||u||_{s_2 - 1}$$

and combining (4.8) with (4.9) yields

$$(4.10) \qquad ||v - v_h||_{\underline{t}_2} \leq ch^{s_2 - \underline{t}_2}(||u||_{s_2 - 1} + ||v||_{s_2}).$$

Finally a standard application of the inverse estimate (3.12) together with the simultaneous approximation property (3.11) yields the assertion (4.3). □

Now we are able to derive the error analysis for the Galerkin method in the case of the operator $\mathcal{A} = \mathcal{A}_0 + \mathcal{K}$.

Theorem 4.2. *Assume that $d_1 \geq -1$, $d_2 \geq 0$ and that the indices s_1 and s_2 satisfy the conditions $\underline{s}_k < s_k \leq \overline{s}_k$, $k = 1,2$. If $U \in H^{\sigma_1} \times H^{s_2}$ with $\sigma_1 = \max\{s_1, s_2 - 1\}$, then for a quasiuniform mesh and with sufficiently small h the Galerkin equation (4.1) has a unique solution $U_h \in S^{d_1} \times S^{d_2}$. Moreover for all $t_1 \leq s_1$, $t_2 \leq s_2$ satisfying $\underline{t}_k \leq t_k < \overline{t}_k = d_k + \frac{1}{2}$, $t_1 < t_2$ the following estimate is valid*

$$(4.11) \qquad ||u - u_h||_{t_1} + ||v - v_h||_{t_2} \leq c[h^{s_1 - t_1}||u||_{s_1} + h^{s_2 - t_2}(||u||_{s_2 - 1} + ||v||_{s_2})].$$

Proof. Since both $\mathcal{A}$ and $\mathcal{A}_0$ are isomorphisms, the operator $I + \mathcal{A}_0^{-1}\mathcal{K} = \mathcal{A}_0^{-1}\mathcal{A}$ is an isomorphism in $H^{t_1} \times H^{t_2}$. Consequently we have

$$(4.12) \qquad ||\mathcal{U} - \mathcal{U}_h||_{t_1,t_2} \leq c||(I + \mathcal{A}_0^{-1}\mathcal{K})(\mathcal{U} - \mathcal{U}_h)||_{t_1,t_2}.$$

The Galerkin equation (3.3) can be written equivalently as

$$(4.13) \qquad (\mathcal{A}_0\mathcal{U}_h|\Psi) = (\mathcal{A}_0(\mathcal{U} + \mathcal{A}_0^{-1}\mathcal{K}(\mathcal{U} - \mathcal{U}_h))|\Psi), \quad \Psi \in S^{d_1} \times S^{d_2}.$$

We decompose the solution $\mathcal{U}_h$ as $\mathcal{U}_h = \mathcal{U}_h^1 + \mathcal{U}_h^2$ where $\mathcal{U}_h^1$, $\mathcal{U}_h^2 \in S^{d_1} \times S^{d_2}$ are Galerkin approximations defined by

$$(4.14) \qquad (\mathcal{A}_0\mathcal{U}_h^k|\Psi) = (\mathcal{A}_0\mathcal{U}^k|\Psi), \quad \Psi \in S^{d_1} \times S^{d_2}$$

for $k = 1,2$. Here we have $\mathcal{U}^1 = \mathcal{U}$ and $\mathcal{U}^2 = \mathcal{A}_0^{-1}\mathcal{K}(\mathcal{U} - \mathcal{U}_h)$. From (4.12) hence follows

$$(4.15) \qquad ||\mathcal{U} - \mathcal{U}_h||_{t_1,t_2} \leq c(||\mathcal{U}^1 - \mathcal{U}_h^1||_{t_1,t_2} + ||\mathcal{U}^2 - \mathcal{U}_h^2||_{t_1,t_2}).$$

Writing $\mathcal{U}_h^k = (u_h^k, v_h^k)$ and $\mathcal{U}^k = (u^k, v^k)$ we obtain by Theorem 4.1 taking $k = 1$ in (4.14)

$$(4.16) \qquad ||u^1 - u_h^1||_{t_1} \leq ch^{s_1 - t_1}||u||_{s_1},$$

$$(4.17) \qquad ||v^1 - v_h^1||_{t_2} \leq ch^{s_2 - t_2}(||u||_{s_2 - 1} + ||v||_{s_2}).$$

Since t_1 satisfies $\underline{t}_1 \leq t_1 < d_1 + \frac{1}{2}$ and $t_1 < t_2$ we can choose the parameter τ_1 such that $t_1 < \tau_1 \leq t_2$ and $\underline{s}_1 < \tau_1 \leq \overline{s}_1$. This follows taking e.g. $\tau_1 = \min\{d_1 + \frac{1}{2}, t_2\}$. We can also choose a $\tau_2 = t_2 + \sigma$ where $0 < \sigma \le 1$ such that τ_2 satisfies $\underline{s}_2 < \tau_2 \le \overline{s}_2$. But then we have by Lemma 4.1 $u^2 \in H^{\tau_1} \cap H^{\tau_2 - 1}$, $v \in H^{\tau_2}$ with

$$(4.18)\qquad ||u^2||_{\tau_1} \le c(||u-u_h||_{t_1} + ||v-v_h||_{\tau_1}) \le c(||u-u_h||_{t_1} + ||v-v_h||_{t_2})$$

and with

$$(4.19)\quad ||u^2||_{\tau_2-1} + ||v^2||_{\tau_2} \le c(||u-u_h||_{t_1} + ||v-v_h||_{\tau_2-1}) \le c(||u-u_h||_{t_1} + ||v-v_h||_{t_2}).$$

Taking $k = 2$ in (4.14) we have by Theorem 4.1 the estimates

$$(4.20)\qquad ||u^2 - u_h^2||_{t_1} \le ch^{\tau_1 - t_1}||u^2||_{\tau_1},$$

$$(4.21)\qquad ||v^2 - v_h^2||_{t_2} \le ch^{\tau_2 - t_2}(||u^2||_{\tau_2-1} + ||v^2||_{\tau_2}).$$

By (4.15)-(4.21) we obtain

$$\begin{aligned} ||u-u_h||_{t_1} + ||v-v_h||_{t_2} \le ch^{s_1-t_1}||u||_{s_1} &+ ch^{s_2-t_2}(||u||_{s_2-1} + ||v||_{s_2}) \\ &+ ch^{\sigma}(||u-u_h||_{t_1} + ||v-v_h||_{t_2}). \end{aligned}$$

where $\sigma = \min\{\tau_1 - t_1,\ \tau_2 - t_2\} > 0$. This yields the assertion (4.11) if the parameter h is small enough. □

Now we return to the original problem of approximating the solution Φ of the clamped plate problem (1.1). By substituting the given boundary data f_1, f_2 and the unknown densities u and v into the representation (1.2) we have

$$(4.22)\qquad \Phi(x) = (Au)(x) + (Bv)(x) + (Sf_2)(x) + (Df_1)(x) - \frac{1}{2\pi}\int_\Gamma f_2 ds$$

for $x \in \Omega$. Having found the approximate densities u_h and v_h we define the biharmonic potential Φ_h by means of

$$(4.23)\qquad \Phi_h(x) = (Au_h)(x) + (Bv_h)(x) + (Sf_2)(x) + (Df_1)(x) - \frac{1}{2\pi}\int_\Gamma f_2 ds$$

for $x \in \Omega$. This function satisfies the biharmonic equation exactly, but the boundary conditions in (1.1) are satisfied only approximately. For the error holds

$$(4.24)\qquad \begin{aligned} \Phi(x) - \Phi_h(x) = &\frac{1}{8\pi}\int_\Gamma (u(y) - u_h(y))|x-y|^2 \ln|x-y| ds_y \\ &+ \frac{1}{8\pi}\int_\Gamma (v(y) - v_h(y))\partial_{n_y}|x-y|^2 \ln|x-y| ds_y. \end{aligned}$$

If $x \in \Omega$ is given, we have by (4.24) for any real numbers $t_1,\ t_2$ the estimate

$$|\Phi(x) - \Phi_h(x)| \le c(||u - u_h||_{t_1} + ||v - v_h||_{t_2}). \tag{4.25}$$

Of particular interest is the maximal convergence rate which can be concluded from Theorem 4.2 for given data. In our concrete application of the clamped plate problem we have $s_1 = s - 1$ if $s_2 = s$. One can then observe from the estimate (4.11) that the maximal rate appears already by choosing $t_1 = t - 1,\ t_2 = t$. Since the mapping $\mathcal{A} : H^{s-1} \times H^s \to H^{s+2} \times H^s$ is an isomorphism, we have

$$||U||_{s-1,s} \le c(||f_1^*||_{s+2} + ||f_2^*||_s),$$

and Theorem 4.2 especially includes the optimal order estimates given by

Theorem 4.3. *We assume that the conditions of Theorem 4.2 are satisfied such that* $u \in H^{s-1} \times H^s$ *where* $\max\{-d_1 - \frac{5}{2},\ -d_2 - \frac{1}{2}\} = \underline{s} < s \le \overline{s} = \min\{d_1 + 2,\ d_2 + 1\}$. *Then for all* $t \le s$, $\underline{t} \le t < \overline{t}$ *with* $\underline{t} = \underline{s} - \frac{1}{2}$, $\overline{t} = \overline{s} - \frac{1}{2}$ *holds the optimal order error estimate*

$$||\mathcal{U} - \mathcal{U}_h||_{t-1,t} \le ch^{s-t}||U||_{s-1,s} \le ch^{s-t}||F^*||_{s+2,s}. \tag{4.26}$$

For the biharmonic problem we have $f_2^* = 0$ which yields by (1.5)

$$||F^*||_{s+2,s} \le c||f_1^*||_{s+2} \le c(||f_1||_{s+2} + ||f_2||_{s-1}). \tag{4.27}$$

A combination of the estimates (4.26), (4.27) gives for the potential the pointwise estimate

$$|\Phi(x) - \Phi_h(x)| \le ch^{s-\underline{t}}||U||_{s-1,s} \le ch^{s-\underline{t}}||F||_{s+2,s-1}. \tag{4.28}$$

Especially the maximal rate for the potential is achieved under the regularity assumption $F \in H^{\overline{s}+2} \times H^{\overline{s}-1}$ which implies

$$|\Phi(x) - \Phi_h(x)| \le ch^{\min\{2d_1+5, 2d_2+2\}}||U||_{\overline{s}-1,\overline{s}} \tag{4.29}$$

for $d_1 \ge -1,\ d_2 \ge 0$. From (4.29) we conclude that the choice of d_1 is less important for the convergence rate than the choice of d_2. For example with the choice $d_1 = d_2 = 0$ our estimate assures the rate $O(h^2)$. This same rate is still retained if $d_1 = -1,\ d_2 = 0$, i.e. if Dirac's distributions are used for approximation of the first component. These rates were observed also in numerical experiments.

4.2. Collocation

We proceed similarly as for the Galerkin method considering first the case $\mathcal{A} = \mathcal{A}_0$. The collocation equations read

$$(4.30a)\qquad (A_0 u_h)(t_k^1) = (A_0 u)(t_k^1), \qquad k = 1, ..., N,$$

$$(4.30b)\qquad (Su_h)(t_k^2) + \frac{1}{2}v_h(t_k^2) = (Su)(t_k^2) + \frac{1}{2}v(t_k^2), \qquad k = 1, ..., N.$$

We have the following

Theorem 4.4. *Assume that $d_1,\ d_2 \geq 0$ and that the indices $s_1,\ s_2$ satisfy the conditions $-\frac{5}{2} < s_1 \leq d_1 + 1,\ \frac{1}{2} < s_2 \leq \min\{d_1 + 2,\ d_2 + 1\}$. If $U \in H^{\sigma_1} \times H^{s_2}$ where $\sigma_1 = \max\{s_1,\ s_2 - 1\}$, then the collocation problem (4.30) has for sufficiently small h a unique solution $U_h = (u_h,\ v_h)$ which satisfies*

$$(4.31)\qquad ||u - u_h||_{t_1} \leq ch^{s_1 - t_1}||u||_{s_1},$$

$$(4.32)\qquad ||v - v_h||_{t_2} \leq ch^{s_2 - t_2}(||u||_{s_2 - 1} + ||v||_{s_2})$$

if $-3 \leq t_1 < d_1 + \frac{1}{2},\quad 0 \leq t_2 < d_2 + \frac{1}{2}$ and $t_1 \leq s_1,\quad t_2 \leq s_2$.

Proof. We argue as in the proof of Theorem 4.1. Estimate (4.31) follows directly from (3.9). Since by (4.30b) the function $v_h \in S^{d_2}$ is the spline interpolation of the function $v + 2S(u - u_h)$, we decompose $v_h = v_h^1 + v_h^2$, where $v_h^1,\ v_h^2 \in S^{d_2}$ are solutions of the collocation problems

$$(4.33)\qquad v_h^1(t_k^2) = v(t_k^2),$$

$$(4.34)\qquad v_h^2(t_k^2) = 2(S(u - u_h))(t_k^2).$$

From (4.33) follows with $v^1 = v$

$$(4.35)\qquad ||v^1 - v_h^1||_0 \leq ch^{s_2}||v||_{s_2}.$$

For the given index $\frac{1}{2} < s_2 \leq \min\{d_1 + 2,\ d_2 + 1\}$ we choose the number κ such that $\frac{1}{2} < \kappa < \min\{s_2,\ d_1 + \frac{3}{2}\}$. This choice allows to estimate by (4.34), (4.31)

$$(4.36)\qquad ||v^2 - v_h^2||_0 \leq ch^{\kappa}||S(u - u_h)||_{\kappa} \leq ch^{\kappa}||u - u_h||_{\kappa - 1} \leq ch^{s_2}||u||_{s_2 - 1}.$$

Combining the estimates (4.36) and (4.37) we get

$$(4.37)\qquad ||v_h - (v + 2S(u - u_h))||_0 \leq ch^{s_2}(||u||_{s_2 - 1} + ||v||_{s_2}).$$

But on the other hand it holds that

$$(4.38) \qquad ||S(u-u_h)||_0 \le c||u-u_h||_{-1} \le ch^{s_2}||u||_{s_2-1}.$$

Thus we obtain from (4.37), (4.38)

$$||v-v_h||_0 \le ch^{s_2}(||u||_{s_2-1}+||v||_{s_2})$$

and the assertion (4.32) follows by a standard argument due to inverse estimates and simultaneous approximation properties. □

For the complete equation with $\mathcal{A}=\mathcal{A}_0+\mathcal{K}$ we obtain

Theorem 4.5. *Assume that $d_1,\ d_2 \ge 0$ and that the indices $s_1,\ s_2$ satisfy the conditions $-\frac{5}{2} < s_1 \le d_1+1,\ \ \frac{1}{2} < s_2 \le \min\{d_1+2,\ d_2+1\}$. If $U \in H^{\sigma_1}\times H^{s_2}$ where $\sigma_1 = \max\{s_1,\ s_2-1\}$, then the collocation problem (3.4) has for sufficiently small h a unique solution $U_h \in S^{d_1}\times S^{d_2}$ such that*

$$(4.39) \qquad ||u-u_h||_{t_1}+||v-v_h||_{t_2} \le c[h^{s_1-t_1}||u||_{s_1}+h^{s_2-t_2}(||u||_{s_2-1}+||v||_{s_2})]$$

if $-3 \le t_1 < d_1+\frac{1}{2},\quad 0 \le t_2 < \min\{d_1+\frac{3}{2}, d_2+\frac{1}{2}\},\quad t_1<t_2$ and $t_1 \le s_1,\quad t_2 \le s_2$.

Proof. We keep the notations used in the proof of Theorem 4.2. Instead of (4.14) we now have $\mathcal{U}_h^k \in S^{d_1}\times S^{d_2}$ such that

$$(4.40) \qquad (\mathcal{A}_0\mathcal{U}_h^k|\Psi) = (\mathcal{A}_0\mathcal{U}^k|\Psi), \quad \Psi \in S^{-1}(\Delta^1)\times S^{-1}(\Delta^2)$$

where $\mathcal{U}^1=\mathcal{U},\quad \mathcal{U}^2=\mathcal{A}_0^{-1}\mathcal{K}(\mathcal{U}-\mathcal{U}_h)$. Taking $k=1$ we have by Theorem 4.4

$$(4.41) \qquad ||u^1-u_h^1||_{t_1} \le ch^{s_1-t_1}||u||_{s_1},$$

$$(4.42) \qquad ||v^1-v_h^1||_{t_2} \le ch^{s_2-t_2}(||u||_{s_2-1}+||v||_{s_2}).$$

Since $t_1<t_2$ we can choose the parameters $\tau_1,\ \tau_2$ such that $t_1<\tau_1<t_2,\quad t_2<\tau_2\le t_2+1$ and that $-\frac{5}{2}<\tau_1\le d_1+1,\quad \frac{1}{2}<\tau_2\le\min\{d_1+2,d_2+1\}$. Then we have $u^2\in H^{\tau_1}\cap H^{\tau_2-1},\quad v^2\in H^{\tau_2}$ and following (4.18)-(4.21) we obtain

$$(4.43) \qquad ||u^2-u_h^2||_{t_1}+||v^2-v_h^2||_{t_2} \le ch^{\sigma}(||u-u_h||_{t_1}+||v-v_h||_{t_2})$$

with $\sigma=\min\{\tau_1-t_1,\tau_2-t_2\}>0$. By combining the estimate (4.15) with inequalities (4.41)-(4.43) we obtain for small h the desired error estimate (4.39). □

Similarly as for the Galerkin method we consider the case with the indices $s_1 = s-1,\ s_2 = s$ and $t_1 = t-1,\ t_2 = t$. We obtain again optimal order convergence

$$\|\mathcal{U} - \mathcal{U}_h\|_{t-1,t} \le ch^{s-t}\|\mathcal{U}\|_{s-1,s} \tag{4.44}$$

where $0 \le t < \min\{d_1 + \frac{3}{2}, d_2 + \frac{1}{2}\},\quad \frac{1}{2} < s \le \min\{d_1 + 2, d_2 + 1\},\quad t \le s$. In particular the maximal rate with $\mathcal{U} \in H^{\bar{s}-1} \times H^{\bar{s}},\quad \bar{s} = \min\{d_1 + 2, d_2 + 1\}$ is

$$\|\mathcal{U} - \mathcal{U}_h\|_{-1,0} \le ch^{\min\{d_1+2,d_2+1\}}\|\mathcal{U}\|_{\bar{s}-1,\bar{s}}, \tag{4.45}$$

which yields for the biharmonic potential

$$|\Phi(x) - \Phi_h(x)| \le c \begin{cases} h^{d_1+2}, & d_2 \ge d_1 + 1, \\ h^{d_2+1}, & d_2 \le d_1 \end{cases} \tag{4.46}$$

if $x \in \Omega$. Our numerical tests have indicated that the above result gives the correct order of the convergence if d_1 and d_2 are odd integers. Instead if one of the integers is even, then the result (4.46) doesn't always yield the true convergence rate. For example if the data is smooth enough our estimate (4.46) assures with $d_1 = d_2 = 0$ and with $d_1 = 1,\ d_2 = 0$ only the convergence rate $O(h)$. However the true rate of the convergence is of the order $O(h^2)$. In order to derive this statement we first prove the following improvement of the estimate (4.44). We are able to show higher order convergence if $d_2 \le d_1$ and d_2 is even, which by convention means that midpoint collocation is used for the second component in equations (3.4). We have

Theorem 4.6. *We assume that the conditions of Theorem 4.5 are valid with* $0 \le d_2 \le d_1$ *such that* d_2 *is an even integer. If* $U \in H^{\bar{s}-1+\tau} \times H^{\bar{s}+\tau}$ *where* $0 \le \tau \le 1$, *then we have for uniform mesh the suboptimal convergence result*

$$\|U - U_h\|_{-1,-\tau} \le ch^{\bar{s}+\tau}\|U\|_{\bar{s}-1+\tau,\bar{s}+\tau} \tag{4.47}$$

where $\bar{s} = d_2 + 1$.

Proof. We use the notations of Theorem 4.5. From equation (4.40) with $k = 1$ follows by Theorem 3.2

$$\|u^1 - u_h^1\|_{-1} \le ch^{\bar{s}+\tau}\|u\|_{\bar{s}-1+\tau}. \tag{4.48}$$

The second component of $\mathcal{U}_h^1 = (u_h^1, v_h^1)$ satisfies

$$v_h^1(t_k^2) = v(t_k^2) + 2S(u^1 - u_h^1)(t_k^2).$$

Therefore we decompose $v_h^1 = v_h^{11} + v_h^{12}$ in S^{d_2} such that

$$v_h^{11}(t_k^2) = v(t_k^2), \quad k = 1, \ldots, N, \tag{4.49}$$

$$v_h^{12}(t_k^2) = 2S(u^1 - u_h^1)(t_k^2), \quad k = 1, \ldots, N. \tag{4.50}$$

By Theorem 3.3 follows from (4.49)

$$\|v^{11} - v_h^{11}\|_{-\tau} \leq ch^{\overline{s}+\tau}\|v\|_{\overline{s}+\tau} \tag{4.51}$$

for $v^{11} = v$. Denoting $v^{12} = 2S(u^1 - u_h^1)$ we obtain from (4.50) and (4.40)

$$\begin{aligned} \|v^{12} - v_h^{12}\|_0 &\leq ch\|S(u^1 - u_h^1)\|_1 \leq ch\|u^1 - u_h^1\|_0 \\ &\leq ch \cdot h^{\overline{s}-1+\tau}\|u\|_{\overline{s}-1+\tau} = ch^{\overline{s}+\tau}\|u\|_{\overline{s}-1+\tau}. \end{aligned} \tag{4.52}$$

Here the last inequality follows from (4.40) with $k = 1$ since we have $\overline{s} - 1 + \tau \leq d_1 + 1$ by $d_2 \leq d_1$. Combining (4.51), (4.52) we have

$$\|v_h^1 - (v^1 + 2S(u^1 - u_h^1))\|_{-\tau} \leq ch^{\overline{s}+\tau}(\|u\|_{\overline{s}-1+\tau} + \|v\|_{\overline{s}+\tau}) \tag{4.53}$$

and since it holds that

$$\|S(u^1 - u_h^1)\|_{-\tau} \leq c\|u^1 - u_h^1\|_{-\tau-1} \leq ch^{\overline{s}+\tau}\|u\|_{\overline{s}-1}, \tag{4.54}$$

we obtain by (4.53), (4.54)

$$\|v^1 - v_h^1\|_{-\tau} \leq ch^{\overline{s}+\tau}(\|u\|_{\overline{s}-1+\tau} + \|v\|_{\overline{s}+\tau}). \tag{4.55}$$

According to (4.48) and (4.55),

$$\|\mathcal{U}^1 - \mathcal{U}_h^1\|_{-1,-\tau} \leq ch^{\overline{s}+\tau}\|U\|_{\overline{s}-1+\tau,\overline{s}+\tau}. \tag{4.56}$$

We have $U_h \in H^{d_1} \times H^{d_2}$ and Lemma 4.1 hence implies that $\mathcal{U}^2 = \mathcal{A}_0^{-1}\mathcal{K}(\mathcal{U} - \mathcal{U}_h) \in H^{\overline{s}-1} \times H^{\overline{s}}$. Equation (4.40) now yields

$$\|\mathcal{U}^2 - \mathcal{U}_h^2\|_{-1,0} \leq ch^{\overline{s}}\|\mathcal{U} - \mathcal{U}_h\|_{\overline{s}-2,\overline{s}-1} \leq ch^{\overline{s}+1}\|\mathcal{U}\|_{\overline{s}-1,\overline{s}} \tag{4.57}$$

where we have applied (4.44) twice and once again Lemma 4.1. Obviously (4.56), (4.57) yield the assertion (4.47). □

Returning to approximation of the potential we have proved if d_2 is even and if $d_2 \leq d_1$ the improved result

$$|\Phi(x) - \Phi_h(x)| \leq ch^{d_2+2} \tag{4.58}$$

for boundary data $F \in H^{d_2+3} \times H^{d_2+1}$.

Our numerical experiments have shown a good agreement with convergence results presented in the paper. For these experiments we refer to the article [4].

References

1. Arnold D.N. and Wendland W.L. (1983), On the Asymptotic Convergence of Collocation Methods, Math. Comp. 41, pp. 349–381.
2. Arnold D.N. and Wendland W.L. (1985), The Convergence of Spline Collocation for Strongly Elliptic Equations on Curves, Numer. Math. 47, pp. 317–341.
3. Christiansen S. and Hougaard P. (1978), An Investigation of a Pair of Integral Equations for the Biharmonic Problem, J. Inst. Math. Appl. 22, pp. 15–27.
4. Costabel M. , Lusikka I. and Saranen J. , Comparison of Three Boundary Element Approaches for the Solution of the Clamped Plate Problem, Boundary Elements IX. 2. Southampton, 1987, pp. 19–34.
5. Costabel M. , Stephan E. and Wendland W.L. (1983), On Boundary Integral Equations of the First Kind for the Bi–Laplacian in a Polygonal Plane Domain, Ann. Scuola Norm. Sup. Pisa, Cl. Sci. (4), 10, pp. 197–241.
6. Costabel M. and Wendland W. L. (1986), Strong Ellipticity of Boundary Integral Operators, J. Reine Angew. Math. 372, pp. 39–63.
7. Dieudonné J. (1978). Eléments d'Analyse Vol. 8, Gauthier–Villars. Paris.
8. Elschner J. and Schmidt G. (1983), On Spline Interpolation in Periodic Sobolev Spaces, Preprint 01/83, Dept. Math. Akademie der Wissenschaften der DDR, Berlin.
9. Fuglede B. (1981), On a Direct Method of Integral Equations for Solving the Biharmonic Dirichlet Problem, ZAMM 61, pp. 449–459.
10. Hsiao G.C. and MacCamy R. (1973), Solution of Boundary Value Problems by Integral Equations of the First Kind, SIAM Rev. 15, pp. 687–705.
11. Hsiao G.C. and Wendland W.L. (1981), The Aubin–Nitsche Lemma for Integral Equations, J. Integral Equations 3, pp. 299–315.
12. Lusikka I. , Ruotsalainen K. and Saranen J. (1986), Numerical Implementation of the Boundary Element Method with Point–Source Approximation of the Potential, Engineering Analysis, Vol. 3, No. 3, pp. 144–153.
13. Ruotsalainen K. and Saranen J. , Some Boundary Element Methods Using Dirac's Distributions as Trial Functions, SIAM J. Numer. Anal. 24, 1987, pp. 816–827.

14. Ruotsalainen K. and Saranen J. , A Dual Method to the Collocation Method, to appear in Math. Meth. Appl. Sci.
15. Ruotsalainen K. and Saranen J. , On the Convergence of the Galerkin Method for Nonsmooth Solutions of Integral Equations, to appear in Numerische Mathematik.
16. Saranen J. , The Convergence of Even Degree Spline Collocation Solution for Potential Problems in Smooth Domains of the Plane, to appear in Numerische Mathematik.
17. Saranen J. and Wendland W.L. (1985), On the Asymptotic Convergence of Collocation Methods with Spline Functions of Even Degree, Math. Comp. 45, pp. 91–108.
18. Wendland W.L. (1981), Asymptotic Convergence of Boundary Element Methods, in Lectures on the Numerical Solution of Partial Differential Equations (Ed. Babuska I. , Liu T.P. and Osborn J.), pp. 435–528, Univ. of Maryland, College Park Md, 1981.
19. Wendland W.L. (1983), Boundary Element Method and their Asymptotic Convergence, in Theoretical Acoustics and Numerical Techniques (Ed. Filippi P.), pp. 135–216, CISM Courses 277, Springer-Verlag, 1983. Wien and New York.

M.Costabel
Department of Mathematics
Technical University
Schlossgartenstrasse 7
6100 Darmstadt
W.-Germany

J.Saranen
Section of Mathematics
Faculty of Technology
University of Oulu
Linnanmaa
90570 Oulu, Finland

Operator Theory:
Advances and Applications, Vol. 41

NONLINEAR LIFTING THEOREMS, INTEGRAL REPRESENTATIONS AND STATIONARY PROCESSES IN ALGEBRAIC SCATTERING SYSTEMS

Mischa Cotlar and Cora Sadosky[1]

To Israel Gohberg, in admiration and friendship,
on his sixtieth birthday

Linear and local nonlinear lifting theorems are given for bounded triplets of Toeplitz forms defined in algebraic scattering systems with two evolutions. These theorems are related to the Nagy-Foias lifting theory and to its local nonlinear extensions. Integral representations of Bochner type are obtained for the lifting forms, as well as representations of Khinchine type for the corresponding stationary and generalized stationary stochastic processes.

INTRODUCTION

This paper is about linear and nonlinear lifting theorems, and corresponding integral representations, for Hankel forms in discrete algebraic scattering systems (a.s.s.) with two evolution operators. Linear lifting theorems in discrete or continuous a.s.s., and their relation to the Nagy-Foias commutant lifting theorem were discussed in [5], [6], [7], and [9]. In the special case of the a.s.s. of trigonometric polynomials in the torus T^2, where the evolution operators are the ordinary shifts, the linear lifting theorem was given a more precise form through an integral representation, called the Generalized Bochner Theorem, GBT, which can be viewed as an expansion into eigenforms of the shifts. This GBT, and its quantized analogue, was used in [9] to provide two-dimensional and operator versions of the theorems of Helson-Szego and Nehari.

[1]Author partially supported by grants from the National Science Foundation (USA).

Here the GBT is extended to more general a.s.s., where the evolution operators are no longer the shifts, and the GBT becomes a generalization of the eigenexpansions of Gel'fand-Kostuchenko, M.G. Krein, and Berezanskii, as developed by Korsunski [cf. [13]).

Local nonlinear lifting theorems in the spirit of [1], and corresponding integral representations for locally nonlinear Hankel forms, are also given.

On the other hand, in studying some moment problems related to the Wightman functionals in quantum mechanics, Berezanskii and Korsunski obtained eigenexpansions for positive symmetric or invariant infinite matrices (B_{jk}), where, for each (j,k), the sesquilinear form B_{jk} is defined in $V^{(j)} \times V^{(k)}$, $V^{(n)} = V \otimes \overset{n}{\cdots} \otimes V$. Accordingly, we discuss also a corresponding GBT for invariant matrices, which can be viewed as another kind of local nonlinear lifting theorem.

Finally, since the classical Bochner-Khinchine and Helson-Szego theorems are basic results in the theory of stationary processes, we give in the last section a formulation of the GBT in terms of stationary processes over an a.s.s., following the pattern of the Gel'fand-Itoh theory of linear processes, as it appears in [11].

The results of this paper, stated here for discrete systems, apply also to a.s.s. with two continuous evolution groups, and, in particular, when the evolution operators are given by difference or differential expressions. In the special case of the "trigonometric" a.s.s. in $\mathbb{R}$, the GBT was discussed in [10]. Quantized versions of these results will be given elsewhere.

1. ALGEBRAIC SCATTERING SYSTEMS (A.S.S.)

Our framework is that of (discrete) two-parameter algebraic scattering system, a.s.s., that is a system $[V;W_1,W_2;\sigma,\tau]$ where V is a vector space, W_1 and W_2 are two vector subspaces of V, and σ,τ are two linear isomorphisms of V, such that $\tau\sigma = \sigma\tau$, and

$$\sigma W_1 \subset W_1,\ \sigma^{-1}W_2 \subset W_2.$$
$$\tau W_1 \subset W_1,\ \tau^{-1}W_2 \subset W_2. \tag{1.1}$$

Since W_2 needs not be either σ or τ invariant, set

$$W_2^{\sigma} = \{f \in W_2: \sigma^n f \in W_2,\ \forall n \in \mathbb{Z}\}$$
$$W_2^{\tau} = \{f \in W_2: \tau^n f \in W_2,\ \forall n \in \mathbb{Z}\}. \tag{1.2}$$

W_1^{σ} and W_1^{τ} are similarly defined.

If σ is the identity on V, the system is called one-parametric, and is denoted by $[V;W_1,W_2;\tau]$.

In the context of a.s.s., a sesquilinear form $B: V \times V \to \mathbb{C}$ is positive if

$$B(f,f) \geq 0,\ f \in V \tag{1.3}$$

and is Toeplitz if it is invariant both with respect to σ and to τ, i.e.:

$$B(\sigma f,\sigma g) = B(f,g) = B(\tau f,\tau g),\ \forall f,g \in V \tag{1.4}$$

The restriction of a Toeplitz form B to $W_1 \times W_2$ is called the Hankel form associated with B. Motivated by the examples below, we are concerned with triplets B_0,B_1,B_2 of Toeplitz forms such that $B_1 \geq 0$, $B_2 \geq 0$ and, for W' and W" subspaces of V,

$$|B_0(f,g)|^2 \leq B_1(f,f)B_2(g,g) \text{ for } (f,g) \in W' \times W''. \tag{1.5}$$

The subordination (1.5) will be expressed as

$$B_0 \leq (B_1,B_2) \text{ in } W' \times W'' \tag{1.5a}$$

and, if $V' = V'' = V$, simply as $B_0 \leq (B_1,B_2)$.

Since two positive forms B_1 and B_2 give rise to two hilbertian seminorms on V, $\|f\|_j = B_j(f,f)^{1/2}$, $j = 1,2$, the relation $B_0 \leq (B_1,B_2)$ expresses the boundedness of the form B_0 with respect to these seminorms. Similarly when $V' = V_1$ and $V'' = V_2$, (1.5a) expresses boundedness in $V_1 \times V_2$, but not of B_0 itself but of its associated Hankel form.

This notion of boundedness is related to positivity as follows. If $B \geq 0$ then, by the Schwarz inequality, $B \leq (B,B)$. Reciprocally, any triplet of forms, B_0,B_1,B_2, in $V \times V$ define a form $\mathcal{B}$: $V^2 \times V^2 \to \mathbb{C}$, given by

$$\mathcal{B}((f,g),\ (f',g')) = B_1(f,f') + B_2(g,g') + B_0(f,g') + B_0(g,f') \qquad (1.6)$$

and $B_0 \leq (B_1,B_2)$ expresses the positivity of $\mathcal{B}$ in $V^2 \times V^2$.

EXAMPLE 1. Setting $V = \{f\colon T \to \mathbb{C}\colon f(t) = \Sigma_{\text{finite}}\ \hat{f}(n)e^{int}\}$, the set of trigonometric polynomials, $\Delta = \{m \in \mathbb{Z},\ m \geq 0\}$, the set of nonnegative integers, $V_1 = \{f \in V\colon \operatorname{supp} \hat{f} \subset \Delta\}$, $V_2 = \{f \in V\colon \operatorname{supp} \hat{f} \subset \Delta^c\}$, the sets of analytic and conjugate analytic polynomials, respectively, $\sigma \equiv I$, the identity, τ = the shift operator, defined by $\tau f(t) = e^{it}f(t)$, we obtain a 1-parameter a.s.s.

EXAMPLE 2. Setting V = the trigonometric polynomials in two variables, $\Delta = \{(m,n) \in \mathbb{Z}^2\colon m \geq 0,\ n \geq 0\}$, $V_1 = \{f \in V\colon \operatorname{supp} \hat{f} \subset \Delta\}$, $V_2 = \{f \in V\colon \operatorname{supp} \hat{f} \subset \Delta^c\}$, and σ and τ the two shifts, defined by $\sigma f(s,t) = e^{is}f(s,t)$, $\tau f(s,t) = e^{it}f(s,t)$, we obtain a 2-parameter a.s.s.

In both examples above V is the direct sum of V_1 and V_2, and there is a unique linear projection $P\colon V \to V$, such that V_1 = range of P, and V_2 = kernel of P. In Example 1, $P = (1/2)(I + iH)$, where H is the Hilbert transform operator, and H acts continuously in a space $L^2(\mu)$ if and only if

$$B(Pf,Pf) \leq cB(f,f), \ \forall f \in V \tag{1.7}$$

where $B(f,g) = \int f\bar{g}d\mu$. Inequality (1.7) is easily seen to be equivalent to

$$B \leq (B_1,B_2) \text{ in } W_1 \times W_2, \text{ for } B_1 = B_2 = (1 - c^{-1})B \tag{1.7a}$$

This, and other related questions, lead to the study of triplets satisfying (1.5a). Similar considerations apply to the double Hilbert transform $f \to H_xH_yf(x,y)$.

REMARK 1. In the case of Example 1, any form $B: V \times V \to \mathbb{C}$ is determined by the associated kernel $K_B: \mathbb{Z} \times \mathbb{Z} \to \mathbb{C}$, defined by

$$K_B(m,n) = B(e^{imt},e^{int}). \tag{1.8}$$

B is positive if and only if K_B is positive definite, and B is Toeplitz iff K_B is a Toeplitz kernel.

K is said to be a generalized Toeplitz kernel (GTK) if there are three Toeplitz kernels, K_0,K_1,K_2, such that $K = K_1$ in $\{m \geq 0, n \geq 0\}$, $K = K_2$ in $\{m < 0, n < 0\}$ and $K = K_0$ in $(m \geq 0, n < 0\}$, with $K(m,n) = \overline{K(n,m)}$. Thus, to give a positive definite GTK is the same as to give a Toeplitz triplet satisfying (1.5a). In the case of Example 1, all the results that follow can be expressed in terms of GTKs.

2. LINEAR AND LOCAL NONLINEAR LIFTING THEOREMS.

The results of this section are based on the following one-parameter lifting theorem.

THEOREM 1. (One-parameter lifting theorem [5],[9]). *If* $[V,W_1,W_2;\tau]$ *is a* 1-*parameter a.s.s., and if* B_0,B_1,B_2 *are three Toeplitz (i.e.,* τ-*invariant) sesquilinear forms satisfying* $B_0 \leq (B_1,B_2)$ *in* $W_1 \times W_2$, *then there exists a Toeplitz form* B *such that* $B \leq (B_1,B_2)$ *and* $B = B_0$ *in* $W_1 \times W_2$. B *is called a* *lifting* *of the Hankel form* $B_0|_{W_1\times W_2}$.

The proof of this theorem was sketched in [5] and given in detail in [9]. This one-parameter lifting theorem is a particular case of the two-parameter lifting theorem below, since for $\sigma \equiv I$, $W_2^{\sigma} = W_2$ and the form B' in Theorem 2 gives the desired lifting of $B_0|_{W_1\times W_2}$ (the other form B" can be ignored in this case).

A full extension of Theorem 1 to multiparametric a.s.s. will not hold in general. This is due to the logical equivalence of that theorem to that of Nagy-Foias (cf. [8]), insuring that every commuting pair T_1, T_2 of contractions in a Hilbert space H has a unitary dilation, i.e., a pair of unitary operators U_1, U_2 in a larger Hilbert space, such that, for j = 1,2, $T_j = P_H\, U|_H$, where P_H is the orthogonal projection on H. It is known, by a counter-example of Parrott, that there exist three commuting contractions T_1,T_2,T_3 for which commuting unitary dilations U_1,U_2,U_3 cannot be found. Thus, there is no hope for a full lifting in a two- parameter a.s.s. Instead, we have the following result, proved in [7], assuming Theorem 1.

THEOREM 2. Two-parameter lifting theorem ([7],[9]). *Let* $[V;W_1,W_2;\sigma,\tau]$ *be a two-parameter a.s.s., and* B_0,B_1,B_2, *three Toeplitz sesquilinear forms satisfying* $B_0 \leq (B_1,B_2)$ *in* $W_1 \times W_2$. *Then there exist two Toeplitz forms,* B' *and* B", *such that*

$$B' \leq (B_1,B_2),\ B'' \leq (B_1,B_2) \tag{2.1}$$

and

$$B' = B_0 \text{ in } W_1 \times W_2^{\sigma},\ B'' = B_0 \text{ in } W_2 \times W_2^{\tau}, \tag{2.2}$$

(B',B") *is called a* *lifting pair* *for the Hankel form* $B_0|_{W_1\times W_2}$.

REMARK 2. While in the one-parameter case the lifting B determines the Hankel form $B_0|_{W_1\times W_2}$, in the two-parameter case the lifting pair (B',B") determines only the pair $(B_0|_{W_1\times W_2^\sigma}, B_0|_{W_1\times W_2^\tau})$. In [9] supplementary conditions on the a.s.s. are given to insure that the pair (B',B") determines the Hankel form $B_0|_{W_1\times W_2}$ (i.e., that $B_0 \le (B_1,B_2)$ in $W_1 \times W_2$ if and only if $B' = B_0$ in $W_1 \times W_2^\sigma$, $B'' = B_0$ in $W_1 \times W_2^\tau$ and $B' \le (B_1,B_2)$, $B'' \le (B_1,B_2)$).

The one-parameter lifting theorem has been shown to be equivalent to the Nagy-Foias commutant lifting theorem (cf. [8]), so that Theorem 2 can be considered as a two-parameter version of the classical Nagy-Foias result (cf. [9]). A local nonlinear version of the Nagy-Foias theorem was recently given in [1] and Theorem 3 below relates to Theorem 2 in the same spirit.

Given the a.s.s. $[V;W_1,W_2;\sigma,\tau]$ and two positive Toeplitz forms, B_1 and B_2, we denote, for j = 1 or 2, by (V,B_j) the space V equipped with the hilbertian seminorm $B_j(f,f)^{1/2}$, and by $\Sigma_r(V)$, $\Sigma_r(W_1)$, the balls of radius r in (V,B_1) and in (W_1,B_1), respectively. Following [12], a mapping ϕ: $\Sigma_r(W_1) \to (W_2,B_2)$ is said <u>holomorphic</u> if the complex function $(z_1,\ldots z_n) \to B_2(\phi(z_1f_1 +\ldots+ z_nf_n),g)$ is holomorphic in a neighborhood of $(1,\ldots 1) \in \mathbb{C}^n$, for all $f_1,\ldots f_n \in W_1$ such that $B_1(f_1 +\ldots+ f_n, f_1 +\ldots+ f_n) < r^2$, and for all $g \in W_1$. Assuming $\phi(0) = 0$, this is equivalent to the existence of a Taylor expansion

$$\phi(f) = \phi_1(f) + \phi_2(f,f) +\ldots+ \phi_n(f,f,\ldots f) +\ldots \tag{2.3}$$

where ϕ_n: $W_1 \times\ldots\times W_1 \to W_2$ is a n-linear map, so that ϕ_n can be identified with a linear operator from the algebraic tensor product $W_1 \otimes\ldots\otimes W_1$ to W_2.

A sequence $\{a_n\}$ is a <u>majorizing sequence</u> for ϕ if $\|\phi_n\| \le a_n$ for all $n \ge 1$. If, in addition, $\limsup a_n^{1/n} = \rho < \infty$, then the expansion (2.3) converges, at least on $\Sigma_{1/\rho}(W_1)$. Similar considerations hold for the function ψ: $\Sigma_r(V) \to (V,B_2)$.

In Theorem 2, B_0 was assumed to be an invariant sesquilinear form satisfying $B_0 \le (B_1,B_2)$ in $W_1 \times W_2$. Thus there exists a continuous operator

$$\phi: (W_1,B_1) \to (W_2,B_2),\ \|\phi\| \le 1, \text{ with } B_0(f,g) = B_2(\phi f,g) \qquad (2.4)$$

In the following theorem, B_0 will be assumed to be a form from $\Sigma_r(W_1) \times W_2$ to $\mathbb{C}$, $B_0(f,g) = B_2(\phi f,g)$ for ϕ: $\Sigma_r(W_1) \to W_2$ a nonlinear holomorphic map, and satisfy the local invariance conditions:

$$B_0(\tau f,g) = B_0(f,\tau^{-1}g),\ B_0(\sigma f,g) = B_0(f,\sigma^{-1}g),$$
$$\forall (f,g) \in \Sigma_r(W_1) \times W_2 \qquad (2.5)$$

(Observe that the τ-invariance of B_1 implies $\tau\Sigma_r(W_1) \subset \Sigma_r(W_1)$; idem for σ.) Instead of $\|\phi\| \le 1$, we shall require the existence of a majorizing sequence with $\rho < \infty$.

THEOREM 3. (Local nonlinear lifting theorem). *Let* $[V{:}W_1,W_2;\ \sigma,\tau]$ *be an a.s.s.,* B_0,B_1,B_2, *three sesquilinear forms,* B_0: $\Sigma_r(W_1) \times W_2 \to \mathbb{C}$, *satisfying* (2.4) *and* (2.5), *for* ϕ: $\Sigma_r(W_1) \to (W_2,B_2)$ *an holomorphic mapping,* $\phi(0) = 0$, *with majorizing sequence* $\{a_n\}$ *and lim sup* $a_n^{1/n} = \rho < \infty$. *Then there exist two holomorphic functions* ψ' *and* ψ'', *from* $\Sigma_{1/\rho}(v)$ *to* (V,B_2), *with the same majorizing sequence* $\{a_n\}$, *such that the forms* B' *and* B'', *defined by*

$$B'(f,g) = B_2(\psi' f,g),\ B''(f,g) = B_2(\psi'' f,g), \qquad (2.6)$$

coincide with B_0 *in* $\Sigma_{1/\rho}(W_1) \times W_2^{\sigma}$ *and* $\Sigma_{1/\rho}(W_1) \times W_2^{\tau}$, *respectively.* B' *and*

B" *satisfy (2.5) in* $\Sigma_r(V) \times V$.

PROOF. Following [1], take $B_1(f_1,f_1) = \ldots = B_1(f_n,f_n) < r/n$ and express ϕ_n as the Taylor coefficient of $z_1 \ldots z_n$ in the expression of $\phi(z_1 f_1 + \ldots + z_n f_n)$:

$$\phi_n(f_1,\ldots f_n) = \frac{1}{n!}\int_0^{2\pi}\ldots\int_0^{2\pi}\phi(e^{it_1}f_1 + \ldots + e^{it_n}f_n)\cdot$$

$$e^{-i(t_1+\ldots+t_n)}dt_1\ldots dt_n \tag{2.7}$$

$$B_2(\phi_n(f_1,\ldots f_n),g) = \frac{1}{n!}\int_0^{2\pi}\ldots\int_0^{2\pi}B_2(\phi(e^{it_1}f_1 + \ldots + e^{it_n}f_n),g)\cdot$$

$$\cdot\, e^{-i(t_1+\ldots+t_n)}dt_1\ldots dt_n \tag{2.7a}$$

By (2.4) and (2.5),

$$B_2(\phi(e^{it_1}f_1 + \ldots + e^{it_n}f_n), \tau^{-1}g) = B_0(e^{it_1}f_1 + \ldots + e^{it_n}f_n, \tau^{-1}g) =$$

$$= B_0(\tau e^{it_1}f_1 + \ldots + \tau e^{it_n}f_n,g) =$$

$$= B_2(\phi(\tau e^{it_1}f_1 + \ldots + \tau e^{it_n}f_n),g),$$

so that (2.7a) gives

$$B_2(\phi_n(\tau f_1,\ldots,\tau f_n),g) = B_2(\phi_n(f_1,\ldots f_n), \tau^{-1}g). \tag{2.8}$$

For each n, consider the spaces $V^{(n)} = V \otimes \ldots \otimes V$, $W_1^{(n)} = W_1 \otimes \ldots \otimes W_1$, $W_2^{(n)} = W_2$ and the forms $B_1^{(n)}(f_1 \otimes \ldots \otimes f_n, g_1 \otimes \ldots \otimes g_n) = B_1(f_1,g_1) \ldots B_1(f_n,g_n)$: $V^{(n)} \times V^{(n)} \to \mathbb{C}$, and $B_2^{(n)} = B_2$. Define the form

$B_0^{(n)}: V_1^{(n)} \times V_2^{(n)} \to \mathbb{C}$ by

$$B_0^{(n)}(f_1 \otimes \ldots \otimes f_n, g) = B_2(\phi_n(f_1, \ldots f_n), g). \tag{2.9}$$

From $\|\phi_n\| \leq a_n$ it is easy to deduce that $B_0^{(n)} \leq a_n(B_1^{(n)}, B_2^{(n)})$ in $V_1^{(n)} \times V_2^{(n)}$, and from (2.5), $B_0^{(n)}(\tau f_1 \otimes \ldots \otimes \tau f_n, g) = B_0^n(f_1 \otimes \ldots \otimes f_n, \tau^{-1}g)$. Similarly for σ. We may, therefore, apply Theorem 2, and get, for each n, a lifting pair $(B'^{(n)}, B''^{(n)})$ for $B_0^{(n)}$, such that both are $\leq a_n(B_1^{(n)}, B_2^{(n)})$.

Thus, $B'^{(n)}$ is given by an operator $\psi'^{(n)}$, and $B''^{(n)}$ by an operator $\psi''^{(n)}$, with $\|\psi'^{(n)}\| \leq a_n$, $\|\psi''^{(n)}\| \leq a_n$. Now we can define ψ' and ψ'' by the Taylor expansions

$$\psi'(f) = \psi'^{(1)}(f) + \psi'^{(2)}(f,f) + \ldots$$

and (2.10)

$$\psi''(f) = \psi''^{(1)}(f) + \psi''^{(2)}(f,f) + \ldots$$

The maps B' and B" defined by

$$B'(f,g) = B_2(\psi'(f),g), \quad B''(f,g) = B_2(\psi'' f,g),$$

satisfy the required conditions. □

3. INTEGRAL REPRESENTATIONS OF THE LIFTINGS

As shown in [9], if $[V;W_1,W_2;\sigma,\tau]$ is the special a.s.s. of Example 2, then, for every triplet B_0,B_1,B_2 of Toeplitz forms satisfying $B_0 \leq (B_1,B_2)$ in $W_1 \times W_2$, there exist three finite measures, ρ_0,ρ_1,ρ_2, in T^2, such that

$$\rho_0(E) \leq \rho_1(E)^{1/2}\, \rho_2(E)^{1/2}, \ \forall E \subset T^2, \text{ Borel set} \tag{3.1}$$

and, for $j = 0,1,2$,

$$B_j(f,g) = \int_{T^2} \Omega(s,t)(f,g)d\rho_j \tag{3.2}$$

with

$$\Omega(s,t)(f,g) = f(s,t)\overline{g(s,t)}, \tag{3.2a}$$

so that $\Omega(s,t)$ is an eigenform of the shifts σ and τ:

$$\Omega(s,t)(\tau f,g) = e^{it}\Omega(s,t)(f,g),\ \Omega(s,t)(\sigma f,g) = e^{is}\Omega(s,t)(f,g). \tag{3.3}$$

For $j = 0$, (3.2) holds only in $V_1 \times V_2^{\sigma}$, and there is another, similar, representation in $V_1 \times V_2^{\tau}$.

This integral representation has been called the Generalized Bochner Theorem, GBT (cf. [8], since the classical Bochner theorem is obtained, in T^2, when $B_0 = B_1 = B_2$. The GBT gives a stronger version of the lifting theorem, providing representations of the liftings through measures, and spectral decompositions of the given triplet of forms. A similar GBT holds for the one-parameter a.s.s. of Example 1, and in [7] the GBT was extended to more general one-parameter a.s.s., where τ can be "reduced" to the classical shift in T.

In this section we extend the GBT to a wider class of a.s.s., where σ and τ are no longer reducible to the ordinary shift, by using eigenforms $\Omega(s,t)$ of these general operators σ and τ. In the case where $B_0 = B_1 = B_2$, $V = V_1 = V_2$, such expansions into general eigenforms were studied by Gel'fand and Kostuchenko, M.G. Krein, Berezanskii, Maurin and other authors (cf. [11], [2]) for invariant and for symmetric forms. Since here we consider only invariant forms, we use results of L.M. Korsunski (specifically, Theorems 1 and 3 of [13]), where the method of Krein-Berezanskii is given for invariant kernels in a form suitable to our purposes.

Recall that $H_+ \subset H_0 \subset H_-$ is a <u>Gelfand triple</u> of Hilbert spaces if H_+ is dense in H_0, the inclusion operator $O: H_+ \to H_0$ is continuous, and the space H_- is constructed as follows. Since, for $f \in H_0$, $u \in H_+$,

$$|(f,0u)| \leq \|0\| \; \|f\|_0 \; \|u\|_+,$$

each $f \in H_0$ can be identified with a continuous antilinear functional in H_+, so that H_0 is contained in the antidual H_+'. Since H_+ is dense in H_+, H_0 is dense in H_+', and we define $H_- = H_+'$.

It will be assumed that an involution $f \to \bar{f}$ is defined in H_+, allowing to transfer sesquilinear forms into bilinear ones: $B(f,g) \to B(f,\bar{g})$. The bilinear form $B(f,\bar{g})$ determines a linear form in the algebraic tensor product $H_+ \otimes H_+$, and, if this linear form is continuous in the Hilbert norm of $H_+ \otimes H_+$ and has norm $\leq c$, then we write $\|B\|_- \leq c$. In such case, there is a bounded operator b in H_+ such that $\|b\| \leq c$ and $B(f,\bar{g}) = (bf,g)$, and B can be identified with an element of $H_- \otimes H_-$.

Fix now an a.s.s., $[V;W_1,W_2;\sigma,\tau]$, such that V is a Hilbert space in a Gelfand triple, i.e.: $V = H_+ \subset H_0 \subset H_-$, and fix a topological vector space $D \subset V$, D dense in V, such that $f \in D$ implies $\bar{f} \in D$, and

$$\sigma(D) \subset D, \; \tau(D) \subset D. \tag{3.4}$$

A family of positive sesquilinear forms $\Omega(s,t)\colon V \times V \to \mathbb{C}$, $(s,t) \in T^2$, is called <u>elementary</u> if each $\Omega(s,t)$ is an eigenform of σ and τ, i.e., satisfies (3.3) for $f,g \in D$, and if $\|\Omega(s,t)\| \leq C$ for all $(s,t) \in T^2$. Thus, we may write

$$\Omega(s,t)(f,g) = \langle \omega(s,t)f,g \rangle_V \tag{3.5}$$

where $\omega(s,t)$ is an operator in V, with norm $\|\omega(s,t)\| \leq C$, for all (s,t), and

$$\Omega(s,t) \in H_- \oplus H_-. \tag{3.5a}$$

A positive form $B: V \times V \to \mathbb{C}$ is of <u>trace class</u> (or <u>nuclear</u>) if $\Sigma_n B(e_n,e_n) < \infty$ for every orthonormal basis (e_n) in V.

THEOREM 4. *Let* $[V;W_1,W_2;\sigma,\tau]$ *and* $D \subset V = H_+ \subset H_0 \subset H_-$ *be as above,* $B_1: V \times V \to \mathbb{C}$ *and* $B_2: V \times V \to \mathbb{C}$, *be two positive Toeplitz forms, both of trace class, and* $B_0: V \times V \to \mathbb{C}$ *be a continuous Toeplitz form satisfying* $B_0 \leq (B_1,B_2)$ *in* $W_1 \times W_2$. *Then there exist two positive scalar measures,* ρ' *and* ρ'', *in* T^2, *two elementary families,* $\Omega'(s,t)$ *and* $\Omega''(s,t)$, *and, for each* $(s,t) \in T^2$, *two positive 2×2 matrices,* $(A'_{ij}(s,t))_{i,j=1,2}$ $(A''_{ij}(s,t))_{i,j=1,2}$, *having operators as elements, such that, setting*

$$d\mu'_{ij} = A'_{ij}(s,t)d\rho', \quad d\mu''_{ij} = A''_{ij}(s,t)d\rho'', \tag{3.6}$$

we have, for $j = 1$ *or* 2, $\mu'_{jj} = \mu''_{jj}$, *and*

$$B_j(f,g) = \int_{T^2} \Omega'(s,t)(f,g)d\mu'_{jj} = \int_{T^2} \Omega''(s,t)(f,g)d\mu''_{jj}, \quad \forall(f,g) \in V \times V$$

$$B_0(f,g) = \int_{T^2} \Omega'(s,t)(f,g)d\mu'_{12}, \qquad \forall(f,g) \in W_1 \times W_2^{\sigma} \tag{3.7}$$

$$B_0(f,g) = \int_{T^2} \Omega''(s,t)(f,g)d\mu''_{12}, \qquad \forall(f,g) \in W_1 \times W_2^{\tau}$$

The integrals converge in the norm $\|\cdot\|_-$, *and the operators* $A'_{ij}(s,t)$ *and* $A''_{ij}(s,t)$ *act in a space varying with* (s,t).

PROOF. Consider first the case $\sigma \equiv I$, so that the a.s.s. is really a one-parameter a.s.s. By the one-parameter lifting theorem, there is a Toeplitz form B satisfying $B \leq (B_1,B_2)$ and $B = B_0$ in $W_1 \times W_2$, so that, by (1.6), the matrix form $\mathcal{B} = \begin{bmatrix} B_1 & B \\ B^* & B_2 \end{bmatrix}$ is positive in $(V \oplus V) \times (V \oplus V)$. Set

$$\mathcal{V} = V \oplus V, \; \mathcal{H}_0 = H_0 \oplus H_0, \; \mathcal{D} = D \oplus D, \tag{3.8}$$

So that, $\mathcal{V} \subset \mathcal{H}_0$, $\mathcal{D} \subset \mathcal{V}$, τ acts on $\mathcal{V}$ and $\mathcal{B}$ is τ-invariant. Moreover, since $B \leq (B_1, B_2)$ and B_1, B_2 are of trace class in V, B_0 is also of trace class in V, and $\mathcal{B} \geq 0$ is of trace class in $\mathcal{V}$. We are therefore in the situation of the above-mentioned Theorem 1 of Korsunski [13], which ensures that, under these conditions, there exists a measure $\rho \geq 0$ in T, and an elementary family of forms $\Omega(t)$: $\mathcal{V} \times \mathcal{V} \to \mathbb{C}$, or, equivalently, an elementary family of 2x2 matrix forms $(\Omega_{ij}(t))_{i,j=1,2}$, $\Omega_{ij}(t): V \times V \to \mathbb{C}$, $t \in T$, such that

$$\mathcal{B} = \begin{bmatrix} B_1 & B \\ B^* & B_2 \end{bmatrix} = \int_T \begin{bmatrix} \Omega_{11}(t) & \Omega_{12}(t) \\ \Omega_{12}(t)^* & \Omega_{22}(t) \end{bmatrix} d\rho(t) \tag{3.9}$$

so that

$$B_j = \int_T \Omega_{jj}(t)d\rho, \; j = 1,2, \; B = \int_T \Omega_{12}(t)d\rho. \tag{3.9a}$$

Setting $\Omega(t) = \Omega_{11}(t) + \Omega_{22}(t) \geq 0$, it is $\Omega_{11}(t) \leq (\Omega(t)), \Omega(t))$, $\Omega_{22}(t) \leq (\Omega(t), \Omega(t))$ and $\Omega_{12}(t) \leq (\Omega(t), \Omega(t))$, so that, for each i,j = 1 or 2, there is an operator $A_{ij}(t)$: $(V, \Omega(t)) \to (V, \Omega(t))$ such that

$$\Omega_{ij}(t)(f,g) = \langle A_{ij}(t)f,g \rangle_{\Omega(t)} = \Omega(t)(A_{ij}(t)f,g). \tag{3.10}$$

If $\omega_{ij}(t)$ and $\omega(t)$ are the operators in V determined by the forms $\Omega_{ij}(t)$ and $\Omega(t)$, then

$$\langle \omega_{ij}(t)f,g \rangle_V = \Omega_{ij}(t)(f,g) = \langle \omega(t) \; A_{ij}(t)f,g \rangle \tag{3.10a}$$

hence, $\omega_{ij}(t) = \omega(t) \; A_{ij}(t)$. Setting $\mathcal{B} = (B_{ij})$ and denoting by b_{ij} the operator in V determined by B_{ij}, (3.9a) reads

$$b_{ij} = \int_T \omega_{ij}(t)d\rho = \int_T \omega(t) \; A_{ij}(t)d\rho. \tag{3.10c}$$

Thus, setting $d\mu_{ij} = A_{ij}(t)d\rho$, i,j = 1,2, we obtain that

$$b_{ij} = \int_T \omega(t)d\mu_{ij} \tag{3.11}$$

and

$$B_1(f,g) = \langle b_{11}f,g\rangle,\ B_{22}(f,g) = \langle b_{22}f,g\rangle \text{ in } V \times V$$

(3.11a)

$$B_0(f,g) = \langle b_{12}f,g\rangle \text{ in } W_1 \times W_2.$$

Formulae (3.11) - (3.11a) give the one-parametric version of the theorem.

The proof of the two-parameter case is entirely similar, using the two-parameter lifting theorem and, also, the fact that Korsunski's theorem hold for every representation $(m,n) \to \sigma^m\tau^n$ of $\mathbb{Z}^2$ into V. □

Theorem 4 gives an integral representation of the liftings in Theorems 1 and 2. A similar representation can be given for the liftings in Theorem 3. In fact, from the proof of Theorem 3 it follows that every triplet B_0,B_1,B_2, satisfying the conditions of that theorem, gives rise to a sequence of spaces $V^{(n)}$, with subspaces $W_1^{(n)}$ and $W_2^{(n)}$, and sesquilinear forms $B_0^{(n)}$, $B_1^{(n)}$, $B_2^{(n)}$, satisfying, for each n, the hypotheses of Theorem 2. Therefore, assuming the additional hypotheses that V be a Hilbert space, $V = H_+ \subset H_0 \subset H_-$, and that B_1 and B_2 be of trace class, we can apply Theorem 4 to the forms $B_0^{(n)}$, $B_1^{(n)}$, $B_2^{(n)}$, for each n. Since the sequence $(B_0^{(n)})$ determines the form B_0, we thus obtain

THEOREM 5. *Let* $[V;W_;,W_2;\sigma,\tau]$ *and* B_0,B_1,B_2 *be as in Theorem 3, and assume that* V *is a Hilbert space in a Gel'fand triple,* $V = H_+ \subset H_0 \subset H_-$, *and that* B_1 *and* B_2 *are of trace class. Then, there exists a pair of sequences* $(\mu'_{ij}(n))$, $(\mu''_{ij}(n))$, *of positive* 2x2 *matrice measures in* T^2, *and a pair of sequences of elementary families,* $\Omega'_n(s,t)$, $\Omega''_n(s,t)$, $(s,t) \in T^2$, *of forms defined in* $V^{(n)} \times V$ $(V^{(n)} = V \oplus\ldots\oplus V)$, *such that*

$$B_0(f,g) = \sum_n \int_{T^2} \Omega_n'(s,t)(f \oplus \ldots \oplus f,g)d\mu_{12}'(n), \ \forall(f,g) \in W_1 \times W_2^\sigma \tag{3.12}$$

$$B_0(f,g) = \sum_n \int_{T^2} \Omega_n''(s,t)(f \oplus \ldots \oplus f,g)d\mu_{12}''(n), \ \forall(f,g) \in W_1 \times W_2^\tau$$

and, for j = 1 *or* 2,

$$B_j(f,g) = \int_{T^2} \Omega'(s,t)(f,g)d\mu_{jj}'(1) \qquad \forall(f,g) \in V \times V \tag{3.12a}$$

$$= \int_{T^2} \Omega''(s,t)(f,g)d\mu_{jj}''(1)$$

In studying some moment problems related to the Wightman functionals in quantum mechanics, Berezanskii [3] and Korsunski [13] obtained integral representations of positive symmetric and invariant infinite matrices $(B_{jk})_{j,k=0}^{\infty}$ of sesquilinear forms B_{jk}: $H_+^{(j)} \times H_+^{(k)} \to \mathbb{C}$, $H_+^{(n)} = H_+ \oplus \ldots \oplus H_+$. We now generalize (a special case of) the result of Korsunski for invariant infinite matrices to triplets of such matrices, satisfying $(B_{jk}^0) \leq ((B_{jk}^1), (B_{jk}^2))$ in an a.s.s. Since the nonlinear version given in Theorem 5 is, as already observed, equivalent to an integral representation for a simple sequence of such forms, Theorem 6 below can be viewed as kind of local nonlinear lifting theorem for "two variables Taylor expansions." Observe that Theorem 6 could reduce to Theorem 5 if we knew that the matrices (B_{jk}^1) and (B_{jk}^2) define trace class operators in the direct sum of the spaces $H_+^{(n)}$. But Theorem 6 assumes only that the diagonal terms B_{jj}^1 and B_{jj}^2 are trace class, so that, in the case when V is the direct sum of the $H_+^{(n)}$'s, Theorem 6 gives a stronger result than Theorem 5.

Let $V = H_+ \subset H_0 \subset H_-$ be a Gel'fand triple of Hilbert spaces, W_1 and W_2 be subspaces of V, and for each n, $V^{(n)} = V \oplus \ldots \oplus V$, let σ_n: $V^{(n)} \to V^{(n)}$, τ_n: $V^{(n)} \to V^{(n)}$ be two commuting isomorphisms. Set

$$V = \{(f_0,f_1,\ldots): \text{ finitely supported sequence, } f_n \in V^{(n)}, \forall n\},$$

$$\sigma(f_0,f_1,\ldots) = (\sigma_0 f_0, \sigma_1 f_1,\ldots),\ \tau(f_0,f_1,\ldots) = (\tau_0 f_0, \tau_1 f_1,\ldots), \tag{3.13}$$

$$W_j = \{(f_0,f_1,\ldots) \in V: f_n \in W_j^{(n)}, \forall n\} \text{ for } j = 1 \text{ or } 2.$$

An infinite matrix $(B_{jk})_{j,k=0}^{\infty}$ of sesquilinear forms, with $B_{jk}: H_+^{(j)} \times H_+^{(k)} \to \mathbb{C}$, is positive if

$$\sum_{j,k=0}^{\infty} B_{jk}(f_j,f_k) \geq 0,\ \forall (f_0,f_1,\ldots) \in V \tag{3.14}$$

and invariant if, for each j,k,

$$B_{jk}(\tau f,\tau g) = B_{jk}(f,g) = B_{jk}(\sigma f,\sigma g) \tag{3.15}$$

for all $(f,g) \in H_+^{(j)} \times H_+^{(k)}$.

A set of positive matrices $\{\Omega_{jk}(s,t): (s,t) \in T^2,\ j,k = 0,1,\ldots\}$ is called elementary if, for each j,k, conditions (3.3) are satisfied (with Ω replaced by Ω_{jk}, and $(f,g) \in H_+^{(j)} \times H_+^{(k)}$), and $\Omega_{jk}(s,t) \in H_-^{(j)} \oplus H_-^{(k)}$, with

$$\sum_{j,k=0}^{\infty} \|\Omega_{jk}(s,t)\|^2_{H_-^{(j)}\oplus H_-^{(k)}} (m_j m_k)^{-1} \leq c < \infty \tag{3.16}$$

for some numerical sequence $\{m_n\}$, $m_n \geq 1$, and for all $(s,t) \in T^2$.

THEOREM 6. *Let* $V = H_+, W_1, W_2, \sigma_n, \tau_n$, $n = 1,2,\ldots$, *be as above, and, for* $i = 0,1,2$, $B_{jk}^i: V^{(j)} \times V^{(k)} \to \mathbb{C}$, (B_{jk}^0), (B_{jk}^1) *and* (B_{jk}^2) *be three invariant continuous matrices. Assume* (B_{jk}^1) *and* (B_{jk}^2) *to be positive in the sense of* (3.14) *and, for each n,* B^1_{nn} *and* B^2_{nn} *to be trace class. Furthermore, assume*

$$|\Sigma_{jk}\, B^0_{jk}(f_j,g_k)| \le (\Sigma_{jk}\, B^1_{jk}(f_j,f_k))^{1/2}\, (\Sigma_{jk}\, B^2_{jk}(g_j,g_k))^{1/2}$$

holds for all $(f_n) \in W_1$ *and* $(g_n) \in W_2$. *Then there exist a pair of positive infinite matrices, having as elements 2x2 matrix measures acting in* T^2, $(\mu'_{\alpha\beta}(j,k))_{\alpha,\beta=1,2}$, $(\mu''_{\alpha\beta}(j,k))_{\alpha,\beta=1,2}$, *and a pair of elementary families*

$\Omega'_{jk}(s,t)$, $\Omega''_{jk}(s,t)$, $(s,t) \in T^2$, *such that, for* $\alpha = 1$ *or* 2,

$$\begin{aligned} B^\alpha_{jk}(f,g) &= \int \Omega'_{jk}(s,t)(f,g)\, d\mu'_{\alpha\alpha}(j,k) \\ &= \int \Omega''_{jk}(s,t)(f,g)\, d\mu''_{\alpha\alpha}(j,k) \end{aligned} \qquad \forall (f,g) \in V^{(j)} \times V^{(k)}$$

and (3.18)

$$B^0_{jk}(f,g) = \int \Omega'_{jk}(s,t)(f,g) d\mu'_{12}(j,k) \qquad \forall (f,g) \in W_1^{(n)} \times (W_2^\sigma)^{(n)}$$

$$B^0_{jk}(f,g) = \int \Omega''_{jk}(s,t)(f,g) d''_{12}(j,k) \qquad \forall (f,g) \in W_1^{(n)} \times (W_2^\tau)^{(n)}$$

SKETCH OF PROOF. The system $[V;W_1,W_2;\sigma,\tau]$ given by (3.13) is an a.s.s. Define, for $\alpha = 1$ and 2, B_α in $V \times V$ by

$$B_\alpha((f_j),\ (g_k)) = \sum_{j,k} B^\alpha_{jk}(f_j,g_k)$$

so that $B_\alpha \ge 0$ in $V \times V$ and is τ- and σ-invariant. Define B_0 similarly. Then B_0 is invariant and, by (3.17),

$$B_0 \le (B_1,B_2) \text{ in } W_1 \times W_2.$$

Applying the lifting theorem 2, one obtains a lifting pair B', B'' for B_0, satisfying $B' \leq (B_1, B_2)$, $B'' \leq (B_1, B_2)$, $B' = B_0$ in $V_1 \times V_2^{\sigma}$ and $B'' = B_0$ in $V_1 \times V_2^{\tau}$. Calling B'_{jk} and B''_{jk} the elements of the matrices B' and B'', we have that

$$\mathcal{B}'_{jk} = \begin{bmatrix} B^1_{jk} & B'_{jk} \\ \cdot & B^2_{jk} \end{bmatrix} \quad \text{and} \quad \mathcal{B}''_{jk} = \begin{bmatrix} B^1_{jk} & B''_{jk} \\ \cdot & B^2_{jk} \end{bmatrix}$$

are positive forms in $(V \oplus V) \times (V \oplus V)$, $\mathcal{B}'_{nn}$ and $\mathcal{B}''_{nn}$ are of trace class, and the matrices $\mathcal{B}'_{jk}$ and $\mathcal{B}''_{jk}$ are positive in the sense of (3.14). We are now in the situation of Theorem 3 of Korsunski [13], which ensures the existence, for $\alpha = 1,2$, of a measure $\rho_\alpha \geq 0$, and of an elementary family $\Omega^{\alpha}_{jk}(s,t) \in (H_-^{(j)} \times H_-^{(j)}) \oplus (H_-^{(k)} \times H_-^{(k)})$, such that

$$\mathcal{B}^{\alpha}_{jk} = \int_{T^2} \Omega^{\alpha}_{jk}(s,t) d\rho_\alpha .$$

From here, the proof follows as that of Theorem 4. □

4. STATIONARY PROCESSES OVER AN A.S.S.

In this section we sketch the extension to stationary processes over an a.s.s. of the classical Bochner-Khinchine theorem, a basic result in the theory of stochastic processes. This requires the translation, in terms of processes, of the generalized Bochner theorem discussed in the previous section.

A function $X: \mathbb{Z} \to H$, where $\mathbb{Z}$ are the integers and H is a Hilbert space, is called an H-process over $\mathbb{Z}$. The positive definite kernel $K: \mathbb{Z} \times \mathbb{Z} \to \mathbb{C}$, defined by $K(m,n) = \langle X(m), X(n)\rangle$, is called the covariance of the process, and the process is called stationary if its covariance is a Toeplitz kernel:

$$K(\tau m, \tau n) = K(m,n) \quad \text{for} \quad \tau m = m + 1, \ \forall m. \tag{4.1}$$

A process X is stationary if and only if there exists a unitary operator U in H such that $UX(n) = X(n + 1)$, $\forall n$. By the Bochner-Khinchine theorem, X is stationary if and only if $K(m,n) = \hat{\mu}(m - n)$, $\forall n,m \in \mathbb{Z}$, for some positive measure μ in T, and if and only if $X(n) = \hat{\nu}(n)$, $\forall n \in \mathbb{Z}$, for some bounded H-valued measure ν that is <u>orthogonally scattered</u>, i.e.:

$$\langle \nu(f), \nu(g) \rangle = 0 \text{ whenever } f\bar{g} = 0 \text{ for } f,g \in C(t). \tag{4.2}$$

Gel'fand and Itoh developed a theory of continuous linear processes over the Schwartz spaces $\mathcal{D}$ (or $\mathcal{S}$), $X: \mathcal{D} \to H$ expounded in [11]. In such case, the covariance is a positive sesquilinear form, $B: \mathcal{D} \times \mathcal{D} \to \mathbb{C}$, continuous in the $\mathcal{D}$ topology, defined by $B(f,g) = \langle X(f), X(g) \rangle$, and the process is stationary if B is τ_t-invariant, where $\tau_t f(x) = f(x + t)$ for $t \in \mathbb{R}$. By the Bochner-Schwartz theorem, $X: \mathcal{D} \to H$ is stationary if and only if $B(f,g) = \int \hat{f} \, \bar{\hat{g}} \, d\mu$, for μ a positive tempered measure in $\mathbb{R}$. In this case, the process itself is the Fourier transform, $X = \hat{\nu}$, of a random measure ν in $\mathbb{R}$, such that

$$E[\nu(\Delta_1) \overline{\nu(\Delta_2)}] = 0 \text{ if } \Delta_1 \cap \Delta_2 = \phi. \tag{4.2a}$$

A process $X: \mathbb{Z} \to H$ is said to be <u>generalized stationary</u> if its covariance is a GTK (see Remark 1 in Section 1), and such processes (as well as the generalized harmonizable ones) were discussed in [4].

Here we discuss briefly the notion of stationary processes over an a.s.s.

Given any a.s.s. $[V;W_1,W_2;\sigma,\tau]$ and any Hilbert space H, an H - <u>process over</u> V is a linear map $X: V \to H$, and the process <u>stationary</u> if its covariance form $B: V \times V \to \mathbb{C}$, given by

$$B(f,g) = \langle X(f), X(g) \rangle_H$$

is Toeplitz, i.e.: τ- and σ- invariant.

The process X will be stationary if and only if there exist a pair of unitary operators in H , U' and U", such that

$$U'X(f) = X(\tau f),\ U''X(f) = X(\sigma f),\ \forall f \in V. \tag{4.3}$$

These definitions do not take into account the subspaces V_1 and V_2 defining the a.s.s., and we shall consider now a notion of stationarity that distinguishes the subspaces.

Every positive sesquilinear form $B: V \times V \to \mathbb{C}$ defines a hilbertian seminorm $B(f,f)^{1/2}$ in V, and gives rise to a Hilbert space H "containing" V as a dense subspace, and to a H-process, $X: V \to H$, having B as its covariance form, and all Hilbert space-valued processes over V arise essentially in this way.

Consider now triplet of forms, $B_1: V \times V \to \mathbb{C}$, $B_2: V \times V \to \mathbb{C}$, $B_0: V_1 \times V_2 \to \mathbb{C}$, satisfying $B_0 \le (B_1,B_2)$ in $V_1 \times V_2$. This triplet determines a positive form $\mathcal{B}: (V_1 \oplus V_2) \times (V_1 \oplus V_2) \to \mathbb{C}$ (cf. (1.6)), and thus gives rise to a Hilbert space-valued process $X_0: V_1 \oplus V_2 \to H_0$, having $\mathcal{B}$ as its covariance form. Since B_1 and B_2 are positive, they give rise to two additional processes, $X_1: V \to H_1$ and $X_2: V \to H_2$, having B_1 and B_2 as covariance forms. Moreover, it follows from this construction that, in an obvious sense,

$$\begin{aligned} &f \in V_1 \text{ implies } X_0\binom{f}{0} = x_1 f \in H_1 \\ \text{and} \qquad & \qquad\qquad\qquad\qquad\qquad\qquad (4.4) \\ &g \in V_2 \text{ implies } X_0\binom{0}{g} = X_2\, g \in H_2 \end{aligned}$$

Accordingly, for any triplet of Hilbert spaces, H_0,H_1,H_2, let us define an (H_0,H_1,H_2)-<u>process over</u> $[V;V_1;V_2;\sigma,\tau]$ to be a triplet $(X_0: V_1 \oplus V_2 \to H_0,\ X_1: V \to H_1,\ X_2: V \to H_2)$ of Hilbert space-valued processes satisfying (4.4).

The covariance of such process is the triplet of forms $(B_0: W_1 \times W_2 \to \mathbb{C}, B_1: V \times V \to \mathbb{C}, B_2: V \times V \to \mathbb{C})$, defined by

$$B_0(f,g) = \langle X_0\binom{f}{0}, X_0\binom{0}{g}\rangle_{H_0},$$

$$B_1(f,g) = \langle X_1 f, X_1 g\rangle_{H_1}, \; B_2(f,g) = \langle X_2 f, X_2 g\rangle_{H_2}, \tag{4.5}$$

so that $B_0 \le (B_1,B_2)$ in $W_1 \times W_2$ holds by (4.4).

Such process is called <u>stationary</u> if B_1 and B_2 are Toeplitz forms and B_0 is a Hankel form, i.e.: $B_0(\tau f,g) = B_0(f,\tau^{-1}g)$, $B_0(\sigma f,g) = B_0(f,\sigma^{-1}g)$.

An (H_0,H_1,H_2)-process over $[V;W_1,W_2;\sigma,\tau]$ is stationary if and only if there exist a pair of commuting unitary operators, U_1' and U_1'', in H_1, another pair of unitary operators, U_2' and U_2'', in H_2, and a pair of isometries, U_0' and U_0'', in H_0, with domains $\mathrm{dom}(U_0') = X_0(W_1 \oplus \sigma^{-1}W_2)$, $\mathrm{dom}(U_0'') = X_0(W_1 \oplus \tau^{-1}W_2)$, such that

$$U_0'X_0\begin{bmatrix} f \\ \sigma^{-1}g \end{bmatrix} = X_0\begin{bmatrix} \sigma f \\ g \end{bmatrix}, \; U_0''X_0\begin{bmatrix} f \\ \tau^{-1}g \end{bmatrix} = X_0\begin{bmatrix} \tau f \\ g \end{bmatrix}, \quad \forall (f,g) \in W_1 \times W_2 \tag{4.6}$$

and

$$U_j'X_j(f) = X_j(\sigma f), \; U_j''X_j(f) = X_j(\tau f), \; \forall f \in V, \; j = 1 \text{ or } 2.$$

The lifting theorem 2 can now be restated in terms of these processes as follows.

THEOREM 2a. (Lifting theorem for stationary processes over an a.s.s.). *If* (X_0,X_1,X_2) *is a stationary* (H_0,H_1,H_2)*-process over* $[V;W_1,W_2;\sigma,\tau]$*, an a.s.s., then there exist two processes,* $\hat{X}': V \times \hat{H}_1$ *and* $\hat{X}'': V \times V \to \hat{H}_2$*, stationary over the a.s.s.* $[V \times V;\, W_1 \times W_1,\, W_2 \times W_2;\, \sigma \times \sigma,$ $\tau \times \tau]$*, such that*

$$\begin{aligned} \hat{X}'\binom{f}{0} &= \hat{X}''\binom{f}{0} = X_1 f \in H_1, \ \forall f \in V \\ \hat{X}'\binom{0}{g} &= X''\binom{0}{g} = X_2 g \in H_2, \ \forall g \in V \\ \hat{X}'\binom{f}{g} &= X_0\binom{f}{g} \in H_0, \ \forall (f,g) \in W_1 \times W_2^{\sigma} \\ \hat{X}''\binom{f}{g} &= X_0\binom{f}{g} \in H_0, \ \forall (f,g) \in W_1 \times W_2^{\tau}. \end{aligned} \tag{4.7}$$

If W_1 and W_2 are complementary subspaces, i.e. if V is the direct sum of W_1 and W_2, then $W_1 \oplus W_2$ can be identified with V, and every linear process $X: V \to H$ is determined by its restrictions to W_1 and to W_2. In such case, $X: V \to H$ is called <u>generalized stationary</u> if there exist two Toeplitz forms, B_1 and B_2, $B_j: V \times V \to \mathbb{C}$, $j = 1$ and 2, and a Hankel form $B_0: W_1 \times W_2 \to \mathbb{C}$, such that the covariance $B(f,g) = \langle X(f), X(g)\rangle_H$ is given by

$$B(f,g) = B_j(f,g), \ \forall (f,g) \in W_j \times W_j, \ j = 1 \text{ or } 2,$$

and (4.8)

$$B(f,g) = B_0(f,g), \ \forall (f,g) \in W_1 \times W_2.$$

A process $X: V \to H$ is <u>generalized stationary</u> if and only if there exists an (H_0,H_1,H_2)-process, stationary over $[V;W_1,W_2;\sigma,\tau]$, such that $X(f + g) = X_0(f,g)$, $\forall (f,g) \in W_1 \times W_2$. Then, X can be identified with X_0, through the isomorphism $V \cong W_1 \oplus W_2$.

In the special case of Example 1, this notion of generalized stationarity coincides with that given in [4]. In the special case of Example 2, the GBT in [9] can also be formulated in terms of generalized stationary processes, as follows.

COROLLARY 1. *Let* $[V;W_1,W_2;\sigma,\tau]$ *be as in Example 2, and let* $X: V \to H$ *be a generalized stationary process. Then there exists a pair of positive* 2x2 *matrix measures,* (μ'_{jk}), (μ''_{jk}), *in* T^2, $\mu'_{jj} = \mu''_{jj}$ *for* $j = 1,2$, *such that the covariance form* B *of* X *satisfies*

$$B(f,g) = \int f\bar{g}d\mu'_{jj} = \int f\bar{g}d\mu''_{jj} \quad \text{if } (f,g) \in W_j \times W_j,\ j = 1,2,$$

$$B(f,g) = \int f\bar{g}d\mu'_{12} \quad \text{if } (f,g) \in W_1 \times W_2^{\sigma} \tag{4.9}$$

$$B(f,g) = \int f\bar{g}d\mu''_{12} \quad \text{if } (f,g) \in W_1 \times W_2^{\tau}.$$

An integral representation of the process itself, as in Khinchine theorem, can also be given, in terms of pairs of mutually orthogonally scattered vector-valued measures, i.e., measures ν_1 and ν_2 satisfying

$$\langle \nu_j(f), \nu_k(g)\rangle = 0 \text{ whenever } f\bar{g} = 0 \text{ for } j,k = 1 \text{ or } 2. \tag{4.10}$$

COROLLARY 2. *Let* $[V;W_1,W_2;\sigma,\tau]$ *be as in Example 2, and let* $X: V \to H$ *be an H-process over V. Then X is generalized stationary if and only if there exist two pairs of mutually orthogonally scattered vector-valued measures* (ν'_1,ν'_2) *and* (ν''_1,ν''_2)*, such that*

$$X(f) = \hat{\nu}'_1(f) = \hat{\nu}''_1(f) \text{ for all } f \in W_1, \tag{4.11}$$
$$X(f) = \hat{\nu}'_2(f) \text{ for } f \in W_2^{\sigma},\ X(f) = \hat{\nu}''_2(f) \text{ for } f \in W_2^{\tau}.$$

IDEA OF THE PROOF. From Corollary 1 it follows that the subspace of H spanned by $X(W_1) + X(W_2^{\sigma})$ is isomorphic to $L^2(T^2,(\mu'_{jk}))$, and that the subspace spanned by $X(W_1) + X(W_2^{\tau})$ is isomorphic to $L^2(T^2,(\mu''_{ij}))$. It is enough to take, for $f \in V$, as $\nu'_1(f)$ the element in $X(W_1) + X(W_2^{\sigma})$ corresponding to $\binom{f}{0} \in L^2(T^2,(\mu'_{jk}))$, and as $\nu'_2(f)$, the one corresponding to $\binom{0}{f} \in L^2(T^2,(\mu'_{jk}))$. Analogous considerations yield ν''_1 and ν''_2. □

In a similar way, Bochner and Khinchine representations of generalized stationary (H_0,H_1,H_2)-processes over $[V;W_1,W_2;\sigma,\tau]$ can be deduced from Theorem 4, assuming that $V = H_+ \subset H_0 \subset H_-$, and that the two positive forms B_1 and B_2 in the covariance triplet B_0,B_1,B_2, are of trace class.

Finally, let us mention that in [6] another type of GBT was stated, for the special case of a one-parameter a.s.s. $[V;W_1,W_2;\tau]$ where $V = W_1 \oplus W_2$,

$$\bigcap_{n=1}^{\infty} \tau^{n} W_1 = \bigcap_{n=1}^{\infty} \tau^{-n} W_2 = \{0\} \tag{4.12}$$

and

$$\bigvee_{n=-\infty}^{\infty} \tau^{n} W_1 = \bigvee_{n=-\infty}^{\infty} \tau^{n} W_2 = V \tag{4.13}$$

In such a case, fix two sets, $E_1 \subset W_1$ and $E_2 \subset W_2$, so that each $f \in V$ has a representation $f = \tau^n e_1 = \tau^m e_2$, $e_1 \in E_1$, $e_2 \in E_2$ and $n = n(f) \in \mathbb{Z}$, with $n \geq 0$ iff $f \in W_1$ and $m < 0$ iff $f \in W_2$. Then, from the Fourier representation theorem ([6], Thm. 3) it is not difficult to deduce the following variant of the Khinchine theorem for this case.

COROLLARY 3. *Let* $[V;W_1,W_2;\tau]$ *satisfy* (4.12) *and* (4.13), *and let* $X: V \to H$ *be a generalized stationary process. For each* $(e_1,e_2) \in E_1 \times E_2$ *let* $H(e_1,e_2)$ *be the subspace of* H *generated by the elements* $X(\tau^{n_1} e_1)$, $X(\tau^{n_2} e_2)$, $n_1,n_2 \in \mathbb{Z}$. *Then, for each* $(e_1,e_2) \in E_1 \times E_2$, *there exists a pair of mutually orthogonally scattered measures,* $\nu_1 = \nu_1(e_1,e_2)$, $\nu_2 = \nu_2(e_1,e_2)$, *defined in* $C(T)$ *and with values in* $H(e_1,e_2)$, *such that*

$$\begin{aligned} &\nu_1(W_1) = H(e_1,e_2) \cap X(W_1), \text{ with } \nu_1(f) = X(f) \text{ for } f \in W_1 \\ &\text{and} \\ &\nu_2(W_2) = H(e_1,e_2) \cap X(W_2), \text{ with } \nu_2(g) = X(g) \text{ for } g \in W_2, \end{aligned} \tag{4.14}$$

while $\nu_1(e_1,e_2)(V)$, *as well as* $\nu_2(e_1,e_2)(V)$ *space the whole of* H.

The converse is also true. Observe that since $X(V) = \cup_{(e_1,e_2)\in E_1\times E_2} H(e_1,e_2)$, the corollary gives an integral representation of the process X.

Corollary 3 can be extended to two parameter a.s.s. $[V;W_1,W_2;\sigma_1,\tau]$, but now conditions (4.12) and (4.13) have to be replaced by

$$\bigcap_{n=1}^{\infty} \sigma^n W_1 = \bigcap_{n=1}^{\infty} \tau^n W_1 = \bigcap_{n=1}^{\infty} \tau^{-n} W_2 = \{0\}, \quad (4.12a)$$

$$\bigvee_{m,n\in\mathbb{Z}} \sigma^m\tau^n W_1 = \bigvee_{m,n\in\mathbb{Z}} \sigma^m\tau^n W_2^\sigma = W_1 + W_2^\sigma \quad (4.13a)$$

and

$$\bigvee_{m,n\in\mathbb{Z}} \sigma^m\tau^n W_1 = \bigvee_{m,n\in\mathbb{Z}} \sigma^m\tau^n W_2^\tau = W_1 + W_2^\tau. \quad (4.13b)$$

Naturally, in this case there will be two pairs of mutually orthogonally scattered measures involved in the representation. While in Theorem 4 there is only one pair of operator-valued matrix measures (μ'_{jk}), (μ^0_{jk}), in the integral representation given in [6], and in Corollary 3, there are as many measures as elements in $E_1 \times E_2$. The expansions in [6] and Corollary 3 are of the classical type, into eigenforms of the shift, for each $(e_1,e_2) \in E_1 \times E_2$, while Theorem 4 gives the expansions in terms of general eigenforms $\Omega_{jk}(s,t)$.

REFERENCES

1. J.A. Ball, C. Foias, W. Helton & A. Tannenbaum, On a local nonlinear commutant lifting theorem, Indiana U. Math. J. 36 (1987), 693-709.

2. Iu M. Berezanskii, "Expansions in Eigenfunctions of Selfadjoint Operators," Transl. of Math. Mon., A.M.S., Providence, 1968.

3. __________, Generalized moment problem, Trans. Moscow Math. Soc. 21 (1970), 47-102. (In Russian.)

4. M. Cotlar & C. Sadosky, Generalized Toeplitz kernels, stationarity and harmonizability, J. Analyse Math. 44 (1985), 117-133.

5. ____________________, Prolongements des formes de Hankel généralisées en formes de Toeplitz, C.R. Acad. Sci. Paris I 305 (1987), 167-170.

6. ____________________, Toeplitz liftings of Hankel forms, in "Function Spaces & Applications, Lund 1986" (H. Wallin, J. Peetre & Y. Sagher, eds.), Springer-Verlag Lect. Notes in Math. 302 (1988), 22-43.

7. ____________________, Integral representations of bounded Hankel forms defined in scattering systems with a multiparametric evolution group, Operator Theory: Adv. & Appl. 35 (1988), 357-375.

8. ____________________, The Generalized Bochner Theorem in algebraic scattering systems, to appear in "Analysis at Urbana" (E. Berkson, ed.), Cambridge University Press.

9. ____________________, Two-parameter lifting theorems and double Hilbert transforms in commutative and non-commutative settings, MSRI #03821-88.

10. M. Cotlar, J. León & M.C. Pereyra, Eigenfunction expansions of covariance kernels, to appear in Acta Cient. Venez.

11. I.M. Gel'fand & N.J. Vilenkin, "Generalized Functions," vol. 4, Fizmatgiz, Moscow, 1961. (In Russian.) English transl.: Academic Press, New York, 1964.

12. E. Hille & R.S. Phillips, "Functional Analysis & Semigroups," A.M.S. Colloquium Publ. 31, Providence, 1957.

13. L.M. Korsunski, Integral representations of positive definite invariant kernels, in "Spectral theory of operators in infinite dimensional analysis," Math. Inst. Kiev (1976), 26-87.

14. K. Yosida, "Functional Analysis," 4th edition, Springer-Verlag, New York, 1974.

M. Cotlar
Fac. de Ciencias
Univ. Central de Venezuela
Caracas 1040, Venezuela

C. Sadosky
Dept. of Mathematics
Howard University
Washington, D.C. 20059, USA

Operator Theory:
Advances and Applications, Vol. 41

CHARACTERISTIC FUNCTIONS OF UNITARY COLLIGATIONS AND OF BOUNDED OPERATORS IN KREIN SPACES

B. Ćurgus, A. Dijksma, H. Langer and H.S.V. de Snoo

Dedicated to Israel Gohberg
on the occasion of his sixtieth birthday

In this note we give necessary and sufficient conditions for a holomorphic operator valued function to coincide weakly with the characteristic function of a bounded operator on a Krein space. We also present a sufficient condition such that the weak isomorphism is an isomorphism, i.e., is bounded.

1. Introduction

Let $\mathfrak{F}$ and $\mathfrak{G}$ be Krein spaces. By $\mathbf{S}(\mathfrak{F},\mathfrak{G})$ we denote the (generalized Schur) class of all functions Θ, defined and holomorphic at $z=0$ and with values in $\mathbf{L}(\mathfrak{F},\mathfrak{G})$, the space of bounded linear operators from $\mathfrak{F}$ to $\mathfrak{G}$ (we write $\mathbf{L}(\mathfrak{F})$ for $\mathbf{L}(\mathfrak{F},\mathfrak{F})$). It is known, see Section 3 below, that any $\Theta\in\mathbf{S}(\mathfrak{F},\mathfrak{G})$ is in some neighborhood of $z=0$ the characteristic function Θ_Δ of a unitary colligation $\Delta=(\mathfrak{K},\mathfrak{F},\mathfrak{G};U)$ with outer spaces $\mathfrak{F},\mathfrak{G}$ and a Krein space $\mathfrak{K}$ as inner space, that is, there exists a unitary mapping U with matrix representation

$$(1.1)\qquad U=\begin{bmatrix} T & F \\ G & H \end{bmatrix}:\begin{bmatrix} \mathfrak{K} \\ \mathfrak{F} \end{bmatrix}\to\begin{bmatrix} \mathfrak{K} \\ \mathfrak{G} \end{bmatrix}$$

such that the relation

$$(1.2)\qquad \Theta(z)=\Theta_\Delta(z)=H+zG(I-zT)^{-1}F$$

holds for all z in this neighborhood of 0. Here, e.g., $\binom{\mathfrak{K}}{\mathfrak{F}}$ stands for the Krein space which is the orthogonal sum $\mathfrak{K}\oplus\mathfrak{F}$ of $\mathfrak{K}$ and $\mathfrak{F}$ (for details see Section 3).

Now let $\mathfrak{K}$ be an arbitrary Krein space and $T\in\mathbf{L}(\mathfrak{K})$. We fix a fundamental symmetry $J_{\mathfrak{K}}$ on $\mathfrak{K}$ and denote by T^* the adjoint of T with respect to the

Hilbert inner product on $\mathfrak{K}$ generated by $J_{\mathfrak{K}}$. We define the operators

$$(1.3)\qquad J_T = \operatorname{sgn}(J_{\mathfrak{K}}-T^*J_{\mathfrak{K}}T),\ D_T = |J_{\mathfrak{K}}-T^*J_{\mathfrak{K}}T|^{1/2}$$

and set $\mathfrak{D}_T = \mathfrak{R}(D_T)^c$, the closure of the range $\mathfrak{R}(D_T)$. Then with T there is associated the unitary colligation $\Delta_T = (\mathfrak{K},\mathfrak{D}_{T^*},\mathfrak{D}_T;U_T)$ with

$$(1.4)\qquad U_T = \begin{bmatrix} T & D_{T^*}|_{\mathfrak{D}_{T^*}} \\ D_T & -J_TL_{T^*} \end{bmatrix} : \begin{bmatrix} \mathfrak{K} \\ \mathfrak{D}_{T^*} \end{bmatrix} \to \begin{bmatrix} \mathfrak{K} \\ \mathfrak{D}_T \end{bmatrix}.$$

Here L_{T^*} is the so-called link operator, introduced by Gr. Arsene, T. Constantinescu and A. Gheondea in [ACG] (for details see Section 4). The characteristic function of the colligation Δ_T:

$$(1.5)\qquad \Theta_T(z) = -J_TL_{T^*} + zD_T(1-zT)^{-1}D_{T^*},$$

defined for all z in a neighborhood of $z=0$, is also called the characteristic function of the operator T (relative to $J_{\mathfrak{K}}$). It is the main purpose of this paper, to prove in Theorem 5.2 necessary and sufficient conditions for a given function $\Theta\in\mathbf{S}(\mathfrak{F},\mathfrak{G})$ to coincide in a neighborhood of $z=0$ with the characteristic function of some $T\in\mathbf{L}(\mathfrak{K})$. In other words, these conditions are equivalent to the fact, that the operators F, G, H and the spaces $\mathfrak{F}$, $\mathfrak{G}$ in (1.1) or (1.2) can be "identified" with the operators $D_{T^*}|_{\mathfrak{D}_{T^*}}$, D_T, $-J_TL_{T^*}$ and the spaces $\mathfrak{D}_{T^*}$, $\mathfrak{D}_T$, respectively, in (1.4) or (1.5), which are all determined by the operator T in the left upper corner of the matrix in (1.4) representing U. If, in particular, the Krein space $\mathfrak{K}$ is a Hilbert space, then $J_{\mathfrak{K}}=I$, $L_T=T$ and the characteristic function in (1.5) becomes the well-known function

$$\Theta_T(z) = -T^*J_{T^*} + zD_T(1-zT)^{-1}D_{T^*}.$$

In this case such necessary and sufficient conditions for the equality $\Theta=\Theta_T$ to hold were proved by J.A. Ball, see [Ba], and we show in Section 6 how his result (formulated below as Theorem 6.1) follows from Theorem 5.2.

In these representation theorems for functions $\Theta\in\mathbf{S}(\mathfrak{F},\mathfrak{G})$ the question arises of the uniqueness of the corresponding colligation Δ or the operator T. As is typical for Krein spaces, the colligation Δ or the Krein space $\mathfrak{K}$ and the operators in it are, in general, only unique up to weak isomorphisms. Recall that two Krein spaces $\mathfrak{K}$, $\mathfrak{K}'$ are called weakly isomorphic, if there is a (closed) linear mapping V from a dense subspace $\mathfrak{D}\subset\mathfrak{K}$ onto a dense subspace of $\mathfrak{K}'$ such that $[Vx,Vy]=[x,y]$, $x,y\in\mathfrak{D}$. In Section 7 (Theorem 7.1) we give a sufficient condition, which assures that such a weak isomorphism V is an isomorphism, i.e., that V is continuous.

At several places in this paper we use an extension of a basic result of M.G. Krein about operators in "spaces with two norms", which is formulated in Section 2. To conclude the outline of the contents of this paper we mention, that we do not repeat the (lengthy) proof of the basic statement (Theorem 3.4), that any $\Theta\in\mathbf{S}(\mathfrak{F},\mathfrak{G})$ is the characteristic function of some unitary colligation Δ, see, e.g., [DLS2]. However, in Proposition 3.6 we prove this result in the particular case of Θ being holomorphic and unitary on an arc of the boundary $\partial\mathbb{D}$ of the open unit disc $\mathbb{D}$ in $\mathbb{C}$. Then the proof in [DLS2] can be simplified considerably.

In the sequel we shall use, as we did above, without further specification the elementary facts and the common notations from the theory of linear operators in Krein and Pontryagin spaces. They can be found in [AI], [Bo], [Cu], [DLS1], [DLS2], [IKL] and [L]. Here we only recall the notation that, if T is a densely defined operator from $\mathfrak{F}$ to $\mathfrak{G}$, T^+ denotes the Krein space adjoint of T: $T^+=J_{\mathfrak{F}}T^*J_{\mathfrak{G}}$, where $J_{\mathfrak{F}}$ and $J_{\mathfrak{G}}$ are fundamental symmetries on $\mathfrak{F}$ and $\mathfrak{G}$, respectively and T^* denotes the adjoint of T with respect to the Hilbert space structures on $\mathfrak{F}$ and $\mathfrak{G}$ induced by these symmetries. We shall use $[.,.]$ as the generic notation for the possibly indefinite inner product and $(.,.)$ in the cases, where we want to emphasize that the inner product is positive definite.

The second author acknowledges the financial support of the "Koninklijke Nederlandse Akademie voor Wetenschappen", K.N.A.W., to present this paper at the International Conference on Operator Theory in Calgary.

2. Extension of a result of M.G. Krein on operators in spaces with two norms

Let $(\mathfrak{H},(.,.))$ be a Hilbert space and let $[.,.]$ be a (possibly indefinite) inner product (hermitian, sesquilinear form) on $\mathfrak{H}$ which is bounded, i.e., there exists a constant $C>0$ such that $|[x,y]|\le C(x,x)^{1/2}(y,y)^{1/2}$, $x,y\in\mathfrak{H}$. By the Riesz' representation theorem there exists a selfadjoint $G\in\mathbf{L}(\mathfrak{H})$ such that $[x,y]=(Gx,y)$, $x,y\in\mathfrak{H}$. G is called the Gram operator. We may write the Hilbert space $\mathfrak{H}$ as the orthogonal sum $\mathfrak{H}=\mathfrak{H}_+\oplus\mathfrak{H}_-\oplus\mathfrak{H}_0$, where

$\mathfrak{H}_\pm = \nu(I \mp \operatorname{sgn} G)$ and $\mathfrak{H}_0 = \nu(G)$. Here the sign function is defined by $\operatorname{sgn}(t) = -1$, 0 or 1, if $t<0$, $t=0$ or $t>0$, respectively, and $\nu(T)$ stands for the null space of the operator T. The spaces $\mathfrak{H}_\pm$ equipped with the inner product $\pm[x,y]$ are pre-Hilbert spaces. We denote their completions by $\mathfrak{K}_\pm$ and retain the notation $\pm[x,y]$ for their inner products. We put $\mathfrak{K} = \mathfrak{K}_+ \oplus \mathfrak{K}_-$, direct sum, and denote by $P_\pm$ the projections of $\mathfrak{K}$ onto $\mathfrak{K}_\pm$ along $\mathfrak{K}_\mp$. Then $(x,y) = [P_+x, P_+y] - [P_-x, P_-y]$ turns $\mathfrak{K}$ into a Hilbert space and $[x,y] = [P_+x, P_+y] + [P_-x, P_-y]$ turns $\mathfrak{K}$ into a Krein space with fundamental symmetry $J_\mathfrak{K} = P_+ - P_-$. We call $\mathfrak{K}$ the Krein space associated with the Hilbert space $\mathfrak{H}$ and the bounded inner product $[.,.]$. Clearly, $\mathfrak{H}_0 = \{x \in \mathfrak{H} \mid [x,y] = 0$ for all $y \in \mathfrak{H}\}$ and the factor space $\hat{\mathfrak{H}} = \mathfrak{H}/\mathfrak{H}_0$, consisting of equivalence classes $\hat{x} = \{x+y \mid y \in \mathfrak{H}_0\}$, can be identified with $\mathfrak{H}_+ \oplus \mathfrak{H}_-$ and is dense in $\mathfrak{K}$.

The following theorem is a generalization of a result announced by M.G. Krein in 1937 and published in [Kr] in 1947 (see also [AI] p. 220 and [Bo] p. 92, p. 98), which was later also proved in a similar form independently by W.T. Reid [Re] (see also [Ha], p.51), P.D. Lax [La] and J. Dieudonné [Di]. For the proof of the theorem, as Krein's proof based on repeated application of the Cauchy–Schwarz inequality, we refer to [DLS2].

THEOREM 2.1. *Let $\mathfrak{H}_1$ and $\mathfrak{H}_2$ be two Hilbert spaces and suppose that on each $\mathfrak{H}_j$ there is given a bounded, and in general indefinite, inner product $[.,.]_j$, $j=1,2$. Furthermore, assume that we are given two operators $U_0 \in \mathbf{L}(\mathfrak{H}_1, \mathfrak{H}_2)$ and $V_0 \in \mathbf{L}(\mathfrak{H}_2, \mathfrak{H}_1)$ such that $[U_0x, y]_2 = [x, V_0y]_1$, $x \in \mathfrak{H}_1$, $y \in \mathfrak{H}_2$. Then the operators $\hat{U}_0 \in \mathbf{L}(\hat{\mathfrak{H}}_1, \hat{\mathfrak{H}}_2)$ and $\hat{V}_0 \in \mathbf{L}(\hat{\mathfrak{H}}_2, \hat{\mathfrak{H}}_1)$ obtained from U_0 and V_0 on the factor spaces $\hat{\mathfrak{H}}_1$ and $\hat{\mathfrak{H}}_2$ in the usual way, can be extended by continuity to operators $U \in \mathbf{L}(\mathfrak{K}_1, \mathfrak{K}_2)$ and $V \in \mathbf{L}(\mathfrak{K}_2, \mathfrak{K}_1)$, respectively, between the associated Krein spaces $\mathfrak{K}_1$ and $\mathfrak{K}_2$ and $[Ux, y]_2 = [x, Vy]_1$, $x \in \mathfrak{K}_1$, $y \in \mathfrak{K}_2$, i.e., $V = U^+$.*

For later reference we amplify two special cases of Theorem 2.1:

COROLLARY 2.2. *Let $\mathfrak{H}$ be a Hilbert space, let $[.,.]$ be a bounded inner product on $\mathfrak{H}$ and let $S_0 \in \mathbf{L}(\mathfrak{H})$ be such that $[S_0x, y] = [x, S_0y]$, $x, y \in \mathfrak{H}$. Then $\hat{S}_0 \in \mathbf{L}(\hat{\mathfrak{H}}_0)$ can be extended by continuity to a bounded selfadjoint operator $S = S^+$ on the associated Krein space $\mathfrak{K}$. If, moreover, $[.,.]$ is positive definite on $\mathfrak{H}$, then $\mathfrak{K}$ is a Hilbert space containing $\mathfrak{H}$ as a dense subspace and S_0 itself can be extended by continuity to a bounded selfadjoint operator $S = S^*$ on $\mathfrak{K}$.*

COROLLARY 2.3. *Let $\mathfrak{H}_1$ and $\mathfrak{H}_2$ be two Hilbert spaces and suppose that on* *each $\mathfrak{H}_j$ there is given a bounded inner product $[.,.]_j$, $j=1,2$. Furthermore, assume that we are given a bijective operator $U_0 \in \mathbf{L}(\mathfrak{H}_1,\mathfrak{H}_2)$ such that $[U_0x,U_0y]_2=[x,y]_1$, $x,y\in\mathfrak{H}_1$. Then $\hat{U}_0\in\mathbf{L}(\hat{\mathfrak{H}}_1,\hat{\mathfrak{H}}_2)$ can be extended by continuity to a unitary operator $U\in\mathbf{L}(\mathfrak{K}_1,\mathfrak{K}_2)$ between the associated Krein spaces $\mathfrak{K}_1$ and $\mathfrak{K}_2$.*

Proof. Apply Theorem 2.1 with $V_0=U_0^{-1}$ and let U and V be the continuous extensions of $\hat{U}_0$ and $\hat{V}_0$ with $V=U^+$. From $[\hat{U}_0x,\hat{U}_0y]_2=[x,y]_1$, $x,y\in\hat{\mathfrak{H}}_1$ it follows by continuity that $[Ux,Uy]_2=[x,y]_1$, $x,y\in\mathfrak{K}_1$, that is $U^+U=I_{\mathfrak{K}_1}$. Similarly, one can show that $V^+V=I_{\mathfrak{K}_2}$ and therefore $UU^+=I_{\mathfrak{K}_2}$. Hence U is unitary.

For more results in the theory of operators in spaces with two norms we refer to [GZ1], [GZ2].

3. Unitary colligations and their characteristic functions, the general representation theorem

Let $\mathfrak{F}$ and $\mathfrak{G}$ be Krein spaces. If $\Theta\in\mathbf{S}(\mathfrak{F},\mathfrak{G})$ we denote by $\mathfrak{D}(\Theta)$ the domain of holomorphy of Θ in $\mathbb{D}=\{z\in\mathbb{C}\mid |z|<1\}$ and by $\mathsf{S}_\Theta^m(z,w)$ the matrix kernel

$$\mathsf{S}_\Theta^m(z,w)=\begin{pmatrix} \dfrac{I-\Theta(w)^+\Theta(z)}{1-\bar{w}z} & \dfrac{\Theta(\bar{z})^+-\Theta(w)^+}{z-\bar{w}} \\ \dfrac{\Theta(\bar{w})-\Theta(z)}{\bar{w}-z} & \dfrac{I-\Theta(\bar{w})\Theta(\bar{z})^+}{1-\bar{w}z} \end{pmatrix}$$

defined for z and w in a neighborhood of 0 contained in $\mathbb{D}$, $z\neq\bar{w}$ with values in $\mathbf{L}\begin{pmatrix}\mathfrak{F}\\ \mathfrak{G}\end{pmatrix}$. Clearly, $\mathsf{S}_\Theta^m(z,w)^+=\mathsf{S}_\Theta^m(z,w)$. A kernel $\mathsf{K}(z,w)$ defined for z,w in some set $\mathfrak{D}\subset\mathbb{C}$ and with values in $\mathbf{L}(\mathfrak{K})$, where $\mathfrak{K}$ is a Krein space, is called positive definite, positive semidefinite, etc., if $\mathsf{K}(z,w)^+=\mathsf{K}(w,z)$, so that all matrices of the form $([\mathsf{K}(z_i,z_j)f_i,f_j])_{i,j=1}^n$, where $n\in\mathbb{N}$, $z_1,\dots,z_n\in\mathfrak{D}$ and $f_1,\dots,f_n\in\mathfrak{K}$ are arbitrary, are hermitian, and the eigenvalues of each of these matrices are positive, nonnegative, etc., respectively. More generally, we say that $\mathsf{K}(z,w)$ has κ positive (negative) squares if $\mathsf{K}(z,w)^+=\mathsf{K}(w,z)$ and all these hermitian matrices have at most κ and at least one has exactly κ positive (negative) eigenvalues. It has infinitely many positive (negative) squares if for each κ at least one of these matrices has not less than κ positive (negative) eigenvalues. A Pontryagin space of (negative) index κ, for example, is a Krein space on which the constant kernel $\mathsf{K}(z,w)=I$ has κ negative squares.

Finally, $\Theta \in \mathbf{S}(\mathfrak{F},\mathfrak{G})$ and $\Theta' \in \mathbf{S}(\mathfrak{F}',\mathfrak{G}')$ are said to coincide, if there are two unitary operators $V \in \mathbf{L}(\mathfrak{F}',\mathfrak{F})$ and $W \in \mathbf{L}(\mathfrak{G}',\mathfrak{G})$ such that $\Theta(z)V = W\Theta'(z)$ for z in some neighborhood of 0 contained in the intersection $\mathfrak{D}(\Theta) \cap \mathfrak{D}(\Theta')$.

A colligation Δ is a quadruple $\Delta = (\mathfrak{K},\mathfrak{F},\mathfrak{G};U)$ consisting of three Krein spaces $\mathfrak{K}$ (the inner or state space), $\mathfrak{F}$ and $\mathfrak{G}$ (the left outer or input space and the right outer or output space, respectively) and a mapping $U \in \mathbf{L}(\mathfrak{K}\oplus\mathfrak{F},\mathfrak{K}\oplus\mathfrak{G})$ (the connecting operator), which we usually write in the form of a 2×2 block matrix

$$U = \begin{pmatrix} T & F \\ G & H \end{pmatrix} : \begin{pmatrix} \mathfrak{K} \\ \mathfrak{F} \end{pmatrix} \to \begin{pmatrix} \mathfrak{K} \\ \mathfrak{G} \end{pmatrix}$$

with bounded operators T (the basic operator), F, G and H; we often write $\Delta = (\mathfrak{K},\mathfrak{F},\mathfrak{G};T,F,G,H)$. The colligation Δ is called unitary, if U is unitary and it is called closely connected if for some small neighborhood $\mathcal{N}$ of 0 contained in $\mathbb{D}$

$$\mathfrak{K} = \bigvee_{z\in\mathcal{N}} (\,\mathfrak{R}((I-zT)^{-1}F) \cup \mathfrak{R}((I-zT^{+})^{-1}G^{+})\,),$$

i.e., $\mathfrak{K}$ is the closed linear span of the elements in the union of the indicated ranges.

An operator $T \in \mathbf{L}(\mathfrak{K})$ will be called simple if there does not exist a nonzero subspace (i.e., linear subset) $\mathfrak{K}_0$ of $\mathfrak{K}$ such that $T\mathfrak{K}_0 = \mathfrak{K}_0$ and $(I - T^{+}T)\mathfrak{K}_0 = \{0\}$. This definition extends the one given in for instance [BDS]. Note that in the definition $\mathfrak{K}_0$ may be degenerated. It is not difficult to check that if T is simple, then so are T^{+} and $T^{*} = J_{\mathfrak{K}}T^{+}J_{\mathfrak{K}}$ for any fundamental symmetry $J_{\mathfrak{K}}$ on $\mathfrak{K}$.

Lemma 3.1. *Let $\Delta = (\mathfrak{K},\mathfrak{F},\mathfrak{G};T,F,G,H)$ be a unitary colligation. Then the following statements are equivalent :*

(*i*) *Δ is closely connected,*

(*ii*) *there does not exist a nonzero subspace $\mathfrak{K}_0$ of $\mathfrak{K}$ such that $T\mathfrak{K}_0 = \mathfrak{K}_0$ and $G\mathfrak{K}_0 = \{0\}$,*

(*iii*) *there does not exist a nonzero subspace $\mathfrak{K}_0$ of $\mathfrak{K}$ such that $T^{+}\mathfrak{K}_0 = \mathfrak{K}_0$ and $F^{+}\mathfrak{K}_0 = \{0\}$.*

Moreover, if T is simple then Δ is closely connected and the converse is true if one of the following four conditions is valid: $\mathfrak{R}(G)^{c} = \mathfrak{G}$, $\mathfrak{G}$ is a Hilbert space, $\mathfrak{R}(F^{+})^{c} = \mathfrak{F}$ or $\mathfrak{F}$ is a Hilbert space.

Proof. For the proof of the first part of the lemma we refer to the proof of Proposition 3.2 in [DLS1], which can easily be adapted from the case where $\mathfrak{F}$ and $\mathfrak{G}$ are Hilbert spaces to the case where they are Krein spaces. In order to prove the second statement, suppose that Δ is not closely connected. Then by the first part there exists a nonzero subspace $\mathfrak{K}_0$ of $\mathfrak{K}$ such that $T\mathfrak{K}_0=\mathfrak{K}_0$ and $G\mathfrak{K}_0=\{0\}$, whence $(I-T^+T)\mathfrak{K}_0=G^+G\mathfrak{K}_0=\{0\}$, so that T is not simple. Conversely, if T is not simple, then there exists a nonzero subspace $\mathfrak{K}_0$ of $\mathfrak{K}$ such that $T\mathfrak{K}_0=\mathfrak{K}_0$ and $G^+G\mathfrak{K}_0=(I-T^+T)\mathfrak{K}_0=\{0\}$. If $\mathfrak{R}(G)^c=\mathfrak{G}$ or if $\mathfrak{G}$ is a Hilbert space this implies that $G\mathfrak{K}_0=\{0\}$ and from *(ii)* it follows that Δ is not closely connected. In a similar way it can be shown that the other two conditions also imply that Δ is not closely connected.

The characteristic function Θ_Δ of a colligation $\Delta=(\mathfrak{K},\mathfrak{F},\mathfrak{G};T,F,G,H)$ is defined by $\Theta_\Delta(z)=H+zG(I-zT)^{-1}F$. This definition stems from M.G. Krein. Clearly, $\Theta_\Delta\in\mathbf{S}(\mathfrak{F},\mathfrak{G})$ and

$$\mathfrak{D}(\Theta_\Delta)=\{\,z\in\mathbb{D}\mid z=0 \text{ or } 1/z\in\rho(T)\,\}.$$

LEMMA 3.2. *If $\Delta=(\mathfrak{K},\mathfrak{F},\mathfrak{G};T,F,G,H)$ is unitary, then*

(i) $\Theta_\Delta(z)-\Theta_\Delta(w)=(z-w)G(I-zT)^{-1}(I-wT)^{-1}F,$

(ii) $I-\Theta_\Delta(w)^+\Theta_\Delta(z)=(1-z\bar{w})F^+(I-\bar{w}T^+)^{-1}(I-zT)^{-1}F,$

(iii) $I-\Theta_\Delta(z)\Theta_\Delta(w)^+=(1-z\bar{w})G(I-zT)^{-1}(I-\bar{w}T^+)^{-1}G^+$

and hence, for $f_j\in\mathfrak{F}$, $g_j\in\mathfrak{G}$, $j=1,2$,

(iv)
$$\left[\, S^m_{\Theta_\Delta}(z,w)\begin{pmatrix} f_1\\ g_1\end{pmatrix},\begin{pmatrix} f_2\\ g_2\end{pmatrix}\right] =$$
$$\left[\,(I-zT)^{-1}Ff_1+(I-zT^+)^{-1}G^+g_1,(I-wT)^{-1}Ff_2+(I-wT^+)^{-1}G^+g_2\,\right].$$

The proof can be given by straightforward calculation, cf. M.S. Brodskii [Br].

COROLLARY 3.3. *Let $\Delta=(\mathfrak{K},\mathfrak{F},\mathfrak{G};T,F,G,H)$ be unitary. (i) If $\mathfrak{K}$ is a Hilbert space or if Δ is closely connected, then*

$$\mathfrak{R}(F^+)^c=\bigvee_{z\in\mathcal{N}}\left(\,\mathfrak{R}(\Theta_\Delta(\bar{z})^+-\Theta_\Delta(0)^+)\cup\mathfrak{R}(\,I-\Theta_\Delta(0)^+\Theta_\Delta(z))\,\right),$$
$$\mathfrak{R}(G)^c=\bigvee_{z\in\mathcal{N}}\left(\,\mathfrak{R}(\Theta_\Delta(0)-\Theta_\Delta(z))\cup\mathfrak{R}(\,I-\Theta_\Delta(0)\Theta_\Delta(\bar{z})^+)\,\right),$$

where $\mathcal{N}$ is some neighborhood of 0 contained in $\mathfrak{D}(\Theta_\Delta)$. (ii) If Δ is closely connected, then

$$\dim \mathfrak{K}_{\pm} = \# \begin{matrix} positive \\ negative \end{matrix} \text{ squares of } S^m_{\Theta_\Delta}(z,w),$$

where $\mathfrak{K} = \mathfrak{K}_+ + \mathfrak{K}_-$ *is any fundamental decomposition of the state space* $\mathfrak{K}$.

An operator V from $\mathfrak{F}$ to $\mathfrak{G}$ is called a weak isomorphism if it has a dense domain $\mathfrak{D}(V) \subset \mathfrak{F}$ and a dense range $\mathfrak{R}(V) \subset \mathfrak{G}$, and is isometric, that is $V^+V = I|_{\mathfrak{D}(V)}$. A weak isomorphism is closable and hence, it may be assumed to be closed. The spaces $\mathfrak{F}$ and $\mathfrak{G}$ are called weakly isomorphic if there exists such a weak isomorphism between them. If V is a weak isomorphism from $\mathfrak{F}$ to $\mathfrak{G}$ then V is bounded (and hence, is an isomorphism between $\mathfrak{F}$ and $\mathfrak{G}$) in each of the following cases: (*i*) $\mathfrak{F}$ and $\mathfrak{G}$ are Pontryagin spaces and (*ii*) there exist fundamental symmetries $J_{\mathfrak{F}}$ and $J_{\mathfrak{G}}$ such that $VJ_{\mathfrak{F}} = J_{\mathfrak{G}}V$ on $\mathfrak{D}(V)$. For a more general result in this direction we refer to Section 7.

Two colligations $\Delta = (\mathfrak{K}, \mathfrak{F}, \mathfrak{G}; T, F, G, H)$ and $\Delta' = (\mathfrak{K}', \mathfrak{F}, \mathfrak{G}; T', F', G', H')$ with the same outer spaces $\mathfrak{F}$ and $\mathfrak{G}$ are called weakly isomorphic if $H = H'$ and there exists a weak isomorphism V from $\mathfrak{K}'$ to $\mathfrak{K}$ such that

$$\begin{bmatrix} T & F \\ G & H \end{bmatrix} \begin{bmatrix} V & 0 \\ 0 & I \end{bmatrix} = \begin{bmatrix} V & 0 \\ 0 & I \end{bmatrix} \begin{bmatrix} T' & F' \\ G' & H' \end{bmatrix} \text{ on } \begin{bmatrix} \mathfrak{D}(V) \\ \mathfrak{F} \end{bmatrix}.$$

If V can be extended to a unitary operator from $\mathfrak{K}'$ onto $\mathfrak{K}$, then of course, Δ and Δ' are called isomorphic or unitarily equivalent.

THEOREM 3.4. *Every* $\Theta \in \mathbf{S}(\mathfrak{F}, \mathfrak{G})$ *can be written as* $\Theta(z) = \Theta_\Delta(z)$ *with* z *in a neighborhood of* 0 *in* $\mathbb{D}$, *for some unitary colligation* $\Delta = (\mathfrak{K}, \mathfrak{F}, \mathfrak{G}; U)$. *Here* Δ *can be chosen to be closely connected in which case it is uniquely determined up to weak isomorphisms and*

$$\dim \mathfrak{K}_{\pm} = \# \begin{matrix} positive \\ negative \end{matrix} \text{ squares of } S^m_{\Theta}(z,w),$$

where $\mathfrak{K} = \mathfrak{K}_+ + \mathfrak{K}_-$ *is any fundamental decomposition of the state space* $\mathfrak{K}$.

For this theorem see T.Ya. Azizov [Az1], [Az2]. Azizov's proof starts with a result of D.Z. Arov [Ar] which states that an arbitrary operator valued mapping, holomorphic at $z = 0$, is the characteristic function of a (not necessarily unitary) colligation and applies C. Davis' statement [Da] that any bounded operator has a unitary dilation in a Krein space. In [DLS2] we have stated and proved a more detailed version of Theorem 3.4 in the sense that we specified beforehand the domain of points z for which the equality $\Theta(z) = \Theta_\Delta(z)$ is to be valid. Our proof is based on Corollary 2.3, but it is quite lengthy and therefore will not be repeated here. However, in order to show how Corollary 2.3 can be applied, we shall prove Theorem 3.4 for Θ's in a special subclass of $\mathbf{S}(\mathfrak{F},\mathfrak{G})$. The assumptions make a much simpler proof possible, than the one in [DLS2] for the general case and lead to the slightly stronger conclusion that the unitary colligation can be chosen to be closely innerconnected and closely outerconnected. The reason for this is formulated in Proposition 3.5 below, which is of some interest of its own. Proposition 3.6 below is the restricted version of Theorem 3.4. Neither of the propositions will be used in the remainder of this paper.

Recall that a unitary colligation $\Delta=(\mathfrak{K},\mathfrak{F},\mathfrak{G};T,F,G,H)$ is closely innerconnected if

$$\mathfrak{K}=\bigvee\{T^nFf\mid n\in\mathbb{N}\cup\{0\},\ f\in\mathfrak{F}\}$$

and closely outerconnected if

$$\mathfrak{K}=\bigvee\{T^{+n}G^+g\mid n\in\mathbb{N}\cup\{0\},\ g\in\mathfrak{G}\}.$$

Clearly, if Δ is closely innerconnected or closely outerconnected, then it is closely connected. The following result gives a sufficient condition for the converse to hold.

Proposition 3.5. *Let $\Delta=(\mathfrak{K},\mathfrak{F},\mathfrak{G};T,F,G,H)$ be a unitary colligation. Assume that the point $0\in\rho(T)$ and that it also belongs to the unbounded component of $\rho(T)$. If Δ is closely connected, then it is closely innerconnected and closely outerconnected. In particular, this holds if $\dim\mathfrak{K}<\infty$ and T is invertible.*

Proof. We have that

$$\bigvee_{z\in\rho(T)}\mathfrak{R}(T-z)^{-1}=\bigvee_{n\in\mathbb{Z}}\mathfrak{R}(T^n)=\bigvee_{n\geq 0}\mathfrak{R}(T^n).$$

The first equality is valid since $0,\infty\in\rho(T)$ and the second one follows from the fact that these points belong to the same component as this implies that T^{-1} can be approximated in the uniform topology by polynomials in T. Let $\mathcal{O}\subset\rho(T)$ be a neighborhood of 0 and ∞, which is symmetric with respect to $\partial\mathbb{D}$. We extend the definition of $\Theta_\Delta(z)$ to all values $z\in\mathbb{C}$ for which $1/z\in\rho(T)$ in the obvious way: $\Theta_\Delta(z)=H+zG(I-zT)^{-1}F$. It is easy to verify that the equalities (*ii*) and (*iii*) of Lemma 3.2 are valid for this extended Θ_Δ. They imply that $\Theta_\Delta(1/z)$ is invertible with inverse $\Theta_\Delta(1/z)^{-1}=\Theta_\Delta(\bar{z})^+$ for all $z\in\mathcal{O}$. It follows from the relation

$$(T-z)^{-1}F\Theta_{\Delta}(\bar{z})^{+} = -(I-zT^{+})^{-1}G^{+}, \qquad z\in\mathcal{O},$$

that

$$\bigvee_{n\geq 0} \Re(T^{+n}G^{+}) \subset \bigvee_{z\in\rho(T)} \Re((T-z)^{-1}F) = \bigvee_{n\geq 0} \Re(T^{n}F).$$

Hence, if Δ is closely connected, then it is closely innerconnected. The same reasoning applied to T^{+}, instead of T easily yields that then Δ is also closely outerconnected. This completes the proof.

PROPOSITION 3.6. *Assume that $\Theta\in\mathbf{S}(\mathfrak{F},\mathfrak{G})$ can be extended holomorphically to a simply connected domain $\mathfrak{D}$ in the extended complex plane $\bar{\mathbb{C}}$, which is symmetric with respect to the unit circle $\partial\mathbb{D}$ and contains neighborhoods of 0,1 and ∞, such that $\Theta(\bar{z})^{-1}=\Theta(1/z)^{+}$. Let ∂ be a closed smooth Jordan curve in $\mathbb{C}$ with interior, exterior denoted by $I(\partial)$, $E(\partial)$, respectively, such that $E(\partial)\cup\partial\subset\mathfrak{D}$, the points 0, 1 and ∞ belong to $E(\partial)$ and there exists a conformal mapping γ from $I(\partial)$ onto $\mathbb{D}$, which can be extended to a continuously differentiable function, also denoted by γ, from $I(\partial)\cup\partial$ onto $\mathbb{D}\cup\partial\mathbb{D}$ with $\gamma'(z)\neq 0$ for all $z\in I(\partial)\cup\partial$. Then, for all $z\in E(\partial)$, $\Theta(z)=\Theta_{\Delta}(z)$ for some unitary colligation Δ, that can be chosen to be closely innerconnected and closely outerconnected, in which case it is uniquely determined up to weak isomorphisms.*

REMARK. According to a theorem of Kellogg, a sufficient condition for the existence of γ is that the angle of the tangent to ∂, considered as a function of the arc length along ∂ satisfies a Lipschitz condition, see [G] Theorem 6, p. 374.

Proof. Put $\mathfrak{D}_{-}=\{z\in\mathbb{C}\,|\,1/z\in\mathfrak{D}\}$ and $\partial_{-}=\{z\in\mathbb{C}\,|\,1/z\in\partial\}$. Then $\mathfrak{D}_{-}$ and ∂_{-} have the same characteristics as $\mathfrak{D}$ and ∂ described in the proposition. We write γ_{-} for the corresponding conformal mapping from $I(\partial_{-})$ onto $\mathbb{D}$. In the proof we shall consider contour integrals $\oint$ which are always over the contour ∂_{-} traversed in the positive direction with respect to $I(\partial_{-})$ and write

$$\bar{\oint} u(z)\,\overline{dz} = \overline{\oint \overline{u(z)}\; dz}.$$

We put $\Psi(z)=\Theta(1/z)$. Then we have Cauchy's formula

$$\Psi(z)=\Psi(\infty)-\frac{1}{2\pi i}\oint \frac{\Psi(w)}{w-z}\,dw, \qquad z\in E(\partial_{-}) \quad \text{(strongly)}.$$

We fix fundamental symmetries $J_{\mathfrak{F}}$ on $\mathfrak{F}$ and $J_{\mathfrak{G}}$ on $\mathfrak{G}$. When we consider $\mathfrak{F}$ and $\mathfrak{G}$

as Hilbert spaces, we mean the linear spaces $\mathfrak{F}$, $\mathfrak{G}$ endowed with the positive definite inner products $[J_{\mathfrak{F}}\cdot,\cdot]_{\mathfrak{F}}$, $[J_{\mathfrak{G}}\cdot,\cdot]_{\mathfrak{G}}$, respectively. By $\mathfrak{H}=H^2_{\mathfrak{F}}(I(\partial_-))$ we denote the Hilbert space of all functions $h:I(\partial_-)\to\mathfrak{F}$ for which $\hat{h}=h\circ\gamma_-^{-1}:\mathbb{D}\to\mathfrak{F}$ belongs to the Hardy class $H^2_{\mathfrak{F}}(\mathbb{D})$. In this class the functions can be extended to $\partial\mathbb{D}$ by their nontangential limits and hence the functions in $\mathfrak{H}$ can also be extended to ∂_-. The inner product is given by

$$(h,k)=\frac{1}{2\pi i}\oint_{\partial\mathbb{D}}[J_{\mathfrak{F}}\hat{h}(w),\hat{k}(w)]_{\mathfrak{F}}\frac{dw}{w}=$$

$$=\frac{1}{2\pi i}\oint[J_{\mathfrak{F}}h(z),k(z)]_{\mathfrak{F}}\frac{\gamma_-'(z)}{\gamma_-(z)}dz.$$

Note that because $0<c\leq|\gamma_-'/\gamma_-|\leq C<\infty$ on ∂_-, the norm corresponding to this inner product is equivalent to the norm

$$h\to(\oint[J_{\mathfrak{F}}h(z),h(z)]_{\mathfrak{F}}|dz|)^{1/2}.$$

This, the fact that Ψ is holomorphic in a neighborhood of ∂_- and the equality $\Psi(w)^+\Psi(z)=I$ when $z\bar{w}=1$, $z,w\in\partial_-$, immediately yield that the inner product $[.,.]$ defined by

$$[h,k]=\frac{1}{4\pi^2}\oint\overline{\oint}\Big[\frac{\Psi(w)^+\Psi(z)-I}{1-z\bar{w}}h(z),k(w)\Big]_{\mathfrak{F}}dz\,\overline{dw},\qquad h,k\in\mathfrak{H},$$

is bounded on $\mathfrak{H}$. We want to apply Corollary 2.3 and define the Hilbert spaces $\mathfrak{H}_1,\mathfrak{H}_2$, the inner products $[.,.]_1$, $[.,.]_2$ and the operator U_0 by

$$\mathfrak{H}_1=\mathfrak{H}\oplus\mathfrak{F},\quad \mathfrak{H}_2=\mathfrak{H}\oplus\mathfrak{G},$$

$$\left[\begin{pmatrix}h\\f_1\end{pmatrix},\begin{pmatrix}k\\f_2\end{pmatrix}\right]_1=[h,k]+[f_1,f_2]_{\mathfrak{F}},\quad \left[\begin{pmatrix}h\\g_1\end{pmatrix},\begin{pmatrix}k\\g_2\end{pmatrix}\right]_2=[h,k]+[g_1,g_2]_{\mathfrak{G}},$$

$$U_0\begin{pmatrix}h\\f\end{pmatrix}=\begin{pmatrix}zh(z)+f\\ \frac{1}{2\pi i}\oint\Psi(z)h(z)dz+\Psi(\infty)f\end{pmatrix},$$

where $h,k\in\mathfrak{H}$, $f_1,f_2,f\in\mathfrak{F}$ and $g_1,g_2\in\mathfrak{G}$. The operator U_0 maps $\mathfrak{H}_1$ onto $\mathfrak{H}_2$ and, as one can easily verify using Cauchy's formula, is invertible with inverse given by

$$U_0^{-1}\begin{pmatrix}k\\g\end{pmatrix}=\begin{pmatrix}\frac{1}{z}k(z)+\frac{1}{z}\Psi(0)^{-1}\big(\frac{1}{2\pi i}\oint\frac{\Psi(z)}{z}k(z)dz-g\big)\\ -\Psi(0)^{-1}\big(\frac{1}{2\pi i}\oint\frac{\Psi(z)}{z}k(z)dz-g\big)\end{pmatrix}.$$

With the exception of the equality $[U_0x,U_0y]_2=[x,y]_1$, the hypotheses of

Corollary 2.3 are easily checked. The equality can be proved in a straightforward manner:

$$\left[U_0\begin{pmatrix}h\\f_1\end{pmatrix}, U_0\begin{pmatrix}k\\f_2\end{pmatrix}\right]_2 =$$

$$= \frac{1}{4\pi^2}\oint\overline{\oint}\left[\frac{\Psi(w)^+\Psi(z)-I}{1-z\bar{w}}(zh(z)+f_1), wk(w)+f_2\right]_{\mathfrak{F}} dz\overline{dw} +$$

$$+\left[\frac{1}{2\pi i}\oint\Psi(z)h(z)dz+\Psi(\infty)f_1, \frac{1}{2\pi i}\oint\Psi(z)k(z)dz+\Psi(\infty)f_2\right]_{\mathfrak{G}}$$

$$= 4 \text{ double integrals } + 4 \text{ inner products}$$

$$= s(h,k)+s(h,f_2)+s(f_1,k)+s(f_1,f_2),$$

where $s(a,b)$ stands for the sum of the double integral and the inner product involving a and b. Using Cauchy's formula we obtain the equalities

$$s(h,k)=[h,k],\quad s(h,f_2)=s(f_1,k)=0,\quad s(f_1,f_2)=[f_1,f_2]_{\mathfrak{F}}.$$

We shall prove the second and fourth equality, as the other two can be proved in a similar way. Concerning the second relation we have

$$s(h,f_2) = \frac{1}{4\pi^2}\oint\overline{\oint}\left[\frac{\Psi(w)^+\Psi(z)-I}{1-z\bar{w}}zh(z), f_2\right]_{\mathfrak{F}} dz\overline{dw} +$$

$$+\left[\frac{1}{2\pi i}\oint \Psi(z)h(z)dz, \Psi(\infty)f_2\right]_{\mathfrak{G}}.$$

The first summand is equal to

$$\frac{1}{2\pi i}\oint\left[h(z), \frac{1}{2\pi i}\oint\frac{\Psi(z)^+\Psi(w)-I}{(\bar{z})^{-1}-w}f_2 dw\right]_{\mathfrak{F}} dz$$

$$=\frac{1}{2\pi i}\oint\left[h(z), \left(\left(\Psi(z)^+\Psi(\bar{z}^{-1})-I\right)-\left(\Psi(z)^+\Psi(\infty)-I\right)\right)f_2\right]_{\mathfrak{F}} dz$$

$$= -\frac{1}{2\pi i}\oint\left[\Psi(z)h(z), \Psi(\infty)f_2\right]_{\mathfrak{G}} dz$$

and, consequently, $s(h,f_2)=0$. Also,

$$s(f_1,f_2) = \frac{1}{4\pi^2}\oint\overline{\oint}\left[\frac{\Psi(w)^+\Psi(z)-I}{1-z\bar{w}}f_1, f_2\right]_{\mathfrak{F}} dz\overline{dw} + [\Psi(\infty)f_1, \Psi(\infty)f_2]_{\mathfrak{G}}$$

and here the first summand equals

$$-\frac{1}{2\pi i}\overline{\oint}\left[\frac{1}{2\pi i}\oint\frac{\Psi(w)^+\Psi(z)-I}{(\bar{w})^{-1}-z}f_1 dz, \frac{1}{w}f_2\right]_{\mathfrak{F}} \overline{dw}$$

$$=\frac{1}{2\pi i}\overline{\oint}\left[(\Psi(w)^{+}\Psi(\infty)-I)f_1,\frac{1}{w}f_2\right]_{\mathfrak{F}}\overline{dw}$$

$$=-\left[f_1,\frac{1}{2\pi i}\oint\frac{\Psi(\infty)^{+}\Psi(w)-I}{w}f_2\,dw\right]_{\mathfrak{F}}$$

$$=-\left[f_1,(\Psi(\infty)^{+}\Psi(\infty)-I)f_2\right]_{\mathfrak{G}},$$

which implies $s(f_1,f_2)=[f_1,f_2]_{\mathfrak{F}}$. Hence, the equality $[U_0x,U_0y]_2=[x,y]_1$ holds. It is not difficult to see that the Krein space associated with $\mathfrak{H}_1$ and $[.,.]_1$ ($\mathfrak{H}_2$ and $[.,.]_2$) is $\mathfrak{K}\oplus\mathfrak{F}$ ($\mathfrak{K}\oplus\mathfrak{G}$, respectively), where $\mathfrak{K}$ is the Krein space associated with $\mathfrak{H}$ and $[.,.]$. Let $U:\mathfrak{K}\oplus\mathfrak{F}\to\mathfrak{K}\oplus\mathfrak{G}$ be the continuous extension of $\hat{U}_0:\hat{\mathfrak{H}}\oplus\mathfrak{F}\to\hat{\mathfrak{H}}\oplus\mathfrak{G}$, which exists and is unitary by Corollary 2.3. Then $\Delta=(\mathfrak{K};\mathfrak{F},\mathfrak{G};U)$ is a unitary colligation and we claim that $\Theta(z)=\Theta_{\Delta}(z)$, $z\in E(\partial_-)$. Indeed, writing

$$U_0=\begin{pmatrix}T_0 & F_0\\ G_0 & H_0\end{pmatrix}:\begin{pmatrix}\mathfrak{H}\\ \mathfrak{F}\end{pmatrix}\to\begin{pmatrix}\mathfrak{H}\\ \mathfrak{G}\end{pmatrix},$$

where $(T_0h)(w)=wh(w)$, $(F_0f)(w)=f$, $G_0h=\frac{1}{2\pi i}\oint\Psi(w)h(w)\,dw$ and $H_0f=\Psi(\infty)f$, we have that for $z\in E(\partial_-)$

$$(H_0+zG_0(I-zT_0)^{-1}F_0)f=\Psi(\infty)f+\frac{z}{2\pi i}\oint\Psi(w)\,\frac{1}{1-zw}f\,dw=$$

$$=\Psi(\infty)f+\frac{1}{2\pi i}\oint\frac{\Psi(w)}{z^{-1}-w}f\,dw=\Psi(\infty)f+(\Psi(1/z)-\Psi(\infty))f=\Theta(z)f$$

and from this the claim can be proved. We leave the details to the reader. It is easy to check that $T_0^nF_0f=w^nf$, $f\in\mathfrak{F}$, which implies that Δ is closely innerconnected. To see this, it is sufficient to show that the polynomials with coefficients in $\mathfrak{F}$ are dense in $\mathfrak{H}=H^2_{\mathfrak{F}}(I(\partial_-))$. Since these polynomials are dense in $H^2_{\mathfrak{F}}(\mathbb{D})$, it follows that when composed with γ_- they form a dense set in $\mathfrak{H}$. Hence, it suffices to prove that for each $n\in\mathbb{N}$ and $f\in\mathfrak{F}$ the function $\gamma_-(z)^nf$ can be approximated in $\mathfrak{H}$ by polynomials in z with coefficients in $\mathfrak{F}$. But this follows from Mergelyan's theorem (see, for instance, [R]), which implies that $\gamma_-(z)^n$ can, in fact, be approximated by polynomials in z, uniformly on $I(\partial_-)\cup\partial$. Hence, Δ is closely innerconnected. Finally, writing U_0^{-1} as

$$U_0^{-1}=\begin{pmatrix}\tilde{T}_0 & \tilde{G}_0\\ \tilde{F}_0 & \tilde{H}_0\end{pmatrix}:\begin{pmatrix}\mathfrak{H}\\ \mathfrak{G}\end{pmatrix}\to\begin{pmatrix}\mathfrak{H}\\ \mathfrak{F}\end{pmatrix},$$

we have that the linear span $\bigvee_{j=1}^{n}\Re(\tilde{T}_0^j\tilde{G}_0)$ is equal to the linear space

$\{\sum_{j=1}^{n} z^{-j} f_j \mid f_j \in \mathfrak{F}\} \subset \mathfrak{H}$ and the above argument (in the end applied to the function $(1/\gamma_-(z))^n/z$ instead of $\gamma_-(z)^n$) gives that Δ is closely outerconnected. We leave to the reader the proof of the uniqueness of Δ up to weak isomorphisms.

We thank Prof. B.L.J. Braaksma for pointing out the above application of Mergelyan's theorem. By similar methods one can obtain Caratheodory type representations of holomorphic operator functions, see, e.g., [DLS3].

4. Characteristic functions of bounded operators in Krein spaces

The definition of the characteristic function of a bounded operator in a Hilbert space goes at least back to M.S. Livšic and V.P. Potapov [LP], A.V. Štraus [St] and Yu.L. Smul'jan [Sm]. In this section we extend it to a bounded operator on a Krein space. The ideas leading to this extension are taken mainly from [ACG], where Arsene, Constantinescu and Gheondeaconsider bounded operators acting from one space to another. For our purpose, however, it suffices to consider a bounded operator T from a Krein space $(\mathfrak{K}, [.,.])$ to itself. Relative to a fixed fundamental symmetry $J_{\mathfrak{K}}$ we define the operators

$$J_T = \operatorname{sgn}(J_{\mathfrak{K}} - T^* J_{\mathfrak{K}} T), \quad D_T = |J_{\mathfrak{K}} - T^* J_{\mathfrak{K}} T|^{1/2},$$

computed using the functional calculus on the Hilbert space $(\mathfrak{K}, [J_{\mathfrak{K}}.,.])$, and we set $\mathfrak{D}_T = \mathfrak{R}(D_T)^c$. Note that $\mathfrak{D}_T$ endowed with the inner product $[J_{\mathfrak{K}} J_T.,.]$ is a Krein space and when we refer to $\mathfrak{D}_T$ as a Krein space it is with respect to this inner product.

Theorem 4.1. *Let $\mathfrak{K}$ be a Krein space, $J_{\mathfrak{K}}$ a fundamental symmetry on $\mathfrak{K}$ and $T \in \mathbf{L}(\mathfrak{K})$. Then*

(*i*) *there exists a uniquely determined operator $L_T \in \mathbf{L}(\mathfrak{D}_T, \mathfrak{D}_{T^*})$ such that $D_{T^*} L_T = T J_{\mathfrak{K}} D_T$ on $\mathfrak{D}_T$,*

(*ii*) *the operator U_T with decomposition*

$$U_T = \begin{bmatrix} T & D_{T^*}|_{\mathfrak{D}_{T^*}} \\ D_T & -J_T L_{T^*} \end{bmatrix} : \begin{bmatrix} \mathfrak{K} \\ \mathfrak{D}_{T^*} \end{bmatrix} \to \begin{bmatrix} \mathfrak{K} \\ \mathfrak{D}_T \end{bmatrix}$$

defines a unitary colligation $\Delta_T = (\mathfrak{K}, \mathfrak{D}_{T^*}, \mathfrak{D}_T; U_T)$ and

(*iii*) *Δ_T is closely connected if and only if T is simple.*

Part (*iii*) of the Theorem 4.1 is a consequence of the foregoing parts and the last statement in Lemma 3.1 as $\mathfrak{R}(D_T)^c = \mathfrak{D}_T$. Parts (*i*) and (*ii*) are copied from [ACG], Proposition 4.1 and Corollary 4.5, respectively. Their proof of part (*i*) applies Corollary 2.2 of this note and in order to show this we briefly repeat it. Take in Corollary 2.2 $\mathfrak{H} = \mathfrak{D}_{T^*}$, provided with the Hilbert inner product $(x,y) = [J_{\mathfrak{K}}x, y]$, $x,y \in \mathfrak{D}_{T^*}$, $[x,y] = (D_{T^*}^2 x, y)$ and $S_0 = J_{T^*}J_{\mathfrak{K}}TJ_TJ_{\mathfrak{K}}T^*|_{\mathfrak{D}_{T^*}}$. Then the hypotheses of Corollary 2.2 are satisfied and, moreover, $[.,.]$ is positive definite. It follows that there exists a constant $C \geq 0$ such that $[Sx,x] \leq C[x,x]$ for all $x \in \mathfrak{K}$, the Hilbert space associated with $\mathfrak{H}$ and $[.,.]$. Hence,

$$(TJ_{\mathfrak{K}}D_T)(TJ_{\mathfrak{K}}D_T)^* = D_{T^*}^2 S_0 \leq C D_{T^*}^2 \text{ on } \mathfrak{D}_{T^*}.$$

The Douglas factorization theorem (see [Fu] p. 124) now implies the existence of L_T with the desired property; its uniqueness follows from the injectivity of D_{T^*}.

We refer to [ACG] for special cases concerning the link operator L_T and the various properties of the operators J_T, D_T and L_T such as

$$L_T^+ = L_{T^*}, \quad D_T^+ = J_{\mathfrak{K}}D_TJ_T, \quad D_T^+J_TL_{T^*} = T^+D_{T^*} \text{ on } \mathfrak{D}_{T^*},$$

if D_T is considered as a (bounded) mapping from the Krein space $\mathfrak{K}$ to the Krein space $\mathfrak{D}_T$.

Theorem 4.1 gives rise to the following definition. We define the characteristic function Θ_T of $T \in \mathbf{L}(\mathfrak{K})$ relative to a fundamental symmetry $J_{\mathfrak{K}}$ to be the mapping

$$\Theta_T(z) = \Theta_{\Delta_T}(z) = -J_TL_{T^*} + zD_T(I - zT)^{-1}D_{T^*}|_{\mathfrak{D}_{T^*}}, \qquad z \in \mathfrak{D}(\Theta_{\Delta_T}),$$

i.e., the characteristic function of the unitary colligation Δ_T. This notion reduces to the usual one, if $\mathfrak{K}$ is a Hilbert space, for then $J_{\mathfrak{K}} = I$, $L_T = T$ and

$$\Theta_T(z) = -T^*J_{T^*} + zD_T(I - zT)^{-1}D_{T^*}|_{\mathfrak{D}_{T^*}}$$

with $J_T = \operatorname{sgn}(I - T^*T)$ and $D_T = |I - T^*T|^{1/2}$.

5. Characteristic functions of bounded operators in Krein spaces (continued)

If in a unitary colligation $\Delta = (\mathfrak{K}, \mathfrak{F}, \mathfrak{G}; U)$ in which U is decomposed as in (1.1) and the null spaces $\nu(F)$ and $\nu(G^+)$ of F and G^+ are orthocomplemented in $\mathfrak{F}$ and $\mathfrak{G}$, then U has the 3x3 block matrix representation

$$U=\begin{pmatrix} T & F_1 & 0 \\ G_1 & H_1 & 0 \\ 0 & 0 & H_2 \end{pmatrix} : \begin{pmatrix} \mathfrak{K} \\ \Re(F^+)^c \\ \nu(F) \end{pmatrix} \to \begin{pmatrix} \mathfrak{K} \\ \Re(G) \\ \nu(G^+) \end{pmatrix},$$

in which $\begin{pmatrix} T & F_1 \\ G_1 & H_1 \end{pmatrix}$ and H_2 are unitary. In this section we want to investigate some sort of equivalence between Δ and the unitary colligation Δ_T corresponding to its basic operator T. A necessary condition for this, even when $\nu(F)$ and $\nu(G^+)$ are not assumed to be orthocomplemented, is that they are equal to $\{0\}$, or, equivalently, that $\Re(F^+)^c=\mathfrak{F}$ and $\Re(G)^c=\mathfrak{G}$, as the corresponding entries in U_T have these properties. With the following definition of equivalence these conditions turn out to be sufficient as well. We shall say that two colligations $\Delta=(\mathfrak{K},\mathfrak{F},\mathfrak{G};T,F,G,H)$ and $\Delta'=(\mathfrak{K},\mathfrak{F}',\mathfrak{G}';T,F',G',H')$ with the same state space $\mathfrak{K}$ and basic operator T coincide weakly if there are two weak isomorphisms V from $\mathfrak{F}'$ to $\mathfrak{F}$ and W from $\mathfrak{G}'$ to $\mathfrak{G}$ such that

$$\begin{pmatrix} T & F \\ G & H \end{pmatrix}\begin{pmatrix} I & 0 \\ 0 & V \end{pmatrix}=\begin{pmatrix} I & 0 \\ 0 & W \end{pmatrix}\begin{pmatrix} T & F' \\ G' & H' \end{pmatrix} \text{ on } \begin{pmatrix} \mathfrak{K} \\ \mathfrak{D}(V) \end{pmatrix}.$$

Similarly, $\Theta\in\mathbf{S}(\mathfrak{F},\mathfrak{G})$ and $\Theta'\in\mathbf{S}(\mathfrak{F}',\mathfrak{G}')$ are said to coincide weakly if for some V and W as above $\Theta(z)V=W\Theta'(z)$ for z in some neighborhood of 0 contained in $\mathfrak{D}(\Theta)\cap\mathfrak{D}(\Theta')$. Clearly, if two colligations Δ and Δ' coincide weakly, then so do their characteristic function. If V and W are continuous, then they are unitary operators from $\mathfrak{F}'$ onto $\mathfrak{F}$ and $\mathfrak{G}'$ onto $\mathfrak{G}$, respectively, and Δ and Δ' as well as Θ_Δ and $\Theta_{\Delta'}$ coincide. A sufficient condition to ensure boundedness of these operators will be formulated and proved in Section 7.

PROPOSITION 5.1. *Let $\Delta=(\mathfrak{K},\mathfrak{F},\mathfrak{G};T,F,G,H)$ be a unitary colligation and let $J_{\mathfrak{K}}$ be a fundamental symmetry on its state space $\mathfrak{K}$. Let Δ_T be the unitary colligation defined for the basic operator T of Δ relative to $J_{\mathfrak{K}}$. Then Δ and Δ_T coincide weakly if and only if $\Re(F^+)^c=\mathfrak{F}$ and $\Re(G)^c=\mathfrak{G}$.*

Proof. Assume $\Re(F^+)^c=\mathfrak{F}$, $\Re(G)^c=\mathfrak{G}$ and define $W:\Re(D_T)\subset\mathfrak{D}_T\to\Re(G)\subset\mathfrak{G}$ by $WD_Tx=Gx$, $x\in\mathfrak{K}$. To see that W is a well defined weak isomorphism we first note that, as Δ is unitary, $I-T^+T=G^+G$ and hence we have that for $x',y'\in\mathfrak{K}$

$$[Gx',Gy']_{\mathfrak{G}}=[(I-T^+T)x',y']_{\mathfrak{K}}=$$

$$=[J_{\mathfrak{K}}J_TD_T^2x',y']_{\mathfrak{K}}=(J_TD_T^2x',y')_{\mathfrak{K}}=(J_TD_Tx',D_Ty')_{\mathfrak{D}_T}.$$

From the density of the ranges $\mathfrak{R}(G)$ and $\mathfrak{R}(D_T)$ in $\mathfrak{G}$ and $\mathfrak{D}_T$, respectively, it follows that the null spaces of G and D_T coincide and this implies that W is well defined. If $x,y\in\mathfrak{R}(D_T)$ and $x=D_Tx'$ and $y=D_Ty'$ for some $x',y'\in\mathfrak{K}$, then the above equalities show that $[Wx,Wy]_{\mathfrak{G}}=[x,y]_{\mathfrak{D}_T}$, i.e., W is isometric. Since $\mathfrak{D}(W)=\mathfrak{R}(D_T)$ and $\mathfrak{R}(W)=\mathfrak{R}(G)$, W is a weak isomorphism. Similarly, it can be shown that the mapping $V:\mathfrak{R}(D_{T^*})\subset\mathfrak{D}_{T^*}\to\mathfrak{R}(F^+)\subset\mathfrak{F}$ defined by

$$VJ_{T^*}D_{T^*}J_{\mathfrak{K}}x=F^+x, \quad x\in\mathfrak{K},$$

is a well defined isometry with a dense domain and, by assumption, a dense range. Therefore, V is a weak isomorphism between $\mathfrak{D}_{T^*}$ and $\mathfrak{F}$. From the defining relation of V and the fact that the Krein space adjoint of

$$D_{T^*}|_{\mathfrak{D}_{T^*}}:\mathfrak{D}_{T^*}\to\mathfrak{K}$$

is given by

$$(D_{T^*}|_{\mathfrak{D}_{T^*}})^+=J_{T^*}D_{T^*}J_{\mathfrak{K}},$$

it easily follows that $FV=D_{T^*}$ on $\mathfrak{D}(V)$. It remains to prove that $HV=-WJ_TL_{T^*}$ on $\mathfrak{D}(V)$. As Δ is unitary $F^+T=-H^+G$ and therefore,

$$V(J_TL_{T^*})^+D_T=V(D_T^+J_TL_{T^*})^+=V(T^+D_{T^*}|_{\mathfrak{D}_{T^*}})^+=$$
$$=VJ_{T^*}D_{T^*}J_{\mathfrak{K}}T=F^+T=-H^+G=-H^+WD_T.$$

This implies that $V(J_TL_{T^*})^+=-H^+W$ on $\mathfrak{R}(D_T)$ and therefore, $(J_TL_{T^*})^+W^+=-V^+H^+$ on $\mathfrak{R}(W)$. Taking adjoints we obtain the desired equality. The proof of the converse is left to the reader.

We now come to the main result of this paper.

THEOREM 5.2. *Let* $\Theta\in\mathbf{S}(\mathfrak{F},\mathfrak{G})$. *If*

$$(5.1)\quad \begin{cases} \mathfrak{F}=\bigvee_{z\in N}\left(\mathfrak{R}(\Theta(\bar z)^+-\Theta(0)^+)\cup\mathfrak{R}(I-\Theta(0)^+\Theta(z))\right),\\ \mathfrak{G}=\bigvee_{z\in N}\left(\mathfrak{R}(\Theta(0)-\Theta(\bar z))\cup\mathfrak{R}(I-\Theta(0)\Theta(\bar z)^+)\right),\end{cases}$$

where N *is some neighborhood of* 0 *contained in* $\mathfrak{D}(\Theta)$, *then* Θ *coincides weakly with a* Θ_T *for some bounded operator* T *on a Krein space* $\mathfrak{K}$. *Here* T *can be chosen to be simple in which case*

$$\dim\mathfrak{K}_\pm=\#\ \begin{matrix}\textit{positive}\\ \textit{negative}\end{matrix}\ \textit{squares of } \mathsf{S}_\Theta^m(z,w),$$

where $\mathfrak{K}=\mathfrak{K}_++\mathfrak{K}_-$ *is any fundamental decomposition of* $\mathfrak{K}$. *If* Θ *coincides weakly with* Θ_T *and if* T *is simple or* $\mathfrak{K}$ *is a Hilbert space, then the equalities* (5.1) *are valid.*

Proof. According to Theorem 3.4 Θ can be represented as the characteristic function Θ_Δ of a closely connected unitary colligation Δ. By Corollary 3.3 (*i*) and (5.1) $\mathfrak{R}(F^+)^c=\mathfrak{F}$ and $\mathfrak{R}(G)^c=\mathfrak{G}$. From Proposition 5.1 it follows that Δ and Δ_T coincide weakly, where T is the basic operator of Δ and hence, so do $\Theta=\Theta_\Delta$ and Θ_{Δ_T}. Moreover, Δ_T is closely connected and therefore, T is simple, see Theorem 4.1 (*iii*). Now assume that V and W are weak isomorphisms such that $\Theta V=W\Theta_T$, where Θ_T is the characteristic function of some simple, bounded operator T on a Krein space $\mathfrak{K}$ relative to a fundamental symmetry $J_{\mathfrak{K}}$. Then

$$\mathsf{S}^m_\Theta(z,w)=\begin{bmatrix} V & 0 \\ 0 & W \end{bmatrix} \mathsf{S}^m_{\Theta_{\Delta_T}}(z,w) \begin{bmatrix} V^+ & 0 \\ 0 & W^+ \end{bmatrix} \quad \text{on} \quad \begin{bmatrix} \mathfrak{R}(V) \\ \mathfrak{R}(W) \end{bmatrix}$$

and the formula concerning $\dim \mathfrak{K}_\pm$ now easily follows from Theorem 4.1 (*iii*) and Corollary 3.3 (*ii*). The last part of the theorem can be proved in the same vein.

6. J.A. Ball's characterization of characteristic functions of bounded operators in Hilbert spaces

The following representation theorem is due to Ball, see [Ba]. Its formulation in terms of Krein spaces stems from B.W. McEnnis, see [Mc].

THEOREM 6.1. *The mapping* $\Theta\in\mathbf{S}(\mathfrak{F},\mathfrak{G})$ *coincides with* $\Theta_T\in\mathbf{S}(\mathfrak{D}_{T^*},\mathfrak{D}_T)$ *for some bounded operator* T *on a Hilbert space if and only if*

(*i*) $I-\Theta(0)^+\Theta(0)$ *is injective on* $\mathfrak{F}$ *and commutes with some fundamental symmetry* $J_{\mathfrak{F}}$ *on* $\mathfrak{F}$,

(*ii*) $I-\Theta(0)\Theta(0)^+$ *is injective on* $\mathfrak{G}$ *and commutes with some fundamental symmetry* $J_{\mathfrak{G}}$ *on* $\mathfrak{G}$ *and*

(*iii*) *the kernel* $\mathsf{S}^m_\Theta(z,w)$ *is positive semidefinite.*

Theorem 6.1 is an extension of a result of D.N. Clark [C] who considered the case where $\Theta(0)$ and hence T are invertible. V.M. Brodskii, I.C. Gohberg and M.G. Krein treated this same case in [BGK]. For the case where $\Theta(0)$ and hence T are contractions we refer to the monograph [Sz.F]. Before we show how Theorem 6.1 can be deduced from Theorem 5.2, we want to make the following remarks.

REMARKS 1. McEnnis proved in [Mc] that Theorem 6.1 remains valid if property (*iii*) is replaced by: the kernels

$$\frac{I-\Theta(w)^{+}\Theta(z)}{1-\bar{w}z} \text{ and } \frac{I-\Theta(\bar{w})\Theta(\bar{z})^{+}}{1-\bar{w}z}$$

(with values in $\mathbf{L}(\mathfrak{F})$ and $\mathbf{L}(\mathfrak{G})$, respectively) are positive semidefinite. Note that these kernels are the elements on the main diagonal of the matrix kernel $S_{\Theta}^{m}(z,w)$. In case $\mathfrak{F}$ and $\mathfrak{G}$ are Hilbert spaces it is known that $S_{\Theta}^{m}(z,w)$ has κ negative squares if and only if one of the kernels above has κ negative squares, see, e.g., [DLS1] Theorem 6.1.

2. If $\Theta\in\mathbf{S}(\mathfrak{F},\mathfrak{G})$ has the property that $I-\Theta(0)^{+}\Theta(0)$ is positive definite on $\mathfrak{F}$ and commutes with a fundamental symmetry $J_{\mathfrak{F}}$ on $\mathfrak{F}$, then the operators $J_{\mathfrak{F}}$ and $J=\operatorname{sgn}(I-\Theta(0)^{+}\Theta(0))$ are equal, where the latter is computed via the functional calculus on $\mathfrak{F}$ with the Hilbert space inner product $[J_{\mathfrak{F}}\cdot,\cdot]$. Indeed, McEnnis showed that J too is a fundamental symmetry on $\mathfrak{F}$ and as it commutes with $J_{\mathfrak{F}}$ the equality $J=J_{\mathfrak{F}}$ follows, cf. [Mc] p. 165. Hence, if $\Theta\in\mathbf{S}(\mathfrak{F},\mathfrak{G})$ has the properties (*i*)–(*iii*) of Theorem 6.1, then $J_{\mathfrak{F}}=\operatorname{sgn}(I-\Theta(0)^{+}\Theta(0))$ and $J_{\mathfrak{G}}=\operatorname{sgn}(I-\Theta(0)\Theta(0)^{+})$.

3. McEnnis also proved that the isomorphisms $V\in\mathbf{L}(\mathfrak{D}_{T^*},\mathfrak{F})$ and $W\in\mathbf{L}(\mathfrak{D}_{T},\mathfrak{G})$ such that $\Theta(z)V=W\Theta_T(z)$ for z near 0 are not only unitary with respect to the indefinite inner products, but are also unitary with respect to the positive definite inner products induced by the symmetries J_{T^*}, J_T, $J_{\mathfrak{F}}$ and $J_{\mathfrak{G}}$. This now follows easily from the previous remark. E.g., $\Theta(0)V=W\Theta_T(0)$ implies that

$$(I-\Theta(0)^{+}\Theta(0))V=V(I-T^{*}T)|_{\mathfrak{D}_{T^*}}$$

(see the argument in the proof of Theorem 6.1 below), whence

$$J_{\mathfrak{F}}V=\operatorname{sgn}(I-\Theta(0)^{+}\Theta(0))V=V\operatorname{sgn}(I-T^{*}T)|_{\mathfrak{D}_{T^*}}=VJ_{T^*}$$

and therefore, V is unitary with respect to the positive definite inner products.

Proof of Theorem 6.1. If Θ coincides with Θ_T, then it has the properties (*i*)–(*iii*), as $\Theta_T\in\mathbf{S}(\mathfrak{D}_{T^*},\mathfrak{D}_T)$ has these properties. This is easy to verify and we skip the details. It remains to prove the converse. Assume Θ satisfies (*i*)–(*iii*), then (5.1) of Theorem 5.2 is valid and Θ coincides weakly with a Θ_T where T is a simple, bounded operator on a Krein space $\mathfrak{K}$:

there exist weak isomorphisms V from $\mathfrak{D}_{T^*}$ to $\mathfrak{F}$ and W from $\mathfrak{D}_T$ to $\mathfrak{G}$ such that $\Theta(z)V = W\Theta_T(z)$ on $\mathfrak{D}(V)$ for z near 0. The proof of Theorem 6.1 is complete when we have shown that V and W are bounded. For, then they are unitary operators and Θ and Θ_T coincide. It suffices to prove the continuity of V, that of W can be established in a similar way. From $\Theta(0)V = W\Theta_T(0)$ we obtain on account of Theorem 4.1(*ii*), that on $\mathfrak{D}(V)$

$$V^+((I-\Theta(0)^+\Theta(0))V = I-(W\Theta_T(0))^+W\Theta_T(0) = I-\Theta_T(0)^+\Theta_T(0)$$

$$= (D_{T^*}|_{\mathfrak{D}_{T^*}})^+(D_{T^*}|_{\mathfrak{D}_{T^*}}) = V^+VJ_{T^*}D_{T^*}J_{\mathfrak{K}}(D_{T^*}|_{\mathfrak{D}_{T^*}}).$$

Hence, since V^+ is injective, we have the intertwining relation $AV=VB$, where we have put $A=I-\Theta(0)^+\Theta(0)$ and $B=J_{T^*}D_{T^*}J_{\mathfrak{K}}(D_{T^*}|_{\mathfrak{D}_{T^*}})$. From the assumptions (*i*) and (*iii*) in Theorem 6.1 concerning Θ it can easily be verified that A has the following properties:

> $A \in \mathbf{L}(\mathfrak{F})$, A is positive definite and commutes with a fundamental symmetry $J_{\mathfrak{F}}$ on $\mathfrak{F}$.

Assumption (*iii*) in Theorem 6.1 implies that $\mathfrak{K}$ is a Hilbert space, hence $J_{\mathfrak{K}} = I$ and $B = J_{T^*}(D_{T^*}|_{\mathfrak{D}_{T^*}})^2 = (I-TT^*)|_{\mathfrak{D}_{T^*}}$. Hence, it follows that B has the same properties as A:

> $B \in \mathbf{L}(\mathfrak{D}_{T^*})$, B is positive definite on the Krein space $\mathfrak{D}_{T^*}$ and commutes with the fundamental symmetry J_{T^*} on $\mathfrak{D}_{T^*}$.

The following proposition with $\mathfrak{G} = \mathfrak{D}_{T^*}$ implies that V has the desired property. In the next section we treat a generalization of this result, see Theorem 7.1.

PROPOSITION 6.2. *Let* $V:\mathfrak{G}\to\mathfrak{F}$ *be a weak isomorphism between the Krein spaces* $\mathfrak{G}$ *and* $\mathfrak{F}$. *Assume there exist positive definite operators* $A \in \mathbf{L}(\mathfrak{F})$ *and* $B \in \mathbf{L}(\mathfrak{G})$ *which commute with a fundamental symmetry on* $\mathfrak{F}$ *and on* $\mathfrak{G}$, *respectively, such that* $AV = VB$ *on* $\mathfrak{D}(V)$. *Then* V *is a unitary operator from* $\mathfrak{G}$ *onto* $\mathfrak{F}$.

Proof. The positive definiteness implies in particular that both operators are selfadjoint, injective and have dense ranges. It follows that the spaces $B\mathfrak{D}(V)$ and $A\mathfrak{R}(V)$ are dense in $\mathfrak{G}$ and $\mathfrak{F}$, respectively, and that V maps $B\mathfrak{D}(V)$ isometrically and bijectively onto $A\mathfrak{R}(V)$. Let $J_{\mathfrak{G}}$ ($J_{\mathfrak{F}}$) be the fundamental symmetry on $\mathfrak{G}$ ($\mathfrak{F}$) which commutes with B (A, respectively). Recall that in $(\mathfrak{G},[\cdot,\cdot]_{\mathfrak{G}})$ $((\mathfrak{F},[\cdot,\cdot]_{\mathfrak{F}}))$ we have denoted by $(\cdot,\cdot)_{\mathfrak{G}}$ $((\cdot,\cdot)_{\mathfrak{F}})$ and $\|\ \|_{\mathfrak{G}}$

$(\| \ \|_{\mathfrak{F}})$ the Hilbert space inner products and corresponding norms on $\mathfrak{G}$ $(\mathfrak{F})$ generated by $J_{\mathfrak{G}}$ ($J_{\mathfrak{F}}$, respectively). From the inequality

$$|[B^{-1}y,x]_{\mathfrak{G}}|^2 = |(J_{\mathfrak{G}}B^{-1}y,x)_{\mathfrak{G}}|^2 \le (J_{\mathfrak{G}}B^{-1}y,y)_{\mathfrak{G}}(J_{\mathfrak{G}}B^{-1}x,x)_{\mathfrak{G}}$$

with $y = BJ_{\mathfrak{G}}x$, one easily obtains the inequality

$$\|x\|_{\mathfrak{G}}^2 \le \|J_{\mathfrak{G}}B\|_{\mathfrak{G}}[B^{-1}x,x]_{\mathfrak{G}}, \qquad x \in \mathfrak{R}(B).$$

We want to apply Corollary 2.3 and define $\mathfrak{H}_1$ as the Hilbert space completion of $\mathfrak{R}(B)$ with respect to the positive definite inner product $(x,y)_1 = [B^{-1}x,y]_{\mathfrak{G}}$ and $\mathfrak{H}_2$ as the Hilbert space completion of $\mathfrak{R}(A)$ with respect to the positive definite inner product $(u,v)_2 = [A^{-1}u,v]_{\mathfrak{F}}$. The above inequality and the relation

$$[x,y]_{\mathfrak{G}} = (Bx,y)_1, \qquad x \in \mathfrak{G}, \quad y \in \mathfrak{H}_1,$$

imply that $\mathfrak{H}_1$ can be identified with a linear manifold in $\mathfrak{G}$, that $B\mathfrak{D}(V)$ is dense in $\mathfrak{H}_1$ and that the inner product $[\cdot,\cdot]_1$, defined as the restriction of $[\cdot,\cdot]_{\mathfrak{G}}$ to $\mathfrak{H}_1$ is bounded on $\mathfrak{H}_1$ with B as the Gram operator. Similarly, it can be shown that $\mathfrak{H}_2$ can be identified with a linear manifold in $\mathfrak{F}$, that $A\mathfrak{R}(V)$ is dense in $\mathfrak{H}_2$ and that the restriction $[\cdot,\cdot]_2$ of $[\cdot,\cdot]_{\mathfrak{F}}$ on $\mathfrak{H}_2$ is a bounded inner product on $\mathfrak{H}_2$ with A as the Gram operator. Since for $x,y \in B\mathfrak{D}(V) \subset \mathfrak{D}(V)$ we have that $Vx, Vy \in A\mathfrak{R}(V)$ and

$$(Vx,Vy)_2 = [A^{-1}Vx,Vy]_{\mathfrak{F}} = [VB^{-1}x,Vy]_{\mathfrak{F}} = [B^{-1}x,y]_{\mathfrak{G}} = (x,y)_1.$$

V on $B\mathfrak{D}(V)$ can be extended by continuity to a unitary and hence bijective mapping $U_0 \in \mathbf{L}(\mathfrak{H}_1,\mathfrak{H}_2)$ and $[Vx,Vy]_{\mathfrak{F}} = [x,y]_{\mathfrak{G}}$ implies that

$$[U_0x,U_0y]_2 = [x,y]_1, \qquad x,y \in \mathfrak{H}_1.$$

We are now in a position to apply Corollary 2.3. Let $\mathfrak{K}_1$, $\mathfrak{K}_2$ and $U \in \mathbf{L}(\mathfrak{K}_1,\mathfrak{K}_2)$ be as in its conclusion. Let $P_{\pm}$ be the orthogonal projection of $\mathfrak{G}$ onto $\mathfrak{G}_{\pm} = \mathfrak{R}(I \mp J_{\mathfrak{G}})$. Then, since B commutes with $J_{\mathfrak{G}}$, we have that $P_{\pm}\mathfrak{H}_2 \subset \mathfrak{H}_2$. It follows that we can identify $\mathfrak{K}_2$ with $\mathfrak{G}$ and, similarly, $\mathfrak{K}_1$ with $\mathfrak{F}$. Finally, it can be verified that $V = U$, which completes the proof.

7. A sufficient condition for the continuity of a weak isomorphism

We begin by recalling some facts from [L]. Let S be a densely defined selfadjoint operator with $\rho(S) \ne \emptyset$ on a Krein space $(\mathfrak{K},[\cdot,\cdot])$ and suppose that the form $[S\cdot,\cdot]$ has a finite number of negative squares on the domain $\mathfrak{D}(S)$.

Then S is definitizable, i.e., $\rho(S)\neq\emptyset$ and there exists a polynomial p with real coefficients such that $[p(S)x,x]\geq 0$ for all $x\in\mathfrak{D}(S^k)$ where k is the degree of p. It follows that S has a spectral function E on $\mathfrak{K}$. A critical point t of E in $\bar{\mathbb{R}}$ (the one point compactification of $\mathbb{R}$), which is also called a critical point of S, is said to be regular if there exists an open neighborhood $\Delta_0\subset\bar{\mathbb{R}}$ of t, in which t is the only critical point, such that the projections $E(\Delta)$, $\bar{\Delta}\subset\Delta_0\backslash\{t\}$, are uniformly bounded. A critical point which is not regular is called singular. The set of singular critical points of S will be denoted by $c_s(S)$.

Let us say that a bounded operator A on a Krein space $(\mathfrak{K},[.,.])$ has property P_κ, if

A is injective, $A=A^+$, the form $[A.,.]$ has κ negative squares and $0\notin c_s(A)$.

If $A\in\mathbf{L}(\mathfrak{K})$ has property P_κ, then $S=A^{-1}$ is a densely defined selfadjoint operator, the form $[S.,.]$ has κ negative squares on $\mathfrak{D}(S)$ and $\infty\notin c_s(S)$. It can be shown that $A\in\mathbf{L}(\mathfrak{K})$ has property P_0 if and only if $A=A^+$, A is positive definite and commutes with some fundamental symmetry on $\mathfrak{K}$, cf. [Cu]. The latter properties are precisely the ones we arrived at at the end of the previous section.

The main theorem of this section is as follows.

THEOREM 7.1. *Let V be a weak isomorphism from the Krein space $\mathfrak{G}$ to the Krein space $\mathfrak{F}$. Suppose that there exist operators $A\in\mathbf{L}(\mathfrak{F})$ and $B\in\mathbf{L}(\mathfrak{G})$ having properties P_κ and $\mathsf{P}_{\kappa'}$, respectively, such that $AV=VB$ on $\mathfrak{D}(V)$. Then V is a unitary operator from $\mathfrak{G}$ onto $\mathfrak{F}$ and $\kappa=\kappa'$.*

In order to prove the theorem we need the following preliminary results. If a subspace $\mathfrak{L}$ of a Krein space $(\mathfrak{K},[.,.])$ is nondegenerate and decomposable, then there exists a fundamental decomposition

(7.1) $\mathfrak{L}=\mathfrak{L}_+ + \mathfrak{L}_-$, orthogonal direct sum,

in which $\mathfrak{L}_+$, $\mathfrak{L}_-$ are positive, negative subspaces of $\mathfrak{K}$, respectively. Hence $\mathfrak{L}$ has a fundamental symmetry $J_\mathfrak{L}$. The topology on $\mathfrak{L}$ induced by the norm $[J_\mathfrak{L}x,x]^{1/2}$ is called a decomposition majorant on $\mathfrak{L}$. Recall that in the case $\mathfrak{L}=\mathfrak{K}$ the decomposition majorant is unique, i.e., independent of the fundamental symmetries on $\mathfrak{K}$, and we shall refer to this topology as the norm topology on $\mathfrak{K}$. A subspace $\mathfrak{L}$ of $(\mathfrak{K},[.,.])$ is termed uniformly decomposable if it admits a decomposition of the form (7.1), such that $\mathfrak{L}_+$, $\mathfrak{L}_-$ are uniformly positive, uniformly negative subspaces of $\mathfrak{K}$, respectively.

LEMMA 7.2. *If $\mathfrak{L}$ is a nondegenerate decomposable subspace of a Krein space $\mathfrak{K}$, then the following statements are equivalent:*

(i) There exists a fundamental symmetry J on $\mathfrak{K}$ such that $J\mathfrak{L}\subset\mathfrak{L}$.

(ii) The subspace $\mathfrak{L}$ is uniformly decomposable.

(iii) There is a decomposition majorant on $(\mathfrak{L},[.,.])$ which is equivalent to the norm topology of $\mathfrak{K}$ restricted to $\mathfrak{L}$.

The implications $(i)\Rightarrow(ii)\Leftrightarrow(iii)$ are easy to prove. The implication $(ii)\Rightarrow(i)$ follows from, e.g., [An] Chapter 1, see also [Bo] Theorem V.9.1. If $\mathfrak{L}$ is a nondegenerate, decomposable and dense subspace of $\mathfrak{K}$ and satisfies (iii) of Lemma 7.2, then $\mathfrak{K}$ is the Krein space completion of $\mathfrak{L}$ with respect to the decomposition majorant referred to in (iii).

Now we recall and supplement some results from [Cu], where criteria are given for the regularity of the crititcal point ∞ of a definitizable (densely defined, selfadjoint) operator, here denoted by S, on a Krein space $(\mathfrak{K},[.,.])$. In these criteria the domain $\mathfrak{D}(S)$ and the set $\mathfrak{D}[JS]$ play an important role, see also Proposition 7.4 below. In what follows J will designate a fundamental symmetry of $\mathfrak{K}$ and $(.,.)$ will stand for the Hilbert inner product $[J.,.]$ on $\mathfrak{K}$. The set $\mathfrak{D}[JS]$ is defined to be the domain $\mathfrak{D}(|JS|^{1/2})$ of $|JS|^{1/2}$, computed in the Hilbert space $(\mathfrak{K},(.,.))$. It coincides with the domain $\mathfrak{D}((|JS|+I)^{1/2})$. In [Cu] it is shown that $\mathfrak{D}[JS]$ is independent of the choice of J. From the fact that $|JS|+I$ is boundedly invertible it easily follows that the inner product spaces $(\mathfrak{D}(S),\ ((|JS|+I).,(|JS|+I).))$ and $(\mathfrak{D}[JS],\ ((|JS|+I)^{1/2}.,(|JS|+I)^{1/2}.))$ are Hilbert spaces.

LEMMA 7.3. *Let S be a selfadjoint operator in a Krein space $(\mathfrak{K},[.,.])$ with $\rho(S)\neq\emptyset$. Then the inner product spaces $(\mathfrak{D}(S),[.,.])$ and $(\mathfrak{D}[JS],[.,.])$ are nondegenerate and decomposable, and have unique decomposition majorants.*

Proof. Since $\mathfrak{D}(S)$ and $\mathfrak{D}[JS]$ are dense in $\mathfrak{K}$, it is easily verified that $[.,.]$ does not degenerate on these subspaces. Furthermore, the Hilbert space topologies on $\mathfrak{D}(S)$ and $\mathfrak{D}[JS]$ described above are both majorants of the inner product $[.,.]$ and therefore, the inner product spaces admit Hilbert majorants. By Theorems IV.5.2 and IV.6.4. in [Bo] they are decomposable and have unique decomposition majorants.

The equivalences $(i)\Leftrightarrow(iii)\Leftrightarrow(iv)$ in the following proposition are extensions to definitizable operators of the equivalences $(i)\Leftrightarrow(v)\Leftrightarrow(vii)$ in [Cu] Theorem 2.5.

PROPOSITION 7.4. *Let S be a densely defined, selfadjoint operator in the Krein space $(\mathfrak{K},[.,.])$ with $\rho(S)\neq\emptyset$. Then the following statements are equivalent:*

(i) *There exists a fundamental symmetry J on $\mathfrak{K}$ such that $J\mathfrak{D}[JS]\subset\mathfrak{D}[JS]$.*

(ii) *Each fundamental decomposition of $(\mathfrak{D}[JS],[.,.])$ is uniform.*

(iii) *The decomposition majorant of $(\mathfrak{D}[JS],[.,.])$ is equivalent to the norm topology of $\mathfrak{K}$ restricted to $\mathfrak{D}[JS]$.*

The three statements obtained from $(i),(ii)$ and (iii) by replacing everywhere $\mathfrak{D}[JS]$ by $\mathfrak{D}(S)$ are also equivalent. If S is definitizable, then all six statements are equivalent to:

(iv) *Infinity is not a singular critical point of S.*

Proof. The equivalence of $(i)-(iii)$ and that of the other three statements follow directly from the preceeding lemmas. The implication $(i)\Rightarrow(iv)$ with $\mathfrak{D}(S)$ instead of $\mathfrak{D}[JS]$ follows from [Cu] Theorem 3.2. Now we prove the converse. Denote the spectral function of S by E and let Δ_∞ be such that $\bar{\mathbb{R}}\backslash\Delta_\infty$ is a bounded open interval containing zero and all the finite critical points of S. Let J_0 be a fundamental symmetry on $\mathfrak{K}$ which commutes with $E(\Delta_\infty)$ and put

$$S_\infty = S|_{E(\Delta_\infty)\mathfrak{D}(S)} + J_0|_{E(\bar{\mathbb{R}}\backslash\Delta_\infty)\mathfrak{K}}.$$

Then the operator S_∞ is boundedly invertible and positive in $\mathfrak{K}$, and $\mathfrak{D}(S_\infty)=\mathfrak{D}(S)$. It is easy to see that $\infty\notin c_s(S)$ if and only if $\infty\notin c_s(S_\infty)$, see [Cu] Corollary 3.3. Now the converse implication follows from the implication $(vii)\Rightarrow(i)$ in [Cu] Theorem 2.5 applied to the operator S_∞, cf. [Cu] Lemma 2.4. Finally, Theorem 3.9 in [Cu] yields that $\infty\notin c_s(S_\infty)$ if and only if $\infty\notin c_s(J_0(J_0S_\infty)^{1/2})$. We also have that

$$\mathfrak{D}[JS]=\mathfrak{D}[J_0S]=\mathfrak{D}[J_0S_\infty]=\mathfrak{D}(J_0(J_0S_\infty)^{1/2}),$$

see [Cu] Remark 1.4. The equivalence of (i) and (iv) now follows when we apply what has already been proved to the operator $J_0(J_0S_\infty)^{1/2}$.

Proof of Theorem 7.1. Let $J_{\mathfrak{G}}$ be a fundamental symmetry on the Krein space $(\mathfrak{G},[.,.]_{\mathfrak{G}})$. Denote by $\mathfrak{H}_{\mathfrak{G}}$ the Hilbert space

$$(\mathfrak{D}[J_{\mathfrak{G}}B^{-1}],\ [J_{\mathfrak{G}}|J_{\mathfrak{G}}B^{-1}|^{1/2}.,|J_{\mathfrak{G}}B^{-1}|^{1/2}.]_{\mathfrak{G}}).$$

As shown in [Cu] Remark 1.7, the inner product $[B^{-1}.,.]_{\mathfrak{G}}$ can be extended from $\mathfrak{D}(B^{-1})$ onto $\mathfrak{D}[J_{\mathfrak{G}}B^{-1}]$. We denote this extension also by $[B^{-1}.,.]_{\mathfrak{G}}$. The space $(\mathfrak{D}[J_{\mathfrak{G}}B^{-1}],[B^{-1}.,.]_{\mathfrak{G}})$ is a Pontryagin space. Provided with its Hilbert majorant it coincides with $\mathfrak{H}_{\mathfrak{G}}$ and $B\mathfrak{D}(V)$ is a dense subspace. The same results are valid if we replace $\mathfrak{G}$ by $\mathfrak{F}$, B by A and $\mathfrak{D}(V)$ by $\mathfrak{R}(V)$. From the intertwining relation $AV=VB$ on $\mathfrak{D}(V)$ it follows that for $x=By\in B\mathfrak{D}(V)\subset\mathfrak{D}(V)$ with $y\in\mathfrak{D}(V)$

$$[A^{-1}Vx,Vx]_{\mathfrak{F}}=[Vy,VBy]_{\mathfrak{F}}=[y,By]_{\mathfrak{G}}=[B^{-1}x,x]_{\mathfrak{G}}.$$

We also have that $V(B\mathfrak{D}(V))=A\mathfrak{R}(V)$ and hence, $V|_{B\mathfrak{D}(V)}$ is a weak isomorphism from the Pontryagin space $\mathfrak{D}[J_{\mathfrak{G}}B^{-1}]$ to the Pontryagin space $\mathfrak{D}[J_{\mathfrak{F}}A^{-1}]$. It follows from [IKL] Theorem 6.3 that $V|_{B\mathfrak{D}(V)}$ is a continuous operator and can be extended by continuity to the unitary operator U_0 from the Pontryagin space $\mathfrak{D}[J_{\mathfrak{G}}B^{-1}]$ to the Pontryagin space $\mathfrak{D}[J_{\mathfrak{F}}A^{-1}]$. Consequently, $\kappa=\kappa'$. The inner product $[.,.]_{\mathfrak{G}}$ $([.,.]_{\mathfrak{F}})$ is continuous on the Hilbert space $\mathfrak{H}_{\mathfrak{G}}$ $(\mathfrak{H}_{\mathfrak{F}})$, since the topology on this space is stronger than the norm topology on the Krein space $\mathfrak{G}$ ($\mathfrak{F}$, respectively). As V is a weak isomorphism from $\mathfrak{G}$ to $\mathfrak{F}$,

$$[Vx,Vy]_{\mathfrak{F}}=[x,y]_{\mathfrak{G}},\qquad x,y\in B\mathfrak{D}(V)$$

and hence, by continuity

$$[U_0x,U_0y]_{\mathfrak{F}}=[x,y]_{\mathfrak{G}},\qquad x,y\in\mathfrak{H}_{\mathfrak{G}}.$$

To finish the proof observe that $0\notin c_s(A)$ $(0\notin c_s(B))$ if and only if $\infty\notin c_s(A^{-1})$ $(\infty\notin c_s(B^{-1})$, respectively). Now, the implication *(iv)* $\Rightarrow$ *(iii)* in Proposition 7.4 implies that the completion of $(\mathfrak{D}[J_{\mathfrak{G}}B^{-1}],[.,.]_{\mathfrak{G}})$ with respect to its unique decomposition majorant is exactly the Krein space $\mathfrak{G}$. In other words $\mathfrak{G}$ is the Krein space associated with $\mathfrak{H}_{\mathfrak{G}}$ and $[.,.]_{\mathfrak{G}}$. Analogously, $\mathfrak{F}$ is the Krein space associated with $\mathfrak{H}_{\mathfrak{F}}$ and $[.,.]_{\mathfrak{F}}$. Because $U_0\in\mathbf{L}(\mathfrak{H}_{\mathfrak{G}},\mathfrak{H}_{\mathfrak{F}})$ and is boundedly invertible, Corollary 2.3 implies that U_0 can be extended by continuity to a unitary operator $U\in\mathbf{L}(\mathfrak{G},\mathfrak{F})$. Since $B\mathfrak{D}(V)$ is dense in $\mathfrak{G}$ and since the operator V is closed, it follows that $V=U\in\mathbf{L}(\mathfrak{G},\mathfrak{F})$. This completes the proof of the theorem.

There is an alternative way to deduce the last conclusion in the proof of Theorem 7.1. To show this we first prove the following simple result.

PROPOSITION 7.5. *Let V be a weak isomorphism from the Krein space $(\mathfrak{G},[.,.]_{\mathfrak{G}})$ to the Krein space $(\mathfrak{F},[.,.]_{\mathfrak{F}})$. Suppose that the subspaces $\mathfrak{D}(V)$ and $\mathfrak{R}(V)$ are uniformly decomposable in $\mathfrak{G}$ and $\mathfrak{F}$, respectively. Furthermore, assume that the inner product space $(\mathfrak{R}(V),[.,.]_{\mathfrak{F}})$ has a unique decomposition majorant. Then V is a unitary operator from $\mathfrak{G}$ to $\mathfrak{F}$.*

Proof. Let $\mathfrak{D}(V)=\mathfrak{D}_{+}+\mathfrak{D}_{-}$ be uniform fundamental decomposition of $(\mathfrak{D}(V),[.,.]_{\mathfrak{G}})$. It is easy to see that $\mathfrak{R}(V)=V(\mathfrak{D}_{+})+V(\mathfrak{D}_{-})$ is a fundamental decomposition of $(\mathfrak{R}(V),[.,.]_{\mathfrak{F}})$. Since $\mathfrak{R}(V)$ has a unique decomposition majorant and is uniformly decomposable, this decomposition is uniform. It follows from [An] Chapter 1, that this uniform decomposition, as well as the uniform decomposition $\mathfrak{D}(V)=\mathfrak{D}_{+}+\mathfrak{D}_{-}$, can be extended to the fundamental decompositions of $\mathfrak{F}$ and $\mathfrak{G}$, respectively. Denote the corresponding fundamental symmetries by $J_{\mathfrak{F}}$ and $J_{\mathfrak{G}}$. Then $J_{\mathfrak{F}}V=VJ_{\mathfrak{G}}$ on $\mathfrak{D}(V)$ and this implies the boundedness of V. This completes the proof.

Now notice that Proposition 7.4 and the fact $\infty\notin c_s(B^{-1})$ imply that $\mathfrak{D}[J_{\mathfrak{G}}B^{-1}]=\mathfrak{D}(U_0)$ satisfies the assumptions in Proposition 7.5. Analogously, Proposition 7.4, Lemma 7.3 and the fact that $\infty\notin c_s(A^{-1})$ imply that $\mathfrak{D}[J_{\mathfrak{F}}A^{-1}]=\mathfrak{R}(U_0)$ satisfies the assumptions in Proposition 7.5. Thus, Proposition 7.5 applied to the weak isomorphism U_0 from $\mathfrak{G}$ to $\mathfrak{F}$ yields the last conclusion of the proof of Theorem 7.1.

In [ACG] it is shown that, if $H\in\mathbf{L}(\mathfrak{K})$ and the form $[(I-H^{+}H).,.]$ has κ negative squares, then the form $[(I-HH^{+}).,.]$ also has κ negative squares. This can be used in the last of the following simple consequences of Theorem 7.1. Let $\Delta=(\mathfrak{K},\mathfrak{F},\mathfrak{G};T,F,G,H)$ and $\Delta'=(\mathfrak{K}',\mathfrak{F}',\mathfrak{G}';T',F',G',H')$ be two unitary colligations.

(*i*) Assume that $\mathfrak{F}=\mathfrak{F}'$, $\mathfrak{G}=\mathfrak{G}'$, and Δ and Δ' are weakly isomorphic. If the operators $I-T^{+}T$ and $I-T'^{+}T'$ have property P_{κ} (for possibly different κ's), then Δ and Δ' are isomorphic (and all κ's are equal).

(*ii*) Assume that $\mathfrak{K}=\mathfrak{K}'$, and Δ and Δ' coincide weakly. If the operators $I-H^{+}H$, $I-HH^{+}$, $I-H'^{+}H'$ and $I-H'H'^{+}$ have property P_{κ} (for possibly different κ's), then Δ and Δ' coincide (and all κ's are equal).

REFERENCES

[ACG] Gr. Arsene, T. Constantinescu, A. Gheondea, "Lifting of operators and prescribed numbers of negative squares", Michigan Math. J., 34 (1987), 201–216.

[AI] T.Ya. Azizov, I.S. Iohvidov, "*Foundations of the theory of linear operators in spaces with indefinite metrics*", (Russian), Moscow, 1986.

[An] T. Ando, "*Linear operators on Krein spaces*", Lecture Notes, Hokkaido University, Sapporo, 1979.

[Ar] D.Z. Arov, "Passive linear stationary dynamic systems", Sibirskii Mat. Zh., 20 (1979), 211–228 (Russian) (English translation: Siberian Math. J., 20 (1979), 149–162).

[Az1] T. Ya. Azizov, "On the theory of extensions of isometric and symmetric operators in spaces with an indefinite metric", Preprint Voronesh University, (1982), deposited paper no. 3420–82 (Russian).

[Az2] T.Ya. Azizov, "Extensions of J-isometric and J-symmetric operators", Funktsional. Anal. i Prilozhen, 18 (1984), 57–58 (Russian) (English translation: Functional Anal. Appl., 18 (1984), 46–48).

[Ba] J.A. Ball, "Models for noncontractions", J. Math. Anal. Appl., 52 (1975), 235–254.

[BDS] P. Bruinsma, A. Dijksma, H.S.V. de Snoo, "Unitary dilations of contractions in Π_κ-spaces", Operator Theory: Adv. Appl., 28 (1988), 27–42.

[BGK] V.M. Brodskii, I.C. Gohberg, M.G. Krein, "On characteristic functions of invertible operators", Acta Sci. Math. (Szeged), 32 (1971), 141–164 (Russian).

[Bo] J. Bognar, *Indefinite inner product spaces*, Springer-Verlag, Berlin–Heidelberg–New York, 1974.

[Br] M.S. Brodskii, "Unitary operator colligations and their characteristic functions", Uspekhi Mat. Nauk, 33:4 (1978), 141–168 (Russian) (English translation: Russian Math. Surveys, 33:4 (1978), 159–191).

[Cl] D.N. Clark, "On models for noncontractions", Acta Sci. Math. (Szeged), 36 (1974), 5–16.

[Cu] B. Ćurgus, "On the regularity of the critical point infinity of definitizable operators", Integral Equations Operator Theory, 8 (1985), 462–488.

[Da] C. Davis, "J-unitary dilation of a general operator", Acta Sci. Math. (Szeged), 31 (1970), 75–86.

[Di] J. Dieudonné, "Quasi-hermitian operators", in Proc. Internat. Symposium Linear Spaces, Jerusalem, 1961, 115 – 122.

[DLS1] A. Dijksma, H. Langer, H.S.V. de Snoo, "Characteristic functions of unitary operator colligations in Π_κ-spaces", Operator Theory: Adv. Appl., 19 (1986), 125–194.

[DLS2] A. Dijksma, H. Langer, H.S.V. de Snoo, "Unitary colligations in Krein spaces and their role in the extension theory of isometries and symmetric linear relations in Hilbert spaces", Functional Analysis II, Proceedings Dubrovnik 1985, Lecture Notes in Mathematics, 1242 (1987), 1–42.

[DLS3] A. Dijksma, H. Langer, H.S.V. de Snoo, "Representations of holomorphic operator functions by means of resolvents of unitary or selfadjoint operators in Krein spaces", Operator Theory: Adv. Appl., 24 (1987), 123–143.

[Fu] P.A. Fuhrmann, *"Linear systems and operators in Hilbert space"*, McGraw–Hill, New York, 1981.

[G] G.M. Goluzin, *"Geometrische Funktionentheorie"*, VEB Deutscher Verlag der Wissenschaften, Berlin, 1957.

[GZ1] I.C. Gohberg, M.K. Zambickii, "Normally solvable operators in spaces with two norms", Bul. Akad. Štiince RSS Moldoven, 6 (1964), 80–84 (Russian).

[GZ2] I.C. Gohberg, M.K. Zambickii, "On the theory of linear operators in spaces with two norms", Ukrain. Mat. Ž., 18 (1966), 11–23 (Russian) (English translation: Amer. Math. Soc. Transl., (2) 85 (1969), 145–163).

[Ha] P.R. Halmos, *"A Hilbert space problem book"*, Graduate texts in Mathematics 19, Springer–Verlag, Berlin–Heidelberg–New York, 1974.

[IKL] I.S. Iohvidov, M.G. Krein, H. Langer, *Introduction to the spectral theory of operators in spaces with an indefinite metric*, Reihe: Mathematical Research 9, Akademie–Verlag, Berlin, 1982.

[Kr] M.G. Krein, "On linear completely continuous operators in functional spaces with two norms", Zbirnik Prac' Inst. Mat. Akad. Nauk. URSR 9 (1947), 104–129 (Ukrainian).

[L] H. Langer, "Spectral functions of definitizable operators in Krein spaces", Functional Analysis, Proceedings Dubrovnik 1981, Lecture Notes in Mathematics, 948 (1982), 1–46.

[La] P.D. Lax, "Symmetrizable linear transformations, Comm. Pure Appl. Math., 7 (1954), 633–647.

[LP] M.S. Livšic, V.P. Potapov, "A theorem on the multiplication of characteristic matrix functions", Dokl. Akad. Nauk. SSSR 72 (1950), 625–628 (Russian).

[Mc] B.W. McEnnis, "Purely contractive analytic functions and characteristic functions of noncontractions", Acta. Sci. Math. (Szeged), 41 (1979), 161–172.

[R] W. Rudin, *Real and complex analysis*, 2nd edition, McGraw–Hill, New York, 1974.

[Re] W.T. Reid, "Symmetrizable completely continuous linear transformations in Hilbert space", Duke Math. J., 18 (1951), 41–56.

[Sm] Yu.L. Smul'jan, "Operators with degenerate characteristic functions", Dokl. Akad. Nauk. SSSR., 93 (1953), 985–988 (Russian).

[St] A.V. Štraus, "Characteristic functions of linear operators", Izv. Akad. Nauk SSSR Ser. Mat., 24 (1960), 43–74 (Russian) (English translation: Amer. Math. Soc. Transl., (2) 40 (1964), 1–37).

[SzF] B. Sz.-Nagy, C. Foiaş, *Harmonic analysis of operators on Hilbert space*, North-Holland, Amsterdam-London, 1970.

Branko Ćurgus
Department of Mathematics
Western Washington University
Bellingham, WA 98225
U.S.A.

Aad Dijksma, Henk de Snoo
Department of Mathematics
University of Groningen
Postbox 800
9700 AV Groningen
Nederland

Heinz Langer
Sektion Mathematik
Technische Universität Dresden
Mommsenstrasse 13
8027 Dresden, D.D.R.

Operator Theory:
Advances and Applications, Vol. 41

DIFFERENTIAL OPERATORS OF FRACTIONAL ORDER AND BOUNDARY VALUE PROBLEMS IN THE COMPLEX DOMAIN

M.M. Djrbashian

0. INTRODUCTION

0.1 In the past ten years at the Institute of Mathematics of the Academy of Sciences of the Armenian SSR quite a number of nonordinary boundary-value problems for differential operators of fractional order have been stated and solved. They were reduced to expansion theorems for eigenfunctions of these problems on proper spaces of functions, set for particular systems of segments in the complex domain.

As a result, a complete theory of particular harmonic analysis was constructed for special finite systems of segments which, in particular, reduces to the classical theory of Fourier series in $L_2(0,\sigma)$ or $L_2(-\sigma,\sigma)$, i.e. for one segment $[0,\sigma)$, or for a system of two segments $[-\sigma,0]\cup[0,\sigma]$, making an angle π.

Each of these systems under consideration is biorthogonal for a proper set of segments in the complex domain. In addition, the most important property of the theory we have developed is that the above-mentioned systems generate from the boundary-value problems in the complex domain on a proper system of segments and for model differential operators of fractional order of Riemann-Liouville type.

Our report offers but a brief review of part of the basic results and methods obtained during the investigations. In addition, the author considers it more expedient to present a complete list of publications at the end of the report instead of supplying the given text with detailed references.

0.2 In the author's investigations, summarized in his monograph [1] in 1966, on the basis of the remarkable asymptotic properties of the Mittag-Leffler type functions

$$E_\rho(Z;\mu) = \sum_{k=0}^{\infty} \frac{Z^k}{\Gamma(\mu+k/\rho)} \quad (\mu > 0,\ \rho > 0) \tag{0.1}$$

a complete theory of generalized harmonic analysis for an arbitrary finite system of rays with origin at the point $z = 0$ in the complex plane was constructed. Thus a principally new analog of the classical Fourier-Plancherel operator for a system of rays was developed.

The given cycle of investigations is based on the three basic theorems of the above-mentioned theory in more general formulation, though here it would be sufficient for us to use their particular cases.

(a) Let us denote by $L\{\psi_1,\ldots,\psi_p\}$ $(p \geq 1)$ the set of rays

ℓ_k: $\arg z = \psi_k$ $(1 \leq k \leq p)$

$0 \leq \psi_1 < \psi_2 < \ldots < \psi_p < \psi_{p+1} = \psi_1 + 2\pi$

This system partitions the z plane into p angular domains with common vertix $z = 0$ and with minimum opening π/ω, where

$$\omega = \max_{1\leq k\leq p} \{\pi(\psi_{k+1}-\psi_k)^{-1}\} \tag{0.2}$$

Then the following theorem (see [1], Theorem 4.4) is valid.

THEOREM I. Let $\rho \geq \omega$, $1/2 < \mu < 1/2 + 1/\rho$ and $g(z)$, $z \in L\{\psi_1,\ldots,\psi_p\}$, be an arbitrary function from the class

$$\int_{L\{\psi_1,\ldots,\psi_p\}} |g(z)^2|z|^{2(\mu-1)} d|z| < +\infty \tag{0.3}$$

1. Defining the set of the Fourier transforms

$$f_k(x) = \frac{1}{\sqrt{2\pi\rho}} \frac{d}{dx} \int_0^{+\infty} \frac{e^{-ixr}-1}{-ir} g(re^{i\psi_k}) r^{\mu-1} dr \in L_2(-\infty,+\infty) \quad (1\leq k\leq p) \tag{0.4}$$

we obtain the inversion formula for $z = re^{i\psi} \in L\{\psi_1,\ldots,\psi_p\}$

$$g(re^{i\psi}) = \frac{r^{1-\mu}}{\sqrt{2\pi\rho}} \sum_{k=1}^{p} \frac{d}{dr} \{r^\mu \int_{-\infty}^{+\infty} E_\rho[(ix)^{1/\rho} r^{1/\rho} e^{i(\psi-\psi_k)};\mu+1](ix)^{\mu-1} f_k(x)dx\} \tag{0.5}$$

2. The Parseval-type equality is true

$$\int_{L\{\psi_1,\ \ldots\ \psi_p\}} |g(z)|^2\ |z|^{2(\mu-1)} d|z| = \rho \sum_{k=1}^{p} \int_{-\infty}^{+\infty} |f_k(x)|^2 dx \tag{0.6}$$

We note that the classical Plancherel theorem for the Fourier integral theory is a particular case of this theorem in the case of one ray $L\{0\}$, or two rays $L\{0,\pi\}$, departing from the zeros.

(b) The inversion of this theorem is the following assertion (see [I], Theorem 4.5).

THEOREM II. *Let $\rho \geq 1/2$, $1/2 < \mu < 1/2 + 1/\rho$ and $f(x)\in L_2(0,+\infty)$ be arbitrary. Then*

1. *The transforms*

$$g^{\pm}(y) = \frac{y^{1-\mu}}{\sqrt{2\pi\rho}} \frac{d}{dy} \{y^{\mu} \int_0^{+\infty} E_{\rho} [e^{\pm i\pi/2\rho} y^{1/\rho} x^{1/\rho}; \mu+1] x^{\mu-1} f(x)dx\} \tag{0.7}$$

from the class $g^{\pm}(y)\, y^{\mu-1}\in L_2(0,+\infty)$, and the inversion formula

$$f(x) = \frac{1}{\sqrt{2\pi\rho}} \{e^{-i\frac{\pi}{2}(i-\mu)} \frac{d}{dx} \int_0^{+\infty} \frac{e^{-ixy}-1}{-iy} g^{+}(y)\, y^{\mu-1}dy + $$

$$+ e^{i\frac{\pi}{2}(1-\mu)} \frac{d}{dx} \int_0^{+\infty} \frac{e^{ixy}-1}{iy} g^{-}(y)\, y^{\mu-1}dy\}, \quad x\in(0+\infty), \tag{0.8}$$

is true.

2. *We have the two-sided inequalities of a form*

$$\int_0^{+\infty} |f(x)|^2 dx \asymp \int_0^{+\infty} |g^{+}(y)|^2\, y^{2(\mu-1)}dy + \int_0^{+\infty} |g^{-}(y)|^2\, y^{2(\mu-1)}dy. \tag{0.9}$$

(c) In the above-mentioned monograph [I] on the basis of Theorems I and II the result of greater generality regarding a parametric representation of any finite order $\rho \geq 1/2$ and normal type $\leq \sigma$ was established such that the square of its module is integrable with respect to the weight $r^{\omega}(-1 < \omega < 1)$ along a given system of rays (see [I], Theorems 6.11-6.16).

Thus, the far-reaching analogs of the classical Wiener-Paley theorem were established.

To formulate the basic result (Theorem 6.13) let us introduce some necessary notations.

Let us assume that $\rho \geq 1/2$ is arbitrary, and $X \geq [2\rho]-1$ is a natural number.

Further we assume that the set of numbers $\{\nu_k\}_0^{X+1}$ is such that

$$-\pi < \nu_0 < \nu_1 < \ldots < \nu_X \le \pi < \nu_{X+1} = \nu_0 + 2\pi,$$
$$\max_{0\le k\le X} \{\nu_{k+1} - \nu_k\} = \pi/\rho \tag{0.10}$$

From the same set of numbers we obtain the sequence of pairs $(\nu_k, \nu_{k+1})_0^X$, and then retaining the mutual order of their succession we select all the pairs $(\nu_{rk}, \nu_{rk+1})_0^p$, for which

$$\nu_{rk+1} - \nu_{rk} = \pi/\rho \quad (k = 0,1,\ldots,p \le X) \tag{0.11}$$

Finally, assuming that $\omega \in (-1,1)$ and $\sigma_k \ge 0$ $(0 \le k \le p)$, let us denote in addition

$$\theta_k = \frac{1}{2}\{\nu_{rk} + \nu_{rk+1}\} \quad (0 \le k \le p) \tag{0.12}$$

Now, to the initial set of numbers (0.10) we associate the class

$$W_\sigma^\rho\ (\omega; \{\nu_k\}, \{\sigma_k\})$$

or entire functions $f(z)$ of order $\rho \ge 1/2$ and of the type $\le \sigma$, satisfying the two conditions:

$$\int_0^{+\infty} |f(te^{-i\nu_k})^2\ t^\omega dt < +\infty \quad (0 \le k \le X) \tag{0.13}$$

$$h(-\theta_k;\ f) \le \sigma_k \le \sigma \qquad (0 \le k \le p)$$

where $h(\nu;\ f)$ is an indicator of the function $f(re^{i\nu})$.

THEOREM III. <u>The class $W_\sigma^\rho(\omega; \{\nu_K\}, \{\sigma_K\})$ coincides with the set of functions $f(z)$, representable in the form</u>

$$f(z) = \sum_{k=0}^{p} \int_0^{\sigma_k} E_\rho\ (e^{i\theta_k} z\, \tau^{1/\rho}; \mu)\ \psi_k(\tau)\tau^{\mu-1} d\tau, \qquad z \in C \tag{0.14}$$

<u>where</u> $\mu = (1 + \rho + \omega)/2\rho$ <u>and</u> $\psi_k(\tau) \in L_2(0, \sigma_k)$, $(0 \le k \le p)$.

<u>Moreover, the functions</u> $\{\psi_k(\tau)\}_0^p$ <u>are unique and almost everywhere are defined from the formulas</u>

$$\frac{i}{\sqrt{2\pi\rho}}\{e^{-i\frac{\pi}{2}\mu}\phi_{r_k+1}(-\tau) - e^{i\frac{\pi}{2}\mu}\phi_{r_k}(\tau)\} = \tilde{\phi}_k(\tau)$$
$$= \begin{cases} \psi_k(\tau), & \tau \in (0, \sigma_k) \\ 0, & \tau \in (\sigma_k, +\infty) \end{cases} \quad (0 \le k \le p), \tag{0.15}$$

<u>where</u>

$$\phi_k(\tau) = \frac{1}{\sqrt{2\pi}} \frac{d}{d\tau} \int_0^{+\infty} f(e^{-i\nu_k} t^{1/\rho})\ \frac{e^{-i\tau t} - 1}{-it}\ t^{\mu-1} dt \tag{0.16}$$
$$(0 \le k \le X)$$

The three given theorems are the basis of our investigations, to be more precise, only their particular cases, when instead of the conditions (0.2) and (0.10) we have

$$\psi_{K+1} - \psi_K = \frac{\pi}{\rho}\ (1 \le k \le p) \text{ and } \nu_{k+1} - \nu_k = \frac{\pi}{\rho}\ (0 \le k \le X),$$

and only four values of the parameter ρ are considered:

$$\rho = 1/2,\ 1,\ 3/2,\ 2. \tag{0.17}$$

(d) Our succinct report is aimed to focus your attention in more general terms on the results we have achieved in constructing an original and new theory of discrete harmonic analysis in the complex domain. As a preliminary we make three important remarks.

1) Theorem I in the case of one ray $L\{0\} = (0,+\infty)$, when $p=1$, $\rho=1/2$, and the parameter $\mu\in(1/2,5/2)$, for particular values $\mu=1$ and $\mu=2$ contains the well-known Plancherel theorems on cos and sin Fourier transforms in the class $L_2(0,+\infty)$.

2) Theorem III in the same case of one ray, when again $p=1$, $\rho=1/2$ and $\mu=3/2+\omega$ $(-1 < \omega < 1)$, for the same particular values $\mu=1$ $(\omega=-1/2)$ and $\mu=2$ $(\omega=1/2)$ contain the famous Wiener-Paley theorems on parametric representation of entire functions of order 1/2 and of the type $\le \sigma$, respectively belonging to the classes

$$W_\sigma^{1/2}(\pm \tfrac{1}{2};\ 0):\ \int_0^{+\infty} |f(t)|^2\, t^{\pm\frac{1}{2}}\, dt < +\infty \tag{0.18}$$

3) Closed in $L_2(0,\sigma)$ and orthogonal on $[0,\sigma]$ the trigonometric systems

$$\left\{ \frac{2}{\sigma} \sin \frac{\pi k}{\sigma} x \right\}_1^\infty \text{ and } \left\{ \frac{1}{\sqrt{\sigma}},\ \left\{ \frac{2}{\sigma} \cos \frac{\pi k}{\sigma} x \right\}_1^\infty \right\} \tag{0.19}$$

are generated by the simplest boundary-value problems on the segment $[0,\sigma]$ for the model Sturm-Liouville operator.

In view of the given remarks we particularly recall that the above-mentioned classical theorems on the series for systems (0.19) are, in their essence, deeply discrete analogs of the Plancherel theorems on cos and sin Fourier transforms in the class $L_2(0,+\infty)$, given in the first remark.

1. A BOUNDARY-VALUE PROBLEM ON THE SEGMENT $[0,\sigma]$ AND A DISCRETE ANALOG OF THEOREM I IN THE CASE $\rho=1/2$

1.1 First we note that the following theorem has played quite an important role not only in Part 1, but also in the whole cycle of our investigations.

THEOREM 1.1. The function

$$\Xi_\sigma(z;\nu) = E_{1/2}(-\sigma^2 z;\ 1+\nu), \qquad \nu > -1, \tag{1.1}$$

is an entire function of order $\rho = 1/2$ and of the type σ, furthermore

1. Under the condition

$$0 \le \nu < 2 \tag{1.2}$$

all the zeros of $\epsilon_\sigma(z;\nu)$ are positive and simple.

2. If $\{r_k\}_1^\infty$ $(0 < r_k < r_{k+1})$ is the sequence of the zeros of the function $\epsilon_\sigma(z;\nu)$, then

a) for $0 < \nu < 1$ or $1 < \nu < 2$

$$\sqrt{r_k} = \frac{\pi}{\sigma}k + \frac{\pi}{2\sigma}(\nu-1) + 0(k^{\nu-2}), \qquad k \to +\infty \tag{1.3}$$

b) for $\nu = 0$ (resp. $\nu = 1$)

$$\sqrt{r_k} = \frac{\pi}{\sigma}k - \frac{\pi}{2\sigma} \quad (\text{resp. } \sqrt{r_k} = \frac{\pi}{\sigma}k) \quad (1 \le k < +\infty) \tag{1.4}$$

1.2 Let us give a definition and some properties of the Riemann-Liouville operator (see [I], Chapter IX).

Let $f(x) \in L(0,\sigma)$ $(0 < \sigma < +\infty)$. Then the integrals of order α $(0 < \alpha < +\infty)$ of $f(x)$ with origin $x = 0$ and endpoint $x = \sigma$, respectively, are the functions

$$D_0^{-\alpha}f(x) \equiv D^{-\alpha}f(x) \equiv \frac{1}{\Gamma(\alpha)}\int_0^x (x-t)^{\alpha-1}f(t)dt, \qquad x \in (0,\ \sigma) \tag{1.5}$$

and

$$D_\sigma^{-\alpha}f(x) \equiv \frac{1}{\Gamma(\alpha)}\int_x^\sigma (t-x)^{\alpha-1}\ f(t)dt, \qquad x \in (0,\ \sigma) \tag{1.6}$$

These functions are again from the class $L(0,\ \sigma)$, moreover, almost everywhere on $(0,\ \sigma)$

$$\lim_{\alpha\to+0} D_0^{-\alpha}f(x) = \lim_{\alpha\to+0} D_\sigma^{-\alpha}f(x) = f(x) \tag{1.7}$$

and, therefore, it is natural to put

$$D_0^{-0}f(x) = D_\sigma^{-0}f(x) = f(x), \qquad x \in (0,\ \sigma) \tag{1.8}$$

Assume again that $\alpha \in (0, +\infty)$ and the integer $p \geq 1$ is defined from the condition $p-1 < \alpha \leq p$.

If for $f(x) \in L(0, \sigma)$ almost everywhere on $(0, \sigma)$ there exist the functions

$$D_0^{\alpha} f(x) = D^{\alpha} f(x) = \frac{d^p}{dx^p} \{D^{-(p-\alpha)} f(x)\} \tag{1.9}$$

$$D_{\sigma}^{\alpha} f(x) = (-1)^p \frac{d^p}{dx^p} \{D_{\sigma}^{-(p-\alpha)} f(x)\} \tag{1.10}$$

then it is accepted to call them the derivative of order α of $f(x)$, respectively, with origin $x = 0$ and endpoint $x = \sigma$.

We note that the functions (1.9) and (1.10) for entire $\alpha = p \geq 1$ coincide with the ordinary derivative $f^{(p)}(X)$.

1.2 (a) Let $\{\gamma_j\}_0^3$ be an arbitrary set of numbers, depending on the conditions

$$0 < \gamma_j \leq 1 \quad (0 \leq j \leq 3), \quad \sum_{j=0}^{3} \gamma_j = 3 \tag{1.11}$$

For the function $y(x)$, given on $(0, \sigma)$, let us formally introduce the operators

$$L_0 y(x) = D^{-(1-\gamma_0)} y(x), \quad L_1 y(x) = D^{-(1-\gamma_1)} D^{\gamma_0} y(x), \tag{1.12}$$

$$L_2 y(x) = D^{-(1-\gamma_2)} D^{\gamma_1} D^{\gamma_0} y(x)$$

$$L_3 y(x) = Zy(x) = D^{-(1-\gamma_3)} D^{\gamma_2} D^{\gamma_1} D^{\gamma_0} y(x)$$

Further, let us consider the functions

$$y_r(x;\lambda) = y_{\mu_r}(x;\lambda) = E_{1/2}(-\lambda x^2; \mu_r)\, x^{\mu_r - 1} \tag{1.13}$$

where

$$\mu_r = \sum_{j=0}^{r} \gamma_j \quad (0 \leq r \leq 2)$$

We assume further that

$$\mu = \mu_0 \text{ for } \mu \in \Delta_0 = (\tfrac{1}{2}, 1]; \ \mu = \mu_1, \text{ for } \mu \in \Delta_1 = (1, 2]; \ \mu = \mu_2 \text{ for } \mu \in \Delta_2 = (2, \tfrac{5}{2}) \tag{1.14}$$

Then the following theorem is true.

THEOREM 1.2. <u>In the class</u> $L_2(0, \sigma)$ <u>the function</u>

$$y_\mu(x;\lambda) = E_{1/2}(-\lambda x^2;\mu)x^{\mu-1}, \qquad \mu \in \Delta_r \qquad (0 \le r \le 2) \tag{1.15}$$

<u>is a unique solution of the following Cauchy-type problem</u>

$$Zy(x) + \lambda y(x) = 0,$$

$$L_k y(x)|_{x=0}\ \delta_{k,r} = \begin{cases} 1, & k=r \\ 0, & k\neq r \end{cases} \qquad (0 \le k \le 2) \tag{1.16}$$

(b) Let us now pass to setting two boundary-value problems, recalling that the parameter ν is in the definition (1.1) of the function $\Xi_\sigma(z;\nu)$, all the zeros of which $\{\lambda_k\}_1^\infty$ are simple and lie on the ray $(0,+\infty)$.

<u>Problem I</u>. Let $\mu \in \Delta_r$ $(0 \le r < 2)$ and the parameter ν satisfies the conditions

$$\max\,(\mu - \tfrac{3}{2}, 0) < \nu < \mu - \tfrac{1}{2} \tag{1.17}$$

Find solutions from the class $L_2(0,\sigma)$ of the Cauchy problem (1.16), satisfying the boundary condition

$$D^{\mu-\nu-1}y(x)|_{x=\sigma} = 0 \tag{1.18}$$

<u>Problem II</u>. Let again $\mu \in \Delta_r$ $(0 \le r < 2)$, but the parameter ν satisfies the conditions

$$\mu - \tfrac{1}{2} < \nu < \min\,\{\mu + \tfrac{1}{2}, 2\} \tag{1.19}$$

Find these solutions from the class $L_2(0,\sigma)$ of the Cauchy problem (1.16), which satisfy the boundary condition

$$D^{1-\nu}\, D^{\mu}\, y(x)|_{x=\sigma} = 0 \tag{1.20}$$

Similarly are also set the boundary-value (I^x) and (II^x) with the replacement of μ by $\tilde{\mu} = 3 + \nu - \mu$ and of the operators D_0 by D_σ.

Using the Riemann-Liouville operators D_σ^α with endpoint at the point $x = \sigma$, the operators L_k $(0 \le k \le 2)$ and Z are defined. A theorem analogous to Theorem 1.2 on the Cauchy-type problem and two related boundary-value problems (I^x) and (II^x) is established with their help.

Omitting the corresponding definitions and formulas, let us formulate the basic theorem established here.

THEOREM 1.3. 1. <u>The boundary-value problems</u> (I) <u>and</u> (I^x) <u>have the same eigenvalues</u> $\{\lambda_k\}_1^\infty$ <u>and, respectively, the following systems of eigenfunctions</u>

$$\{y_\mu(x;\lambda_K)\}_1^\infty = \{E_{1/2}(-\lambda_K x^2;\mu)x^{\mu-1}\}_1^\infty, \tag{1.21}$$

$$\{Z_{\tilde{\mu}}(x;\lambda_k)\}_1^\infty = \{E_{1/2}(-\lambda_k(\sigma-x)^2;\tilde{\mu})(\sigma-x)^{\tilde{\mu}-1}\}_1^\infty \tag{1.22}$$

2. <u>The systems (1.21) and (1.22) are biorthogonal on $(0,\sigma)$ as regards</u>

$$\int_0^\sigma y_\mu(x;\lambda_k)Z_{\tilde{\mu}}(x;\lambda_n)dx = \begin{cases} 0, & k\neq n \\ C_n, & k=n \end{cases} \quad (k,\ n=1,2,3\ldots) \tag{1.23}$$

<u>where</u> $C_n \neq 0$ $(n \geq 1)$ <u>are definite constants</u>.

1.3 The following basic theorem, which completes the solution of our boundary-value problems (I)-(I^x), is established.

THEOREM 1.4. <u>Under the condition of the parameters (1.17) and</u> $Z_{\tilde{\mu}}(x;\lambda_k) = C_k^{-1} Z_{\tilde{\mu}}(x;\lambda_k)$, <u>any function</u> $\psi(x) \in L_2\ (0,\ \sigma)$ <u>in its metric is expanded into series of two types</u>

$$\psi(x) = \sum_{k=1}^\infty a_k\, y_\mu\ (x;\lambda_k), \qquad \psi(x) = \sum_{k=1}^\infty b_k\, Z_{\tilde{\mu}}(x;\lambda_k) \tag{1.24}$$

<u>where</u>

$$a_k = \int_0^\sigma \psi(t)\ Z_{\tilde{\mu}}(t;\lambda_k)dt, \qquad b_k = \int_0^\sigma \psi(t)\ y_\mu(t;\lambda_k)dt \quad (k=1,2,\ldots) \tag{1.25}$$

<u>Furthermore, the two-sided inequalities are valid</u>

$$||\{a_K\}_1^\infty||_{2(\mu-1)} \asymp ||\{b_k\}_1^\infty||_{2(\mu-1)} \asymp \int_0^\sigma |\psi(t)|^2 dt\ ^{1/2} \tag{1.26}$$

<u>where</u>

$$||\{d_K\}_1^\infty||_X = \left(\sum_{k=1}^\infty (1+k)^{2X}|d_k|^2\right)^{1/2} \tag{1.27}$$

For the boundary-value problems (II) and (II^x) two theorems analogous to 1.3 and 1.4 are true with the difference that they have one more eigenfunction $X^{\mu-1}/\Gamma(\mu)$ and $(\sigma-X)^{\mu-1}/\Gamma(\mu)$ each, corresponding to the eigenvalue $\lambda_0 = 0$, which is added to the systems (1.21) and (1.22), respectively.

Some particular cases of our boundary-value problems have led us to new, simpler biorthogonal systems on $[0,\sigma]$, also involving the orthogonal trigonometric systems (0.19).

2. BOUNDARY-VALUE PROBLEMS IN THE COMPLEX DOMAIN FOR OPERATOR OF RIEMANN-LIOUVILLE TYPE OF ORDER 2/3.

The boundary-value problems given in Part 1 were stated not in the complex domain, but only on the segment $[0,\sigma]$, and the theorems formulated there were, essentially, the discrete analogs of Theorem I only for the value of the parameter $\rho=1/2$. They, in particular, contained theorems on the Fourier series in the class $L_2(0,\sigma)$ for sin and cos trigonometric systems (0.19).

An analogous fact is true in the case $\rho=1$. Then the corresponding boundary-value problem for a proper differential operator of fractional order is formulated on the segment $[-\sigma,\sigma]$ of the real axis, whereas in the established expansion theorems the functions belong to the class $L_2(-\sigma,\sigma)$. Thus, the discrete analogs of Theorem I were established in the case $\rho=1$ and $1/2 < \mu < 3/2$. As a particular case, for $\mu=1$ they contain a well-known theorem on completeness and basisness in $L_2(-\sigma,\sigma)$ of the system of Fourier functions.

$$\left\{ \frac{1}{\sqrt{2\pi}} e^{i \frac{\pi x}{\sigma} k} \right\}_{-\infty}^{+\infty}$$

In view of the above said, we found it necessary to provide a detailed description of one boundary-value problem in the complex domain that has led us to the construction of the discrete analog of Theorem I in the case $\rho = 3/2$, $1/2 < \mu < 1/2 + 2/3$.

2.1 (a) Together with the Riemann-Liouville operator $D^{-\alpha}$ on the Riemann surface

$$G^{\infty} = \{z,\ |\mathrm{Arg}\ z| < +\infty,\ 0 < |z| < +\infty\}$$

of the function Lnz let us define the operator $I^{-\alpha}$. Namely, assuming that $\alpha \in (0, +\infty)$ and $\phi(z)$ is analytic on G^{∞} or in the domain $G \subset G^{\infty}$, star-shaped regarding the origin $Z = 0 \in G$, defining

$$I_{3/2}^{-\alpha}\, \phi(z) \equiv I^{-\alpha}\phi(z) = \frac{3}{2\Gamma(\alpha)} \int_0^z (z^{3/2}-\zeta^{3/2})^{\alpha-1}\, \phi(\zeta)d\zeta, \qquad z \in G \tag{2.1}$$

Furthermore, we assert the integration in (2.1) along the interval $(0, z) \subset G$. Analogous to the operator D^{α} for $\alpha \in (0, +\infty)$ the operator I^{α} may be defined for the same α values. However, we confine ourselves to the case $\alpha \in (0, 1)$, putting

$$I^{\alpha}\phi(z) = \frac{2}{3}\frac{d}{dz}\{I^{-(1-\alpha)}\phi(z)\}, \qquad z \in G \tag{2.2}$$

Let us note the simplest properties of the introduced operators; for $z \in G$

$$I^{-0}\phi(z) = z^{-1/2}\phi(z), \qquad I^{-1}\phi(z) = \frac{3}{2}\int_0^z \phi(\zeta)d\zeta = \frac{3}{2}D^{-1}\phi(z)$$

$$I^{+0}\phi(z) = \phi(z), \qquad I^{1}\phi(z) = \frac{2}{3}\frac{d}{dz}\{z^{-1/2}\phi(z)\} \tag{2.3}$$

(b) Let us now define via $L_2^* = L_2^*(G^{\infty})$ a set of functions $\psi(z)$, measurable on any interval $(0, w_0)$ $(w_0 \in G^{\infty})$ for which

$$\int_{(0, w_0)} |\psi(w)|^2 |w|^{-1/2} |dw| < +\infty \tag{2.4}$$

Now we introduce the function

$$Y_{\mu}(z;\lambda) = E_{3/2}(\lambda z;\mu)\ z^{3/2\ \mu-1} \tag{2.5}$$

and denote the proposition.

THEOREM 2.1. <u>Under the condition</u>

$$2/3 \le \mu \le 1 \tag{2.6}$$

<u>for the Cauchy-type problem</u>

$$I^{2/3}Y(w) = \frac{w^{3/2\ \mu-2}}{\Gamma(\mu-2/3)} + \lambda\, Y(w), \qquad w \in G^{\infty} \tag{2.7}$$

$$I^{-(1-\mu)}Y(w)\Big|_{w=0} = 1$$

<u>in the class</u> L_2^* <u>there is a unique solution</u>

$$Y(w) = Y_{\mu}(w;\lambda) = E_{3/2}(\lambda w;\mu)\ w^{3/2\ \mu-1} \tag{2.8}$$

(b) We introduce for consideration also the operator

$$L_{\mu}^* = I^{-(\mu-2/3)}I^{\mu} \qquad (2/3 \le \mu \le 1) \tag{2.9}$$

Then the following theorem is valid.

THEOREM 2.2. <u>If</u> $2/3 \le \mu \le 1$, <u>then together with the problem (2.7) the function</u> $Y_{\mu}(w;\lambda)$ <u>is also a unique solution in the class</u> $L_2^*(G^{\infty})$ <u>of the homogeneous Cauchy-type problem</u>

$$w^{1/2}\, L_{\mu}^*\, Y(w) = \lambda\, Y(w), \qquad w \in G^{\infty} \tag{2.10}$$

$$I^{-(1-\mu)}\, Y(w)\Big|_{w=0} = 1$$

2.2 Let

$$\nabla(\sigma) = \bigcup_{j=-1}^{1} \nabla_j(\sigma), \; \nabla_j(\sigma) = \{z=re^{i\frac{2\pi}{3}j}; \; 0 \le r \le \sigma^{2/3}\} \quad (j = -1, 0, 1) \qquad (2.11)$$

be a set of three segments uniform in length, which equals to $\sigma^{2/3}$, departing from the origin $z = 0$ at equal angles $2\pi/3$.

Further, by Theorem 1.1 an entire function $E_{1/2}(w; 1+\nu)$ of order 1/2 and of type 1 for the values of the parameter $\nu \in [0,2)$ has only simple and negative roots $\{-r_k\}_1^\infty$ $(r_k < r_{k+1})$.

Therefore, all the roots of the entire function

$$E_{1/2}(\sigma^2 z^3; 1 + \nu) \qquad (2.12)$$

are also simple and lie on three rays of the complex plane

$$L_j = \left(0, e^{i\frac{2\pi}{3}(j+\frac{1}{2})}\right) \qquad (j = -1, 0, 1)$$

After numbering the zeros of the function (2.12) separately on each ray L_j, we obtain three different sequences

$$\{\lambda_k^{(j)}\}_1^\infty \subset L_j, \qquad |\lambda_k^{(j)}| < |\lambda_{k+1}^{(j)}| \qquad (j = -1, 0, 1)$$

It is obvious that

$$\lambda_k^{(j)} = e^{i\frac{2\pi}{3}(j+\frac{1}{2})} \sqrt[3]{r_k} \qquad (k = 1, 2, \ldots; \; j = -1, 0, 1)$$

Let us introduce, finally, the unified numbering for the entire set of the zeros of our function (2.12), putting

$$\lambda_{3k-2} = \lambda_k^{(-1)}, \; \lambda_{3k-1} = \lambda_k^{(0)}, \; \lambda_{3k} = \lambda_k^{(1)} \qquad (k = 1, 2, \ldots)$$

Then the sequence of the complex numbers $\{\lambda_k\}_1^\infty$ coincides with the set of all zeros of the function (2.12).

(c) Now let us pass to formulating two boundary-value problems for the set of segments $\nabla(\sigma)$ in the complex domain.

Problem A. Find such values of the parameter λ, for which in the class $L_2^*(G^\infty)$ on the set of segments $\nabla(\sigma)$ there exists a non-trivial solution to the Cauchy-type problem (2.10)

$$z^{1/2} I^{-(\mu-2/3)} I^{\mu} Y(z) = \lambda Y(z), \qquad z \in G^\infty$$
$$I^{-(1-\mu)} Y(z)|_{z=0} = 1,$$

such as that at the endpoints $\nabla(\sigma)$: $a_j = \sigma^{2/3} e^{i\frac{2\pi}{3}j}$ $(j = -1, 0, 1)$,

these solutions satisfy the boundary condition of the form

$$X(\lambda,\mu,\nu) = \sum_{j=-1}^{1} e^{-i\pi\nu j} \{I^{(\nu-\mu+1/3)}Y(z)\Big|_{z=a_j}\} = 0 \tag{2.13}$$

<u>Problem $\tilde{A}$</u>. Find the values of the parameter λ, for which in the class $L_2^*(G^\infty)$ on $\nabla(\sigma)$ there exists a non-trivial solution of the same Cauchy-type problem (2.10), which satisfies the boundary condition of the form

$$\tilde{X}(\lambda,\mu,\nu) = \sum_{j=-1}^{1} e^{-i\pi\nu j} e^{i\frac{2\pi}{3}j} \{I^{(\nu-\mu-1/3)}Y(z)\Big|_{z=a_j}\} = 0 \tag{2.14}$$

Adding to the sequence $\{\lambda_k\}_1^\infty$ of the zeros of the function (2.12) the number $\lambda_0 = 0$, we arrive at the solubility theorem of both boundary-value problems in their explicit form.

THEOREM 2.3. 1. <u>If</u>

$$2/3 \le \mu \le 1, \qquad \nu \in \Delta = (\mu - \frac{1}{6}, \mu + \frac{1}{6}),$$

<u>then the boundary problem A has a countable number of eigenvalues</u> $\{\lambda_k\}_0^\infty$ <u>and the corresponding eigenfunctions</u>

$$E_n(z) = E_{3/2}(\lambda_n z;\mu)\, z^{3/2\,\mu-1} \in L_2^*(G^\infty) \qquad (n = 0, 1, 2, \ldots) \tag{2.15}$$

2. <u>The system (2.15) is complete in the class</u> $L_2^{-1/2}\{\nabla(\sigma)\}$ <u>of the functions</u> $X(z)$, $z \in \nabla(\sigma)$, <u>depending on the condition</u>

$$\int_{\nabla(\sigma)} |X(z)|^2 \, |z|^{-1/2} \, |dz| < +\infty \tag{2.16}$$

THEOREM 2.4. 1. <u>If</u>

$$2/3 \le \mu \le 1, \qquad \nu \in \tilde{\Delta} = (\mu + \frac{1}{6}, \mu + \frac{1}{2}),$$

<u>then the boundary-value problem</u> $(\tilde{A})$ <u>has the same eigenvalues</u> $\{\lambda_k\}_0^\infty$ <u>and the corresponding eigenfunctions</u> $\{E_n(z)\}_0^\infty$. <u>Except the function</u> $E_0(z)$ <u>for the eigenvalue</u> $\lambda_0 = 0$ <u>(two-multiple root</u> $\tilde{X}(\lambda,\mu,\nu)$) <u>one more associated function is added</u>

$$E_0^*(z) = \frac{\partial}{\partial\lambda} E_{3/2}(\lambda z;\mu) z^{3/2\,\mu-1}\Big|_{\lambda=0} = \frac{z^{3/2\,\mu}}{\Gamma(\mu+2/3)} \tag{2.17}$$

2. <u>The system</u>

$$\{E_0^*(z),\ [E_n(z)]_0^\infty\} \tag{2.18}$$

<u>of eigenfunctions and associated functions of the problem</u> $(\tilde{A})$ <u>is complete in the class</u> (2.16) <u>of the functions</u> $X(z)$.

2.3 We complete this paragraph by formulating the basic theorems on the basisness of the systems $\{E_n(z)\}_0^\infty$ and $\{E_0^*(z), [E_n(z)]_0^\infty\}$ in the class of the functions $L_2^{-1/2}\{\nabla(\sigma)\}$.

a. First, we have to formulate a theorem on the systems, biorthogonal on the system of segments $\nabla(\sigma)$ in the complex domain.

1. For $\nu \in \Delta = (\mu - \frac{1}{6}, \mu + \frac{1}{6})$, $2/3 \le \mu \le 1$, let us introduce a pair of systems

$$\{E_n(z)\}_0^\infty \text{ and } \{F_n(z)\}_0^\infty, \quad z \in \nabla(\sigma), \tag{2.19}$$

putting for $Z = r\,\alpha^j \in \nabla(\sigma)$ $(j = -1, 0, 1)$ and $\alpha = e^{i\frac{2\pi}{3}}$

$$\begin{aligned} E_n(z) &= E_{3/2}(\lambda_n z;\mu)\, z^{3/2\,\mu-1} \\ F_n(z) &= \frac{3}{2}\alpha^{3/2\,\mu-1}\,\omega_{n,j}(r^{3/2})\, r^{-1/2} \end{aligned} \qquad (n = 0, 1, 2, \dots) \tag{2.20}$$

where

$$\omega_{0,j}(\tau) = \frac{\sigma^{-\nu}\,\Gamma(1+\nu)}{3\,\Gamma(1+\nu-\mu)}(\sigma - \tau)^{\nu-\mu},$$

$$\omega_{n,j}(\tau) = \frac{\sigma^{-\nu}\,\Gamma(1+\nu)}{3\,\Gamma(1+\nu-\mu)} E_{3/2}(\alpha^{-j}\lambda_n(\sigma-\tau)^{2/3};\, 1+\nu-\mu)\,(\sigma-\tau)^{\nu-\mu} \quad (n = 1, 2, \dots) \tag{2.21}$$

2. For $\nu \in \tilde{\Delta} = (\mu + 1/6, \mu + 1/2)$, $2/3 \le \mu \le 1$, we introduce another pair of systems

$$\{\tilde{E}_0^*(z), [\tilde{E}_n(z)]_0^\infty\} \text{ and } \{\tilde{F}_0^*(z), [\tilde{F}_n(z)]_0^\infty\}, \qquad z \in \nabla(\sigma), \tag{2.22}$$

putting for $z = r\alpha^j \in \nabla(\sigma)$

$$\begin{aligned} \tilde{E}_0^*(z) &= \frac{z^{3/2\,\mu}}{\Gamma(\mu+2/3)}, \\ \tilde{F}_0^*(z) &= \frac{3}{2}\alpha^{(3/2\mu-1)j}\,\tilde{\omega}_{0,j}^*(r^{3/2})\, r^{-1/2}, \qquad z \in \nabla_j(\sigma), \end{aligned} \tag{2.23}$$

and then

$$\tilde{E}_n(z) = E_n(z), \qquad \tilde{F}_n(z) = F_n(z) \tag{2.24}$$

where

$$\tilde{\omega}_{0,j}^*(\tau) = \alpha^j \frac{\sigma^{-\nu}\,\Gamma(1+\nu)}{3\,\Gamma(1/3+\nu-\mu)}(\sigma-\tau)^{\Upsilon-\mu-2/3} \quad (j = -1, 0, 1) \tag{2.25}$$

Finally, let us define on $\nabla(\sigma)$ the set $L_2^{(1/2)}[\nabla(\sigma)]$ of functions $\psi(z)$, for which

$$||\psi;\nabla(\sigma)|| = \int_{\nabla(\sigma)} |\psi(z)|^2 |z|^{1/2} |dz| ^{1/2} < +\infty \tag{2.26}$$

Furthermore, we define, as usual, for the pair of functions $\psi_k(z) \in L_2^{(1/2)}[\nabla(\sigma)]$ $(k = 1, 2)$ their scalar product, putting

$$[\psi_1, \psi_2] = \int_{\nabla(\sigma)} \psi_1(z) \overline{\psi_2(z)} |z|^{1/2} |dz| \tag{2.27}$$

The following theorem is valid.

THEOREM 2.5. 1. <u>For</u> $\nu \in \Delta$ <u>the systems of functions</u> (2.19) <u>are biorthogonal on</u> $\nabla(\sigma)$ <u>in the metric</u> (2.27), <u>i.e.</u>

$$[E_n, F_m] = \delta_{n,m} \quad (n, m = 0, 1, 2, \ldots) \tag{2.28}$$

2. <u>For</u> $\nu \in \tilde{\Delta}$ <u>the systems of functions</u> (2.22) <u>are biorthogonal in the metric</u> (2.27), <u>i.e.</u>

$$\begin{aligned} &[\tilde{E}_0^*, \tilde{F}_0^*] = 1, \ [\tilde{E}_0^*, \tilde{F}_m] = 0 \\ &[\tilde{E}_n, \tilde{F}_0^*] = 0, \ [\tilde{E}_n, \tilde{F}_m] = \delta_{n,m} \end{aligned} \qquad (n, m = 0, 1, 2, \ldots) \tag{2.29}$$

(b) With the above introduced systems for further formulation it is expedient to consider such pairs of systems.

1. For $\nu \in \Delta$, $2/3 \le \mu \le 1$, the systems

$$\{e_n(z)\}_0^\infty \text{ and } \{T_n(z)\}_0^\infty, \quad z \in \nabla(\sigma), \tag{2.30}$$

where

$$e_n(z) = (1+n)^{\mu-1} E_n(z), \qquad T_n(z) = (1+n)^{1-\mu} F_n(z) \tag{2.31}$$

2. For $\nu \in \tilde{\Delta}$, $2/3 \le \mu \le 1$ the systems

$$\{\tilde{e}_0^*(z), [\tilde{e}_n(z)]_0^\infty\} \text{ and } \{\tilde{T}_0^*(z), [\tilde{T}_n(z)]_0^\infty\} \tag{2.32}$$

where

$$\begin{aligned} &\tilde{e}_0^*(z) = \tilde{E}_0^*(z), \qquad \tilde{T}_0^*(z) = \tilde{F}_0^*(z), \\ &\tilde{e}_n(z) = (1+n)^{\mu-1} \tilde{E}_n(z), \qquad \tilde{T}_n(z) = (1+n)^{1-\mu} \tilde{F}_n(z) \end{aligned} \tag{2.33}$$

The following basic theorem is established.

THEOREM 2.6. 1. <u>Let</u> $\nu \in \Delta = (\mu - \frac{1}{6}, \mu + \frac{1}{6})$, $2/3 \le \mu \le 1$ <u>and</u> $\{E_n(z)\}_0^\infty$, $z \in \nabla(\sigma)$, <u>be the systems of eigenfunctions of the boundary-value</u>

problem (A).

Then after suitable normalization according to the formulas (2.31) the functional systems (2.30), biorthogonal on $\nabla(\sigma)$, form a Riesz basis in the space of functions

$$L_2^{(-1/2)}[\nabla(\sigma)]: \int_{\nabla(\sigma)} |\phi(z)|^2 \ |z|^{-1/2} \ |dz| < +\infty$$

2. Let $\nu \in \tilde{\Delta} = (\mu + \frac{1}{6}, \mu + \frac{1}{2})$, $2/3 \le \mu \le 1$ and

$$\{\tilde{E}_0^*(z) \ , \ [\tilde{E}_n(z)]_0^\infty \}$$

be a system of eigenfunctions and an associated function of the problem $(\tilde{A})$. Then the systems of the functions (2.32), biorthogonal on $\nabla(\sigma)$, form a Riesz basis in the same space $L_2^{(-1/2)} [\nabla(\sigma)]$.

3. A BOUNDARY-VALUE PROBLEM ON THE FINITE CROSS IN THE COMPLEX DOMAIN AND A RIESZ BASIS GENERATED BY IT.

3.1 Let us proceed to formulating the Cauchy-type problem and a boundary-value problem for the cross of the type

$$e_+(\surd\sigma) = [-\sigma^{1/2}, \sigma^{1/2}] \cup [-i\sigma^{1/2}, i\sigma^{1/2}] \tag{3.1}$$

in the complex domain. As a result, we arrive at the systems of eigenfunctions for a special differential operator of order 1/2. These systems form a Riesz basis in the class L_2 for the functions on the cross $e_+(\surd\sigma)$ and are deep analogs of the Fourier system $\left\{\frac{1}{\surd 2\pi} e^{i\frac{\pi x}{\sigma}k}\right\}_{-\infty}^{+\infty}$ for the class $L_2(-\sigma, \sigma)$.

(a) For any function F(z), given on the Riemann surface of the function Lnz, we introduce the following modification of the Riemann-Liouville integral

$$D^{-\alpha} F(z) = \frac{2}{\Gamma(\alpha)} \int_0^z (z^2 - \zeta^2)^{\alpha-1} F(\zeta)d\zeta, \qquad z \in G^\infty \tag{3.2}$$

where $\alpha \in (0, +\infty)$ is arbitrary.

Let us also define the analog of the differential operator D^α, putting for the entire $p \ge 1$ and $p-1 < \alpha \le p$.

$$D^{\alpha}F(z) = \frac{d^P}{d\zeta^P} D^{-(p-\alpha)} [\zeta^{-1/2} F(\zeta^{1/2})] \Big|_{\zeta=z^2} \tag{3.3}$$

For $0 < \alpha < 1$ we obtain the representation

$$D^{\alpha}F(z) = (2z)^{-1} \frac{d}{dz} D^{-(1-\alpha)} F(z) \tag{3.4}$$

and formulas

$$D^{-0}F(z) = D^{+0}F(z) = z^{-1} F(z)$$

Finally, on G^{∞} we introduce for consideration the function

$$\Xi_{\mu}(z;\ \lambda) = E_2(\lambda z;\ \mu)z^{2\mu-1} \tag{3.5}$$

and the operator[x]

$$\tilde{L}_{\mu}\ \psi(z) \equiv D^{1/2}\ \psi(z) - \frac{z^{2\mu-3}}{\Gamma(\mu-1/2)} \tag{3.6}$$

assuming always that the parameter $\mu \in (1/2,\ 1)$.

THEOREM 3.1. <u>Under the condition</u> $1/2 < \mu < 1$ <u>the Cauchy-type problem</u>

$$\tilde{L}_{\mu}\ \psi(z) = \lambda\ z^{-1}\ \psi(z), \qquad z \in G^{\infty}, \tag{3.7}$$

$$D^{(-1-\mu)}\ \psi(z)\Big|_{z=0} = 1$$

<u>in the class</u> $L_2^{-1}(G^{\infty})$ <u>has a unique solution</u>

$$\psi(z) \equiv \Xi_{\mu}(z;\ \lambda) = E_2(\lambda z;\ \mu)z^{2\mu-1} \tag{3.8}$$

(b) Prior to setting a boundary-value problem, we note an important identity

$$\phi_{\mu,\nu}(\lambda) = \sum_{k=-1}^{2} e^{-i\pi\nu k} \{D^{-(\nu-\mu)}\ \Xi_{\mu}\ (z;\ \lambda)\}_{z=\sqrt{\sigma}\, e^{i\frac{\pi}{2}k}}$$

$$= 4\sigma\lambda^2\ E_{1/2}(\sigma^2\lambda^4;\ 1+\nu) \tag{3.9}$$

which contains a peculiar linear combination, depending on the solutions of the Cauchy-type problem (3.7) at the endpoints on the cross $e_+\{\sqrt{\sigma}\}$ under consideration.

Now let us assume that

$$1/2 < \mu < 1,\ \mu < \nu < \mu + 1/2 \tag{3.10}$$

and λ is an arbitrary complex parameter.

---

x The operator $\tilde{L}_{\mu}$ may also assume the homogeneous form.

Let us note, finally, that by Theorem 1 all the zeros of the function $\phi_{\mu,\nu}(\lambda)$, except for the two-multiple zero $\tilde{\lambda}_0 = 0$, are simple and lie on the rays

$$\arg \lambda = \pm \frac{\pi}{4} \text{ and } \arg \lambda = \pm \frac{3\pi}{4}$$

Numbering them as to the non-increase of their modules and arguments, we arrive at the sequence $\{\tilde{\lambda}_j\}_{-\infty}^{+\infty}$, symmetrical in relation to the origin $\lambda_0=0$.

THEOREM 3.2. 1. <u>All eigenvalues of the boundary-value problem</u>

$$\tilde{L}_\mu \psi(z) = \lambda z^{-1} \psi(z), \qquad z \in e_+\{\surd\sigma\} \tag{3.7}$$

$$D^{-(1-\mu)} \psi(z)\Big|_{z=0} = 1$$

$$\phi_{\mu,\nu}(\lambda) = 4\sigma\lambda^2 E_{1/2} (\sigma^2\lambda^4;\ 1+\nu) = 0 \tag{3.11}$$

<u>coincide with the sequence of complex numbers</u> $\{\tilde{\lambda}_j\}_{-\infty}^{+\infty}$.

2. <u>The problem</u> (3.7)-(3.9) <u>has a countable set of eigenfunctions</u>

$$\{E_2(\tilde{\lambda}_n z;\ \mu)\ z^{2\mu-1}\}_{-\infty}^{+\infty}, \qquad z \in e_+\{\surd\sigma\}, \tag{3.12}$$

$$\left\{ \frac{\partial}{\partial\lambda} E_2 \left[e^{i\frac{\pi}{4}} \lambda z;\ \mu\right] z^{2\mu-1} \right\}_{\lambda=\lambda_0=0} = \frac{e^{i\frac{\pi}{4}} \cdot z^{2\mu}}{\Gamma(\mu+1/2)} \tag{3.12}$$

(c) Except for the system

$$\{E_0^*(z),\ [E_n(z)]_{-\infty}^{+\infty}\} \tag{3.13}$$

the existence of the second system

$$\{F_0^*(z),\ [F_n(z)]_{-\infty}^{+\infty}\} \in L_2\{e_+\{\surd\sigma\}\} \tag{3.14}$$

biorthogonal on the entire cross $e_+\{\surd\sigma\}$ of the z plane with the system (3.13), is proved.

The basic theorem of the given boundary-value problem is as follows.

THEOREM 3.3. <u>Let</u> $\phi(z)$ <u>be any function of the class</u>

$$L_2^*[e_+\{\surd\sigma\}]:\ ||\phi||_* = \left(\int_{e_+\{\surd\sigma\}} |\phi(z)|^2\ |z|^{-1}\ |dz| \right)^{1/2} < +\infty \tag{3.15}$$

<u>Then the following assertions are valid:</u>

1. <u>In the accepted metric</u> (3.15) <u>the expansion is true</u>

$$\phi(z) = C_0^* E_0^*(z) + \sum_{n=-\infty}^{+\infty} C_n E_n(z), \qquad z \in e_+\{\surd\sigma\} \tag{3.16}$$

$$C_0^* = \int_{e_+\{\surd\sigma\}} \phi(z)\, \overline{F_0^*(\zeta)}\, |d\zeta|, \quad C_n = \int_{e_+\{\surd\sigma\}} \phi(\zeta)\, \overline{F_n(\zeta)}\, |d\zeta| \tag{3.17}$$

$(n = 0, \pm1, \pm2, \ldots)$

2. <u>The two-sided inequalities are valid</u>

$$||\phi||_* \asymp \{|C_0^*|^2 + \sum_{k=-\infty}^{+\infty} (1+|k|)^{2(1-\mu)} |C_k|^2 \}^{1/2} \tag{3.18}$$

Thus the results of this paragraph represent far-reaching analogs of the well-known approach, by means of which the ortho-normal Fourier system complete in $L_2(-\sigma, \sigma)$ in the complex form is generated

$$\left\{ \frac{1}{\sqrt{2\pi}} e^{i \frac{\pi x}{\sigma} k} \right\}_{-\infty}^{+\infty} \tag{3.19}$$

This system, however, is a solution of the segment $[-\sigma, \sigma]$ of the following simplest boundary-value problem

$$y'(x) = \lambda\, y(x), \qquad y(-\sigma) - y(\sigma) = 0 \tag{3.20}$$

REFERENCES

1. M.M. Djrbashian. Integral Transforms and Representations of Functions in the Complex Domain M., Nauka, 1966.

2. M.M. Djrbashian, S.G. Raphaelian. On Entire Functions of Exponential Type in Weight Classes L_p; Dokl. Akad. Nauk Armenian SSR. 53, No. 1 (1981), 29-36.

3. M.M. Djrbashian. Basisness of Biorthogonal Systems Generated by Boundary-value Problems for Differential Operators of Fractional Order. Dokl. Akad. Nauk USSR. 264, No. 5 (1981).

4. S.G. Raphaelian. On Basisness of Some Systems of Entire Functions. Dokl. Akad. Nauk Armenian SSR. 50, No. 4 (1980), 198-204.

5. M.M. Djrbashian. Interpolating and Spectral Expansions Associated with Differential Operators of Fractional Order; Izv. Armenian Acad. Sci. 19, No. 2 (1984), 81-181.

6. M.M. Dzrbashjan. Interpolation and Spectral Expansions Associated with Differential Operators of Fractional Order; Alfred Haar Memorial Conference, Budapest (Hungary), 1985, 303-322.

7. M.M. Djrbashian, S.G. Raphaelian. Interpolation Theorems and Expansions with Respect to Fourier Type Systems; Journ. of Approximation Theory; 50, No. 4 (1987), 297-329.

8. M.M. Djrbashian, S.G. Raphaelian. Special Boundary-value Problems Generating the Fourier-type Systems. Izv. Armenian Acad. Sci. Series Mathematica. 23, No. 2 (1988).

9. M.M. Djrbashian, S.G. Raphaelian. Interpolating Expansions in Classes of Entire Functions and Riesz Bases Generated by Them. Izv. Armenian Acad. Sci. Series Mathematica. 22, No. 1 (1987), 23-63.

10. M.M. Djrbashian. On a Boundary-value Problem in the Complex Domain. Izv. Armenian Acad. Sci. Series Mathematica. 22, No. 6 (1987), 523-542.

11. M.M. Djrbashian. On Basisness of Systems of Eigenfunctions Generated by One Boundary-value Problem in the Complex Domain. Izv. Armenian Acad. Sci. Series Mathematica (In Print).

Note: The reports, published in the journal Mathematica Izv. Armenian Acad. Sci., are being translated and published in English in the Soviet Journal of Contemporary Mathematical Analysis (Allerton Press Inc.). These reports are published in the corresponding volumes of the above journal.

M.M. Djrbashian
Institute of Mathematics
Armenian Academy of Sciences
Yerevan
USSR

Operator Theory:
Advances and Applications, Vol. 41

ON REPRODUCING KERNEL SPACES, J UNITARY MATRIX FUNCTIONS, INTERPOLATION AND DISPLACEMENT RANK

Harry Dym*

To Israel Gohberg: teacher, colleague and valued friend, with admiration and affection, on being sixty years young.

In this paper structured reproducing kernel Hilbert spaces are used to solve matrix versions of a number of classical interpolation problems. Enroute a characterization of a class of such spaces which originates with de Branges is reformulated in terms of matrix equations of the Liapunov and Stein type in the finite dimensional case. Some generalizations to indefinite inner product spaces are also formulated. Finally it is shown that every invertible Hermitean matrix with displacement rank m is the Gram matrix of a "chain" of $m \times 1$ vector valued functions in a suitably defined reproducing kernel Pontryagin space.

CONTENTS

* The author would like to acknowledge with thanks Renee and Jay Weiss for endowing the chair which supported this research.

1. INTRODUCTION

In this article we shall use two special classes of reproducing kernel Hilbert spaces (which originate in the work of de Branges [dB] and de Branges-Rovnyak [dBR1], respectively) to solve matrix versions of a number of classical interpolation problems. Enroute we shall reinterpret de Branges' characterization of the first of these spaces, when it is finite dimensional, in terms of matrix equations of the Liapunov and Stein type and shall subsequently draw some general conclusions on rational $m \times m$ matrix valued functions which are "J unitary" a.e. on either the circle or the line. We shall also make some connections with the notation of displacement rank which has been introduced and extensively studied by Kailath and a number of his colleagues as well as the one used by Heinig and Rost [HR].

The first of the two classes of spaces alluded to above is distinguished by a reproducing kernel of the special form

$$K_\omega(\lambda) = \frac{J - U(\lambda)JU(\omega)^*}{\rho_\omega(\lambda)} , \tag{1.1}$$

in which J is a constant $m \times m$ signature matrix and U is an $m \times m$ J inner matrix valued function over Δ_+, where Δ_+ is equal to either the open unit disc $\mathbb{D}$ or the open upper half plane $\mathbb{C}_+$ and $\rho_\omega(\lambda)$ is defined in the table below.

The second class of spaces is distinguished by reproducing kernels of the special form

$$\Lambda_\omega(\lambda) = \frac{I_p - S(\lambda)S(\omega)^*}{\rho_\omega(\lambda)} , \tag{1.2}$$

where S is a $p \times q$ matrix valued function which is contractive and analytic in Δ_+ and $\rho_\omega(\lambda)$ is defined in the table below.

To clarify the preceding statements, let us recall that a signature matrix is a constant $m \times m$ matrix which is both selfadjoint and unitary (with respect to the standard inner product in $\mathbb{C}^m$). The eigenvalues of a signature matrix J are thus both real and of modulus one and therefore J is unitarily equivalent to

$$J_{pq} = \begin{bmatrix} I_p & 0 \\ 0 & -I_q \end{bmatrix} , \qquad p + q = m . \tag{1.3}$$

An $m \times m$ matrix valued function U is said to be J inner over Δ_+ if it is meromorphic in Δ_+ and

$$U(\lambda)JU(\lambda)^* \leq J \tag{1.4}$$

at every point λ of analyticity of U in Δ_+ with equality (of its nontangential limits) a.e. on the boundary (i.e., $\mathbb{T}$ or $\mathbb{R}$). The better known class of matrix valued inner functions corresponds to the special choice $J = I_m$. In this instance (1.4) forces U to be analytic in Δ_+. In the general case U may have poles in Δ_+, as the example

$$U(\lambda) = \begin{bmatrix} 1 & 0 \\ 0 & 1/\lambda \end{bmatrix},$$

which is J_{11} inner over $\mathbb{D}$, clearly indicates.

We shall refer to reproducing kernel Hilbert space with reproducing kernels of the form (1.1) [resp. (1.2)] as $\mathcal{H}(U)$ [resp. $\mathcal{H}(S)$] spaces. Actually $\mathcal{H}(U)$ spaces are meaningful for U's which are meromorphic and J contractive in Δ_+. However, we shall restrict ourselves here to the smaller class of J inner U.

The paper itself is organized as follows: In Section 2 we first recall some definitions and some general elementary properties of reproducing kernel Hilbert spaces and their kernels, and present a number of examples. We then focus our attention on $\mathcal{H}(U)$ spaces for J inner U. These, as de Branges [dB] was the first to show (for $\Delta_+ = \mathbb{C}_+$) are reproducing kernel Hilbert spaces of $m \times 1$ vector valued meromorphic functions in Δ_+ which are characterized by two extra conditions: "R_α invariance" and a special identity; see Theorem 2.1. The role of these two conditions (which seemed somewhat mysterious in the early days) are fully explained in the remainder of Section 2 and Sections 3 and 4, for finite dimensional spaces. It turns out that R_α invariance determines the structure of the elements in the space: they must be expressible in terms of certain "chains", whereas the identity determines the inner product. In particular it is shown that the requisite identity is met if and only if the Gram matrix of an appropriately chosen basis for the space satisfies a certain matrix equation. The verification involves a series of detailed calculations which occupy Sections 3 and 4 and which can be skipped without too much loss of continuity. General conclusions and a reformulation of the basic Theorem 2.1 in the even more general setting of finite dimensional indefinite inner product spaces are then presented in Section 5. There are a number of points of contact here with the work of Ball and Ran [BR], though the methods are entirely different. Section 5 also includes some general conclusions on rational $m \times m$ matrix valued functions which are J unitary a.e. on either the circle or the line. Section 6 is a short diversion on J inner functions which assembles together a number of their properties which prove useful in the analysis of the interpolation problems which are treated herein. The latter are formulated and solved in Sections 7-9, which treat in turn left interpolation, right interpolation and two-sided interpolation.

The basic theme in Sections 7–9 is to show first that the stated interpolation problem is solvable if and only if a certain matrix $\mathbb{P}$ which is based on the data of the problem is positive semidefinite and then, secondly, to identify the set of all solutions in terms of a linear fractional transformation based on an associated J inner function U when $\mathbb{P} > 0$. The methods rest on the theory of $\mathcal{H}(U)$ spaces. In Section 9, $\mathcal{H}(S)$ spaces are also used heavily. These methods can also be adapted to identify the solutions to these interpolation problems when $\mathbb{P} \geq 0$ and singular. However, we shall discuss that elsewhere. Some indications may be found in Section 7 of [D], which treats the case of one chain. Section 10 is devoted to a more detailed analysis of the two sided interpolation problem with just two chains. This has the advantage of keeping the notation under control and at the same time it includes all the essential ideas.

Finally, in Section 11, we show that every positive definite matrix with displacement rank equal to m can be interpreted as the Gram matrix of a chain in an allied $\mathcal{H}(U)$ space based on an $m \times m$ J inner function U, and some generalizations thereof. This serves to explain a number of the properties of this class of matrices which have been established earlier by purely algebraic methods.

To sum up: This article is largely devoted to the thesis that reproducing kernel Hilbert spaces of the type alluded to above are a natural and effective tool for dealing with a wide class of matrix interpolation problems. This theme is developed at greater length in [D], to which this article may serve both as an introduction and a partial extension. It is only fair to emphasize that there are many other approaches to interpolation, of which perhaps the most far-reaching to date is that of Ball and Helton. For an introduction to their methods, and additional sources, the expositions [H] and [B2] are suggested.

The specific interpolation problems which are solved here are treated by other methods in the paper [BR] by Ball and Ran. They make heavy use of the Ball-Helton machinery, in juxtaposition with the theory of the zero pole structure of rational matrix valued functions which is developed in [GKLR]. A friendlier exposition of [BR] is promised in a forthcoming monograph by Ball, Gohberg and Rodman [BGR3], portions of which have appeared in [BGR1] and [BGR2].

Finally some words on notation: The symbols $\mathbb{C}$, $\mathbb{C}^{j\times k}$ and $\mathbb{R}$ will denote the complex numbers, the space of complex $j \times k$ matrices and the real numbers, respectively, whereas $\mathbb{C}^j$ is short for $\mathbb{C}^{j\times 1}$; $\mathbb{C}_+$ [resp. $\mathbb{C}_-$] stands for the open upper [resp. lower] half plane, $\mathbb{T}$ for the unit circle, $\mathbb{D}$ for its interior: $\mathbb{D} = \{\lambda \in \mathbb{C} : |\lambda| < 1\}$ and $\mathbb{E}$ for its exterior with respect to the extended complex plane $\hat{\mathbb{C}} = \mathbb{C} \cup \{\infty\} : \mathbb{E} = \{\lambda \in$

$\hat{\mathbb{C}} : 1 < |\lambda| \leq \infty\}$.

If A is a set, then $\bar{A}$ denotes its closure. If A is a matrix or operator, then A^* stands for its adjoint with respect to the standard inner product unless indicated otherwise.

The symbol R_α denotes the generalized backwards shift:

$$(R_\alpha f)(\lambda) = \frac{f(\lambda) - f(\alpha)}{\lambda - \alpha} .$$

Just as in [AD] and [D], we shall adopt a flexible notation which permits us to treat problems in $\mathbb{D}$ and problems in $\mathbb{C}_+$ more or less simultaneously by using Δ_+ to denote either $\mathbb{D}$ or $\mathbb{C}_+$ and then invoking the following table:

Δ_+	$\mathbb{D}$	$\mathbb{C}_+$
$\rho_\omega(\lambda)$	$1 - \lambda\omega^*$	$-2\pi i\ (\lambda - \omega^*)$
∂	$\mathbb{T}$	$\mathbb{R}$
$< f, g >$	$\frac{1}{2\pi}\int_0^{2\pi} g(e^{i\theta})^* f(e^{i\theta}) d\theta$	$\int_{-\infty}^{\infty} g(\lambda)^* f(\lambda) d\lambda$
$\bar{\Delta}_+$	$\bar{\mathbb{D}}$	$\mathbb{C}_+ \cup \mathbb{R}$
Δ_-	$\mathbb{E}$	$\mathbb{C}_-$

We shall also deal with the following spaces:

$\mathcal{S}_{p\times q}(\Delta_+)$ denotes the set of $p \times q$ matrix valued functions $S(\lambda)$ which are analytic and contractive (i.e., $S(\lambda)^* S(\lambda) \leq I_q$) in Δ_+.

$H^p(\Delta_+)$ denotes the space of scalar valued Hardy functions of class p over Δ_+, $1 \leq p \leq \infty$.

$H^p_{k\times j}(\Delta_+)$ denotes the space of $k \times j$ matrix valued functions each entry of which belongs to $H^p(\Delta_+)$.

$H^p_k(\Delta_+)$ is short for $H^p_{k\times 1}(\Delta_+)$.

The symbol $\underline{\underline{p}}$ is used to denote the orthogonal projection of the space $L^2_k(\partial)$ of square summable $k \times 1$ vector valued functions on ∂ onto $H^2_k(\Delta_+)$ and $\underline{\underline{q}}' = I - \underline{\underline{p}}$.

2. REPRODUCING KERNELS

A Hilbert space $\mathcal{H}$ of $m \times 1$ vector valued functions which are analytic in some open nonempty subset Ω of $\mathbb{C}$ is said to be a reproducing kernel Hilbert space if there exists an $m \times m$ matrix valued function $K_\omega(\lambda)$ on $\Omega \times \Omega$ such that for every choice of $\omega \in \Omega$, $v \in \mathbb{C}^m$ and $f \in \mathcal{H}$

(1) $K_\omega v \in \mathcal{H}$ (as a function of λ) and

(2) $< f, K_\omega v >_{\mathcal{H}} = v^* f(\omega)$.

It is readily checked that a reproducing kernel Hilbert space $\mathcal{H}$ has exactly one reproducing kernel $K_\omega(\lambda)$ and that

$$K_\alpha(\beta) = K_\beta(\alpha)^*$$

for every choice of α and β in Ω; see e.g., Chapter 2 of [D] for the verification of these details and more.

We next list and briefly discuss a number of examples.

EXAMPLE 1. *Let* $\Omega = \Delta_+$. *Then* $\mathcal{H} = H_m^2(\Delta_+)$, *is a reproducing kernel Hilbert space with respect to the standard inner product (see the table in Section 1)*

$$< f, g >_{\mathcal{H}} = < f, g > ,$$

with reproducing kernel

$$K_\omega(\lambda) = \frac{I_m}{\rho_\omega(\lambda)} .$$

DISCUSSION. The verification that this choice of $K_\omega(\lambda)$ is a reproducing kernel for H_m^2 is just Cauchy's formula for H^2 spaces.

EXAMPLE 2. *Let* $\Omega = \Delta_-$. *Then* $\mathcal{H} = (H_m^2(\Delta_+))^\perp$, *is a reproducing kernel Hilbert space with respect to the same inner product as in Example 1 but with reproducing kernel*

$$K_\omega(\lambda) = -\frac{I_m}{\rho_\omega(\lambda)} .$$

DISCUSSION. In checking the details it is important to bear in mind that now ω and λ belong to Δ_-.

EXAMPLE 3. *Let $f_1, \ldots, f_n$ be any linearly independent set of $m \times 1$ vector valued functions which are analytic in some open nonempty subset Ω of $\mathbb{C}$. Let*

$$P = [p_{ij}] \ , \qquad i,j = 1, \ldots, n \ ,$$

be any positive definite matrix and let

$$G = [g_{ij}] \ , \qquad i,j = 1, \ldots, n \ ,$$

be its inverse. Then

$$\mathcal{H} = \operatorname{span}\{f_1, \ldots, f_n\}$$

with

$$< f_j, f_i >_{\mathcal{H}} = p_{ij} \ , \qquad i,j = 1, \ldots, n \ ,$$

is a reproducing kernel Hilbert space with reproducing kernel

$$K_\omega(\lambda) = \sum_{i,j=1}^{n} f_i(\lambda) g_{ij} f_j(\omega)^* \ . \tag{2.1}$$

DISCUSSION. The verification is a straightforward calculation.

EXAMPLE 4. *For $S \in \mathcal{S}_{p\times q}(\Delta_+)$ and $f \in H_p^2(\Delta_+)$, let*

$$\mu(f) = \sup_{g \in H_q^2} \{\|f + Sg\|^2 - \|g\|^2\} \ .$$

Then

$$\mathcal{H}(S) = \{f \in H_p^2 : \ \mu(f) < \infty\}$$

is a reproducing kernel Hilbert space with inner product based on

$$\|f\|^2_{\mathcal{H}(S)} = \mu(f)$$

and reproducing kernel

$$\Lambda_\omega(\lambda) = \frac{I_p - S(\lambda)S(\omega)^*}{\rho_\omega(\lambda)} \ . \tag{2.2}$$

DISCUSSION. This space and the verification of its advertised properties is due to de Branges and Rovnyak [dBR1] and [dBR2]; for additional discussion, see also Chapter 2 of [D]. It is instructive to check that if S is isometric on $\mathbb{T}$, i.e., if $S(e^{i\theta})^*S(e^{i\theta}) = I_q$ for a.e. θ, then

$$\mathcal{H}(S) = H_p^2 \ominus SH_q^2 \quad \text{and} \quad \mu(f) = < f, f > \ .$$

EXAMPLE 5. *If J is an $m \times m$ signature matrix and U is J inner over Δ_+, then there is an associated reproducing kernel Hilbert space $\mathcal{H}(U)$ with reproducing kernel*

$$K_\omega(\lambda) = \frac{J - U(\lambda)JU(\omega)^*}{\rho_\omega(\lambda)} \ . \tag{2.3}$$

DISCUSSION. Notice that we haven't specified the space or the inner product. For more concrete descriptions see Theorems 2.4 and 2.7 of [D]. If U is a J inner function which belongs to $H^\infty_{m\times m}$, then

$$\mathcal{H}(U) = \{f \in H_m^2 : < Jf, Ug >= 0 \ \text{ for every } \ g \in H_m^2\}$$

endowed with the J inner product.

EXAMPLE 6. *Let $\Omega = \mathbb{D}$. Then the Bergman space $\mathcal{H}$ of $m \times 1$ vector valued analytic functions with*

$$< f, g >_{\mathcal{H}} = \frac{1}{\pi}\int_0^{2\pi}\int_0^1 g(re^{i\theta})^* f(re^{i\theta}) r dr d\theta$$

is a reproducing kernel Hilbert space with reproducing kernel

$$K_\omega(\lambda) = \frac{I_m}{\rho_\omega(\lambda)^2} \ .$$

For present purposes (and in fact many others), spaces with kernels of the form (2.3) will turn out to be of special interest. Fortunately, there is a beautiful characterization of reproducing kernel Hilbert spaces of $m \times 1$ vector valued analytic functions with reproducing kernels of the form (2.3) with a U which is J inner over $\mathbb{C}_+$, which is due to de Branges [dB]. It contained an extra technical condition which was later shown to be superfluous by Rovnyak [R]. The characterization for spaces with kernels based on U's which are J inner over $\mathbb{D}$ was worked out by Ball [B1]. A unified approach to both settings may be found in [AD1].

For ease of exposition we shall restrict ourselves here to a finite dimensional version because this will be adequate for the applications we have in mind.

THEOREM 2.1. *Let $\mathcal{H} = span\{f_1, \ldots, f_n\}$ be a reproducing kernel Hilbert space of $m \times 1$ vector valued functions which are analytic in some open nonempty subset Ω of Δ_+. Then its reproducing kernel $K_\omega(\lambda)$ can be expressed in the form*

$$K_\omega(\lambda) = \frac{J - U(\lambda)JU(\omega)^*}{\rho_\omega(\lambda)} \tag{2.4}$$

where U is J inner over Δ_+ if and only if

(1) $\mathcal{H}$ is R_α invariant for every point $\alpha \in \Omega$, and

(2) the first [resp. the second] of the following two identities

$$< f, g >_{\mathcal{H}} + \alpha < R_\alpha f, g >_{\mathcal{H}} + \beta^* < f, R_\beta g >_{\mathcal{H}} - (1 - \alpha\beta^*) < R_\alpha f, R_\beta g >_{\mathcal{H}} \\ = g(\beta)^* J f(\alpha) \tag{2.5}$$

$$< R_\alpha f, g >_{\mathcal{H}} - < f, R_\beta g >_{\mathcal{H}} - (\alpha - \beta^*) < R_\alpha f, R_\beta g >_{\mathcal{H}} \\ = 2\pi i \; g(\beta)^* J f(\alpha) \tag{2.6}$$

holds for every choice of α, β in Ω and f, g in $\mathcal{H}$ if $\Delta_+ = \mathbb{D}$ [resp. $\Delta_+ = \mathbb{C}_+$].

Moreover, U is uniquely determined by the space up to a J unitary constant factor on the right and can in fact be specified by the formula

$$U(\lambda) = I_m - \rho_\beta(\lambda) \sum_{i,j=1}^{n} f_i(\lambda) g_{ij} f_j(\beta)^* J \ , \tag{2.7}$$

where g_{ij} is the ij entry of the inverse to the Gram matrix $[< f_j, f_i >_{\mathcal{H}}]$ based on $f_1, \ldots, f_n$ (as in Example 3) and β is any point on ∂ at which the f_j (which turn out to be rational!) are all analytic. (Changing the β just introduces an extra J unitary constant factor on the right of (2.7).)

Everything except for the specific formula (2.7) is contained in one form or another in the references cited just above. For a proof in the style of de Branges' original argument but in the more general setting of Pontryagin spaces, see Theorems 6.9 and 6.12 of [AD2]. A recursive proof is outlined in Theorems 2.6, 4.1 and the Supplementary Notes to Chapter 4 of [D].

Our next objective is to clarify the role of the two conditions in Theorem 2.1.

The first condition: R_α invariance, determines the form of the elements in the space. Thus if $f \in \mathcal{H}$ and $\mathcal{H}$ is one-dimensional, then the R_α invariance guarantees the existence of a constant μ such that

$$(R_\alpha f)(\lambda) = \frac{f(\lambda) - f(\alpha)}{\lambda - \alpha} = \mu f(\lambda) .$$

But this in turn implies that

$$f(\lambda) = \frac{f(\alpha)}{1 + \mu\alpha - \mu\lambda} ,$$

which in turn is readily seen to be of the form

$$f(\lambda) = \begin{cases} v/(\lambda - \omega) & \text{if} \quad \mu \neq 0 \\ v & \text{if} \quad \mu = 0 \end{cases}$$

for some choice of $\omega \in \mathbb{C}$ and $v \in \mathbb{C}^m$.

More generally, if dim $\mathcal{H} = n > 1$, then $\mathcal{H}$ consists of chains of the form

$$\frac{v_1}{\lambda - \omega} , \frac{v_1}{(\lambda - \omega)^2} + \frac{v_2}{\lambda - \omega}, \ldots, \frac{v_1}{(\lambda - \omega)^k} + \ldots + \frac{v_k}{\lambda - \omega}$$

or

$$v_1 , \lambda v_1 + v_2, \ldots, \lambda^{k-1} v_1 + \ldots + v_k .$$

Typically $\mathcal{H}$ will contain many such chains of different lengths and with different poles. Roughly speaking, each chain corresponds to a Jordan cell in the Jordan decomposition of the operator R_α on the finite dimensional space $\mathcal{H}$ with respect to a suitably chosen basis. In particular (as noted parenthetically in the statement of the theorem) it follows that the elements of such a finite dimensional R_α invariant space are automatically rational. Consequently, *in the setting of Theorem 2.1, the reproducing formula*

$$v^* f_i(\omega) =< f_i, \frac{J - UJU(\omega)^*}{\rho_\omega} v >_{\mathcal{H}} \tag{2.8}$$

is valid in the whole complex plane $\mathbb{C}$ *except for at most n points corresponding to the (not necessarily distinct) poles of* $f_1, \ldots, f_n$. This is verified most simply by invoking the identity

$$\frac{J - U(\lambda)JU(\omega)^*}{\rho_\omega(\lambda)} = \sum_{i,j=1}^{n} f_i(\lambda) g_{ij} f_j(\omega)^* \tag{2.9}$$

in the inner product in (2.8), where g_{ij} is defined as in Example 3 and the identity (2.9) itself is an important byproduct of the uniqueness of the reproducing kernel.

The role of the second condition in Theorem 2.1 is to determine the inner product. Thus, for example, if

$$f(\lambda) = \frac{v}{\rho_\omega(\lambda)} \ ,$$

then

$$(R_\alpha f)(\lambda) = \begin{cases} \frac{\omega^*}{\rho_\omega(\alpha)} f(\lambda) & \text{if} \quad \Delta_+ = \mathbb{D} \\ \frac{1}{\omega^* - \alpha} f(\lambda) & \text{if} \quad \Delta_+ = \mathbb{C}_+ \ , \end{cases} \tag{2.10}$$

and it follows from (2.5) and (2.6) with $\alpha = \beta$ and $f = g$ that in both cases

$$\rho_\omega(\omega) < f, f >_{\mathcal{H}} = v^* J v \ . \tag{2.11}$$

This in turn leads to the following conclusions:

(1) If $\omega \in \Delta_+$ (and hence $f \in H_m^2(\Delta_+)$), then v is strictly J positive and

$$< f, f >_{\mathcal{H}} = \frac{v^* J v}{\rho_\omega(\omega)} = < Jf, f > \ . \tag{2.12}$$

(2) If $\omega \in \Delta_-$ (and hence $f \in H_m^2(\Delta_+)^\perp$), then v is strictly J negative and

$$< f, f >_{\mathcal{H}} = - < Jf, f > \ . \tag{2.13}$$

(3) If $\omega \in \partial$, then v is J neutral and (unless there are other constraints) $< f, f >_{\mathcal{H}}$ can be set equal to δ for any choice of $\delta > 0$.

Formulas (2.12) and (2.13) are indicative of the general fact that if f and $g \in \mathcal{H}(U) \cap H_m^2$, then

$$< f, g >_{\mathcal{H}} = < Jf, g > \ , \tag{2.14}$$

whereas, if f and $g \in \mathcal{H}(U) \cap (H_m^2)^\perp$, then

$$< f, g >_{\mathcal{H}} = < -Jf, g > \ ; \tag{2.15}$$

see Corollaries 1 and 2 to Theorem 2.7 in [D] (which though formulated for $J = J_{pq}$ are easily adapted to any signature matrix J). These conclusions will also emerge from the more detailed analysis of the implications of (2.5) and (2.6) on the inner product of chains which is carried out in Sections 3 and 4.

We next establish a pair of converse statements, the point of which is to emphasize that in certain circumstances (which occur quite commonly) the identities which form such an integral part of Theorem 2.1 are automatically satisfied and hence do not even need to be mentioned. This is the reason that they are not as well known as they should be.

THEOREM 2.2. *If f and $g \in H^2_m(\Delta_+)$, then identity (2.5) [resp. (2.6)] holds for every choice of α, β in $\mathbb{D}$ [resp. $\mathbb{C}_+$] for*

$$< f, g >_{\mathcal{H}} = < Jf, g > ,$$

the J inner product.

PROOF. If $\Delta_+ = \mathbb{C}_+$, then the left hand side of (2.6) with the J inner product is readily seen to reduce to

$$< \frac{Jf}{\lambda - \alpha}, g > - < Jf, \frac{g}{\lambda - \beta} > -(\alpha - \beta^*) < \frac{Jf}{\lambda - \alpha}, R_\beta g >$$

$$= < J\left\{\frac{1}{\lambda - \alpha} - \frac{1}{\lambda - \beta^*} - \frac{(\alpha - \beta^*)}{(\lambda - \alpha)(\lambda - \beta^*)}\right\} f, g > +(\alpha - \beta^*) < \frac{Jf}{\lambda - \alpha}, \frac{g(\beta)}{\lambda - \beta} >$$

$$= 0 + 2\pi i \; g(\beta)^* Jf(\alpha) ,$$

where Cauchy's formula for H^2_m is used for the last evaluation. This establishes (2.6).

The proof for $\Delta_+ = \mathbb{D}$ goes through in more or less the same way:

$$\begin{aligned} < J(f + \alpha R_\alpha f), g + \beta R_\beta g > &= < J\frac{\lambda f}{\lambda - \alpha}, \frac{\lambda g}{\lambda - \beta} > - < J\frac{\lambda f}{\lambda - \alpha}, \frac{\beta g(\beta)}{\lambda - \beta} > \\ &= < \frac{Jf}{\lambda - \alpha}, \frac{g}{\lambda - \beta} > - \frac{\alpha\beta^* g(\beta)^* Jf(\alpha)}{\rho_\beta(\alpha)} \\ &= < \frac{Jf}{\lambda - \alpha}, R_\beta g > + g(\beta)^* Jf(\alpha) \\ &= < JR_\alpha f, R_\beta g > + g(\beta)^* Jf(\alpha) , \end{aligned}$$

which leads easily to (2.5). ∎

THEOREM 2.3. *If f and $g \in (H^2_m)^\perp$, then identity (2.5) [resp. (2.6)] holds for every choice of α, β in $\mathbb{E} \cap \mathbb{C}$ [resp. $\mathbb{C}_-$] for*

$$< f, g >_{\mathcal{H}} = < -Jf, g > ,$$

the $-J$ inner product.

PROOF. The proof is much the same as for Theorem 2.2 and is therefore left to the reader. ∎

For rational functions of the type to be considered below the identities (2.5) and (2.6) are in fact valid for every choice of α and β in $\mathbb{C}$ except for at most a finite number of points, i.e., except for the poles of f and g, respectively.

3. EQUATIONS FOR THE GRAM MATRIX WHEN $\Delta_+ = \mathbb{D}$

In this section we shall prepare the way for showing that the identity (2.5) is equivalent to a matrix equation for the Gram matrix of a suitably chosen basis of chains. For ease of future reference we shall carry out the calculations in the more general setting of a space $\mathcal{M}$ with a possibly indefinite inner product $\langle\ ,\ \rangle_{\mathcal{M}}$; the work is just the same.

Throughout this section we shall fix $\rho_\omega(\lambda) = 1 - \lambda\omega^*$. Then every chain of length n, except for the chain

$$\frac{v_1}{\lambda}, \frac{v_1}{\lambda^2} + \frac{v_2}{\lambda}, \ldots, \frac{v_1}{\lambda^n} + \ldots + \frac{v_n}{\lambda} \tag{3.1}$$

can be expressed in the form

$$\begin{aligned} f_1(\lambda) &= \frac{v_1}{\rho_\omega(\lambda)} \\ f_j(\lambda) &= \lambda^{j-1}\frac{v_1}{\rho_\omega(\lambda)^j} + \lambda^{j-2}\frac{v_2}{\rho_\omega(\lambda)^{j-1}} + \ldots + \frac{v_j}{\rho_\omega(\lambda)}, \qquad j = 2, \ldots, n, \end{aligned} \tag{3.2}$$

for some choice of $\omega \in \mathbb{C}$ (and, of course, $v_1, \ldots, v_n \in \mathbb{C}^m$).

LEMMA 3.1. *Every chain of the form (3.2) is analytic at zero. Furthermore:*

(1) $f_j(0) = v_j$,

(2) $f_j(\lambda) = \frac{\lambda}{\rho_\omega(\lambda)} f_{j-1}(\lambda) + \frac{v_j}{\rho_\omega(\lambda)}$,

(3) $(R_\alpha f_j)(\lambda) = \frac{1}{\rho_\omega(\alpha)}\{\omega^* f_j(\lambda) + f_{j-1}(\lambda) + \alpha(R_\alpha f_{j-1})(\lambda)\}$, (3.3)

(4) $(R_0 f_j)(\lambda) = \omega^* f_j(\lambda) + f_{j-1}(\lambda)$, (3.4)

for $j = 1, \ldots, n$, with the understanding that $f_0 = 0$.

PROOF. (1) and (2) are selfevident and (3) has already been checked for $j = 1$. For $j = 2, \ldots, n$, (3) may be obtained by invoking (2) and the general identity

$$(R_\alpha gh)(\lambda) = (R_\alpha g)(\lambda)h(\lambda) + g(\alpha)(R_\alpha h)(\lambda) \tag{3.5}$$

with $g(\lambda) = \lambda/\rho_\omega(\lambda)$ and $h(\lambda) = f_{j-1}(\lambda)$. (4) is immediate from (3). ∎

We shall also have need for chains of the form

$$
\begin{aligned}
g_1 &= \frac{w_1}{\lambda-\mu} \\
g_j(\lambda) &= \frac{1}{\lambda-\mu} g_{j-1}(\lambda) + \frac{w_j}{\lambda-\mu}, \qquad j = 2,\dots,n\,.
\end{aligned}
\tag{3.6}
$$

LEMMA 3.2. *If g_j, $j = 1,\dots,n$, is a chain of the form (3.6), then*

$$
(R_\beta g_j)(\lambda) = \frac{1}{\mu-\beta}\{g_j(\lambda) - (R_\beta g_{j-1})(\lambda)\}\,, \tag{3.7}
$$

with the understanding that $g_0 = 0$.

PROOF. By direct calculation

$$
(R_\beta g_1)(\lambda) = \frac{1}{\mu-\beta} g_1(\lambda)\,.
$$

The corresponding formula for general j is obtained much as in Lemma 3.1, by applying (3.5) to formula (3.6) for g_j with $g(\lambda) = 1/(\lambda-\mu)$ and $h(\lambda) = g_{j-1}(\lambda)$. ∎

Chains of these two particular forms: (3.2) and (3.6), will play an important role in the interpolation problems which will be studied in the sequel.

The symbol Z will be used to denote a square upper triangular matrix with ones on the first super diagonal only and zeros elsewhere:

$$
Z = \begin{bmatrix} 0 & 1 & \dots & 0 & 0 \\ 0 & 0 & \dots & 0 & 0 \\ \vdots & & & & \vdots \\ 0 & 0 & \dots & 0 & 1 \\ 0 & 0 & \dots & 0 & 0 \end{bmatrix} \quad \text{and} \quad Z_\alpha = \alpha I + Z\,. \tag{3.8}
$$

The size of Z will, if not already clear from the context, be explained separately but will not be indicated in the notation.

THEOREM 3.1. *If $\rho_\omega(\lambda) = 1 - \lambda\omega^*$,*

$$
f_t(\lambda) = \begin{cases} v_t/\rho_\omega(\lambda) & \text{for} \quad t = \tau+1 \\ \{\lambda f_{t-1}(\lambda) + v_t\}/\rho_\omega(\lambda) & \text{for} \quad t = \tau+2,\dots,\tau+n \end{cases}
$$

and

$$
g_s(\lambda) = \begin{cases} w_s/\rho_\mu(\lambda) & \text{for} \quad s = \sigma+1 \\ \{\lambda g_{s-1}(\lambda) + w_s\}/\rho_\mu(\lambda) & \text{for} \quad s = \sigma+2,\dots,\sigma+k \end{cases}
$$

are chains in a space $\mathcal{M}$ *with inner product* $< \ , \ >_{\mathcal{M}}$ *(definite or not) and if*

$$V = [v_{\tau+1} \dots v_{\tau+n}] \quad \text{and} \quad W = [w_{\sigma+1} \dots w_{\sigma+k}] \ ,$$

then the $k \times n$ *matrix* B *with entries*

$$b_{st} = < f_t, g_s >_{\mathcal{M}}$$

is a solution of the matrix equation

$$B - (\mu^* I_k + Z)^* B(\omega^* I_n + Z) = W^* J V$$

if and only if

$$< f_t + \alpha R_\alpha f_t, g_s + \beta R_\beta g_s >_{\mathcal{M}} - < R_\alpha f_t, R_\beta g_s >_{\mathcal{M}} = g_s(\beta)^* J f_t(\alpha) \tag{3.9}$$

for $s = \sigma+1, \dots, \sigma+k, \quad t = \tau+1, \dots, \tau+n$ *and every choice of* α, β *in* $\mathbb{C}$ *with* $\alpha\omega^* \neq 1$ *and* $\beta\mu^* \neq 1$.

PROOF. Suppose first that (3.9) holds. Then, with the special choice $\alpha = \beta = 0$, it implies that

$$< f_t, g_s >_{\mathcal{M}} - < R_0 f_t, R_0 g_s >_{\mathcal{M}} = g_s(0)^* J f_t(0) \ ,$$

and hence, in view of Lemma 3.1, that

$$< f_t, g_s >_{\mathcal{M}} - < \omega^* f_t + f_{t-1}, \mu^* g_s + g_{s-1} >_{\mathcal{M}} = w_s^* J v_t \ ,$$

with the understanding that $f_\tau = 0$ and $g_\sigma = 0$. But this is the same as

$$b_{st} - \omega^* \mu b_{st} - \mu b_{s,t-1} - \omega^* b_{s-1,t} - b_{s-1,t-1} = w_s^* J v_t \ , \tag{3.10}$$

for $s = \sigma + 1, \dots, \sigma + k$ and $t = \tau + 1, \dots, \tau + n$ with the understanding that $b_{ij} = 0$ if $i = \sigma$ or $j = \tau$. This in turn is equivalent to the stated matrix equation.

It remains to show the converse: that if B is a solution of the stated matrix equation, then (3.9) holds. This is carried out in steps with the help of the formulas

$$R_\alpha f_t = \frac{1}{\rho_\omega(\alpha)} \{\omega^* f_t + f_{t-1} + \alpha R_\alpha f_{t-1}\} \tag{3.11}$$

$$f_t + \alpha R_\alpha f_t = \frac{f_t}{\rho_\omega(\alpha)} + \frac{\alpha}{\rho_\omega(\alpha)} \{f_{t-1} + \alpha R_\alpha f_{t-1}\} \tag{3.12}$$

$$R_\beta g_s = \frac{1}{\rho_\mu(\beta)}\{\mu^* g_s + g_{s-1} + \beta R_\beta g_{s-1}\} \tag{3.13}$$

$$g_s + \beta R_\beta g_s = \frac{g_s}{\rho_\mu(\beta)} + \frac{\beta}{\rho_\mu(\beta)}\{g_{s-1} + \beta R_\beta g_{s-1}\} \tag{3.14}$$

which follow from Lemma 3.1. They are rewritten here in the form which will be used in the proof, at the cost of some redundancy, to ease the burden on the reader, if such there be. The convention $g_\sigma = f_\tau = 0$ is in force, as usual.

STEP 1. *(3.9) is valid for* $s = \sigma + 1$ *and* $t = \tau + 1$.

PROOF OF STEP 1. For this choice of s and t, it is readily checked with the help of (3.11) and (3.13) that the left hand side of (3.9) is equal to

$$\frac{(1-\omega^*\mu)b_{st}}{\rho_\omega(\alpha)\rho_\mu(\beta)^*} = \frac{w_s^* J v_t}{\rho_\omega(\alpha)\rho_\mu(\beta)^*} ,$$

where the right hand side of the last equality follows from (3.10). This completes the step since the right hand side of the last equality is equal to $g_s(\beta)^* J f_t(\alpha)$ for this choice of s and t by the definition of the chain.

STEP 2. *(3.9) is valid for* $s = \sigma + 1$ *and* $t = \tau + 2, \ldots, \tau + n$.

PROOF OF STEP 2. Let L_{st} denote the left hand side of (3.9). Then, it follows readily from (3.12) that

$$L_{st} = \frac{\alpha}{\rho_\omega(\alpha)} L_{s,t-1} + < \frac{\alpha}{\rho_\omega(\alpha)} R_\alpha f_{t-1} - R_\alpha f_t, R_\beta g_s >_{\mathcal{M}} + \frac{1}{\rho_\omega(\alpha)} < f_t, g_s + \beta R_\beta g_s >_{\mathcal{M}} .$$

Thus, by (3.11),

$$L_{st} = \frac{1}{\rho_\omega(\alpha)}\{\alpha L_{s,t-1} - < \omega^* f_t + f_{t-1}, R_\beta g_s >_{\mathcal{M}} + b_{st} + < f_t, \beta R_\beta g_s >_{\mathcal{M}}\} . \tag{3.15}$$

Formula (3.15) is valid for any choice of s and t. If $s = \sigma + 1$, as in the present case, then $R_\beta g_s = \mu^* g_s / \rho_\mu(\beta)$ and the right hand side of (3.15) reduces to

$$\frac{1}{\rho_\omega(\alpha)}\{\alpha L_{s,t-1} + \frac{1}{\rho_\mu(\beta)^*}(b_{st}(1-\mu\omega^*) - \mu b_{s,t-1})\} = \frac{1}{\rho_\omega(\alpha)}\{\alpha L_{s,t-1} + g_s(\beta)^* J v_t\} ,$$

by (3.10) and the definition of g_s for $s = \sigma + 1$. The rest follows by induction on t: If the assertion is true for $L_{s,t-1}$, then, by the last line,

$$L_{st} = \frac{1}{\rho_\omega(\alpha)}\{\alpha g_s(\beta)^* J f_{t-1}(\alpha) + g_s(\beta)^* J v_t\} = g_s(\beta)^* J f_t(\alpha) ,$$

as needed. The induction starts with Step 1.

STEP 3 *is to establish the formula*

$$b_{st} + < R_\alpha f_t, (\alpha - \mu)^* g_s - g_{s-1} >_{\mathcal{M}} = w_s^* J f_t(\alpha)$$

for $s = \sigma + 1, \ldots, \sigma + k$ *and* $t = \tau + 1, \ldots, \tau + n$.

PROOF OF STEP 3. The formula is first verified for $t = \tau + 1$ with the help of (3.11) and (3.10). The rest follows by straightforward induction on t: Let E_{st} denote the given left hand side. Then, by (3.11),

$$\begin{aligned} E_{st} &= b_{st} + \frac{1}{\rho_\omega(\alpha)} < \omega^* f_t + f_{t-1}, (\alpha - \mu)^* g_s - g_{s-1} >_{\mathcal{M}} + \frac{\alpha}{\rho_\omega(\alpha)} \{E_{s,t-1} - b_{s,t-1}\} \\ &= w_s^* J \left\{ \frac{\alpha}{\rho_\omega(\alpha)} f_{t-1}(\alpha) + \frac{v_t}{\rho_\omega(\alpha)} \right\} , \end{aligned}$$

by (3.10) and the induction hypothesis. This completes the proof since the term in curly brackets is equal to $f_t(\alpha)$.

STEP 4. *(3.9) is valid for the remaining choices of* s *and* t.

PROOF OF STEP 4. In view of Steps 1 and 2, which verify (3.9) for $s = \sigma + 1$ and $t = \tau + 1, \ldots, \tau + n$, it suffices to show that if (3.9) is valid for any choice of $s - 1$ and t, then it is also valid for s and t. Again let L_{st} denote the left hand side of (3.9). Then, by (3.14),

$$\begin{aligned} L_{st} &= < f_t + \alpha R_\alpha f_t, \{g_s + \beta(g_{s-1} + \beta R_\beta g_{s-1})\}/\rho_\mu(\beta) >_{\mathcal{M}} - < R_\alpha f_t, R_\beta g_s >_{\mathcal{M}} \\ &= \frac{1}{\rho_\mu(\beta)^*} \{\beta^* L_{s-1,t} + b_{st} + < \alpha R_\alpha f_t, g_s >_{\mathcal{M}}\} + < R_\alpha f_t, \frac{\beta}{\rho_\mu(\beta)} R_\beta g_{s-1} - R_\beta g_s >_{\mathcal{M}} . \end{aligned}$$

But now, by (3.13), the last inner product is equal to

$$- < R_\alpha f_t, \{\mu^* g_s + g_{s-1}\}/\rho_\mu(\beta) >_{\mathcal{M}}$$

and hence

$$L_{st} = \frac{1}{\rho_\mu(\beta)^*} \{\beta^* L_{s-1,t} + b_{st} + < R_\alpha f_t, (\alpha - \mu)^* g_s - g_{s-1} >_{\mathcal{M}}\} .$$

Thus, by Step 3 and the induction hypothesis,

$$L_{st} = \{\frac{\beta}{\rho_\mu(\beta)} g_{s-1}(\beta)\}^* J f_t(\alpha) + \{\frac{w_s}{\rho_\mu(\beta)}\}^* J f_t(\alpha) = g_s(\beta)^* J f_t(\alpha) ,$$

as needed.

This completes the proof of both Step 4 and the theorem. We remark that the assumptions on α and β insure that the terms $\rho_\mu(\beta)$ and $\rho_\omega(\alpha)$, which occur repeatedly throughout the proof are nonzero. ∎

The next two theorems deal with chains of the form (3.6). There is some overlap with Theorem 3.1 because every such chain can also be expressed in the form (3.2) except if $\mu = 0$. Nevertheless, it is convenient to have the results available in the indicated form.

THEOREM 3.2. *If* $\rho_\omega(\lambda) = 1 - \lambda\omega^*$,

$$f_t(\lambda) = \begin{cases} v_t/\rho_\omega(\lambda) & for \quad t = \tau + 1 \\ \{\lambda f_{t-1}(\lambda) + v_t\}/\rho_\omega(\lambda) & for \quad t = \tau + 2, \ldots, \tau + n \end{cases}$$

and

$$g_s(\lambda) = \begin{cases} w_s/(\lambda - \mu) & for \quad s = \sigma + 1 \\ \{g_{s-1}(\lambda) + w_s\}/(\lambda - \mu) & for \quad s = \sigma + 2, \ldots, \sigma + k \end{cases}$$

are chains in a space $\mathcal{M}$ *with inner product* $< \ , \ >_\mathcal{M}$ *(definite or not) and if* V *and* W *are as in Theorem 3.1, then the* $k \times n$ *matrix* B *with entries*

$$b_{st} = < f_t, g_s >_\mathcal{M}$$

is a solution of the matrix equation

$$B(\omega^* I_n + Z) - (\mu I_k + Z)^* B = W^* J V$$

if and only if (3.9) holds for the present set of f_t *and* g_s *for every choice of* α *and* β *in* $\mathbb{C}$ *with* $\alpha\omega^* \neq 1$ *and* $\beta \neq \mu$.

PROOF. Suppose first that (3.9) holds. Then, since the second chain will have a pole at zero if $\mu = 0$, we can only allow $\alpha = 0$. With this choice of α and with $\beta \neq \mu$, (3.9) reduces to

$$< f_t, g_s >_\mathcal{M} + \beta^* < f_t, R_\beta g_s >_\mathcal{M} - < R_0 f_t, R_\beta g_s >_\mathcal{M} = g_s(\beta)^* J v_t \ .$$

If $s > \sigma + 1$, then also

$$< f_t, g_{s-1} >_\mathcal{M} + \beta^* < f_t, R_\beta g_{s-1} >_\mathcal{M} - < R_0 f_t, R_\beta g_{s-1} >_\mathcal{M} = g_{s-1}(\beta)^* J v_t \ .$$

Multiplying the first of these equations by $(\mu^* - \beta^*)$ and then adding it to the second, and invoking (3.7), leads to

$$(\mu-\beta)^* < f_t, g_s >_{\mathcal{M}} + \beta^* < f_t, g_s >_{\mathcal{M}} - < R_0 f_t, g_s >_{\mathcal{M}} \\ + < f_t, g_{s-1} >_{\mathcal{M}} = \{(\mu-\beta)g_s(\beta) + g_{s-1}(\beta)\}^* J v_t \ ,$$

which, with the aid of (3.4), is readily seen to reduce to

$$\mu^* b_{st} - \omega^* b_{st} - b_{s,t-1} + b_{s-1,t} = -w_s^* J v_t \ . \tag{3.16}$$

But this is the same as the asserted matrix equation.

The converse is proved in steps (much as in Theorem 3.1) with the help of formulas (3.11), (3.12) and

$$(R_\beta g_s)(\lambda) = \frac{1}{\mu-\beta}\{g_s(\lambda) - (R_\beta g_{s-1})(\lambda)\} \tag{3.17}$$

$$g_s(\lambda) + \beta(R_\beta g_s)(\lambda) = \frac{1}{\mu-\beta}\{\mu g_s(\lambda) - \beta(R_\beta g_{s-1})(\lambda)\} \ , \tag{3.18}$$

for $s = \sigma+1, \ldots, \sigma+k$. The last two formulas follow from Lemma 3.2 and are subject to the understanding that $g_\sigma = 0$.

STEP 1. *(3.9) is valid for* $s = \sigma + 1$ *and* $t = \tau + 1$.

PROOF OF STEP 1. By (3.11), (3.12), (3.17) and (3.18), the left hand side of (3.9) is equal to

$$\frac{(\mu^* - \omega^*)b_{st}}{\rho_\omega(\alpha)(\mu-\beta)^*}$$

for this choice of s and t. The rest drops out easily from (3.16).

STEP 2. *(3.9) is valid for* $s = \sigma + 1$ *and* $t = \tau + 1, \ldots, \tau + k$.

PROOF OF STEP 2. The proof is by induction on t, the first step of which is justified by Step 1. Let L_{st} denote the right hand side of (3.9). Then (3.15) is still valid because the present chain $f_{\tau+1}, \ldots, f_{\tau+n}$ is the same as in Theorem 3.1. But now $R_\beta g_s = g_s/(\mu - \beta)$ for $s = \sigma + 1$ and hence (3.15) reduces to

$$L_{st} = \frac{1}{\rho_\omega(\alpha)}\left\{\alpha L_{s,t-1} + \frac{(\mu^* - \omega^*)b_{st} - b_{s,t-1}}{(\mu-\beta)^*}\right\} \ ,$$

which, by the induction hypothesis and (3.16), can be reexpressed as

$$L_{st} = g_s(\beta)^* J \left\{\frac{\alpha}{\rho_\omega(\alpha)} f_{t-1}(\alpha) + \frac{v_t}{\rho_\omega(\alpha)}\right\} = g_s(\beta)^* J f_t(a) \ ,$$

for $s = \sigma + 1$, as needed.

STEP 3 *is to verify the formula*

$$< f_t + \alpha R_\alpha f_t, \mu g_s + g_{s-1} >_{\mathcal{M}} - < R_\alpha f_t, g_s >_{\mathcal{M}} = -w_s^* J f_t(\alpha)$$

for $s = \sigma + 1, \ldots, \sigma + k$ *and* $t = \tau + 1, \ldots, \tau + n$.

PROOF OF STEP 3. With the help of (3.11) and (3.12) the left hand side is readily seen to reduce to

$$\frac{(\mu^* - \omega^*)b_{st} + b_{s-1,t}}{\rho_\omega(\alpha)}$$

for $t = \tau + 1$. An application of (3.16) serves to complete the proof for this choice of t.

The proof for general t is by induction on t. To this end let E_{st} denote the left hand side of the stated formula. Then, by (3.12) and (3.11),

$$\begin{aligned} E_{st} &= \frac{\alpha}{\rho_\omega(\alpha)} E_{s,t-1} + < \frac{\alpha}{\rho_\omega(\alpha)} R_\alpha f_{t-1} - R_\alpha f_t, g_s >_{\mathcal{M}} + \frac{1}{\rho_\omega(\alpha)} < f_t, \mu g_s + g_{s-1} >_{\mathcal{M}} \\ &= \frac{1}{\rho_\omega(\alpha)} \{\alpha E_{s,t-1} - < \omega^* f_t + f_{t-1}, g_s >_{\mathcal{M}} + < f_t, \mu g_s + g_{s-1} >_{\mathcal{M}}\} \\ &= \frac{1}{\rho_\omega(\alpha)} \{\alpha E_{s,t-1} + (\mu^* - \omega^*) b_{st} + b_{s-1,t} - b_{s,t-1}\} . \end{aligned}$$

By the induction hypothesis and (3.16) the last formula can be reexpressed as

$$\begin{aligned} E_{st} &= -w_s^* J \left\{ \frac{\alpha}{\rho_\omega(\alpha)} f_{t-1}(\alpha) + \frac{v_t}{\rho_\omega(\alpha)} \right\} \\ &= -w_s^* J f_t(\alpha) , \end{aligned}$$

as needed.

STEP 4 *is to verify (3.9) for the remaining choices of* s *and* t.

PROOF OF STEP 4. By Steps 1 and 2, (3.9) is valid for $s = \sigma + 1$ and $t = \tau + 1, \ldots, \tau + n$. The proof for general s is by induction on s. Let L_{st} denote the left hand side of (3.9). Then, by (3.18) and (3.17),

$$\begin{aligned} L_{st} &= \frac{1}{(\mu - \beta)^*} \{-L_{s-1,t} + < f_t + \alpha R_\alpha f_t, \mu g_s + g_{s-1} >_{\mathcal{M}}\} \\ &\qquad - < R_\alpha f_t, \frac{1}{\mu - \beta} R_\beta g_{s-1} + R_\beta g_s >_{\mathcal{M}} \\ &= \frac{1}{(\mu - \beta)^*} \{-L_{s-1,t} + < f_t + \alpha R_\alpha f_t, \mu g_s + g_{s-1} >_{\mathcal{M}} - < R_\alpha f_t, g_s >_{\mathcal{M}}\} . \end{aligned}$$

By Step 3 and the induction hypothesis, the last equation can be expressed in the form

$$\begin{aligned} L_{st} &= \frac{1}{(\mu-\beta)^*}\{-g_{s-1}(\beta)^* J f_t(\alpha) - w_s^* J f_t(\alpha)\} \\ &= g_s(\beta)^* J f_t(\alpha) \ , \end{aligned}$$

as needed. This completes the proof of both Step 4 and the theorem. ∎

THEOREM 3.3. *If* $\rho_\omega(\lambda) = 1 - \lambda\omega^*$,

$$f_t(\lambda) = \begin{cases} v_t/(\lambda-\omega) & \text{for} \quad t = \tau+1 \\ \{f_{t-1}(\lambda) + v_t\}/(\lambda-\omega) & \text{for} \quad t = \tau+2,\ldots,\tau+n \end{cases}$$

and

$$g_s(\lambda) = \begin{cases} w_s/(\lambda-\mu) & \text{for} \quad s = \sigma+1 \\ \{g_{s-1}(\lambda) + w_s\}/(\lambda-\mu) & \text{for} \quad s = \sigma+2,\ldots,\sigma+k \end{cases}$$

are chains in a space $\mathcal{M}$ *with inner product* $< \ , \ >_{\mathcal{M}}$ *(definite or not) and if* V *and* W *are as in Theorem 3.1, then the* $k \times n$ *matrix* B *with entries*

$$b_{st} = < f_t, g_s >_{\mathcal{M}}$$

is a solution of the matrix equation

$$(\mu I_k + Z)^* B(\omega I_n + Z) - B = W^* J V$$

if and only if (3.9) holds for $s = \sigma+1,\ldots,\sigma+k$, $t = \tau+1,\ldots,\tau+n$ *and every choice of* α *and* β *in* $\mathbb{C}$ *except for* $\alpha = \omega$ *and* $\beta = \mu$.

PROOF. Since both ω and μ may be equal to zero, it is necessary to work with (3.9) in its full glory. We shall make extensive use of (3.17) and (3.18) and their counterparts for the present f_t chain:

$$(R_\alpha f_t)(\lambda) = \frac{1}{\omega-\alpha}\{f_t(\lambda) - (R_\alpha f_{t-1})(\lambda)\} \tag{3.19}$$

and

$$f_t(\lambda) + \alpha(R_\alpha f_t)(\lambda) = \frac{1}{\omega-\alpha}\{\omega f_t(\lambda) - \alpha(R_\alpha f_{t-1})(\lambda)\} \tag{3.20}$$

for $t = \tau+1,\ldots,\tau+n$, with the understanding that $f_\tau = 0$.

Let L_{st} denote the left hand side of (3.9) and let us suppose for the moment that $s > \sigma+1$ and $t > \tau+1$. Then it is readily checked with the help of (3.17) that $(\mu-\beta)^* L_{st} + L_{s-1,t}$ is equal to

$$< f_t + \alpha R_\alpha f_t, \mu g_s + g_{s-1} >_{\mathcal{M}} - < R_\alpha f_t, g_s >_{\mathcal{M}} = -w_s^* J f_t(\alpha) \ .$$

But now, upon multiplying the last expression through by $\omega - \alpha$ and adding it to the same expression but with f_t replaced by f_{t-1} and invoking (3.19), it is readily seen that

$$< \omega f_t + f_{t-1}, \mu g_s + g_{s-1} >_{\mathcal{M}} - < f_t, g_s >_{\mathcal{M}} = w_s^* J v_t$$

or equivalently that

$$\omega\mu^* b_{st} + \omega b_{s-1,t} + \mu^* b_{s,t-1} + b_{s-1,t-1} - b_{st} = w_s^* J v_t \ . \tag{3.21}$$

The same conclusions prevail with $t = \tau + 1$ and/or $s = \sigma + 1$, with the understanding that $b_{st} = 0$ if $s = \sigma$ and/or $t = \tau$. But this is the same as the asserted equation.

The converse is again proved in steps, much as in the preceding two theorems:

STEP 1. *(3.9) is valid for $s = \sigma + 1$ and $t = \tau + 1$.*

PROOF OF STEP 1. By (3.17) and (3.19) the left hand side of (3.9) is equal to

$$\frac{(\omega\mu^* - 1)b_{st}}{(\omega - \alpha)(\mu - \beta)^*}$$

which in turn is readily identified as $g_s(\beta)^* J f_t(\alpha)$ for this choice of s and t by (3.21).

STEP 2. *(3.9) is valid for $s = \sigma + 1$ and $t = \tau + 1, \ldots, \tau + n$.*

PROOF OF STEP 2. The proof is by induction on t, with Step 1 serving to start things off. By (3.19) and (3.20) the left hand side of (3.9):

$$\begin{aligned} L_{st} &= \frac{1}{\alpha - \omega}\{L_{s,t-1} - < \omega f_t + f_{t-1}, g_s + \beta R_\beta g_s >_{\mathcal{M}}\} \\ &\quad - < \frac{1}{\omega - \alpha} R_\alpha f_{t-1} + R_\alpha f_t, R_\beta g_s >_{\mathcal{M}} \\ &= \frac{1}{\alpha - \omega}\{L_{s,t-1} - < w f_t + f_{t-1}, g_s + \beta R_\beta g_s >_{\mathcal{M}} + < f_t, R_\beta g_s >_{\mathcal{M}}\} \end{aligned}$$

If $s = \sigma + 1$, then $R_\beta g_s = g_s/(\mu - \beta)$ and so the last formula reduces to

$$L_{st} = \frac{1}{\alpha - \omega}\left\{L_{s,t-1} + \frac{(1 - \omega\mu^*)b_{st} - \mu^* b_{s,t-1}}{(\mu - \beta)^*}\right\}$$

for this choice of s, which, with the help of (3.21) and the induction hypothesis, is readily seen to equal $g_s(\beta)^* J f_t(\alpha)$, as needed.

STEP 3 *is to verify the formula*

$$< R_\alpha f_t, g_s >_{\mathcal{M}} - < f_t + \alpha R_\alpha f_t, \mu g_s + g_{s-1} >_{\mathcal{M}} = w_s^* J f_t(\alpha)$$

for $s = \sigma + 1, \ldots, \sigma + k$ *and* $t = \tau + 1, \ldots, \tau + n$.

PROOF OF STEP 3. Let E_{st} denote the left hand side of the asserted formula. Then if $t = \tau + 1$ it is readily checked that

$$E_{st} = \frac{1}{\omega - \alpha}\{(1 - \omega\mu^*)b_{st} - \omega b_{s-1,t}\}$$

which, by (3.21), is equal to $w_s^* J f_t(\alpha)$ for this choice of t. The proof for general t proceeds by induction. By (3.19) and (3.20)

$$E_{st} = \frac{1}{\alpha - \omega}\{E_{s,t-1} + (\omega\mu^* - 1)b_{st} + \omega b_{s-1,t} + \mu^* b_{s,t-1} + b_{s-1,t-1}\}$$

which in turn is readily seen to reduce to what it should by the induction hypothesis and (3.21).

STEP 4 *is to verify (3.9) for the remaining choices of* s *and* t.

PROOF OF STEP 4. The proof is by induction on s, the first step of which is justified by Step 2. By (3.18) and (3.17)

$$L_{st} = \frac{1}{(\beta - \mu)^*}\{L_{s-1,t} + E_{st}\}$$

which in turn is equal to $g_s(\beta)^* J f_t(\alpha)$, as needed, by the induction hypothesis and Step 3.

This completes the proof of both Step 4 and the theorem. ∎

4. EQUATIONS FOR THE GRAM MATRIX WHEN $\Delta_+ = \mathbb{C}_+$

In this section we shall work out the matrix equations for the "Gram matrix" of certain chains which correspond to the identity (2.6).

If $\rho_\omega(\lambda) = -2\pi i\ (\lambda - \omega^*)$, then every chain of length n except for the chain

$$v_1, \lambda v_1 + v_2, \ldots, \lambda^{n-1} v_1 + \ldots + v_n \tag{4.1}$$

can be expressed in the form

$$\begin{aligned} f_1(\lambda) &= \frac{v_1}{\rho_\omega(\lambda)} \\ f_j(\lambda) &= (-2\pi i\ )^{j-1}\frac{v_1}{\rho_\omega(\lambda)^j} + (-2\pi i\ )^{j-2}\frac{v_2}{\rho_\omega(\lambda)^{j-1}} + \ldots + \frac{v_j}{\rho_\omega(\lambda)}\ , \quad j = 2, \ldots, n\ . \end{aligned} \tag{4.2}$$

THEOREM 4.1. *If* $\rho_\omega(\lambda) = -2\pi i\ (\lambda - \omega^*)$,

$$f_t(\lambda) = \begin{cases} v_t/\rho_\omega(\lambda) & \text{for} \quad t = \tau + 1 \\ \{(-2\pi i\ )f_{t-1}(\lambda) + v_t\}/\rho_\omega(\lambda) & \text{for} \quad t = \tau + 2, \ldots, \tau + n \end{cases}$$

and

$$g_s(\lambda) = \begin{cases} w_s/\rho_\mu(\lambda) & \text{for} \quad s = \sigma + 1 \\ \{-2\pi i\ g_{s-1}(\lambda) + w_s\}/\rho_\mu(\lambda) & \text{for} \quad s = \sigma + 2, \ldots, \sigma + k \end{cases}$$

are chains in a space $\mathcal{M}$ *with an inner product* $<\ ,\ >_{\mathcal{M}}$ *(definite or not) and if*

$$V = [v_{\tau+1} \ldots v_{\tau+n}] \qquad \text{and} \qquad W = [w_{\sigma+1} \ldots w_{\sigma+n}]\ ,$$

then the $k \times n$ *matrix* B *with entries*

$$b_{st} =< f_t, g_s >_{\mathcal{M}}$$

is a solution of the matrix equation

$$B(\omega^* I_n + Z) - (\mu^* I + Z)^* B = \frac{1}{2\pi i} W^* J V$$

if and only if

$$< R_\alpha f_t, g_s + \beta R_\beta g_s >_{\mathcal{M}} - < f_t + \alpha R_\alpha f_t, R_\beta g_s >_{\mathcal{M}} = 2\pi i\ g_s(\beta)^* J f_t(\alpha) \qquad (4.3)$$

for $s = \sigma + 1, \ldots, \sigma + k$ *and* $t = \tau + 1, \ldots, \tau + n$ *and every choice of* α *and* β *in* $\mathbb{C}$ *with* $\alpha \neq \omega^*$ *and* $\beta \neq \mu^*$.

PROOF. We shall make extensive use of the formulas

$$f_t(\lambda) = (R_\alpha f_{t-1})(\lambda) + (\omega^* - \alpha) R_\alpha f_t \qquad (4.4)$$

and

$$g_s(\lambda) = (R_\beta g_{s-1})(\lambda) + (\mu^* - \beta) R_\beta g_s\ , \qquad (4.5)$$

which are valid for $t = \tau + 1, \ldots, \tau + n$ and $s = \sigma + 1, \ldots, \sigma + k$, respectively, with the understanding that $f_\tau = g_\sigma = 0$. The formulas themselves are obtained from Lemma 3.2.

Suppose first that (4.3) holds and choose $\alpha = \beta \in \mathbb{R}$ therein. Then it reduces to

$$< R_\alpha f_t, g_s >_{\mathcal{M}} - < f_t, R_\alpha g_s >= 2\pi i\ g_s(\alpha)^* J f_t(\alpha)\ , \qquad (4.6)$$

which, when multiplied by $(\omega^* - \alpha)$ and added to (4.6) with f_t replaced by f_{t-1} yields

$$< f_t, g_s >_{\mathcal{M}} - < (\omega^* - \alpha)f_t + f_{t-1}, R_\alpha g_s >_{\mathcal{M}} = g_s(\alpha)^* J v_t \ , \tag{4.7}$$

thanks to (4.4). Now multiply (4.7) through by $\mu - \alpha$ and then add to it (4.7) with g_s replaced by g_{s-1} to obtain

$$\begin{aligned}(\mu - \alpha) < f_t, g_s >_{\mathcal{M}} - &< (\omega^* - \alpha)f_t + f_{t-1}, g_s >_{\mathcal{M}} \\ + < f_t, g_{s-1} >_{\mathcal{M}} &= \{(\mu^* - \alpha)g_s(\alpha) + g_{s-1}(\alpha)\}^* J v_t \\ &= -\frac{1}{2\pi i} w_s^* J v_t \ ,\end{aligned}$$

via (4.5). But this is readily seen to reduce to

$$(\omega^* - \mu)b_{st} + b_{s,t-1} - b_{s-1,t} = \frac{1}{2\pi i} w_s^* J v_t \ . \tag{4.8}$$

It remains to check that the conclusion is also valid if $s = \sigma$ and/or $t = \tau$, with the usual conventions, but that is straightforward. Thus the proof that (4.3) implies the stated matrix equation is complete.

The converse is obtained in steps, much as in the proofs of Theorems 3.1–3.3. To this end, let L_{st} denote the left hand side of (4.3).

STEP 1. *(4.3) is valid for* $s = \sigma + 1$ *and* $t = \tau + 1$.

PROOF OF STEP 1. It follows readily from (4.4) and (4.5) that for this choice of s and t

$$L_{st} = \frac{(\omega^* - \mu)b_{st}}{(\omega^* - \alpha)(\beta - \mu^*)^*} \ .$$

The rest is immediate from (4.8) and the definitions of f_t and g_s.

STEP 2. *(4.3) is valid for* $s = \sigma + 1$ *and* $t = \tau + 1, \ldots, \tau + n$.

PROOF OF STEP 2. The proof is by induction on t. First, with the help of (4.4), it is readily checked that

$$L_{st} = \frac{1}{\alpha - \omega^*}\{L_{s,t-1} - < f_t, g_s + \beta R_\beta g_s >_{\mathcal{M}} + < \omega^* f_t + f_{t-1}, R_\beta g_s >_{\mathcal{M}} \ .$$

But for $s = \sigma + 1$, $R_\beta g_s = g_s/(\mu^* - \beta)$ and therefore the preceding formula reduces to

$$L_{st} = \frac{1}{\alpha - \omega^*}\left\{L_{s,t-1} + \frac{(\omega^* - \mu)b_{st} + b_{s,t-1}}{\mu - \beta^*}\right\}$$

which in turn is easily identified with $2\pi i\, g_s(\beta)^* J f_t(\alpha)$ for this choice of s, by the induction hypothesis and (4.8).

STEP 3 *is to verify the formula*

$$b_{st} - < R_\alpha f_t, (\mu - \alpha)^* g_s + g_{s-1} >_{\mathcal{M}} = w_s^* J f_t(\alpha)$$

for $s = \sigma + 1, \ldots, \sigma + k$ *and* $t = \tau + 1, \ldots, \tau + n$.

PROOF OF STEP 3. Let E_{st} denote the left side of the asserted formula. Then, since $R_\alpha f_t = f_t/(\omega^* - \alpha)$ for $t = \tau + 1$, it is readily checked with the help of (4.8) that

$$E_{st} = \frac{(\omega^* - \mu)b_{st} - b_{s-1,t}}{\omega^* - \alpha} = \frac{w_s^* J v_t}{2\pi i\ (\omega^* - \alpha)}$$

for this choice of t, as needed. The proof for general t is established by induction: By (4.4),

$$\begin{aligned} E_{st} &= b_{st} + \frac{1}{\alpha - \omega^*}\{< f_t - R_\alpha f_{t-1}, (\mu - \alpha)^* g_s + g_{s-1} >_{\mathcal{M}}\} \\ &= b_{st} + \frac{1}{\alpha - \omega^*}\{(\mu - \alpha)b_{st} + b_{s-1,t} + E_{s,t-1} - b_{s,t-1}\} \\ &= \frac{1}{\alpha - \omega^*}\{E_{s,t-1} + (\mu - \omega^*)b_{st} + b_{s-1,t} - b_{s,t-1}\}\ , \end{aligned}$$

which in turn is readily seen to be equal to $w_s^* J f_t(\alpha)$ by (4.8) and the induction hypothesis.

STEP 4 *is to verify (4.3) for the remaining choices of* s *and* t.

PROOF OF STEP 4. The proof is by induction on s, using Step 2 as a springboard: By (4.5), and a little computation,

$$L_{st} = \frac{1}{\beta^* - \mu}\{L_{s-1,t} + E_{st}\}\ .$$

The rest is plain by the induction hypothesis and Step 3. This completes the proof of both Step 4 and the theorem. ∎

THEOREM 4.2. *If* $\rho_\omega(\lambda) = -2\pi i\ (\lambda - \omega^*)$,

$$f_t(\lambda) = \begin{cases} v_t/\rho_\omega(\lambda) & \textit{for} \quad t = \tau + 1 \\ \{-2\pi i\ f_{t-1}(\lambda) + v_t\}/\rho_\omega(\lambda) & \textit{for} \quad t = \tau + 1, \ldots, \tau + n \end{cases}$$

and

$$g_s(\lambda) = \begin{cases} w_s/2\pi i & \textit{for} \quad s = \sigma + 1 \\ \lambda g_{s-1}(\lambda) + w_s/2\pi i & \textit{for} \quad s = \sigma + 2, \ldots, \sigma + k \end{cases}$$

are chains in a space $\mathcal{M}$ with an inner product $<\ ,\ >_{\mathcal{M}}$ (definite or not) and if

$$V = [v_{\tau+1} \cdots v_{\tau+n}] \quad \textit{and} \quad W = [w_{\sigma+1} \cdots w_{\sigma+k}] ,$$

then the $k \times n$ matrix B with st entry

$$b_{st} = < f_t, g_s >_{\mathcal{M}}$$

is a solution of the matrix equation

$$(OI_k + Z)^* B(\omega^* I_n + Z) - B = \frac{1}{2\pi i} W^* J V$$

if and only if (4.3) holds for this choice of f_t, $t = \tau + 1, \ldots, \tau + n$, and g_s, $s = \sigma + 1, \ldots, \sigma + k$, and every choice of α and β in $\mathbb{C}$ except for $\alpha = \omega^*$.

PROOF. The proof rests heavily on the formulas (4.4) and

$$(R_\beta g_s)(\lambda) = \beta(R_\beta g_{s-1})(\lambda) + g_{s-1}(\lambda) , \tag{4.9}$$

where, as usual, $g_\sigma = 0$.

Suppose first that (4.3) holds, and, for this part of the proof, take $\alpha = \beta \in \mathbb{R}$. Then (4.7) remains valid for the present choice of f_t and g_s also. Moreover, much as before, let us subtract α times (4.7) with g_s replaced by g_{s-1} from (4.7) to obtain

$$< f_t, g_s - \alpha g_{s-1} >_{\mathcal{M}} - < (\omega^* - \alpha) f_t + f_{t-1}, g_{s-1} >_{\mathcal{M}} = \{g_s(\alpha) - \alpha g_{s-1}(\alpha)\}^* J v_t ,$$

thanks to (4.9). But this is the same as

$$b_{st} - \omega^* b_{s-1,t} - b_{s-1,t-1} = -\frac{1}{2\pi i} w_s^* J v_t , \tag{4.10}$$

which in turn, with the usual conventions, is equivalent to the stated matrix equation.

The proof of the converse proceeds in steps. Let L_{st} denote the left hand side of (4.3) for the current choice of g_s and f_t.

STEP 1. *(4.3) is valid for $s = \sigma + 1$ and $t = \tau + 1$.*

PROOF OF STEP 1. For this choice of s and t,

$$L_{st} = \frac{b_{st}}{\omega^* - \alpha} ,$$

by (4.4) and (4.9). The rest is immediate from (4.10).

STEP 2. *(4.3) is valid for $s = \sigma + 1$ and $t = \tau + 1, \ldots, \tau + n$.*

PROOF OF STEP 2. Since $R_\beta g_s = 0$ for $s = \sigma + 1$,

$$L_{st} = < R_\alpha f_t, g_s >_{\mathcal{M}}$$

for this choice of s. The proof is now easily completed by induction on t: By (4.4),

$$L_{st} = \frac{1}{\omega^* - \alpha}\{b_{st} - L_{s,t-1}\} ,$$

and the rest is plain by the induction hypothesis and (4.10).

STEP 3.

$$b_{s-1,t} - < R_\alpha f_t, g_s - \alpha^* g_{s-1} >_{\mathcal{M}} = w_s^* J f_t(\alpha)$$

for $s = \sigma + 1, \ldots, \sigma + k$ and $t = \tau + 1, \ldots, \tau + n$.

PROOF OF STEP 3. Let E_{st} denote the left hand side of the asserted formula. If $t = \sigma + 1$, then it is readily seen that

$$E_{st} = \frac{b_{st} - \omega^* b_{s-1,t}}{\alpha - \omega^*}$$

and hence, by (4.10), that the asserted formula is correct for this choice of t. The proof for general t is by induction: By (4.4),

$$\begin{aligned} E_{st} &= b_{s-1,t} - \frac{1}{\omega^* - \alpha}\{E_{s,t-1} + b_{st} - \alpha b_{s-1,t} - b_{s-1,t-1}\} \\ &= \frac{1}{\alpha - \omega^*}\{E_{s,t-1} + b_{st} - \omega^* b_{s-1,t} - b_{s-1,t-1}\} , \end{aligned}$$

which is easily seen to be equal to $w_s^* J f_t(\alpha)$, by (4.10) and the induction hypothesis.

STEP 4 *is to verify (4.3) for the remaining choices of s and t.*

PROOF OF STEP 4. The proof is by induction on s, using Step 2 as a springboard. By (4.9),

$$L_{st} = \beta^* L_{s-1,t} - E_{st}$$

and the rest is plain by Step 3 and the induction hypothesis.

This completes the proof of both Step 4 and the theorem. ∎

THEOREM 4.3. *If $\rho_\omega(\lambda) = -2\pi i\ (\lambda - \omega^*)$,*

$$f_t(\lambda) = \begin{cases} v_t/2\pi i & \text{for} \quad t = \tau + 1 \\ \lambda f_{t-1}(\lambda) + v_t/2\pi i & \text{for} \quad t = \tau + 2, \ldots, \tau + n \end{cases}$$

and

$$g_s(\lambda) = \begin{cases} w_s/2\pi i & \text{for} \quad s = \sigma + 1 \\ \lambda g_{s-1}(\lambda) + w_s/2\pi i & \text{for} \quad s = \sigma + 2, \ldots, \sigma + k \end{cases}$$

are chains in a space $\mathcal{M}$ with an inner product $< \ , \ >_{\mathcal{M}}$ (definite or not) and if

$$V = [v_{\tau+1} \cdots v_{\tau+n}] \qquad \text{and} \qquad W = [w_{\sigma+1} \cdots w_{\sigma+n}] \ ,$$

then the $k \times n$ matrix B with st entry

$$b_{st} = < f_t, g_s >_{\mathcal{M}}$$

is a solution of the matrix equation

$$(OI_k + Z)^* B - B(OI_n + Z) = \frac{1}{2\pi i} W^* J V$$

if and only if (4.3) holds for this choice of f_t, $t = \tau + 1, \ldots, \tau + n$, and g_s, $s = \sigma + 1, \ldots, \sigma + k$, for every choice of α and β in $\mathbb{C}$.

PROOF. The proof rests heavily on the formulas (4.9) and its counterpart for the current choice of the f_t:

$$(R_\alpha f_t)(\lambda) = \alpha(R_\alpha f_{t-1})(\lambda) + f_{t-1}(\lambda) \ , \tag{4.11}$$

with the understanding that $f_\tau = 0$.

Suppose first that (4.3) holds and that $\alpha = \beta \in \mathbb{R}$, in which case it reduces to (4.6). Then, subtracting α times (4.6) with f_t replaced by f_{t-1} from (4.6) yields

$$< f_{t-1}, g_s >_{\mathcal{M}} - < f_t - \alpha f_{t-1}, R_\alpha g_s >_{\mathcal{M}} = g_s(\alpha)^* J v_t \ , \tag{4.12}$$

thanks to (4.11). Next, subtract α times (4.12) with g_s replaced by g_{s-1} from (4.12) to get

$$< f_{t-1}, g_s - \alpha g_{s-1} >_{\mathcal{M}} - < f_t - \alpha f_{t-1}, g_{s-1} >_{\mathcal{M}} = -\frac{1}{2\pi i} w_s^* J v_t \ .$$

But this is the same as

$$b_{s-1,t} - b_{s,t-1} = \frac{1}{2\pi i} w_s^* J v_t \tag{4.13}$$

for $s = \sigma + 1, \ldots, \sigma + k$ and $t = \tau + 1, \ldots, \tau + n$ with the understanding that $b_{st} = 0$ if either $s = \sigma$ or $t = \tau$. Therefore B is a solution of the stated matrix equation.

Now, suppose conversely that B is a solution of the stated matrix equation and let L_{st} denote the left hand side of (4.3) for the current choice of g_s and f_t.

STEP 1. *(4.3) is valid for $s = \sigma + 1$ and $t = \tau + 1$.*

PROOF OF STEP 1. For this choice of s and t, $R_\alpha f_t = R_\beta g_s = 0$ and hence $L_{st} = 0$, whereas the right hand side of (4.3) is equal to

$$2\pi i \; g_s(\beta)^* J f_t(\alpha) = -\frac{1}{2\pi i} w_s^* J v_t \; .$$

But this also vanishes by (4.13) and hence all is well.

STEP 2. *(4.3) is valid for $s = \sigma + 1$ and $t = \tau + 1, \ldots, \tau + n$.*

PROOF OF STEP 2. The proof is by induction on t, using Step 1 for the start. By (4.11),

$$L_{st} = \alpha L_{s,t-1} + b_{s,t-1} \; ,$$

since $R_\beta g_s = 0$ for $s = \sigma + 1$. The rest follows easily by the induction hypothesis and (4.13).

STEP 3.

$$< f_t + \alpha R_\alpha f_t, g_{s-1} >_{\mathcal{M}} - < R_\alpha f_t, g_s >_{\mathcal{M}} = w_s^* J f_t(\alpha)$$

for $s = \sigma + 1, \ldots, \sigma + k$ and $t = \tau + 1, \ldots, \tau + n$.

PROOF OF STEP 3. The formula is easily seen to be valid via (4.13) when $t = \tau + 1$, because then $R_\alpha f_t = 0$. The proof for general t is by induction: Let E_{st} denote the left hand side of the stated formula. Then, by (4.11),

$$E_{st} = \alpha E_{s,t-1} + b_{s-1,t} - b_{s,t-1} \; ,$$

which in turn is readily seen to equal $w_s^* J f_t(\alpha)$, by the induction hypothesis and (4.13).

STEP 4 *is to verify (4.3) for the remaining choices of s and t.*

PROOF OF STEP 4. The proof is by induction on s, using Step 2 to start, as usual: By (4.9),

$$L_{st} = \beta^* L_{s-1,t} - E_{st}$$

and the rest is plain by Step 3 and the induction hypothesis. ∎

5. CONSOLIDATED MATRIX EQUATIONS AND A BASIC REFORMULATION

In this section we shall consolidate the conclusions drawn in the preceding two sections into two unified statements and shall then use these statements to reformulate Theorem 2.1 in the language of matrix equations. To this end it is convenient to introduce the following supplementary notation (which will intervene again in the solution of the interpolation problem below):

$$\varphi_{\omega,j}(\lambda) = \begin{cases} \lambda^j/\rho_\omega(\lambda)^{j+1} & \text{if} \quad \Delta_+ = \mathbb{D} \\ (-2\pi i\ )^j/\rho_\omega(\lambda)^{j+1} & \text{if} \quad \Delta_+ = \mathbb{C}_+ \ , \end{cases}$$

$$\psi_{\omega,j}(\lambda) = \begin{cases} 1/(\lambda-\omega)^{j+1} & \text{if} \quad \Delta_+ = \mathbb{D} \\ 1/\{2\pi i\ (\lambda-\omega)^{j+1}\} & \text{if} \quad \Delta_+ = \mathbb{C}_+ \ , \end{cases}$$

$$\theta_j(\lambda) = \lambda^j/2\pi i \quad .$$

The symbols $\Phi_{\omega,\ell}(\lambda)$, $\Psi_{\omega,\ell}(\lambda)$, $\Theta_\ell(\lambda)$, based on the corresponding capital letter, will be used to denote $\ell \times \ell$ upper triangular Toeplitz matrices with top rows

$$[\varphi_{\omega,0} \ \dots \ \varphi_{\omega,\ell-1}] \ , \quad [\psi_{\omega,0} \ \dots \ \psi_{\omega,\ell-1}] \ \text{ and } \ [\theta_0 \ \dots \ \theta_{\ell-1}] \ ,$$

respectively. Then every chain of length ℓ which intervened in the preceding two sections can be described in terms of the columns of at least one (and "mostly" two) of the following $m \times \ell$ matrix valued functions:

$$V\Phi_{\omega,\ell}(\lambda) \ , \quad V\Psi_{\omega,\ell}(\lambda) \ , \quad V\Theta_\ell(\lambda) \ ,$$

where

$$V = [v_1 \ \dots \ v_\ell]$$

is an $m \times \ell$ constant matrix with columns $v_1, \dots, v_\ell$.

THEOREM 5.1. *Let $\mathcal{M}$ be an indefinite inner product space of $m \times 1$ vector valued functions which contains the columns*

$$[f_1 \ \dots \ f_\nu] = [F_1 \ \dots \ F_n] \ , \qquad \nu = \ell_1 + \dots + \ell_n \ ,$$

of the $m \times \ell_j$ matrix valued functions

$$F_j(\lambda) = \begin{cases} V_j\Phi_{\omega_j,\ell_j}(\lambda) \ , & j = 1, \dots, k \\ -V_j\Psi_{\omega_j,\ell_j}(\lambda) \ , & j = k+1, \dots, n \ , \end{cases}$$

where the V_j *are* $m \times \ell_j$ *constant matrices and* Φ_{ω_j,ℓ_j} *and* Ψ_{ω_j,ℓ_j} *are defined as above for* $\Delta_+ = \mathbb{D}$. *Then the identity*

$$< f_t + \alpha R_\alpha f_t, f_s + \beta R_\beta f_s >_{\mathcal{M}} - < R_\alpha f_t, R_\beta f_s >_{\mathcal{M}} = f_s(\beta)^* J f_t(\alpha) \qquad (5.1)$$

holds for $s,t = 1,\dots,\nu$ *and every choice of* α, β *in the common domain of analyticity* Ω *of* $f_1,\dots,f_\nu$ *if and only if the* $\nu \times \nu$ *matrix* $\mathbb{P}$ *with st entry*

$$p_{st} = < f_t, f_s >_{\mathcal{M}} , \qquad s,t = 1,\dots,\nu ,$$

is a solution of the matrix equation

$$M^*\mathbb{P}M - N^*\mathbb{P}N = V^*JV , \qquad (5.2)$$

where

$$M = I_{\ell_1} \oplus \dots \oplus I_{\ell_k} \oplus Z_{\omega_{k+1}} \oplus \dots \oplus Z_{\omega_n} ,$$

$$N = Z_{\omega_1^*} \oplus \dots \oplus Z_{\omega_k^*} \oplus I_{\ell_{k+1}} \oplus \dots \oplus I_{\ell_n} ,$$

M_{jj} *and* N_{jj} *are* $\ell_j \times \ell_j$, $j = 1,\dots,n$, *and* $V = [V_1 \ \dots \ V_n]$.

PROOF. Suppose first that $\mathbb{P}$ is a solution of (5.2). Then

$$M_{ii}^*\mathbb{P}_{ij}M_{jj} - N_{ii}^*\mathbb{P}_{ij}N_{jj} = V_i^*JV_j , \qquad i,j = 1,\dots,n , \qquad (5.3)$$

where M_{ij}, N_{ij} and $\mathbb{P}_{ij}$ are the $\ell_i \times \ell_j$ subblocks in the natural partitioning induced by the chains. The validity of (5.2) for α, β in Ω then drops out easily by successive applications of Theorems 3.1, 3.2 and 3.3, according as (1) $i,j = 1,\dots,k$, (2) $i = k+1,\dots,n$ and $j = 1,\dots,k$ or (3) $i,j = k+1,\dots,n$, respectively.

Conversely, if (5.1) holds for every choice of α, β in Ω, then, for any specific choice of f_t and f_s, it also holds by analytic continuation for every choice of α in the domain of analyticity of f_t and every choice of β in the domain of analyticity of g_s. The rest is a straightforward application of Theorems 3.1–3.3. ∎

THEOREM 5.2. *Let* $\mathcal{M}$ *be an indefinite inner product space of* $m \times 1$ *vector valued functions which contains the columns*

$$[f_1 \ \dots \ f_\nu] = [F_1 \ \dots \ F_n] ,$$

of the $m \times \ell_j$ *matrix valued functions*

$$F_j(\lambda) = \begin{cases} V_j\Phi_{\omega_j,\ell_j}(\lambda) , & j = 1,\dots,k \\ -V_j\Psi_{\omega_j,\ell_j}(\lambda) , & j = k+1,\dots,r , \\ V_j\Theta_{\ell_j}(\lambda) , & j = r+1,\dots,n , \end{cases}$$

where the V_j *are* $m \times \ell_j$ *constant matrices and* Φ_{ω_j,ℓ_j}, Ψ_{ω_j,ℓ_j} *and* Θ_{ℓ_j} *are defined as above for* $\Delta_+ = \mathbb{C}_+$. *Then the identity*

$$< R_\alpha f_t, f_s + \beta R_\beta f_s >_{\mathcal{M}} - < f_t + \alpha R_\alpha f_t, R_\beta f_s >_{\mathcal{M}} = 2\pi i \ f_s(\beta)^* J f_t(\alpha) \tag{5.4}$$

holds for $s,t = 1,\ldots,\nu$ *and every choice of* α, β *in the common domain of analyticity* Ω *of* $f_1,\ldots,f_\nu$ *if and only if the* $\nu \times \nu$ *matrix* $\mathbb{P}$ *with st entry*

$$p_{st} =< f_t, f_s >_{\mathcal{M}} \ , \qquad s,t = 1,\ldots,\nu \ ,$$

is a solution of the matrix equation

$$M^*\mathbb{P}N - N^*\mathbb{P}M = \frac{1}{2\pi i} V^*JV \ , \tag{5.5}$$

where

$$M = I_{\ell_1} \oplus \ldots \oplus I_{\ell_k} \oplus I_{\ell_{k+1}} \oplus \ldots \oplus I_{\ell_r} \oplus Z \oplus \ldots \oplus Z \ ,$$

$$N = Z_{\omega_1^*} \oplus \ldots \oplus Z_{\omega_\kappa^*} \oplus Z_{\omega_{k+1}} \oplus \ldots \oplus Z_{\omega_r} \oplus I_{\ell_{r+1}} \oplus \ldots \oplus I_{\ell_n} \ ,$$

M_{jj} *and* N_{jj} *are* $\ell_j \times \ell_j$, $j = 1,\ldots,n$, *and* $V = [V_1 \ \ldots \ V_n]$.

PROOF. The proof is much the same as the proof of Theorem 5.1 except that now the burden rests on Theorems 4.1–4.3. When applying them, it is important to bear in mind that in the present setting $\varphi_{\omega,j}(\lambda) = -\psi_{\omega^*,j}(\lambda)$. ∎

Theorems 5.1 and 5.2 may be used to reformulate Theorem 2.1 in the language of matrix equations. In fact, since Sections 3 and 4 are formulated in terms of general indefinite inner product spaces it is now easy to make a more general formulation for finite dimensional reproducing kernel Pontryagin spaces with reproducing kernels of the special form (2.3) based on a rational $m \times m$ matrix valued function U which is J unitary on ∂ except for possibly finitely many points. The theory of such spaces (and their infinite dimensional brothers) is worked out in [AD2], where they are called $\mathcal{K}(U)$ spaces.

THEOREM 5.3. *Suppose that the system of equations (5.2) [resp. (5.5)] admits an invertible Hermitean solution* $\mathbb{P}$ *and let* $\mathcal{M}$ *now denote the span of the functions* $f_1,\ldots,f_\nu$ *specified in Theorem 5.1 [resp. 5.2] with indefinite inner product*

$$< f_j, f_i >_{\mathcal{M}} = p_{ij} \ , \qquad i,j = 1,\ldots,\nu \ . \tag{5.6}$$

Then $\mathcal{M}$ *is a* $\mathcal{K}(U)$ *space, i.e. it is a reproducing kernel Pontryagin space with reproducing kernel of the form (2.3) based on an* $m \times m$ *rational matrix valued function* U *which is*

J unitary on $\mathbb{T}$ *[resp.* $\mathbb{R}$*] (except for at most finitely many points) and which may be specified by the recipe*

$$U(\lambda) = I_m - \rho_\beta(\lambda)F(\lambda)\mathbb{P}^{-1}F(\beta)^*J \tag{5.7}$$

for some choice of $\beta \in \partial$*, where*

$$F(\lambda) = [f_1(\lambda) \ \dots \ f_\nu(\lambda)] \ .$$

If $\mathbb{P} > 0$*, then* U *is* J *inner and* $\mathcal{M}$ *is an* $\mathcal{H}(U)$ *space.*

Conversely, if $\mathcal{M}$ *is a reproducing kernel Pontryagin space with basis* $f_1, \dots, f_\nu$ *and reproducing kernel of the form (2.3) with* U *rational and* J *unitary on* $\mathbb{T}$ *[resp.* $\mathbb{R}$*] except for at most finitely many points, then the Gram matrix with* ij *entry given by (5.6) is an invertible solution of (5.2) [resp. (5.5)]. If* $\mathcal{M}$ *is a Hilbert space, then* U *is* J *inner and* $\mathbb{P}$ *is positive definite.*

PROOF. Suppose first that $\mathbb{P}$ is a positive definite solution of (5.2) [resp. (5.5)]. Then, by Theorem 5.1 [resp. 5.2] the requisite identity (2.5) [resp. (2.6)] is met and hence, since $\mathcal{M}$ is automatically R_α invariant by construction, $\mathcal{M}$ is an $\mathcal{H}(U)$ space by Theorem 2.1. In just the same way it follows from Theorem 6.12 of [AD2] that if $\mathbb{P}$ is a Hermitean invertible solution of (5.2) [resp. (5.5)], then $\mathcal{M}$ is a $\mathcal{K}(U)$ space with U rational and J unitary a.e. on $\mathbb{T}$ [resp. $\mathbb{R}$]. Formula (5.7) is just a different way of writing (2.7).

Conversely, if $\mathcal{M}$ is a $\mathcal{K}(U)$ space, with a U which is J unitary a.e. on $\mathbb{T}$ [resp. $\mathbb{R}$], then by Theorem 6.9 of [AD2], the indefinite inner product version of the identity (2.5) [resp. (2.6)] holds and hence by Theorem 5.1 [resp. (5.2)] the indicated Gram matrix $\mathbb{P}$ with entries given by (5.6) is an invertible Hermitean solution of (5.2) [resp. (5.5)]. If $\mathcal{M}$ is a Hilbert space, then U is J inner and $\mathbb{P}$ is positive definite. In this instance, Theorem 2.1 could have been used directly. ∎

An immediate corollary (which is essentially a restatement in a more striking way) is:

THEOREM 5.4. *If* $\mathcal{M}$ *is a reproducing kernel Pontryagin space with the basis* $f_1, \dots, f_\nu$ *which is specified in Theorem 5.1 [resp. 5.2] and Gram matrix* $\mathbb{P}$ *(defined as in (5.6)), then* $\mathcal{M}$ *is a* $\mathcal{K}(U)$ *space with* U *rational and* J *unitary a.e. on* $\mathbb{T}$ *[resp.* $\mathbb{R}$*] if and only if* $\mathbb{P}$ *is a solution of (5.2) [resp. (5.5)].*

It is perhaps well to remark that the formulations in terms of equations (5.2) and (5.5), though elegant, make things look harder than they really are. Because of the block diagonal form of M and N, equations (5.2) and (5.5) readily reduce to an uncoupled set of equations, one for each of the subblocks $\mathbb{P}_{ij}$ of $\mathbb{P}$. Indeed, these are precisely the matrix equations which are exhibited in the theorems of Section 3 and 4, which is no surprise since this was our starting point. The equations under consideration fall into one of the following two types:

$$B - (Z_\alpha)^* B Z_\beta = \cdots \tag{5.8}$$

or

$$(Z_\alpha)^* B - Z_\beta B = \cdots . \tag{5.9}$$

It is well known that (5.8) [resp. (5.9)] is uniquely solvable if and only if $\alpha^*\beta \neq 1$ [resp. $\alpha^* \neq \beta$], as is apparent by reexpressing the equations as

$$(1 - \alpha^*\beta)B = Z^* B Z + \cdots \tag{5.10}$$

and

$$(\alpha^* - \beta)B = ZB - Z^* B + \cdots , \tag{5.11}$$

respectively, both of which can easily be solved by iterations since the Z's are nilpotent. This is selfevident for (5.10) and just a shade harder for (5.11); see e.g. Russell [Ru] for help with the latter if need be.

The same conclusions are also apparent from the corresponding detailed recursion formulas (with an appropriate adaptation of the notation): (3.10) [resp. (3.16)], the point being that the equation is uniquely solvable if and only if the coefficient of b_{st} is nonzero. Moreover, it will turn out in the applications to interpolation theory which will be treated below, that the solution to the requisite equation will often be known from other considerations:

THEOREM 5.5. *If in Theorem 3.1 [resp. 4.1]* ω *and* μ *belong to* Δ_+*, then the* $k \times n$ *matrix* B *with entries*

$$b_{st} = < Jf_t, g_s >$$

based on the chains given there is the one and only solution of the indicated matrix equation.

PROOF. Under the stated conditions f_t and g_s both belong to H^2_m. Therefore, by Theorem 2.2 and Theorem 3.1 [resp. 4.1], B is a solution of the matrix equation stated in the latter. The rest is immediate from the fact that under the given conditions these equations (one for $\mathbb{D}$ and one for $\mathbb{C}_+$) have only one solution. ∎

THEOREM 5.6. *If in Theorem 3.3 [resp. 4.1]* ω *and* μ *belong to* $\mathbb{D}$ *[resp.* ω *and* μ *belong to* $\mathbb{C}_-$*], then the* $k \times n$ *matrix* B *with entries*

$$b_{st} = \langle -Jf_t, g_s \rangle$$

based on the chains given there is the one and only solution of the indicated matrix equation.

PROOF. Under the stated conditions f_t and g_s both belong to $(H_m^2)^\perp$. Therefore, by Theorem 2.3 and Theorem 3.3 [resp. 4.1], B is a solution of the matrix equation stated in the latter. The rest is immediate from the fact that under the given conditions these equations (one for $\mathbb{D}$ and one for $\mathbb{C}_+$) have only one solution. ∎

Finally we remark that formula (5.7) can be reexpressed in realization theory form by taking advantage of the auxiliary identities

$$\Phi_{\omega,\ell}(\lambda)^{-1} = \begin{cases} I_\ell - \lambda Z_{\omega^*} & \text{if} \quad \Delta_+ = \mathbb{D} \\ -2\pi i \, (\lambda I_\ell - Z_{\omega^*}) & \text{if} \quad \Delta_+ = \mathbb{C}_+ \, , \end{cases} \tag{5.12}$$

$$\Psi_{\omega,\ell}(\lambda)^{-1} = \begin{cases} \lambda I_\ell - Z_\omega & \text{if} \quad \Delta_+ = \mathbb{D} \\ 2\pi i \, (\lambda I_\ell - Z_\omega) & \text{if} \quad \Delta_+ = \mathbb{C}_+ \, , \end{cases} \tag{5.13}$$

and

$$\Theta_\ell(\lambda)^{-1} = 2\pi i \, (I_\ell - \lambda Z) \, . \tag{5.14}$$

Formulas (7.5), (7.6) and (8.5) will serve as illustrative examples. They also suggest connections with the work of Alpay and Gohberg [AG] and Sakhnovich [S1], [S2], but these will not be pursued here.

6. DIVERSION ON J INNER FUNCTIONS

In this section we shall assemble some facts about J inner functions U for use in the sequel. For ease of exposition we shall assume that U is rational and shall fix $J = J_{pq}$ throughout this section and shall accordingly write U in the block form

$$U(\lambda) = \begin{bmatrix} A(\lambda) & B(\lambda) \\ C(\lambda) & D(\lambda) \end{bmatrix}$$

where A is $p \times p$, B is $p \times q$, C is $q \times p$ and D is $q \times q$.

Let Ω denote the set of points in Δ_+ at which U is analytic. It follows from the 22 block of the inequality

$$U(\lambda)^* J U(\lambda) \leq J \, ,$$

which is valid for every point $\lambda \in \Omega$, that

$$B(\lambda)^*B(\lambda) + I_q \le D(\lambda)^*D(\lambda)$$

and hence that D is invertible in Ω. Thus

$$\sigma_1 = BD^{-1} \quad \text{and} \quad \varepsilon_2^{-1} = D^{-1}$$

are analytic in Ω and furthermore are subject to the inequality

$$\sigma_1(\lambda)^*\sigma_1(\lambda) + \{D(\lambda)^*\}^{-1}\{D(\lambda)\}^{-1} \le I_q$$

for $\lambda \in \Omega$. But this in turn implies that both σ_1 and D^{-1} are pole free in Δ_+ and hence are analytic and contractive in all of Δ_+:

$$\sigma_1 \in \mathcal{S}_{p\times q} \quad \text{and} \quad \varepsilon_2^{-1} \in \mathcal{S}_{q\times q} \ .$$

In much the same way the inequality $U^*JU \le J$ in Ω leads to the auxiliary conclusion that

$$\sigma_2 = D^{-1}C \in \mathcal{S}_{q\times p}$$

Since D is invertible in Ω, U can be expressed in the form

$$U = \begin{bmatrix} I_p & \sigma_1 \\ 0 & I_q \end{bmatrix} \begin{bmatrix} \varepsilon_1 & 0 \\ 0 & \varepsilon_2 \end{bmatrix} \begin{bmatrix} I_p & 0 \\ \sigma_2 & I_q \end{bmatrix} \tag{6.1}$$

in Ω, with

$$\varepsilon_1 = A - BD^{-1}C \in \mathcal{S}_{p\times p}$$

and, as noted earlier,

$$\varepsilon_2 = D \ .$$

Formula (6.1), even if unfamiliar, is readily seen to be correct by direct computation; a proof that $\varepsilon_1 \in \mathcal{S}_{p\times p}$ may be found in Chapter 1 of [D].

Next we observe that if $S_o \in \mathcal{S}_{p\times q}$, then

$$CS_o + D = D(I_q + \sigma_2 S_o)$$

is invertible in Ω and hence the linear fractional transformation

$$T_U[S_o] = (AS_o + B)(CS_o + D)^{-1}$$

is well defined at every point in Ω for every choice of $S_o \in \mathcal{S}_{p\times q}$. In fact, since $T_U[S_o]$ is contractive in Ω, it can be identified with an element in $\mathcal{S}_{p\times q}$:

THEOREM 6.1. *If U is J inner and $S_o \in \mathcal{S}_{p\times q}$, then*

(1) $T_U[S_o] \in \mathcal{S}_{p\times q}$

(2) $T_U[S_o] - T_U[0] = \varepsilon_1 S_o(I_q + \sigma_2 S_o)^{-1}\varepsilon_2^{-1}$

(3) If also V is J inner, then

$$T_{UV}[S_o] = T_U[T_V[S_o]] \ .$$

PROOF. (1) and (3) are straightforward (they can also be found in Chapter 3 of [D]), whereas (2) is an easy consequence of the formula

$$\begin{aligned}(AS_o + B)(CS_o + D)^{-1} - BD^{-1} &= \{(AS_o + B) - BD^{-1}(CS_o + D)\}(CS_o + D)^{-1}\\ &= \varepsilon_1 S_o(CS_o + D)^{-1} \ .\end{aligned}$$

∎

THEOREM 6.2. *Let U be a J inner function which belongs to $L^\infty_{m\times m}(\partial)$ and suppose further that*

$$\underline{\underline{p}}\ U^* J f = 0 \quad \textit{for some} \quad f = \begin{bmatrix} g \\ h \end{bmatrix}$$

with components $g \in H^2_p$ and $h \in H^2_q$. Then

$$\underline{\underline{p}}\ \varepsilon_1^* g = 0 \tag{6.2}$$

and

$$\underline{\underline{p}}\ \sigma_1^* g = h \ . \tag{6.3}$$

PROOF. Since $\sigma_2 \in \mathcal{S}_{q\times p}$ and $\varepsilon_2^{-1} \in \mathcal{S}_{q\times q}$,

$$G = \begin{bmatrix} I_p & 0 \\ -\sigma_2 & I_q \end{bmatrix} \begin{bmatrix} I_p & 0 \\ 0 & \varepsilon_2^{-1} \end{bmatrix}$$

clearly belongs to $H^\infty_{m\times m}$ and therefore

$$\underline{\underline{p}}\ G^* \underline{\underline{q}}\,' = 0 \ .$$

Thus, for $f \in H_m^2$ as above,

$$\underline{\underline{p}}\, G^* \underline{\underline{p}}\, U^* J f = \underline{\underline{p}}\, G^* U^* J f$$

$$= \underline{\underline{p}} \begin{bmatrix} \varepsilon_1^* & 0 \\ 0 & I_q \end{bmatrix} \begin{bmatrix} I_p & 0 \\ \sigma_1^* & -I_q \end{bmatrix} \begin{bmatrix} g \\ h \end{bmatrix}$$

and the rest is a straightforward calculation. ∎

THEOREM 6.3. *Let U be a J inner function which belongs to $L_{m\times m}^\infty(\partial)$ and suppose further that*

$$\underline{\underline{q}}\,' U^{-1} f = 0 \quad \textit{for some} \quad f = \begin{bmatrix} g \\ h \end{bmatrix} ,$$

with components $g \in (H_p^2)^\perp$ and $h \in (H_q^2)^\perp$. Then

$$\underline{\underline{q}}\,' \sigma_1 h = g \tag{6.4}$$

and

$$\underline{\underline{q}}\,' \varepsilon_2^{-1} h = 0 . \tag{6.5}$$

PROOF. Since $\sigma_2 \in \mathcal{S}_{q\times p}$ and $\varepsilon_1 \in \mathcal{S}_{p\times p}$,

$$F = \begin{bmatrix} \varepsilon_1 & 0 \\ 0 & I_q \end{bmatrix} \begin{bmatrix} I_p & 0 \\ \sigma_2 & I_q \end{bmatrix}$$

belongs to $H_{m\times m}^\infty$ and hence

$$\underline{\underline{q}}\,' F \underline{\underline{p}} = 0 .$$

Thus, for $f \in (H_m^2)^\perp$ as above,

$$\underline{\underline{q}}\,' F \underline{\underline{q}}\,' U^{-1} f = \underline{\underline{q}}\,' F U^{-1} f$$

$$= \underline{\underline{q}}\,' \begin{bmatrix} I_p & 0 \\ 0 & \varepsilon_2^{-1} \end{bmatrix} \begin{bmatrix} I_p & -\sigma_1 \\ 0 & I_q \end{bmatrix} \begin{bmatrix} g \\ h \end{bmatrix}$$

and the rest drops out by a routine calculation (the identity $\varepsilon_1 A^* = I_p$ a.e. on ∂ guarantees the existence of ε_1^{-1} a.e. on ∂). ∎

We remark that the conclusions of the last two theorems are obtained by other methods in a less restrictive setting in Theorem 2.7 (and its corollaries) of [D].

7. LEFT INTERPOLATION

In this section we shall begin to explain how to apply the preceding theory to solve matrix versions of some classical interpolation problems in $\mathcal{S}_{p\times q}$, starting with "left" interpolation.

The data for the general left interpolation problem is a set of points $\omega_1,\ldots,\omega_n$ in Δ_+, a set of vectors $\xi_1,\ldots,\xi_n$ in $\mathbb{C}^p$ such that

$$\frac{\xi_1}{\rho_{\omega_1}},\ldots,\frac{\xi_n}{\rho_{\omega_n}}$$

are linearly independent and a second set of vectors

$$\eta_{11},\ldots,\eta_{1\ell_1};\ \eta_{21},\ldots,\eta_{2\ell_2};\cdots;\eta_{n1},\ldots,\eta_{n\ell_n}$$

in $\mathbb{C}^q$. The objective is to find conditions on the data which insure the existence of an $S\in\mathcal{S}_{p\times q}$ such that

$$\xi_j^*\frac{S^{(t-1)}(\omega_j)}{(t-1)!}=\eta_{jt}^*\ ,\qquad t=1,\ldots,\ell_j\ ,\ \ j=1,\ldots,n\ ,$$

and to describe the set of all such solutions when these conditions are met.

The special case in which $\ell_j=1$ for $j=1,\ldots,n$ is the left Nevanlinna Pick problem, whereas the special case in which $n=1$ is the left Carathéodory-Féjer problem. The former (as should become clear in a moment) involves n chains of length one whereas the latter involves one chain of length n. The general left interpolation problem includes both of these as special cases. It involves n chains of length $\ell_1,\ldots,\ell_n$, respectively, which are defined in terms of the data as follows:

$$v_{j1}=\begin{bmatrix}\xi_j\\ \eta_{j1}\end{bmatrix}\ ,\qquad v_{jt}=\begin{bmatrix}0\\ \eta_{jt}\end{bmatrix}\ ,\qquad t=2,\ldots,\ell_j\ ,$$

$$V_j=[v_{j1}\ \ldots\ v_{j\ell_j}]\ ,$$

$$F_j=V_j\Phi_{\omega_j,\ell_j}\ ,\qquad j=1,\ldots,n\ .$$

The chains are the columns of F_j, which we index by the rule

$$[f_1\ \ldots\ f_\nu]=[F_1\ \ldots\ F_n]\ ,$$

where $\nu=\ell_1+\ldots+\ell_n$. The relevance of these particular chains to the interpolation problem under study rests on the evaluation:

$$(\underline{\underline{p}}\ \{S^*\varphi_{\omega,j}\xi\})(\lambda)=\sum_{i=0}^{j}\varphi_{\omega,i}(\lambda)\left\{\frac{S^{(j-i)}(\omega)}{(j-i)!}\right\}^*\xi\ ,\tag{7.1}$$

which is valid for any choice of $\omega \in \Delta_+$, $S \in \mathcal{S}_{p\times q}$ and $\xi \in \mathbb{C}^p$; for a proof of (7.1) see Lemma 6.1 of [D]. Indeed, upon writing

$$f_t = \begin{bmatrix} g_t \\ h_t \end{bmatrix} , \qquad t = 1, \ldots, \nu ,$$

with components $g_t \in H_p^2$ and $h_t \in H_q^2$ it is readily checked that

LEMMA 7.1. *$S \in \mathcal{S}_{p\times q}$ is a solution of the left interpolation problem if and only if*

$$\underline{\underline{p}}\ S^* g_t = h_t$$

for $t = 1, \ldots, \nu$.

THEOREM 7.1. *The left interpolation problem admits a solution if and only if the $\nu \times \nu$ matrix $\mathbb{P}$ with st entry*

$$p_{st} = \langle Jf_t, f_s \rangle , \qquad s, t = 1, \ldots, \nu ,$$

is positive semidefinite.

PROOF. Suppose first that the left interpolation problem admits a solution $S \in \mathcal{S}_{p\times q}$. Then, by Lemma 7.1,

$$f_t = \begin{bmatrix} g_t \\ \underline{\underline{p}}\ S^* g_t \end{bmatrix} , \qquad t = 1, \ldots, \nu ,$$

and thus

$$\begin{aligned} \sum_{i,j=1}^{\nu} c_i^* p_{ij} c_j &= \sum_{i,j=1}^{\nu} \langle Jc_j f_j, c_i f_i \rangle \\ &= \langle J \left\{ \sum_{j=1}^{\nu} c_j f_j \right\} , \left\{ \sum_{i=1}^{\nu} c_i f_i \right\} \rangle \\ &= \left\| \sum_{j=1}^{\nu} c_j g_j \right\|^2 - \left\| \underline{\underline{p}}\ S^* \sum_{j=1}^{\nu} c_j g_j \right\|^2 \\ &\geq 0 \end{aligned}$$

for any choice of the constants $c_1, \ldots, c_\nu$. This proves the necessity of the condition $\mathbb{P} \geq 0$.

Suppose next that $\mathbb{P} > 0$ and let

$$\mathcal{M} = \operatorname{span}\{f_s : \ s = 1, \ldots, \nu\}$$

endowed with the J inner product:

$$< f_t, f_s >_{\mathcal{M}} = < Jf_t, f_s > = p_{st} \ .$$

Then, since $\mathbb{P} > 0$, $\mathcal{M}$ is clearly a reproducing kernel Hilbert space. It is also R_α invariant for every choice of $\alpha \in \Delta_+$ by construction and furthermore, by Theorem 2.2, the assigned inner product meets the requisite identities. Therefore, by Theorem 2.1, $\mathcal{M}$ is an $\mathcal{H}(U)$ space. Moreover, it is clear from the recipe (2.7) and the form of $f_1, \ldots, f_\nu$ that U is rational with all its poles in Δ_-. Thus, upon invoking formula (2.3), which expresses the reproducing kernel for $\mathcal{M}$ in terms of U, it follows that

$$v^* f_i(\omega) = < Jf_i, \frac{J - UJU(\omega)^*}{\rho_\omega} v >$$

for $i = 1, \ldots, \nu$ and every choice of $\omega \in \Delta_+$ and $v \in \mathbb{C}^m$. Since both of the summands in the right hand entry of the inner product belong to H_m^2, the inner product can be split into

$$< Jf_i, \frac{Jv}{\rho_\omega} > - < Jf_i, \frac{UJU(\omega)^*}{\rho_\omega} v > = v^* f_i(\omega) - < U^* Jf_i, \frac{JU(\omega)^*}{\rho_\omega} v > \ .$$

But this in turn implies that

$$< U^* Jf_i, \frac{JU(\omega)^* v}{\rho_\omega} > = 0$$

for every choice of $\omega \in \Delta_+$ and $v \in \mathbb{C}^m$ and hence, since $U(\omega)$ is invertible except for at most finitely many points (use the fact that $|\det U(\omega)| = 1$ for $\omega \in \partial$ to verify this), it follows readily from Example 1 that

$$\underline{\underline{p}}\, U^* Jf_i = 0 \ . \tag{7.2}$$

Therefore, by Theorem 6.2,

$$\underline{\underline{p}}\, \varepsilon_1^* g_i = 0 \tag{7.3}$$

and

$$\underline{\underline{p}}\, \sigma_1^* g_i = h_i \tag{7.4}$$

for $i = 1, \ldots, \nu$. Consequently, by Lemma 7.1, (7.4) exhibits σ_1 as a solution to the left interpolation problem, when $\mathbb{P} > 0$.

The sufficiency of the condition $\mathbb{P} \geq 0$ is established by a limiting argument. ∎

We remark that in the Nevanlinna Pick case it is easy to evaluate the (Pick) matrix $\mathbb{P}$ directly as the $n \times n$ matrix with st entry

$$p_{st} = \frac{v_s^* J v_t}{\rho_{\omega_t}(\omega_s)} , \qquad s,t = 1,\ldots,n ,$$

where $v_s = v_{s1}$ is the first (and only) column of the matrix V_s. For the general left interpolation problem the evaluation is more complicated. However, in view of Theorems 3.1 and 4.1, $\mathbb{P}$ can be identified as the unique solution of the matrix equation

$$\begin{aligned} \mathbb{P} - N^*\mathbb{P}N &= V^*JV && \text{if} \quad \Delta_+ = \mathbb{D} \\ \mathbb{P}N - N^*\mathbb{P} &= \tfrac{1}{2\pi i} V^*JV && \text{if} \quad \Delta_+ = \mathbb{C}_+ , \end{aligned}$$

respectively, where

$$N = Z_{\omega_1^*} \oplus \ \ldots \ \oplus Z_{\omega_n^*}$$

and

$$V = [V_1 \ \ldots \ V_n] .$$

Moreover, with the help of (5.12), formula (5.7) for U can be reexpressed as

$$U(\lambda) = I_m - \rho_\beta(\lambda) V (I - \lambda N)^{-1} \mathbb{P}^{-1} (I - \beta^* N^*)^{-1} V^* J \tag{7.5}$$

if $\Delta_+ = \mathbb{D}$ and as

$$U(\lambda) = I_m - \rho_\beta(\lambda) V \{2\pi i \ (\lambda I - N)\}^{-1} \mathbb{P}^{-1} \{-2\pi i \ (\beta^* I - N^*)\}^{-1} V^* J \tag{7.6}$$

if $\Delta_+ = \mathbb{C}_+$, when $\mathbb{P}$ is invertible.

We turn next to a description of the set of all solutions to the left interpolation problem when $\mathbb{P} > 0$.

LEMMA 7.2. *If $\mathbb{P} > 0$ and U is as in Theorem 7.1, then $T_U[S_o]$ is a solution of the right interpolation problem for every choice of $S_o \in \mathcal{S}_{p\times q}$.*

PROOF. We have already shown that if $\mathbb{P} > 0$, then

$$T_U[0] = BD^{-1} = \sigma_1$$

is a solution. Next let $S = T_U[S_o]$ for some choice of $S_o \in \mathcal{S}_{p\times q}$. Then, by Theorem 6.1,

$$S - \sigma_1 = \varepsilon_1 F$$

with

$$F = S_o(I_q + \sigma_2 S_o)^{-1}\varepsilon_2^{-1} \in H^\infty_{p\times q} ,$$

since σ_2 is strictly contractive on ∂ (use the fact that $U(\infty)$ exists and is J unitary to get this for $\partial = \mathbb{R}$) in the present setting. Therefore,

$$\begin{aligned} \underline{\underline{p}}\ S^* g_i &= \underline{\underline{p}}\ \sigma_1^* g_i + \underline{\underline{p}}\ F^*\varepsilon_1^* g_i \\ &= h_i + \underline{\underline{p}}\ F^* \underline{\underline{p}}\ \varepsilon_1^* g_i \\ &= h_i , \end{aligned}$$

by (7.4) and (7.3). Thus, by Lemma 7.1, $T_U[S_o]$ is a solution of the left interpolation problem for every choice of $S_o \in \mathcal{S}_{p\times q}$. ∎

The converse of the last lemma is also true as follows from (the more general) Theorem 9.4, which will be established in Section 9. The statements have been separated somewhat in order to both minimize repetition and to keep the logic clear, because Lemma 7.2 is used to prove Theorem 9.3 which in turn is used to prove part of Theorem 9.4. Be that as it may, Lemma 7.2 and Theorem 9.4 imply:

THEOREM 7.2. *If* $\mathbb{P} > 0$ *and* U *is as in Theorem 7.1, then*

$$\{T_U[S_o] : \ S_o \in \mathcal{S}_{p\times q}\}$$

is a complete list of the solutions to the left interpolation problem.

We remark that formula (7.2) in conjunction with the evaluation (7.1) (which in the present setting is applicable to U as well as to $S \in \mathcal{S}_{p\times q}$, since $U \in H^\infty_{m\times m}$) exhibits the f_i as the carriers of the left zero data of U in the sense of [GKLR].

8. RIGHT INTERPOLATION

The data for the general right interpolation problem in $\mathcal{S}_{p\times q}$ is a set of points $\omega_1, \ldots, \omega_n$ in Δ_+, a set of vectors $\eta_1, \ldots, \eta_n$ in $\mathbb{C}^q$ such that

$$\frac{\eta_1}{\lambda - \omega_1}, \ldots, \frac{\eta_n}{\lambda - \omega_n}$$

are linearly independent and a second set of vectors

$$\xi_{11}, \ldots, \xi_{1\ell_1}; \ \ \xi_{21}, \ldots, \xi_{2\ell_2}; \ldots; \xi_{n1}, \ldots, \xi_{n\ell_n} ;$$

in $\mathbb{C}^p$. The objective is to find an $S \in \mathcal{S}_{p\times q}$ such that

$$\frac{S^{(t-1)}(\omega_j)}{(t-1)!}\eta_j = \xi_{jt} \quad \text{for} \quad t = 1,\ldots,\ell_j\ , \qquad j = 1,\ldots,n\ ,$$

and to describe the set of all such solutions when these conditions are met.

To set up the chains for this problem we introduce the $m \times 1$ vectors

$$v_{j1} = \begin{bmatrix} \xi_{j1} \\ \eta_j \end{bmatrix}\ , \quad v_{jt} = \begin{bmatrix} \xi_{jt} \\ 0 \end{bmatrix}\ , \qquad t = 2,\ldots,\ell_j\ ,$$

the $m \times \ell_j$ matrices

$$V_j = [v_{j1} \ \ldots \ v_{j\ell_j}]$$

and the $m \times \ell_j$ matrix valued functions

$$F_j = V_j \Psi_{\omega_j,\ell_j}\ ,$$

for $j = 1,\ldots,n$. The columns of F_j form a chain of length ℓ_j of the form (3.6) which we index by the rule

$$[f_1 \ \ldots \ f_\nu] = [F_1 \ \ldots \ F_n]\ ,$$

where $\nu = \ell_1 + \ldots + \ell_n$. Notice that since $\omega_j \in \Delta_+,\ f_t \in (H_m^2)^\perp$ for $t = 1,\ldots,\nu$.

The relevance of these chains to the interpolation problem at hand rests on the evaluation

$$(\underline{\underline{q}}\,'\{S\psi_{\omega,j}\eta\})(\lambda) = \sum_{i=0}^{j} \psi_{\omega,i}(\lambda)\frac{S^{(j-i)}(\omega)}{(j-i)!}\eta\ , \qquad j = 0,1,\ldots\ , \tag{8.1}$$

which is valid for every choice of $S \in \mathcal{S}_{p\times q},\ \omega \in \Delta_+$ and $\eta \in \mathbb{C}^q$. Indeed, upon writing

$$f_t = \begin{bmatrix} g_t \\ h_t \end{bmatrix}\ , \qquad t = 1,\ldots,\nu\ ,$$

with $g_t \in (H_p^2)^\perp$ and $h_t \in (H_q^2)^\perp$ it is readily checked with the aid of (8.1) that:

LEMMA 8.1. *$S \in \mathcal{S}_{p\times q}$ is a solution of the right interpolation problem if and only if*

$$g_t = \underline{\underline{q}}\,' S h_t$$

for $t = 1,\ldots,\nu$.

THEOREM 8.1. *The right interpolation problem admits a solution if and only if the $\nu \times \nu$ matrix $\mathbb{P}$ with st entry*

$$p_{st} =< -Jf_t, f_s > , \qquad s,t = 1,\ldots,\nu ,$$

is positive semidefinite.

PROOF. Suppose first that the right interpolation problem admits a solution $S \in \mathcal{S}_{p\times q}$. Then, by Lemma 8.1,

$$f_t = \begin{bmatrix} \underline{\underline{q}}\,'Sh_t \\ h_t \end{bmatrix} , \qquad t = 1,\ldots,\nu ,$$

and thus

$$\sum_{i,j=1}^{\nu} c_i^* p_{ij} c_j = - < J\sum_{j=1}^{\nu} c_j f_j, \sum_{i=1}^{\nu} c_i f_i >$$

$$= - \left\| \underline{\underline{q}}\,'S\sum_{j=1}^{\nu} c_j h_j \right\|^2 + \left\| \sum_{j=1}^{\nu} c_j h_j \right\|^2 \geq 0$$

for any choice of constants $c_1,\ldots,c_\nu$. This proves the necessity of the condition $\mathbb{P} \geq 0$.

Suppose next that $\mathbb{P} > 0$ and let

$$\mathcal{M} = \text{span}\{f_s : \ s = 1,\ldots,\nu\}$$

endowed with the $-J$ inner product:

$$< f_t, f_s >_{\mathcal{M}} =< -Jf_t, f_s >= p_{st} \ .$$

Then, since $\mathbb{P} > 0$, $\mathcal{M}$ is clearly a reproducing kernel Hilbert space. It is also R_α invariant for every choice of α in $\mathbb{C}$ except for the points $\omega_1,\ldots,\omega_n$. Moreover, thanks to Theorem 2.3, this inner product satisfies the requisite identities. Therefore, by Theorem 2.1, $\mathcal{M}$ is an $\mathcal{H}(U)$ space. Moreover, it is clear from the recipe (2.7) and the form of the f_i that U is rational with all its poles in Δ_+.

Next, by (2.8) and the discussion thereof,

$$v^* f_i(\omega) =< -Jf_i, \frac{J - UJU(\omega)^*}{\rho_\omega} v >$$

for $i = 1,\ldots,\nu$ and every choice of $\omega \in \mathbb{C}\backslash\{\omega_1,\ldots,\omega_n\}$ and $v \in \mathbb{C}^m$. Since both of the summands in the right hand entry of the inner product belong to $(H_m^2)^\perp$ when $\omega \in \Delta_-$, the inner product can be split as

$$- < Jf_i, \frac{Jv}{\rho_\omega} > + < Jf_i, \frac{UJU(\omega)^*}{\rho_\omega} v >= v^* f_i(\omega) + < U^* Jf_i, \frac{JU(\omega)^*}{\rho_\omega} v >$$

for such ω. Therefore

$$< JU^*Jf_i, \frac{U(\omega)^*}{\rho_\omega} v >= 0$$

for every choice of $\omega \in \Delta_-$ and $v \in \mathbb{C}^m$. This implies that U^*Jf_i is orthogonal to $(H_m^2)^\perp$ with respect to the standard inner product and hence that

$$\underline{\underline{q}}\,' JU^*Jf_i = \underline{\underline{q}}\,' U^{-1} f_i = 0 \tag{8.2}$$

for $i = 1, \ldots, \nu$. Thus, by Theorem 6.3,

$$\underline{\underline{q}}\,' \sigma_1 h_i = g_i \ , \qquad i = 1, \ldots, \nu \ , \tag{8.3}$$

and

$$\underline{\underline{q}}\,' \varepsilon_2^{-1} h_i = 0 \ , \qquad i = 1, \ldots, \nu \ . \tag{8.4}$$

In view of Lemma 8.1, the former exhibits σ_1 as a solution of the right interpolation problem when $\mathbb{P} > 0$.

The sufficiency of the condition $\mathbb{P} \geq 0$ is obtained by limiting arguments. ∎

We remark that if $\ell_j = 1$, for $j = 1, \ldots, n$, then the entries in $\mathbb{P}$ can be evaluated explicitly as

$$p_{st} = -\frac{v_s^* J v_t}{\rho_{\omega_s}(\omega_t)} \ ,$$

in which $v_s = v_{s1}$ is the first (and only) column of V_s. For chains with arbitrary lengths ℓ_j, the calculations are more involved. By Theorems 3.3 and 4.1, respectively, $\mathbb{P}$ can, however, be characterized as the unique solution of the matrix equation

$$\begin{aligned} M^*\mathbb{P}M - \mathbb{P} &= V^*JV & \text{if} \quad \Delta_+ &= \mathbb{D} \\ \mathbb{P}M - M^*\mathbb{P} &= \tfrac{1}{2\pi i} V^*JV & \text{if} \quad \Delta_+ &= \mathbb{C}_+ \ , \end{aligned}$$

where

$$M = Z_{\omega_1} \oplus \ \ldots \ \oplus Z_{\omega_n}$$

and

$$V = [V_1 \ \ldots \ V_n] \ .$$

Theorem 4.1 is applicable to obtain the latter because, for $\Delta_+ = \mathbb{C}_+$, $\varphi_{\omega,j}(\lambda) = -\psi_{\omega^*,j}(\lambda)$. Moreover, with the help of (5.13), formula (5.7) for U can be reexpressed as

$$U(\lambda) = I_m - \rho_\beta(\lambda)|\delta|^2 V(\lambda I - M)^{-1}\mathbb{P}^{-1}(\beta^* I - M^*)^{-1} V^* J \ , \tag{8.5}$$

when $\mathbb{P}$ is invertible. Here $\delta = 1$ if $\Delta_+ = \mathbb{D}$ and $\delta = 1/2\pi i$ if $\Delta_+ = \mathbb{C}_+$.

LEMMA 8.2. *If* $\mathbb{P} > 0$ *and* U *is as in Theorem 8.1, then* $T_U[S_o]$ *is a solution of the right interpolation problem for every choice of* $S_o \in \mathcal{S}_{p\times q}$.

PROOF. We have already shown that if $\mathbb{P} > 0$, then

$$T_U[0] = BD^{-1} = \sigma_1$$

is a solution. Next let $S = T_U[S_o]$ for some choice of $S_o \in \mathcal{S}_{p\times q}$. Then, by Theorem 6.1,

$$S - \sigma_1 = \varepsilon_1 S_o(I_q + \sigma_2 S_o)^{-1}\varepsilon_2^{-1} .$$

Moreover, since in the present setting σ_2 is strictly contractive on ∂ (just as in Lemma 7.2),

$$\underline{\underline{q}}\,'\varepsilon_1 S_o(I_q + \sigma_2 S_o)^{-1}\underline{\underline{p}} = 0$$

and therefore, by (8.3) and (8.4),

$$\underline{\underline{q}}\,'Sh_i = \underline{\underline{q}}\,'\sigma_1 h_i + \underline{\underline{q}}\,'\varepsilon_1 S_o(I_q + \sigma_2 S_o)^{-1}\underline{\underline{q}}\,'\varepsilon_2^{-1}h_i = g_i ,$$

for $i = 1, \ldots, \nu$. This proves that $T_u[S_o]$ is an interpolant. ∎

The fact that conversely every interpolant is of the indicated form is a consequence of (the more general) Theorem 9.4 which will be established in the next section. We have separated the statements because Lemma 8.2 is used in the proof of Theorem 9.3 which in turn is used in the proof of part of Theorem 9.4. The two together lead at once to:

THEOREM 8.2. *If* $\mathbb{P} > 0$ *and* U *is as in Theorem 8.1, then*

$$\{T_U[S_o] : \; S_o \in \mathcal{S}_{p\times q}\}$$

is a complete list of the solutions to the right interpolation problem.

Finally, to complete the section, we remark that formula (8.2) in conjunction with the evaluation (8.1) (which is applicable to U^{-1} as well as to $S \in \mathcal{S}_{p\times q}$, since in the present setting $U^{-1} \in H^\infty_{m\times m}$) exhibit the f_i as the carriers of the right zero data of U^{-1} alias the right pole data of U in the sense of [GKLR].

9. TWO-SIDED INTERPOLATION

The data for the general two-sided problem is the set of "triples":

$$\omega_j;\ \xi_j;\ \eta_{j1},\ldots,\eta_{j\ell_j}\ , \qquad j=1,\ldots,k\ ,$$

$$\omega_j;\ \xi_{j1},\ldots,\xi_{j\ell_j};\ \eta_j\ , \qquad j=k+1,\ldots,n\ ,$$

where $\omega_j \in \Delta_+$, ξ_j and ξ_{ji} belong to $\mathbb{C}^p$, η_{ji} and η_j belong to $\mathbb{C}^q$ and it is assumed that $\frac{\xi_1}{\rho_{\omega_1}},\ldots,\frac{\xi_k}{\rho_{\omega_k}}$ are linearly independent, and $\frac{\eta_{k+1}}{\lambda-\omega_{k+1}},\ldots,\frac{\eta_n}{\lambda-\omega_n}$ are linearly independent. The objective is to find conditions on the data which ensure the existence of an $S \in \mathcal{S}_{p\times q}$ such that

$$\frac{\xi_j^* S^{(t-1)}(\omega_j)}{(t-1)!} = \eta_{jt}^*\ , \qquad t=1,\ldots,\ell_j\ ,\ j=1,\ldots,k\ ,$$

and

$$\frac{S^{(t-1)}(\omega_j)\eta_j}{(t-1)!} = \xi_{jt}\ , \qquad t=1,\ldots,\ell_j,\ j=k+1,\ldots,n\ ,$$

and to describe the set of all such solutions when these conditions are met.

Define

$$v_{j1} = \begin{bmatrix} \xi_j \\ \eta_{j1} \end{bmatrix}\ , \qquad v_{jt} = \begin{bmatrix} 0 \\ \eta_{jt} \end{bmatrix}\ , \qquad t=2,\ldots,\ell_j,\ \ j=1,\ldots,k\ ,$$

$$v_{j1} = \begin{bmatrix} \xi_{j1} \\ \eta_j \end{bmatrix}\ , \qquad v_{jt} = \begin{bmatrix} \xi_{jt} \\ 0 \end{bmatrix}\ , \qquad t=2,\ldots,\ell_j,\ \ j=k+1,\ldots,n\ ,$$

$$V_j = [v_{j1}\ \ldots\ v_{j\ell_j}]\ , \qquad j=1,\ldots,n\ ,$$

$$F_j = \begin{cases} V_j\Phi_{\omega_j,\ell_j}\ , & j=1,\ldots,k \\ -V_j\Psi_{\omega_j,\ell_j}\ , & j=k+1,\ldots,n \end{cases}$$

and

$$[f_1\ \ldots\ f_\nu] = [F_1\ \ldots\ F_n]\ ,$$

where $\nu = \ell_1+\ldots+\ell_n$ and $\mu = \ell_1+\ldots+\ell_k$; the latter will intervene later. The chains are chosen as indicated because an application of the evaluations (7.1) and (8.1) leads easily to the conclusion that if

$$f_i = \begin{bmatrix} g_i \\ h_i \end{bmatrix}\ , \qquad i=1,\ldots,\nu\ ,$$

with components g_i of size $p\times 1$ and h_i of size $q\times 1$, then:

LEMMA 9.1. *$S \in \mathcal{S}_{p\times q}$ is a solution of the general two-sided interpolation problem if and only if*

$$\underline{\underline{p}}S^*g_j = h_j \quad for \quad j = 1, \ldots, \mu$$

and

$$\underline{\underline{q}}'Sh_j = g_j \quad for \quad j = \mu + 1, \ldots, \nu$$

(with the selfevident conventions if the problem is only one-sided).

The strategy is to identify

$$\mathcal{M} = \text{span}\{f_1, \ldots, f_\nu\}$$

endowed with an appropriate choice of

$$p_{st} = < f_t, f_s >_{\mathcal{M}} , \qquad s, t = 1, \ldots, \nu ,$$

as an $\mathcal{H}(U)$ space. This is complicated because of the presence of terms with, say, $f_t \in H_m^2$ and $f_s \in (H_m^2)^\perp$ to which no simple rule is applicable. The most that can be said is that $\mathbb{P}$ must be a positive definite solution of a certain matrix equation, which will be specified below. It is obtained with the aid of the theory of $\mathcal{H}(S)$ spaces, which were introduced in Example 4 of Section 2. For the convenience of the reader, if such there be, we list a number of the properties of $\mathcal{H}(S)$ spaces which will be useful in the sequel.

THEOREM 9.1. *Let $\mathcal{H}(S)$ denote the reproducing kernel Hilbert space with reproducing kernel $\Lambda_\omega(\lambda)$ which was introduced in Example 4 of Section 2 for $S \in \mathcal{S}_{p\times q}(\Delta_+)$ and let $X = [I_p \quad -S]$ and $J = J_{pq}$. Then:*

(1) $R_\omega S\eta \in \mathcal{H}(S)$ for every choice of $\omega \in \Delta_+$ and $\eta \in \mathbb{C}^q$.

(2) $X\underline{\underline{p}}JX^*\varphi_{\omega,j}\xi = (R_\omega^*)^j\Lambda_\omega\xi$, $j = 0, 1, \ldots$, for every choice of $\omega \in \Delta_+$ and $\xi \in \mathbb{C}^p$, where the $*$ on the right indicates the adjoint in $\mathcal{H}(S)$.

(3) If $u_j \in H_p^2$, then $X\underline{\underline{p}}JX^*u_j \in \mathcal{H}(S)$ and

$$< X\underline{\underline{p}}JX^*u_j, X\underline{\underline{p}}JX^*u_i >_{\mathcal{H}(S)} = < J\underline{\underline{p}}X^*u_j, \underline{\underline{p}}X^*u_i > .$$

(4)

$$\left\| \sum_{j=1}^n R_{\alpha_j}^j S\eta_j \right\|_{\mathcal{H}(S)}^2 \leq - < J\sum_{j=1}^n g_j, \sum_{j=1}^n g_j > ,$$

for any choice of points $\alpha_j \in \Delta_+$ *and vectors* $\eta_j \in \mathbb{C}^q$, *where*

$$g_j = \begin{bmatrix} \underline{\underline{q}}' S u_j \\ u_j \end{bmatrix} \quad \text{and} \quad u_j = \frac{\eta_j}{(\lambda - \alpha_j)^j} \; .$$

(5) *If* U *is* J *inner, then* $S = T_U[S_o]$ *for some choice of* $S_o \in \mathcal{S}_{p\times q}$ *if and only if* (1) $Xf \in \mathcal{H}(S)$ *and* (2) $\|Xf\|_{\mathcal{H}(S)} \leq \|f\|_{\mathcal{H}(U)}$ *for every* $f \in \mathcal{H}(U)$.

PROOF. Verifications of (1) and (4) may be found in Theorem 2.3 of [D], (2) in Lemma 6.2 of [D] and (3) in Corollary 1 to Theorem 2.2 of [D]. (5) is a synopsis of Theorems 2.3 and 2.4 of [AD2], which in turn are adapted in part from de Branges and Rovnyak [dBR2]. The latter also contains a proof of (1) for $\Delta_+ = \mathbb{D}$ and $\omega = 0$. ∎

THEOREM 9.2. *Let* $S \in \mathcal{S}_{p\times q}$ *be a solution of the two-sided interpolation problem, let* $X = [I_p \quad -S]$ *and let* $\mathbb{P}_{ij}$ *denote the* $\ell_i \times \ell_j$ *matrix with entries*

$$p_{st} = < Xf_t, Xf_s >_{\mathcal{H}(S)} \; ,$$

where

$$f_s, \quad s = \sigma + a, \quad \sigma = \ell_0 + \ldots + \ell_{i-1}, \quad a = 1, \ldots, \ell_i, \quad i = 1, \ldots, k \; , \quad \text{and}$$

$$f_t, \quad t = \tau + b, \quad \tau = \ell_0 + \ldots + \ell_{j-1}, \quad b = 1, \ldots, \ell_j, \quad j = k+1, \ldots, n \; ,$$

denote the columns of $V_i \Phi_{\omega_i, \ell_i}$ *and* $-V_j \Psi_{\omega_j, \ell_j}$, *respectively. Then* $\mathbb{P}_{ij}$ *is a solution of the matrix equation*

$$\mathbb{P}_{ij} Z_{\omega_j} - (Z_{\omega_i^*})^* \mathbb{P}_{ij} = \delta V_i^* J V_j \; , \tag{9.1}$$

where $\delta = 1$ *if* $\Delta_+ = \mathbb{D}$ *and* $\delta = 1/2\pi i$ *if* $\Delta_+ = \mathbb{C}_+$ *and its entries*

$$p_{st} = \begin{cases} \delta \xi_i^* (R_{\omega_j}^b S)^{(a-1)}(\omega_i)\eta_j/(a-1)! & \text{if} \quad \omega_i \neq \omega_j \\ \delta \xi_i^* S^{(a+b-1)}(\omega_i)\eta_j/(a+b-1)! & \text{if} \quad \omega_i = \omega_j \; . \end{cases} \tag{9.2}$$

PROOF. Since S is a solution of the two-sided interpolation problem,

$$Xf_s = X\underline{\underline{p}} J X^* \varphi_{\omega_i, a-1} \xi_i = (R_{\omega_i}^*)^{a-1} \Lambda_{\omega_i} \xi_i$$

for $s = \sigma + a, \quad a = 1, \ldots, \ell_i$, by Lemma 7.1 and Theorem 9.1, and

$$Xf_t = -X\underline{\underline{q}}' \begin{bmatrix} S \\ I_q \end{bmatrix} \psi_{\omega_j, b-1} \eta_j = \delta R_{\omega_j}^b S \eta_j$$

for $t = \tau + b, \quad b = 1, \ldots, \ell_j$, by Lemma 8.1 and direct calculation. Therefore

$$p_{st} = \delta < R_{\omega_j}^b S\eta_j, (R_{\omega_i}^*)^{a-1}\Lambda_{\omega_i}\xi_i >_{\mathcal{H}(S)}$$
$$= \delta < R_{\omega_i}^{a-1} R_{\omega_j}^b S\eta_j, \Lambda_{\omega_i}\xi_i >_{\mathcal{H}(S)} ,$$

which readily yields (9.2).

Now, if $a = b = 1$ and $\omega_i \neq \omega_j$, then, since S is a solution of the interpolation problem, it follows directly from formula (9.2) that

$$(\omega_i - \omega_j)p_{st} = -\delta v_s^* J v_t \ . \tag{9.3}$$

On the other hand, if $a = b = 1$ and $\omega_i = \omega_j$, then (9.3) still holds because in this instance

$$v_s^* J v_t = \xi_i^* \xi_{j1} - \eta_{i1}^* \eta_j$$
$$= \xi_i^* \{S(\omega_j) - S(\omega_i)\}\eta_j = 0 \ .$$

Next, if $a = 1$ and $b > 1$, and $\omega_i \neq \omega_j$, then (9.2) implies that

$$(\omega_i - \omega_j)p_{st} = \delta\xi_i^*(R_{\omega_j}^b S)(\omega_i)\eta_j$$
$$= \delta\xi_i^* \left\{ (R_{\omega_j}^{(b-1)} S)(\omega_i) - \frac{S^{(b-1)}(\omega_j)}{(b-1)!} \right\} \eta_j$$
$$= p_{s,t-1} - \delta\xi_i^* \frac{S^{(b-1)}(\omega_j)}{(b-1)!}\eta_j \ .$$

Another look at (9.2) reveals that this final formula is also valid if $\omega_i = \omega_j$ and hence, since S is a solution of the interpolation problem, we see that the formula

$$(\omega_i - \omega_j)p_{st} - p_{s,t-1} = -\delta v_s^* J v_t \ , \quad a = 1, \quad b = 1, \ldots, \ell_j \ , \tag{9.4}$$

holds whether or not $\omega_i = \omega_j$.

Next, we turn to the case $a > 1$ and $b \geq 1$. Then, by the "resolvent identity"

$$R_\omega - R_\mu = (\omega - \mu)R_\omega R_\mu \ ,$$

which is readily seen to be valid by direct calculation,

$$(\omega_i - \omega_j)p_{st} = \delta < (R_{\omega_i})^{a-2}(R_{\omega_i} - R_{\omega_j})(R_{\omega_j})^{b-1} S\eta_j, \ \Lambda_{\omega_i}\xi_i >_{\mathcal{H}(S)}$$

$$= \begin{cases} p_{s,t-1} - p_{s-1,t} & \text{if} \quad b > 1 \\ \delta\xi_i^* \frac{S^{(a-1)}(\omega_i)}{(a-1)!}\eta_j - p_{s-1,t} & \text{if} \quad b = 1 \ . \end{cases}$$

The former yields

$$(\omega_i - \omega_j)p_{st} - p_{s,t-1} + p_{s-1,t} = -\delta\ v_s^* J v_t \tag{9.5}$$

for $s > \sigma + 1$ and $t > \tau + 1$ (since $v_s^* J v_t = 0$ for this choice of s and t) as does the latter, since $p_{s,t-1} = 0$ and

$$\delta \xi_i^* \frac{S^{(a-1)}(\omega_i)}{(a-1)!} \eta_j = \delta \eta_{ia}^* \eta_j = -\delta v_s^* J v_t \ ,$$

if $b = 1$ and $a > 1$.

Thus formula (9.5) is seen to hold for all choices of s, t with the usual understanding that $p_{st} = 0$ if either $s = \sigma$ or $t = \tau$. This completes the proof, since (9.5) is equivalent to (9.1). ∎

COROLLARY. *The numbers specified in (9.2) when $\omega_i \neq \omega_j$ are the same for every solution S of the two-sided interpolation problem.*

PROOF. For every solution S the indicated numbers are the entries in a solution of the matrix equation (9.1). But if $\omega_i \neq \omega_j$, then (as noted in Section 5) this equation has only one solution. ∎

THEOREM 9.3. *The two-sided interpolation problem in $\mathbb{D}$ admits a solution $S \in \mathcal{S}_{p\times q}(\mathbb{D})$ if and only if the matrix equation*

$$M^* \mathbb{P} M - N^* \mathbb{P} N = V^* J V \tag{9.6}$$

with

$$M = I_{\ell_1} \oplus \ldots \oplus I_{\ell_k} \oplus Z_{\omega_{k+1}} \oplus \ldots \oplus Z_{\omega_n} \ ,$$

$$N = Z_{\omega_1^*} \oplus \ldots \oplus Z_{\omega_k^*} \oplus I_{\ell_{k+1}} \oplus \ldots \oplus I_{\ell_n} \ ,$$

and

$$V = [V_1 \ \ldots \ V_n]$$

admits a positive semidefinite solution $\mathbb{P}$.

The two-sided interpolation problem in $\mathbb{C}_+$ admits a solution $S \in \mathcal{S}_{p\times q}(\mathbb{C}_+)$ if and only if the matrix equation

$$\mathbb{P} N - N^* \mathbb{P} = \frac{1}{2\pi i} V^* J V \tag{9.7}$$

with

$$N = Z_{\omega_1^*} \oplus \ldots \oplus Z_{\omega_k^*} \oplus Z_{\omega_{k+1}} \oplus \ldots \oplus Z_{\omega_n}$$

and V as above admits a positive semidefinite solution $\mathbb{P}$.

PROOF. Suppose first that $S \in \mathcal{S}_{p\times q}(\Delta_+)$ is a solution of the two-sided interpolation problem, let $\Lambda_\omega(\lambda)$ denote the reproducing kernel for the corresponding $\mathcal{H}(S)$ space and let $X = [I_p \quad -S]$. Then, by Lemma 7.1 and Theorem 9.1,

$$XF_j = \left[I_p \; R^*_{\omega_j} \; \cdots \; (R^*_{\omega_j})^{\ell_j-1}\right] \Lambda_{\omega_j}\xi_j \ , \quad j = 1, \ldots, k \ ,$$

where the $*$ denotes the adjoint in $\mathcal{H}(S)$ and, by Lemma 8.1 and direct evaluation,

$$XF_j = \delta\left[R_{\omega_j}S\eta_j \; \cdots \; R^{\ell_j}_{\omega_j}S\eta_j\right] \ , \quad j = k+1, \ldots, n \ ,$$

where $\delta = 1$ if $\Delta_+ = \mathbb{D}$ and $\delta = 1/2\pi i$ if $\Delta_+ = \mathbb{C}_+$. Thus, by Theorem 9.1, $Xf_j \in \mathcal{H}(S)$ for $j = 1, \ldots, \nu$.

Let $\mathbb{P}$ denote the $\nu \times \nu$ Hermitean matrix with ij entry

$$p_{ij} = \begin{cases} < Jf_j, f_i > \ , & i,j = 1, \ldots, \mu \ , \\ < Xf_j, Xf_i >_{\mathcal{H}(S)} \ , & i = 1, \ldots, \mu \ , \quad j = \mu+1, \ldots, \nu \ , \\ < -Jf_j, f_i > \ , & i,j = \mu+1, \ldots, \nu \ . \end{cases}$$

Our first objective is to show that $\mathbb{P}$ is a positive semidefinite solution of the stated equations. We proceed in steps.

STEP 1. $p_{ij} =< Xf_j, Xf_i >_{\mathcal{H}(S)}$ *for* $i,j = 1, \ldots, \mu$.

PROOF OF STEP 1. The $f_1, \ldots, f_\mu$ are the columns of $[F_1 \; \ldots \; F_k]$, where

$$F_j = \underline{\underline{p}}\begin{bmatrix} I_p \\ S^* \end{bmatrix} [\varphi_{\omega_j,0} \; \cdots \; \varphi_{\omega_j}, \ell_j - 1]\xi_j \ ,$$

$j = 1, \ldots, k$. This means that f_j is of the form

$$f_j = \underline{\underline{p}}JX^*u_j$$

with $u_j \in H^2_p$ for $j = 1, \ldots, \mu$. Thus, by (3) of Theorem 9.1,

$$< Xf_j, Xf_i >_{\mathcal{H}(S)} =< Jf_j, f_i > \ .$$

STEP 2.

$$\left\| \sum_{i=\mu+1}^{\nu} c_i Xf_i \right\|^2_{\mathcal{H}(S)} \leq \ < -J \sum_{i=\mu+1}^{\nu} c_i f_i, \sum_{j=\mu+1}^{\nu} c_j f_j > \ .$$

PROOF OF STEP 2. By Lemma 9.1, $f_{\mu+1},\ldots,f_\nu$ are the columns of $[F_{k+1} \ \cdots \ F_n]$, where

$$F_j = -\underline{\underline{q}}' \begin{bmatrix} S \\ I_q \end{bmatrix} [\psi_{\omega_j,0},\ldots,\psi_{\omega_j,\ell_j-1}]\eta_j \ , \qquad j = k+1,\ldots,n \ ,$$

and, as has already been noted above,

$$XF_j = \delta[R_{\omega_j}S\eta_j \ \ldots \ R_{\omega_j}^{\ell_j}S\eta_j] \ , \qquad j = k+1,\ldots,n \ .$$

The desired inequality drops out from (4) of Theorem 9.1.

STEP 3. $\mathbb{P} \geq 0$.

PROOF OF STEP 3. By Steps 1 and 2,

$$\begin{aligned} \sum_{i,j=1}^{\nu} c_i^* p_{ij} c_j &= \sum_{i,j=1}^{\mu} c_i^* p_{ij} c_j + 2Re \sum_{i=1}^{\mu} \sum_{j=\mu+1}^{\nu} c_i^* p_{ij} c_j + \sum_{i,j=\mu+1}^{\nu} c_i^* p_{ij} c_j \\ &\geq \sum_{i,j=1}^{\nu} c_i^* < Xf_j, Xf_i >_{\mathcal{H}(S)} c_j \\ &= \| X \sum_{j=1}^{\nu} c_j f_j \|_{\mathcal{H}(S)}^2 \geq 0 \ . \end{aligned}$$

STEP 4. $\mathbb{P}$ *is a solution of (9.6) [resp. (9.7)] if* $\Delta_+ = \mathbb{D}$ *[resp.* $\mathbb{C}_+$*].*

PROOF OF STEP 4. Let $\mathbb{P}_{ij}$ denote the $\ell_i \times \ell_j$ block of $\mathbb{P}$ with entries

$$p_{st}, \ s = \sigma+1,\ldots,\sigma+\ell_i, \quad t = \tau+1,\ldots,\tau+\ell_j \ ,$$

where $\sigma = \ell_0 + \ldots + \ell_{i-1}$, $\tau = \ell_0 + \ldots + \ell_{j-1}$ and $\ell_0 = 0$ and suppose first that $\Delta_+ = \mathbb{D}$. Then, using the same subdivision for M and N, $\mathbb{P}$ is a solution of (9.6) if and only if

$$M_{ii}^* \mathbb{P}_{ij} M_{jj} - N_{ii}^* \mathbb{P}_{ij} N_{jj} = V_i^* J V_j \tag{9.8}$$

for $i,j = 1,\ldots,n$. By Theorems 2.2 and 3.1 [resp. 2.3 and 3.3] this is clearly the case for $i,j = 1,\ldots,k$ [resp. $i,j = k+1,\ldots,n$], since $f_1,\ldots,f_\mu \in H_m^2$ and $f_{\mu+1},\ldots,f_\nu \in (H_m^2)^\perp$. It suffices therefore to check that (9.8) is met for $i = 1,\ldots,k$ and $j = k+1,\ldots,n$. In this case, however, (9.8) reduces to (9.1), which is met by Theorem 9.2.

The proof for $\Delta_+ = \mathbb{C}_+$ goes through in much the same way. This completes the proof of Step 4 and the proof of the necessity of the given conditions on $\mathbb{P}$.

STEP 5. *If* $\mathbb{P}$ *is a positive semidefinite solution of the matrix equation (9.6) or (9.7), according as* $\Delta_+ = \mathbb{D}$ *or* $\Delta_+ = \mathbb{C}_+$*, then the two sided interpolation problem is solvable.*

PROOF OF STEP 5. Suppose first that $\mathbb{P} > 0$ and let $\mathcal{M}$ denote the span of $f_1, \ldots, f_\nu$ endowed with the inner product

$$< f_t, f_s >_{\mathcal{M}} = p_{st} \ , \qquad s,t = 1, \ldots, \nu \ ,$$

where p_{st} denotes the st entry of $\mathbb{P}$. Then, by Theorem 5.3, $\mathcal{M}$ is an $\mathcal{H}(U)$ space based on the J inner function U given in formula (5.7). Moreover, U admits a pair of factorizations:

$$U = U_1 U_2 = U_3 U_4 \ ,$$

where

$$\mathcal{H}(U_1) = \text{span}\{f_1, \ldots, f_\mu\} \quad \text{and} \quad \mathcal{H}(U_3) = \text{span}\{f_{\mu+1}, \ldots, f_\nu\}$$

are the subspaces of $\mathcal{H}(U)$ which are generated by the indicated subsets of $f_1, \ldots, f_\nu$ which correspond in turn to left interpolation conditions and right interpolation conditions, respectively. In particular, it follows from (7.2) and (8.2), that

$$\underline{\underline{p}} U_1^* J f_j = 0 \quad \text{for} \quad j = 1, \ldots, \mu \qquad \text{and}$$

$$\underline{\underline{q}}' U_3^{-1} f_j = 0 \quad \text{for} \quad j = \mu+1, \ldots, \nu \ .$$

Thus, by Lemmas 7.2 and 8.2 and the fact that $T_{U_i}[\ ]$ maps $\mathcal{S}_{p\times q}$ into itself,

$$\sigma_1 = T_U[0] = T_{U_1}[T_{U_2}[0]] = T_{U_3}[T_{U_4}[0])$$

is a solution of the two-sided interpolation problem (9.1). This completes the proof of the step and the theorem if $\mathbb{P} > 0$. The case $\mathbb{P} \geq 0$ is handled by limits. The choice

$$V_j^\varepsilon = \begin{bmatrix} (1+\varepsilon)I_p & 0 \\ 0 & (1+\varepsilon)^{-1}I_q \end{bmatrix} V_j \ , \qquad j = 1, \ldots, k \ ,$$

and

$$V_j^\varepsilon = \begin{bmatrix} (1+\varepsilon)^{-1}I_p & 0 \\ 0 & (1+\varepsilon)I_q \end{bmatrix} V_j \ , \qquad j = k+1, \ldots, n \ ,$$

(with $\varepsilon > 0$) for the perturbed data seems to be particularly convenient since it does not affect $V_i^* J V_j$ for $i = 1, \ldots, k$ and $j = k+1, \ldots, n$. ∎

THEOREM 9.4. *If* $\mathbb{P} > 0$ *and* U *are as in Step 5 of Theorem 9.3, then*

$$\{T_U[S_o] : \ S_o \in \mathcal{S}_{p\times q}\}$$

is a complete list of the solutions to the two-sided interpolation problem.

PROOF. The formulas

$$T_U[S_o] = T_{U_1}[T_{U_2}[S_o]] = T_{U_3}[T_{U_4}[S_o]]$$

based on the factorizations of U given in Step 5 of Theorem 9.3 clearly exhibit $T_U[S_o]$ as a solution of the two-sided interpolation problem for every choice of $S_o \in \mathcal{S}_{p\times q}$.

To prove the converse, let S be any solution of the two-sided interpolation problem and let $X = [I_p \quad -S]$. Then, as follows readily from the analysis in the proof of Theorem 9.3, $Xf_i \in \mathcal{H}(S)$ for $i = 1, \dots, \nu$ and

$$\left\| \sum_{i=1}^{\nu} c_i X f_i \right\|_{\mathcal{H}(S)}^2 \leq \left\| \sum_{i=1}^{\nu} c_i f_i \right\|_{\mathcal{H}(U)}^2$$

for every choice of the complex constants $c_1, \dots, c_\nu$. Therefore, by (5) of Theorem 9.1, $S = T_U[S_o]$ for some choice of $S_o \in \mathcal{S}_{p\times q}$, as needed. ∎

Finally we remark (just as in Section 5) that although the formulations in terms of the unified equations (9.6) and (9.7) is elegant, it is easier to work with the equations for the individual $\ell_i \times \ell_j$ subblocks $\mathbb{P}_{ij}$ of $\mathbb{P}$. In the present setting it follows easily from Theorems 5.5 and 5.6 and the fact that the corresponding block equations have unique solutions that the entries in every solution $\mathbb{P}$ of (9.6)/(9.7) are given in part by the recipe

$$p_{st} = \begin{cases} < Jf_t, f_s > & \text{for} \quad s, t = 1, \dots, \mu \\ < -Jf_t, f_s > & \text{for} \quad s, t = \mu + 1, \dots, \nu \,. \end{cases}$$

The entries in the remaining blocks $\mathbb{P}_{ij}$ are uniquely specified by Theorem 9.2 when $\omega_i \neq \omega_j$ in (9.1). If $\omega_i = \omega_j$ in (9.1), then it turns out that $\mathbb{P}_{ij}$ is a Hankel matrix and that its entries are only partially specified. The prototype for this case is considered in more detail in the next section.

10. AN EXAMPLE AND SOME IMPLICATIONS

At first approach the general setting of Section 9 may well be a little overwhelming. Accordingly, in this section we shall explore in greater detail an example with just two chains. This is both merciful and instructive since it contains the core of what is going on. In particular, we shall investigate the existence of an $S \in \mathcal{S}_{p\times q}(\Delta_+)$ such that

$$\xi_1^* \frac{S^{(t-1)}(\omega_1)}{(t-1)!} = \eta_{1t}^* \,, \quad t = 1, \dots, \ell_1 \,,$$

and

$$\frac{S^{(t-1)}(\omega_2)}{(t-1)!}\eta_2 = \xi_{2t} \ , \qquad t = 1, \ldots, \ell_2 \ ,$$

where ω_1 and ω_2 are points in Δ_+ and both ξ_1 and η_2 are presumed to be nonzero vectors.

Following the notation introduced earlier, we set

$$V_1 = [v_1 \ldots v_{\ell_1}] = \begin{bmatrix} \xi_1 & 0 & \ldots & 0 \\ \eta_{11} & \eta_{12} & \ldots & \eta_{1\ell_1} \end{bmatrix},$$

$$V_2 = [v_{\ell_1+1} \ldots v_{\ell_1+\ell_2}] = \begin{bmatrix} \xi_{21} & \xi_{22} & \ldots & \xi_{2\ell_2} \\ \eta_2 & 0 & \ldots & 0 \end{bmatrix},$$

$$[f_1 \ldots f_{\ell_1}] = F_1 = V_2 \Phi_{\omega_1,\ell_1} \ ,$$

$$[f_{\ell_1+1} \ldots f_{\ell_1+\ell_2}] = F_2 = -V_2 \Psi_{\omega_2,\ell_2} \ ,$$

and take $\mu = \ell_1$ and $\nu = \ell_1 + \ell_2$.

By Theorem 9.3, this interpolation problem is solvable in $\mathbb{D}$ if and only if (9.6) admits a positive semidefinite solution $\mathbb{P}$. In the present setting (9.6) reduces to the three separate equations

$$\mathbb{P}_{11} - (Z_{\omega_1^*})^* \mathbb{P}_{11} Z_{\omega_1^*} = V_1^* J V_1 \ , \tag{10.1}$$

$$\mathbb{P}_{12} Z_{\omega_2} - (Z_{\omega_1^*})^* \mathbb{P}_{12} = \delta V_1^* J V_2 \ , \tag{10.2}$$

$$(Z_{\omega_2})^* \mathbb{P}_{22} Z_{\omega_2} - \mathbb{P}_{22} = V_2^* J V_2 \ , \tag{10.3}$$

where $\delta = 1$ if $\Delta_+ = \mathbb{D}$ and $\delta = 1/2\pi i$ if $\Delta_+ = \mathbb{C}_+$. (We have kept the δ, because then the discussion of (10.2) will be applicable to $\Delta_+ = \mathbb{C}_+$ also.)

In view of Theorems 5.5 and 5.6 the solutions of (10.1) and (10.3) can be identified as the matrices with entries as indicated below:

$$\mathbb{P}_{11}: \quad p_{st} = \langle J f_t, f_s \rangle \ , \qquad s,t = 1, \ldots, \ell_1 \ ,$$

$$\mathbb{P}_{22}: \quad p_{st} = \langle -J f_t, f_s \rangle \ , \quad s,t = \ell_1 + 1, \ldots, \ell_1 + \ell_2 \ .$$

The same conclusions hold for the solutions of the corresponding block equations of (9.7).

Equation (10.2) is uniquely solvable if and only if $\omega_1 \neq \omega_2$. If $\omega_1 \neq \omega_2$ and the resulting

$$\mathbb{P} = \begin{bmatrix} \mathbb{P}_{11} & \mathbb{P}_{12} \\ \mathbb{P}_{21} & \mathbb{P}_{22} \end{bmatrix}$$

is positive definite, then, by Theorem 5.3, the space $\mathcal{M} = \text{span}\{f_1, \ldots, f_\nu\}$, endowed with the inner product

$$\langle f_t, f_s \rangle_{\mathcal{M}} = p_{st} \ , \qquad s,t = 1, \ldots, \nu \ ,$$

is an $\mathcal{H}(U)$ space and, by Theorem 9.4,

$$\{T_U[S_o]:\ S_o \in \mathcal{S}_{p\times q}\}$$

is a complete list of the solutions to the stated interpolation problem. In this instance it follows from Theorem 9.2 that

$$p_{i,\ell_1+j} = \delta\xi_1^* \frac{(R_{\omega_2}^j S)^{(i-1)}(\omega_1)}{(i-1)!}\eta_2$$

for $i = 1,\ldots,\ell_1,\ j = 1,\ldots,\ell_2$. Moreover, by the corollary to Theorem 9.2, these numbers are the same for every solution S to the two-sided interpolation problem.

Suppose next that $\omega_1 = \omega_2$. Then, as follows from (9.5), the entries of every solution $\mathbb{P}_{12}$ to equation (10.2) are subject to the recursion

$$p_{s,t-1} - p_{s-1,t} = \delta v_s^* J v_t$$

for $s = 1,\ldots,\mu\ ,\quad t = \mu+1,\ldots,\nu$, with the understanding that $p_{st} = 0$ if either $s = 0$ or $t = \mu$. Therefore, we must have

$$v_1^* J v_{\mu+1} = 0 \tag{10.4}$$

and, because of the special form of V_1 and V_2,

$$v_s^* J v_t = 0$$

for $s = 2,\ldots,\mu$ and $t = \mu+2,\ldots,\nu$. This leads to the conclusion that $\mathbb{P}_{12}$ must be an $\ell_1 \times \ell_2$ Hankel matrix:

$$\mathbb{P}_{12} = \begin{bmatrix} c_1 & c_2 & c_3 & \cdots & c_{\ell_2} \\ c_2 & c_3 & & & \\ c_3 & & & & \\ \vdots & & & & \vdots \\ c_{\ell_1} & & & \cdots & c_{\nu-1} \end{bmatrix} \qquad (\nu = \ell_1 + \ell_2)\ ,$$

with

$$c_j = \begin{cases} \delta v_1^* J v_{\mu+1+j}\ , & j = 1,\ldots,\ell_2 - 1\ , \\ -\delta v_{1+j}^* J v_{\mu+1}\ , & j = 1,\ldots,\ell_1 - 1\ . \end{cases} \tag{10.5}$$

In particular these two recipes must agree in the overlap and therefore, in view of (10.4),

$$v_{1+j}^* J v_{\mu+1} + v_1^* J v_{\mu+1+j} = 0 \tag{10.6}$$

for $j = 0, \ldots, \gamma - 1$, where

$$\gamma = \min(\ell_1, \ell_2) .$$

The last γ entries: c_j for $j = \max(\ell_1, \ell_2), \ldots, \ell_1 + \ell_2 - 1$, are unspecified. This leads to the following conclusion:

THEOREM 10.1. *The two-sided interpolation problem stated at the beginning of this section with* $\omega_1 = \omega_2$ *is solvable if and only if*

(1) $v_{1+j}^* J v_{\mu+1} + v_1^* J v_{\mu+1+j} = 0$ *for* $j = 0, \ldots, \gamma - 1$, *and*

(2) there exists a choice of γ *constants:* c_j, $j = \max(\ell_1, \ell_2), \ldots, \ell_1 + \ell_2 - 1$ *such that the* $\nu \times \nu$ *Hermitean matrix with st entry*

$$\begin{aligned} p_{st} &= < Jf_t, f_s > , & s, t &= 1, \ldots, \ell_1 , \\ p_{st} &= < -Jf_t, f_s > , & s, t &= \ell_1 + 1, \ldots, \ell_1 + \ell_2 , \\ p_{st} &= c_{s+t-1-\ell_1} , & s &= 1, \ldots, \ell_1, \quad t = \ell_1 + 1, \ldots, \ell_1 + \ell_2 , \end{aligned}$$

where the c_j *are specified by (10.5) for* $j = 1, \ldots, \max(\ell_1 - 1, \ell_2 - 1)$, *is positive semidefinite.*

This is Theorem 6.3 of [D].

It is well to note that the conditions (10.6) are natural consistency conditions: If S is a solution of the problem, then

$$v_1^* J v_{\ell_1+1+j} = \xi_1^* \xi_{2,1+j} = \xi_1^* \frac{S^{(j)}(\omega_1)}{j!} \eta_2 ,$$

whereas

$$-v_{j+1}^* J v_{\ell_1+1} = \eta_{1,j+1}^* \eta_2 = \xi_1^* \frac{S^{(j)}(\omega_1)}{j!} \eta_2 ,$$

and hence clearly the two must agree for $j = 0, \ldots, \gamma - 1$, and

$$c_j = \delta \xi_1^* \frac{S^{(j)}(\omega_1)}{j!} \eta_2 , \qquad j = 1, \ldots, \max(\ell_1 - 1, \ell_2 - 1) .$$

The preceding theorem can be sharpened:

THEOREM 10.2. *There exists an* $S \in \mathcal{S}_{p\times q}(\Delta_+)$ *which*

(1) solves the interpolation problem stated at the beginning of this section with $\omega = \omega_1 = \omega_2$ *and for which*

(2) $\delta \xi_1^* \frac{S^{(j)}(\omega)}{j!} \eta_2 = c_j , \qquad j = 1, \ldots, \ell_1 + \ell_2 - 1 ,$

if and only if the $\nu \times \nu$ *Hermitean matrix* $\mathbb{P}$ *with entries*

$$\begin{aligned} p_{st} &= \langle Jf_t, f_s \rangle \quad , \quad s,t = 1,\ldots,\ell_1 , \\ p_{st} &= \langle -Jf_t, f_s \rangle \quad , \quad s,t = \ell_1+1,\ldots,\ell_1+\ell_2 , \\ p_{s,\ell_1+t} &= c_{s+t-1} \quad , \quad s = 1,\ldots,\ell_1 , \quad t = 1,\ldots,\ell_2 , \end{aligned}$$

where the c_j *are specified by (10.5) for* $j = 1,\ldots,\max(\ell_1-1,\ell_2-1)$, *is positive semidefinite.*

PROOF. Suppose first that S is a solution of both (1) and (2) and let B denote the $\nu \times \nu$ matrix with ij entry

$$b_{ij} = \langle Xf_j, Xf_i \rangle_{\mathcal{H}(S)} , \quad i,j = 1,\ldots,\nu ,$$

where $X = [I_p \quad -S]$. Then by Theorem 9.2, which supplies the identification

$$b_{i,\ell_1+j} = \delta\xi_1^* \frac{S^{(i+j-1)}(\omega)}{(i+j+1)!}\eta_2 = c_{i+j-1} = p_{i,\ell_1+j}$$

for $i = 1,\ldots,\ell_1$ and $j = 1,\ldots,\ell_2$, and the first three steps in the proof of Theorem 9.3,

$$\mathbb{P} \geq B \geq 0 .$$

Since S is an interpolant, the preceding identification also serves to show that c_j is specified by (10.5) for $j = 1,\ldots,\max(\ell_1-1,\ell_2-1)$.

Suppose next that $\mathbb{P} > 0$ for some choice of $c_{\nu-\gamma},\ldots,c_{\nu-1}$ (the other entries in $\mathbb{P}$ are completely specified by the data as indicated in the second half of the statement of the theorem; the bookkeeping is straight because if $\gamma = \min(\ell_1,\ell_2)$, then $\nu - \gamma = \max(\ell_1,\ell_2)$). Then, since $\mathbb{P}$ is a solution of the matrix equation (9.6) or (9.7), according as $\Delta_+ = \mathbb{D}$ or $\Delta_+ = \mathbb{C}_+$, the space

$$\mathcal{M} = \operatorname{span}\{f_1,\ldots,f_\nu\}$$

endowed with the inner product

$$\langle f_j, f_i \rangle_{\mathcal{M}} = p_{ij} , \quad i,j = 1,\ldots,\nu ,$$

is an $\mathcal{H}(U)$ space by Theorem 5.3 and therefore, by Theorem 9.4, $S = T_U[S_o]$ is a solution of problem (1) in the statement of the theorem. Moreover, for any such solution S,

$$\begin{aligned} f_i &= \underline{\underline{p}}\, X^* J \varphi_{\omega_1,i-1}\xi_1 \quad , \quad i = 1,\ldots,\ell_1 , \\ f_{\ell_1+j} &= -\underline{\underline{q}}' \begin{bmatrix} S \\ I_q \end{bmatrix} \psi_{\omega_2,j-1}\eta_2 \quad , \quad j = 1,\ldots,\ell_2 , \end{aligned}$$

and hence, by Corollary 3 to Theorem 2.7 of [D],

$$p_{i,\ell_1+j} = < f_{\ell_1+j}, f_i >_{\mathcal{H}(U)} = < \sigma_1 \psi_{\omega_2,j-1}\eta_2, \varphi_{\omega_1,i-1}\xi_1 > ,$$

where $\sigma_1 = T_U[0]$, as usual. The inner product on the right can be reexpressed as a simple contour integral over the circle or (the limit of) a semicircle based on the line, according as $\Delta_+ = \mathbb{D}$ or $\Delta_+ = \mathbb{C}_+$. In particular, if $\omega_1 = \omega_2$ as in the present case, then it is readily checked that

$$p_{i,\ell_1+j} = \delta\xi_1^* \frac{\sigma_1^{(i+j-1)}(\omega)}{(i+j-1)!}\eta_2$$

for $i = 1, \ldots, \ell_1$ and $j = 1, \ldots, \ell_2$. Thus σ_1 is also a solution of (2) in the statement. This completes the proof of the theorem if $\mathbb{P} > 0$. The case $\mathbb{P} \geq 0$ is handled by limits; see the proof of Step 5 of Theorem 9.3 for some hints. ∎

Theorem 10.2 can of course be extended to the more general setting of many chains. An even more general formulation may be found in Ball and Ran [BR], who obtained it by entirely different methods. As noted earlier, yet another approach will be presented in the forthcoming monograph [BGR3]. Indeed, the present Theorem 10.2 was motivated by what the author remembered of a lecture given by J. Ball at the Calgary conference wherein he described some of the BGR work.

11. DISPLACEMENT RANK

An $n \times n$ matrix B is said to have displacement rank m (with respect to Z) if

$$\text{rank}(B - Z^*BZ) = m \ . \tag{11.1}$$

This class of matrices was introduced by Kailath, Kung and Morf [KKM] (though it should be noted that their Z is our Z^*) and further investigated by Kailath in numerous collaborations with assorted combinations of colleagues and students, because of its wide occurrence in signal processing problems. The paper [LK] by Lev-Ari and Kailath, and its bibliography, will serve to give some idea of the scope. The purpose of this section is to show that every invertible Hermitean matrix with displacement rank m can be identified as the Gram matrix of a suitably defined chain in a reproducing kernel space of the type considered above. For ease of exposition we begin with the positive definite case.

THEOREM 11.1. *Every $n \times n$ positive definite matrix B with displacement rank m is the Gram matrix of a chain $f_1, \ldots, f_n$ of $m \times 1$ vector valued polynomials in a* *reproducing kernel Hilbert space $\mathcal{H}(U)$ based upon an $m \times m$ matrix valued polynomial U which is J inner over $\mathbb{D}$.*

PROOF. If B is positive definite (on $\mathbb{C}^n$), then B is automatically Hermitean. Therefore

$$B - Z^*BZ ,$$

being a Hermitean matrix of rank m, admits a representation of the form

$$B - Z^*BZ = C^*DC ,$$

where

$$D = \begin{bmatrix} I_p & 0 & 0 \\ 0 & -I_q & 0 \\ 0 & 0 & 0 \end{bmatrix} \quad \text{with } p + q = m .$$

Thus, upon expressing

$$C = [c_1 \ldots c_n]$$

as an array of $n \times 1$ column vectors and setting

$$V = [v_1 \ldots v_n] = [I_m \quad 0]C ,$$

it is readily seen that

$$C^*DC = V^*JV$$

and hence that B is a solution of the matrix equation

$$B - Z^*BZ = V^*JV . \tag{11.2}$$

Now let

$$f_1 = v_1 , \quad f_2 = \lambda v_1 + v_2, \ldots, f_n = \lambda^{n-1}v_1 + \lambda^{n-2}v_2 + \ldots + v_n .$$

Then, by Theorem 5.5 (or by direct calculation), the $n \times n$ matrix $\mathbb{P}$ with ij entry

$$p_{ij} = \langle Jf_j, f_i \rangle , \qquad i, j = 1, \ldots, n ,$$

(for $\Delta_+ = \mathbb{D}$) is a solution of (11.2). Therefore, since B is positive definite and equation (11.2) is uniquely solvable, it follows that $\mathbb{P} = B$ and hence that $\mathbb{P}$ is positive definite. Thus, by Theorems 2.1 and 3.1, the space

$$\mathcal{M} = \text{span}\{f_1, \ldots, f_n\} ,$$

endowed with the inner product

$$< f_j, f_i >_{\mathcal{M}} = < Jf_j, f_i > = p_{ij}$$

is an $\mathcal{H}(U)$ space and B is the Gram matrix of the chain $f_1, \ldots, f_n$. Finally, formula (2.7) exhibits U as a matrix polynomial. ∎

THEOREM 11.2. *Every $n \times n$ invertible Hermitean matrix B with displacement rank m is the Gram matrix of a chain $f_1, \ldots, f_n$ of $m \times 1$ vector valued polynomials in a reproducing kernel Pontryagin space $\mathcal{K}(U)$ based upon an $m \times m$ matrix valued polynomial U which is J unitary on $\mathbb{T}$.*

PROOF. The proof is almost exactly the same as that of Theorem 11.1, except that now we need to invoke the Pontryagin space analogue from [AD2] of Theorem 2.1. The needed ingredients are incorporated in Theorem 5.3. ∎

The interplay between the underlying reproducing kernel spaces and the given matrix B, as expressed in the preceding two theorems, gives additional insight into the structural properties of B. Thus, for example, the fact that the Schur complement of an invertible upper left hand corner of a Gram matrix for a single chain, is the Gram matrix for a shorter chain of the same form, which is established in Theorem 8.2 of [AD2], serves to explain why the Schur complement of an invertible upper left hand corner of an invertible matrix with displacement rank m is of displacement rank no larger than m. A matrix proof of this fact may be found in Bitmead and Anderson [BA].

We take this opportunity to point out a misprint in the aforementioned theorem of [AD2]: $\mathbb{P}^{-1}$ should be replaced by $\hat{\mathbb{P}}^{-1}$ where $\hat{\mathbb{P}}$ designates the upper left hand $k \times k$ corner of $\mathbb{P}$.

The preceding two theorems can be generalized in many ways. For example Z can be replaced by Z_ω in both (11.1) and the statements of the last two theorems. The corresponding U, which is still given by (2.7), will then be rational. It is also possible to deal with non-Hermitean matrices with displacement rank m by passing to two chains and invoking Theorem 3.2 in place of Theorem 3.1:

THEOREM 11.3. *If B is an $n \times n$ matrix with displacement rank m, then there exist a pair of $m \times 1$ vector valued polynomial chains $f_1, \ldots, f_n$ and $g_1, \ldots, g_n$, such that the ij entry of B is given by*

$$b_{ij} = < Jf_j, g_i > , \qquad i, j = 1, \ldots, n ,$$

where the inner product is taken on $\mathbb{T}$.

PROOF. If B has displacement rank m, then there exist a pair of $m \times n$ matrices $W = [w_1 \dots w_n]$ and $V = [v_1 \dots v_n]$ and an $m \times m$ signature matrix J such that

$$B - Z^*BZ = W^*JV \ .$$

Let

$$f_1 = v_1, \quad f_2 = \lambda v_1 + v_2, \dots, f_n = \lambda^{n-1} v_1 + \dots + \lambda^{n-2} v_2 + \dots + v_n \ ,$$

$$g_1 = w_1, \quad g_2 = \lambda w_1 + w_2, \dots, g_n = \lambda^{n-1} w_1 + \lambda^{n-2} w_2 + \dots + w_n \ ,$$

and finish by invoking Theorems 2.2 and 3.2. ∎

A Hermitean invertible matrix B can also be interpreted as the Gram matrix of a suitably defined chain in a reproducing kernel space over $\mathbb{C}_+$ by appropriately modifying the definition of displacement rank:

THEOREM 11.4. *If B is an $n \times n$ Hermitean invertible matrix with*

$$rank\ 2\pi i\ \{(Z_{\omega^*})^* B - B Z_{\omega^*}\} = m \tag{11.3}$$

for some $\omega \in \mathbb{C}$, then B is the Gram matrix of a chain $f_1, \dots, f_n$ of $m \times 1$ vector valued rational functions in a reproducing kernel Pontryagin space $\mathcal{K}(U)$ which is based on an $m \times m$ rational matrix function U which is J unitary on $\mathbb{R}$.

PROOF. By the argument used in the proof of Theorem 11.1, there exists an $m \times m$ matrix V with columns $v_1, \dots, v_n$ such that

$$(Z_{\omega^*})^* B - B Z_{\omega^*} = \frac{1}{2\pi i} V^* J V \ .$$

Let $f_1, \dots, f_n$ denote the columns of $V\Phi_{\omega,n}$ (for $\Delta_+ = \mathbb{C}_+$). Then, by Theorem 5.3, the space $\mathcal{M}$ =span$\{f_1, \dots, f_n\}$ endowed with the inner product

$$< f_j, f_i >_{\mathcal{M}} = b_{ij} \ , \qquad i, j = 1, \dots, n \ ,$$

is a $\mathcal{K}(U)$ space. ∎

The definition of displacement rank introduced in (11.3) resembles the one used by Heinig and Rost [HR].

More can be said, but the hour is late and the manuscript is already both overly long and long overdue.

12. REFERENCES

[AD1] D. Alpay and H. Dym, *Hilbert spaces of analytic functions, inverse scattering, and operator models I*, Integral Equations and Operator Theory **7** (1984), 589–641.

[AD2] —, *On applications of reproducing kernel spaces to the Schur algorithm and rational J unitary factorization*, in I. Schur Methods in Operator Theory and Signal Processing (I. Gohberg, ed.), Operator Theory: Advances and Applications, **OT18**, Birkhäuser Verlag, Basel, 1986, pp. 89-159.

[AG] D. Alpay and I. Gohberg, *Unitary rational matrix functions*, in Topics in Interpolation Theory of Rational Matrix-valued Functions (I. Gohberg, ed.), Operator Theory: Advances and Applications **OT33**, Birkhäuser Verlag, Basel, 1988, pp.175–222.

[B1] J.A. Ball, *Models for non contractions*, J. Math. Anal. Appl. **52** (1975), 235-254.

[B2] —, *Nevanlinna-Pick interpolation: Generalizations and applications*, in Surveys of Some Recent Results in Operator Theory (J.B. Conway and B.B. Morrel, eds.), Pitman Research Notes in Mathematics, Longman, in press.

[BA] R.R. Bitmead and B.D.O. Anderson, *Asymptotically fast solution of Toeplitz and related systems of linear equations*, Linear Algebra Appl. **34** (1980), 117–124.

[BGR1] J.A. Ball, I. Gohberg and L. Rodman, *Realization and interpolation of rational matrix functions*, in Topics in Interpolation Theory of Rational Matrix-valued Functions (I. Gohberg, ed.), Operator Theory: Advances and Applications **OT33**, Birkhäuser Verlag, Basel, 1988, pp.1-72.

[BGR2] —, *Two-sided Nudelman interpolation problem for rational matrix functions*, preprint, 1988.

[BGR3] —, *Interpolation Problems for Matrix Valued Functions, Part I: Rational Functions*, Monograph in preparation.

[BR] J.A. Ball and A.C.M. Ran, *Local inverse spectral problems for rational matrix functions*, Integral Equations and Operator Theory **10** (1987), 349–415.

[D] H. Dym, *J Contractive Matrix Functions, Reproducing Kernel Hilbert Spaces and Interpolation*, CBMS Lecture Notes, in press.

[dB] L. de Branges, *Some Hilbert spaces of analytic functions I*, Trans. Amer. Math. Soc. **106** (1963), 445–468.

[dBR1] L. de Branges and J. Rovnyak, *Square Summable Power Series*, Holt, Rinehart and Winston, New York, 1966.

[dBR2] —, *Canonical Models in Quantum Scattering Theory*, in Perturbation Theory and its Applications in Quantum Mechanics (C. Wilcox, ed.), Wiley, New York, 1966, pp. 295-392.

[GKLR] I. Gohberg, M.A. Kaashoek, L. Lerer and L. Rodman, *Minimal divisors of rational matrix function with prescribed zero and pole structure*, in Topics in Operator Theory Systems and Networks (H. Dym and I. Gohberg, eds.), Operator Theory: Advances and Applications **OT12**, Birkhäuser Verlag, Basel, 1984, pp.241–275.

[H] J.W. Helton, *Operator Theory, Analytic Functions, Matrices and Electrical Engineering*, Regional Conference Series in Mathematics, **68**, Amer. Math. Soc., Providence, R.I., 1987.

[HR] G. Heinig and K. Rost, *Algebraic Methods for Toeplitz-like Matrices and Operators*, Birkhäuser Verlag, Basel, 1984.

[LK] H. Lev-Ari and T. Kailath, *Triangular factorization of structured Hermitean matrices*, in I. Schur Methods in Operator Theory and Signal Processing (I. Gohberg, ed.), Operator Theory: Advances and Applications **OT18**, Birkhäuser-Verlag, Basel, 1986, pp.301–324.

[KKM] T. Kailath, S.-Y. Kung and M. Morf, *Displacement ranks of matrices and linear equations*, J. Math. Anal. and Appl. **68** (1979), 395–407.

[R] J. Rovnyak, *Characterization of spaces* $\mathcal{K}(M)$, unpublished manuscript.

[Ru] D.L. Russell, *Mathematics of Finite-Dimensional Control Systems*, Lecture Notes in Pure and Applied Mathematics, Vol. 43, Marcel Dekker, New York, 1979.

[S1] L.A. Sakhnovich, *Factorization problems and operator identities*, Russian Math. Surveys **41** (1986), 1-64.

[S2] A.L. Sakhnovich, *On a class of extremal problems*, Math. USSR Izvestiya, **30** (1988), 411–418.

Department of Theoretical Mathematics
The Weizmann Institute of Science
Rehovot 76100, ISRAEL

Operator Theory:
Advances and Applications, Vol. 41

ON ASYMPTOTIC TOEPLITZ AND HANKEL OPERATORS

Avraham Feintuch

Dedicated to Professor Israel Gohberg on the occasion of his sixtieth birthday.

An operator T is asymptotic Toeplitz if for S the unilateral shift, the sequence $\{S^{*n}TS^n\}$ converges. It is asymptotic Hankel if for J_n the permutation isometry on the subspace determined by the first n coordinate vectors, the sequence $\{J_nTS^{n+1}\}$ converges. The relationship between these notions is studied and operator analogues of the A.A.K. distance formulae in terms of the s numbers of a Hankel operator are obtained.

1 Introduction

The theory of Toeplitz and Hankel operators has been studied in great detail and has major importance in numerous areas of applied mathematics, especially in the theory of linear systems and control. Both these classes of operators can be represented either as "pieces" of a multiplication operator of equivalently as matrices of a particular structure on a Hilbert sequence space. In control theory these representations correspond to frequency domain and time domain representations of a *time invariant linear system.* While the Toeplitz operator "is the system" the associated Hankel operator containes a great deal of information about the system. For example, its minimal dimension is determined by the rank of the Hankel operator. All these facts are classical ([4], [7]).

For time-varying systems, the situation is much more complicated. These systems have a natural description as linear operators in the time domain, and while there is a frequency domain theory developed in ([8], [11]) for such operators, the transfer function suggested there is usually an operator valued function which is extremely difficult to analyze.

In this paper, we study a class of operators which are asymptotic Toeplitz and Hankel operators. The idea of an asymptotic Toeplitz operator was introduced and first studied in ([3]). Let S denote the unilateral shift and for an operator T consider the sequence $\{S^{*n}TS^n\}$. If this sequence converges strongly, then T was referred to by Barria and Halmos ([3]) as asymptotic Toeplitz. In particular every element of the uniformly closed algebra generated by the Toeplitz operators (and / or the Hankel operators) is asymptotic Toeplitz in this sense ([3], Theorem 4). Here we consider asymptotic Toeplitz operators in the weak and uniform sense as well and associate with T another sequence which when it converges, converges to the corresponding Hankel operator. We give a characterization of both uniform and weak asymptotic Toeplitz and Hankel operators and show that operators in the uniformly closed algebra generated by the Toeplitz operators are asymptotic Hankel operators. The significance of these objects is illustrated in a distance formula which is an operator theoretic analogue of results of Adamjan, Arov and Krein ([1]). Some related results in control theory are studied in ([5]).

2 Terminology and Notation.

The set of all one-sided square summable sequences is denoted by h_2; that is $\{x_k\}_{k=0}^{\infty}$ is in h_2 if $\sum_0^\infty |x_k|^2 < \infty$. Sequences in h_2 will be represented as vectors starting from the zeroth co-ordinate; that is, $x \in h_2$ will be written as $x = \langle x_0, x_1, x_2, \ldots \rangle$. The truncation projections on h_2 will be denoted by P_i, $0 \leq i \leq \infty$. Then

$$\begin{aligned} P_i x &= P_i\langle x_0, x_1, x_2, \ldots, x_i, x_{i+1}, \ldots\rangle \\ &= \langle x_0, x_1, \ldots, x_i, 0, 0, \ldots\rangle. \end{aligned}$$

These are, of course, orthogonal proections on h_2 which converge strongly to the identity I on h_2.

The unilateral shift on h_2 is denoted by S. Recall that T on h_2 is

Toeplitz if and only if $S^*TS = T$. The standard orthonormal basis on h_2 is denoted by $\{e_i\}_{i=0}^{\infty}$ and corresponds to the co-ordinate representation given above. On the subspace $P_n h_2$ spanned by $\{e_0, \ldots, e_n\}$, define the unitary operator J_n by

$$J_n e_i = e_{n-i} \qquad\qquad 0 \le i \le n.$$

J_n is called the permutation operator of order n. We extend J_n to h_2 by defining it to be zero on $(I - P_n)h_2$ and we will denote this operator on h_2 by J_n as well. Note that $J_n = J_n P_n = P_n J_n$.

By $\mathbf{L}^\infty$ we mean the algebra of essentially-bounded complex-valued functions defined on the unit circle $\mathbf{T}$. $\varphi \in \mathbf{L}^\infty$ induces a Toeplitz operator T_φ on h_2 as follows. Let $\{a_i\}_{i=-\infty}^{\infty}$ be the sequence of Fourier coefficients of φ. T_φ will be the operator whose matrix representation with respect to the standard orthonormal basis is given by

$$\begin{bmatrix} a_0 & a_{-1} & a_{-2} & \cdots \\ a_1 & a_0 & a_{-1} & \cdots \\ a_2 & a_1 & a_0 & \cdots \\ \cdot & \cdot & \cdot & \\ \cdot & \cdot & \cdot & \\ \cdot & \cdot & \cdot & \end{bmatrix}$$

We also associate with φ a Hankel operator H_φ on h_2. Its matrix representation is given by

$$H_\varphi = \begin{bmatrix} a_{-1} & a_{-2} & a_{-3} & \cdots \\ a_{-2} & a_{-3} & a_{-4} & \cdots \\ a_{-3} & a_{-4} & a_{-5} & \cdots \\ \cdot & \cdot & \cdot & \\ \cdot & \cdot & \cdot & \\ \cdot & \cdot & \cdot & \end{bmatrix}$$

Note that if h_{ij} denotes the (i,j)-th entry in H_φ then $h_{ij} = a_{-(i+j+1)}$. Nehari's Theorem ([10], p.4) states that all Hankel operators are determined in this way.

$\mathbf{H}^\infty$ denotes the subalgebra of $\mathbf{L}^\infty$ consisting of functions whose negative Fourier coefficients are zero. Note that $\varphi \in \mathbf{H}^\infty$ if and only if the matrix representation of T_φ is lower triangular if and only if $H_\varphi = 0$.

$\mathbf{R}_n$ denotes the rational functions with at most n poles (counting multiplicities) all of which are in the interior of $\mathbf{T}$. $\mathbf{C}$ will denote the algebra of continuous functions on $\mathbf{T}$. Note that the following inclusion relations hold:

$$\mathbf{H}^\infty \subset \mathbf{H}^\infty + \mathbf{R}_1 \subset \mathbf{H}^\infty + \mathbf{R}_2 \subset \ldots \subset \mathbf{H}^\infty + \mathbf{C}.$$

Kronecker's Lemma ([9], p.183) characterizes those $\varphi \in \mathbf{L}^\infty$ for which H_φ is finite rank. Rank $H_\varphi \leq n$ if and only if $\varphi \in H^\infty + \mathbf{R}_n$. Hartman's Theorem ([10], p.6) characterizes those $\varphi \in \mathbf{L}^\infty$ for which H_φ is compact. This is the case if and only if $\varphi \in H^\infty + \mathbf{C}$.

$\mathcal{C}$ will denote the (weakly closed) algebra of bounded linear operators in h_2 which have a lower triangular matrix representation with respect to the standard basis. $\mathcal{C}$ is a nest algebra and has a system theoretic interpretation as the algebra of physically realizable (= causal) stable linear systems (see [6]). $\mathcal{K}$ will denote the uniformly closed two-sided ideal of compact operators on h_2. $\mathcal{C} + \mathcal{K}$ is a uniformly closed algebra ([2]). It is characterized as $\{T \in \mathcal{L}(h_2) : \|P_i T(I - P_i)\| \to 0\}$ and is also referred to as the algebra of quasi-triangular operators with respect to the sequence $\{P_i\}$.

$\mathcal{F}_n$ will denote the operators on h_2 of rank less than or equal to n. Each operator in $\mathcal{F}_n$ can be expressed as a sum of rank one operators and of course

$$\mathcal{F}_1 \subset \mathcal{F}_2 \subset \ldots \subset \mathcal{K}$$

Finally, for an operator T on h_2, the n-th s-number of T is

$$s_n(T) = \inf\{\|T + F\| : F \in \mathcal{F}_n\}$$

and the essential norm of T is given by

$$\|T\|_e = \inf\{\|T + K\| : K \in \mathcal{K}\}.$$

3 Asymptotic Toeplitz operators and associated asymptotic Hankels.

For $T \in \mathcal{L}(h_2)$ consider the sequences $\{S^{*n}TS^n\}$ and $\{H_n(T) = J_n T S^{n+1}\}$. It is clear that if $\{S^{*n}TS^n\}$ converges in the weak (strong, uniform) topology then its limit is a Toeplitz operator T_φ for some $\varphi \in L^\infty$. Following

[3] we will refer to the function φ as the symbol of T and will denote it as $sym(T)$. It is not immediately obvious, however, that when $H_n(T)$ converges that its limit is a Hankel operator.

LEMMA 3.1 *If $T \in \mathcal{L}(h_2)$ has matrix representation $[t_{ij}]_{i,j=0}^{\infty}$ with respect to the standard basis and $\{H_n(T)\}$ converges weakly, its limit is a Hankel operator.*

Proof: Denote $w - \lim_{n\to\infty} H_n(T)$ by $H(T)$. Then for each $i, j \geq 0$, $(H_n(T)e_i, e_j) \to (H(T)e_i, e_j)$ as $n \to \infty$. But

$$(H_n(T)e_i, e_j) = \begin{cases} 0 & j \leq n \\ t_{n-j,n+i+1} & j > n \end{cases}$$

For n sufficiently large this is just the sequence $\{t_{n,n+i+j+1}\}$ which by hypothesis converges to a number which depends only on the sum $i+j$. Denote this limit by $t_{-(i+j)}$. Thus the matrix representation of $H(T)$ is

$$H(T) = \begin{bmatrix} t_{-1} & t_{-2} & t_{-3} & \cdots \\ t_{-2} & t_{-3} & t_{-4} & \cdots \\ t_{-3} & t_{-4} & t_{-5} & \cdots \\ \cdot & \cdot & \cdot & \\ \cdot & \cdot & \cdot & \\ \cdot & \cdot & \cdot & \end{bmatrix}$$

which is of course a Hankel matrix. ∎

Two natural questions arise at this stage;

(1) Is an asymptotic Toeplitz operator also asymptotic Hankel ?

(2) Suppose T is both asymptotic Toeplitz and Hankel. Do the symbols correspond ?

The answers to these questions are related. We show (Theorem 5.1) that weak asymptotic Toeplitz implies weak asymptotic Hankel, and that in this case the sympols are the same. Thus when T is both asymptotic Toeplitz and Hankel in the uniform and strong sense, the symbols must be the same.

With respect to the first question, the weak topology case is special. In both the uniform and strong topologies we shall see that asymptotic Toeplitz does not imply asymptotic Hankel.

4 Uniform Asymptotic Toeplitz and Hankel Operators.

The uniform asymptotic Toeplitz operators are easily characterized.

THEOREM 4.1 $\{S^{*n}TS^n\}$ *converges uniformly if and only if* $T = T_\varphi + K$, $K \in \mathcal{K}$, $\varphi \in \mathbf{L}^\infty$.

Proof: The necessity is immediate. If $T = T_\varphi + K$ then $S^{*n}T_\varphi S^n = T_\varphi$ and $\|S^{*n}KS^n\| \to 0$ by the compactness of K.

For sufficiency, note that if $S^{*n}TS^n$ converges uniformly to T_φ then $\|S^{*n}(T - T_\varphi)S^n\| \to 0$ as $n \to \infty$.

It is easily checked that $\|S^{*n}(T - T\varphi)S^n\| = \|(I - P_{n-1})(T - T_\varphi)(I - P_{n-1})\|$. The first is the norm of the matrix of $T - T_\varphi$ after the first n rows and columns are removed and the second is the norm of the same matrix when these rows and columns are replaced by zeroes. Thus $\|(I - P)_{n-1})(T - T_\varphi)(I - P_{n-1})\| \to 0$ as $n \to \infty$ and this implies that $T - T_\varphi \in \mathcal{K}$. ∎

It is obvious that uniformly asymptotic Toeplitz does not imply uniformly asymptotic Hankel. Indeed,

$$\begin{aligned} \|H_n(T)\| &= \|J_nTS^{n+1}\| \\ &= \|J_nP_nT(I - P_n)S^{n+1}\| \\ &= \|P_nT(I - P_n)\|. \end{aligned}$$

Thus for K compact $\|H_n(K)\| \to 0$ as $n \to \infty$. However, for a Toeplitz operator T_φ, $H_n(T_\varphi)$ does not generally converge uniformly to $H_n(T_\varphi)$. This is a consequence of the following lemma.

LEMMA 4.2 $\{H_n(T_\varphi)\}$ *converges uniformly if and only if* H_φ *is compact.*

Proof: For a Toeplitz operator T_φ, $H_n(T_\varphi) = P_nH_\varphi$ which is finite rank and $H_n(T_\varphi)$ converges strongly to H_φ. If convergence is uniform, it is also to H_φ and thus H_φ is compact. On the other hand if H_φ is compact, $H_n(T_\varphi) = P_n H_\varphi$ converges uniformly. ∎.

We will characterize all uniformly asymptotic Hankel operators. We begin with those for which $\|H_n(T)\| \to 0$. As observed, this is equivalent to $\|P_n T(I - P_n)\| \to 0$. But this is exactly $\mathcal{C} + \mathcal{K}$ ([2]) the algebra of quasi-triangular operators with respect to $\{P_n\}$.

THEOREM 4.3 *T is uniformly asmptotic Hankel if and only if $T = T_\varphi + U$ where $\varphi \in \mathbf{H}^\infty + \mathbf{C}$ and $U \in \mathcal{C} + \mathcal{K}$.*

Proof: Suppose $\{H_n(T)\}$ converges uniformly to $H(T)$. Since $H_n(T)$ is finite rank, $H(T)$ is compact. By Lemma 3.1, $H(T)$ is a Hankel operator. It follows from Hartman's theorem ([10]) that $H(T) = H(T_\varphi)$ where $\varphi = \mathbf{H}^\infty + \mathbf{C}$. Then

$$\|H_n(T) - H_n(T_\varphi)\| \leq \|H_n(T) - H(T)\| + \|H(T) - H_n(T_\varphi)\|.$$

By Lemma 4.2, $\|H(T) - H_n(T\varphi)\| \to 0$ as $n \to \infty$. Thus $\|H_n(T - T_\varphi)\| \to 0$. This implies that $T - T_\varphi \in \mathcal{C} + \mathcal{K}$.

On the other hand suppose $T = T_\varphi + U$ where $\varphi \in \mathbf{H}^\infty + \mathbf{C}$ and $U \in \mathcal{C} + \mathcal{K}$. Then $\|H_n(T - T_\varphi)\| = \|H_n(U)\| \to 0$ as $n \to \infty$. Since $\|H_n(T) - H(T_\varphi)\| \leq \|H_n(T - T_\varphi)\| + \|H_n(T_\varphi) - H(T_\varphi)\|$, and since $\|H_n(T_\varphi) - H(T_\varphi)\| \to 0$ as $n \to \infty$ by Lemma 4.2, it follows that $\|H_n(T) - H(T_\varphi)\| \to 0$ as $n \to \infty$. ∎

Can Lemma 4.2 be extended to asymptotic Hankel operators ? That is, suppose $H(T)$ is compact; does it follow that $\{H_n(T)\}$ converges uniformly? We will see in section 7 that this is in fact the case.

5 The Weak Asymptotic Case.

THEOREM 5.1 *Let $[t_{ij}]_{i,j=0}^{\infty}$ be the matrix representation of T with respect to the standard basis. Then*

(i) *T is weak asymptotic Hankel if and only if for each $j \geq 1$, the sequence $\{t_{i,i+j}\}$ converges and the sequence of limits $\{t_{-j}\}_{j=1}^{\infty}$ is the sequence of negative Fourier coefficients of some $\varphi \in \mathbf{L}^\infty$.*

(ii) *T is weakly asymptotic Toeplitz if and only if for each $-\infty < j < \infty$ the sequence $\{t_{i,i+j}\}$ converges and the sequence of limits $\{t_{-j}\}_{j=-\infty}^{\infty}$ is the sequence of Fourier coefficients of some $\varphi \in \mathbf{L}^\infty$.*

Proof: The necessity of (i) is contained in Lemma 3.1. For sufficiency, consider the Hankel operator H given by

$$H(T) = \begin{bmatrix} t_{-1} & t_{-2} & t_{-3} & \cdots \\ t_{-2} & t_{-3} & t_{-4} & \cdots \\ t_{-3} & t_{-4} & t_{-5} & \cdots \\ \cdot & \cdot & \cdot & \\ \cdot & \cdot & \cdot & \\ \cdot & \cdot & \cdot & \end{bmatrix}$$

Then, $(H_n(T)e_i, e_j) = t_{n-j,n+i+1}$ for $n \geq j$ and by a change of indexes this can be written as $t_{n,n+i+j+1}$ which by hypothesis converges to $t_{-(i+j)} = (He_i, e_j)$ as $n \to \infty$. It follows that $H_n(T)$ converges weakly to H.

The proof of (ii) is similar. It simply involves all diagonals while (i) involved only the upper diagonals. ∎

COROLLARY 5.2 *T weakly asymptotic Toeplitz implies T weakly asymptotic Hankel.*

6 The strong asymptotic case.

Strong asymptotic Toeplitz operators were studied in [3]. This is the most important case both for Toeplitz and Hankel operators. Every operator in the Toeplitz algebra $\mathcal{T}$ (the uniformly closed algebra generated by the Toeplitz operators) is strong asymptotic Toeplitz ([3], Theorem 4). In fact every operator in $\mathcal{T}$ is of the form $T = T_\varphi + Q$ where φ is $sym(T)$ and Q is in the commutator ideal of $\mathcal{T}$.

What about strongly asymptotic Hankels ? We have observed that every Toeplitz is a strongly asymptotic Hankel, $H_n(T_\varphi) = P_n H_\varphi$ which converges strongly to H_φ. We show that every $T \in \mathcal{T}$ is a strongly asymptotic Hankel. We begin with a simple lemma.

LEMMA 6.1 *If A is a strongly asymptotic Toeplitz and Hankel with zero symbol, then BA is a strongly asymptotic Hankel, for any operator B.*

Proof: We assume that $H_n(A) \to 0$ and $S^{*n}AS^n \to 0$. Now

$$\begin{aligned} H_n(BA) &= J_nBAS^{n+1} \\ &= J_nB(I-P_n)AS^{n+1} + J_nBP_nAS^{n+1} \\ &= J_nBS^{n+1}S^{*n+1}AS^{n+1} + J_nBJ_nJ_nAS^{n+1} \\ &= H_n(B)S^{*n+1}AS^{n+1} + J_nBJ_nH_n(A) \end{aligned}$$

Now $\|H_n(B)\|$ and $\|J_nBJ_n\|$ are uniformly bounded by $\|B\|$. Thus $H_n(BA) \to 0$. ∎

LEMMA 6.2 *If $T = T_{\varphi_1}T_{\varphi_2}\dots T_{\varphi_n}$ with $\varphi_1, \varphi_2, \dots, \varphi_n \in \mathbf{L}^\infty$, then T is strong asymptotic Hankel and $H(T) = H_{\varphi_1\varphi_2\dots\varphi_n}$.*

Proof: Since $H(T_{\varphi_1\varphi_2\dots\varphi_n}) = H_{\varphi_1\varphi_2\dots\varphi_n}$, it suffices to show that $H_i(T_{\varphi_1}\dots T_{\varphi_n} - T_{\varphi_1\dots\varphi_n})$ converges to zero as $i \to \infty$. Using an idea of ([3]), we write $T_{\varphi_1}\dots T_{\varphi_n} - T_{\varphi_1\dots\varphi_n}$ as a telescoping sum:

$$\begin{aligned} T_{\varphi_1}T_{\varphi_2}\dots T_{\varphi_n} - T_{\varphi_1\varphi_2\dots\varphi_n} &= T_{\varphi_1}T_{\varphi_2\dots\varphi_n} - T_{\varphi_1(\varphi_2\dots\varphi_n)} + \\ &+ T_{\varphi_1}(T_{\varphi_2}T_{\varphi_3\dots\varphi_n} - T_{\varphi_2(\varphi_3\dots\varphi_n)}) + \\ &+ T_{\varphi_1}T_{\varphi_2}T_{\varphi_3}T_{\varphi_4\dots\varphi_n} - T_{\varphi_3(\varphi_4\dots\varphi_n)} + \\ &+ \dots \\ &+ T_{\varphi_1}T_{\varphi_2}\dots T_{\varphi_{n-2}}(T_{\varphi_{n-1}}T_{\varphi_n} - T_{\varphi_{n-1}\varphi_n}) \end{aligned}$$

Each term is of the form $A(T_\psi T_\varphi - T_{\psi\varphi})$. It thus suffices to show that for any operator A and $\psi, \varphi \in \mathbf{L}^\infty$, $H_i(A[T_\psi T_\varphi - T_{\psi\varphi}]) \to 0$ as $i \to \infty$. Using Lemma 6.1, we check that

(1) $H_i(T_\psi T_\varphi - T_{\psi\varphi}) \to 0$ as $i \to \infty$.

(2) $S^{*i}(T_\psi T_\varphi - T_{\psi\varphi})S^i \to 0$ as $i \to \infty$.

Since $T_\varphi T_\psi - T_{\varphi\psi}$ is the product of the two Hankels, (2) follows from Lemma 3, in ([3]). For (1), note that

$$H_i(T_\psi T_\varphi) = J_iT_\psi T_\varphi S^{i+1}$$

$$
\begin{aligned}
&= (J_iT_\psi S^{i+1})(S^{*i+1}T_\varphi S^{i+1}) + (J_iT_\psi J_i)(J_iT_\varphi S^{i+1}) \\
&\doteq H_i(T_\psi)T_\varphi + J_iT_\psi J_iP_iH_\varphi \\
&= P_iH_\psi T_\varphi + J_iT_\psi J_iP_iH_\varphi .
\end{aligned}
$$

As $i \to \infty$, this converges strongly to $H_\psi T_\varphi + T_\psi H_\varphi = H_{\psi\varphi}$. Since $H_i(T_{\psi\varphi}) = P_iH_{\psi\varphi}$, this proves (1), and the proof of the lemma is complete. ∎

THEOREM 6.3 *Each $T \in \mathcal{T}$ is a strongly asymptotic Hankel.*

Proof: It now suffices to show that if $\{T_n\} \in \mathcal{T}$ and $\|T_n - T\| \to 0$, then the existence of $H(T_n)$ implies the existence of $H(T)$. We show that for each $x \in h_2$, $\{H_n(T)x\}$ is a Cauchy sequence. Fix $x \in h_2$ and let $\varepsilon > 0$. Choose k so that $\|T - T_k\| < \varepsilon/4\|x\|$. Then

$$
\begin{aligned}
\|[H_n(T) - H_m(T)]x\| &= \|H_n(T - T_k)x + [H_n(T_k)x - H_m(T_k)]x + \\
&\quad + H_m(T_k - T)x\| \\
&\leq \|H_n(T - T_k)x\| + \|[H_n(T_k)x - H_m(T_k)]x\| + \\
&\quad + \|H_m(T_k - T)x\| \\
&\leq 2\|T - T_k\|\|x\| + \|H_n(T_k) - H_n(T_k)x]\|.
\end{aligned}
$$

Since $T_k \in \mathcal{T}$, there exists N such that for $n, m > N$,

$$\|[H_n(T_k) - H_m(T_k)]x\| < \varepsilon/2.$$

The rest is immediate. ∎

It is natural to ask whether every strongly asymptotic Toeplitz operator is also strongly asymptotic Hankel. This is the case in the weak topology and is not so in the uniform case. We show that strong asymptotic Toeplitz does not imply strong asymptotic Hankel. First, note that for any operator T and $x \in h_2$,

$$
\begin{aligned}
\|TS^nx\|^2 &= \|P_{n-1}TS^nx\|^2 + \|(I - P_{n-1})TS^nx\|^2 \\
&= \|J_{n-1}TS^nx\|^2 + \|S^{*n}TS^nx\|^2
\end{aligned}
$$

Thus if $\|S^{*n}TS^nx\| \to 0$ and $\|H_n(T)x\| \to 0$ it follows that $\|TS^nx\| \to 0$.

We give an example of an operator T for which $S^{*n}TS^n$ converges strongly to zero but TS^n doesn't converge. Since if $H_n(T)$ converges strongly it must converge to the Hankel of zero (namely zero), it follows that $H_n(T)$ doesn't converge strongly either. The example is based on ([3], Example 13). Let U_k be the square matrix of size $2k$ defined as follows: all entries are 0 except the first k in the last column and they are equal to $1/\sqrt{k}$ ($k = 1,2,3,\ldots$). Let T be the operator whose matrix is the direct sum of the U_k's. As shown in ([3], Example 13) $S^{*n}TS^n \to 0$ strongly as $n \to \infty$. But $\{S^{*n}T^*TS^n\}$ is not strongly convergent. It follows that $\{TS^n\}$ doesn't converge strongly to zero.

7 Distance Formulas for strongly asymptotic Hankels.

Recall that a classical resalt of Nehari ([9]) gives distance the from $v \in \mathbf{L}^\infty$ to the subalgebra $\mathbf{H}^\infty$ as the norm of $\mathbf{H}_v$. We begin with the operator analogue of this result.

THEOREM 7.1 *If T is strongly asymptotic Hankel then* $d(T, \mathcal{C}+\mathcal{K}) = \|H(T)\|$.

Proof: By ([2])

$$\begin{aligned} d(T, \mathcal{C}+\mathcal{K}) &= \lim_{n\to\infty} \sup \|P_n T(I-P_n)\| \\ &= \lim_{n\to\infty} \sup \|H_n(T)\| \end{aligned}$$

Since $H_n(T)x \to H(T)x$ for all $x \in h_2$ and $\|H_n(T)\| \leq \|T\|$, it follows that $\|H_n(T)\| \to \|H(T)\|$. This completes the proof. ∎

Remark This result requires amplification. Why is $\mathcal{C}+\mathcal{K}$ the natural operator analogue of $\mathbf{H}^\infty$ (at least in this context)? $\mathbf{H}^\infty$ is the algebra of Toeplitz operators whose associated Hankel operator is zero. $\mathcal{C}+\mathcal{K}$ is the algebra of uniformly asymptotic Hankel operators whose symbol is zero.

This raises the following question. For $\varphi \in \mathbf{L}^\infty$, it is easy to see (Lemma 4.2), that $\{H_n(T_\varphi)\}$ converges uniformly if and only if H_φ is compact. Is this true for strongly asymptotic Hankel operators as well?

In Theorem 4.3 we have characterized the set of uniform asymptotic Hankel operators as $\{T_\varphi + U, \varphi \in \mathbf{H}^\infty + \mathbf{C}, U \in \mathcal{C} + \mathcal{K}\}$. Denote this set by $\mathcal{U}_\infty$ and for each non-negative integer n, let $\mathcal{U}_n = \{T \in \mathcal{U}_\infty :$ rank $H(T) < n\}$. Then

(1) $\mathcal{C} + \mathcal{K} = \mathcal{U}_0 \subset \mathcal{U}_1 \subset \mathcal{U}_2 \subset \ldots \subset \mathcal{U}_\infty$.

(2) $\{T_v : v \in \mathbf{H}^\infty + \mathbf{R}_n\} \subset \mathcal{U}_n$ for all $n \geq 0$.

THEOREM 7.2 *For all* $n \geq 1$,

$$\mathcal{U}_n = \{T_\varphi + U : \varphi \in \mathbf{H}^\infty + \mathbf{R}_n, U \in \mathcal{C} + \mathcal{K}\}.$$

Proof: Suppose $\{H_n(T)\}$ converges uniformly to $H(T)$ and rank $H(T) \leq n$. By Kronecker's Lemma ([9]) $H(T) = H(T_\varphi)$ with $\varphi \in \mathbf{H}^\infty + \mathbf{R}_n$. Then

$$\|H_n(T) - H_n(T_\varphi)\| \leq \|H_n(T) - H(T)\| + \|H(T) - H_n(T_\varphi)\|.$$

By Lemma 4.2, $\|H(T) - H_n(T_\varphi)\| \to 0$ as $n \to \infty$. Thus $\|H_n(T - T_\varphi)\| \to 0$ as $n \to \infty$ which implies that $T - T_\varphi \in \mathcal{C} + \mathcal{K}$.

On the other hand, suppose that $T = T_\varphi + U$ where $\varphi \in \mathbf{H}^\infty + \mathbf{R}_n$. and $U \in \mathcal{U}_0$. Then $\|H_n(T - T_\varphi)\| = \|H_n(U)\| \to 0$ as $n \to \infty$. By Theorem 4.3 $T \in \mathcal{U}_\infty$. Since rank $H(T) =$ rank $H(T_\varphi) \leq n$, this implies that $T \in \mathcal{U}_n$ and the proof is complete. ∎

Recall that the k-th s-number of an operator A on h_2 is $s_k(A) = \inf\{\|A + F\| : F \in \mathcal{F}_k\}$.

THEOREM 7.3 *For* $T \in \mathcal{T}$, $d(T, \mathcal{U}_k) = s_k(H(T))$.

Proof: Since for $u \in \mathcal{U}_k$,

$$\|H(T) - H(U)\| = \|H(T - U)\| \leq \|T - U\|$$

it follows that

$$\begin{aligned} d(T, \mathcal{U}_k) &= \inf\{\|T - U\| : U \in \mathcal{U}_k\} \\ &\geq \inf\{\|H(T) - H(U)\| : U \in \mathcal{U}_k\} \\ &\geq \inf\{\|H(T) - F\| : F \in \mathcal{F}_k\} \\ &= s_k(H(T)). \end{aligned}$$

For the opposite inequality recall ([9], p. 305) that there exists a Hankel operator H_v of rank $\leq k$ such that

$$s_k(H(T)) = \|H(T) - H_v\|.$$

Thus

$$\begin{aligned} s_k(H(T)) &= \|H(T - T_v)\| \\ &= d(T - T_v, \mathcal{C} + \mathcal{K}) \\ &= d(T, T_v + \mathcal{C} + \mathcal{K}) \\ &\leq d(T, \mathcal{U}_k) \end{aligned}$$

since $T_v + \mathcal{C} + \mathcal{K} \subseteq \mathcal{U}_k$ by Theorem 7.2. This completes the proof. ∎

COROLLARY 7.4 *If $T = T_v$ for $v \in \mathbf{L}^\infty$, then $d(T, \mathcal{U}_k) = d(v, \mathbf{H}^\infty + \mathbf{R}_k)$.*

The next result is an analogoue of Hartman's Theorem ([9]).

THEOREM 7.5 *For $T \in \mathcal{T}$, $d(T, \mathcal{U}_\infty) = \|H(T)\|_e$ (see [9] p. 311).*

Proof: This follows from the fact that

$$\begin{aligned} d(T, \mathcal{U}_\infty) &= \lim_{n \to \infty} d(T, \mathcal{U}_n) \\ &= \lim_{n \to \infty} s_n(H(T)) \\ &= \|H(T)\|_e. \end{aligned}$$

∎

COROLLARY 7.6 *For $v \in \mathbf{L}^\infty$, $d(T_v, \mathcal{U}_\infty) = D(v, \mathbf{H}^\infty + \mathbf{C})$.*

As another consequence of Theorem 7.2 we obtain an extension of Lemma 4.2 to strongly asymptotic Hankels.

COROLLARY 7.7 *If T is strongly asymptotic Hankel and $H(T)$ is compact, then*

$$\lim_{n \to \infty} \|H_n(T) - H(T)\| = 0.$$

Bibliography

[1] V.M. Adamjan, D.Z. Arov and M.G. Krein, "Analytic properties of Schmidt pairs for a Hankel operator and the generalized Schur-Takagi problem", Math USSR Sbornik 15(1971), 31-73.

[2] W. Arveson, "Interpolation problems in nest algebras", J. Func. Anal. 20(1975), 208-233.

[3] J. Barria and P.R. Halmos, "Asymptotic Toeplitz operators", Trans. A.M.S. 273(1982) 621-630.

[4] R. Brockestt, "Finite Dimensional Linear Systems", John Wiley and Sons, New York, 1970.

[5] A. Feintuch, "Stabilization and Sensitivity for eventually time-invariant systems", to appear Linear Algebra and Applications.

[6] A. Feintuch and R. Saeks, "System Theory: A Hilbert Space Approach", Academic Press, New york, 1982.

[7] J. Kailath, "Linear Systems", Prentice Hall, Englewood Cliffs, N.J. 1980.

[8] E. Kamen, P. Khargonekar and K. Polla, "A transfer function approach to linear time-varying discrete-time systems", SIAM J. on Control and Opt. 23(1980), 550-565.

[9] N.K. Nikol'skii, "Treatise On The Shift Operator", Springer Verlag, Berlin, 1986.

[10] S.E. Power, "Hankel Operators On Hilbert Space", Research Notes in Mathematics 64, Pitman, Boston 1982.

[11] R. Saeks and G. Knowels, "The Arveson Frequency Response and System Theory", Int. Jour. Control, 42(1985), 639-650.

Avraham Feintuch
Department of Mathematics
Ben Gurion University
Beer-Sheva
Israel

Operator Theory:
Advances and Applications, Vol. 41

ITERATIVE COMMUTANT LIFTING FOR SYSTEMS WITH RATIONAL SYMBOL

Ciprian Foias   Allen Tannenbaum

This paper is dedicated to our dear friend and colleague Professor Israel Gohberg on the occasion of his sixtieth birthday.

Among Professor Gohberg's many impressive achievements is the solid and fruitful connection he established among the fields of operator theory, linear algebra, and systems theory. Many researchers have now since followed in this pioneering trail led by Israel Gohberg.

The present paper is concerned with giving an explicit way of carrying out the iterative commutant lifting procedure described in [9] for systems with rational symbol. Thus for such systems, our technique may be applied to yield the optimal controller for a nonlinear weighted sensitivity optimization problem.

This research was supported in part by grants from the Research Fund of Indiana University, Department of Energy DE-FG02-86ER25020, NSF (ECS-8704047), NSF (DMS-8811084), and the Air Force Office of Scientific Research (AFOSR-88-0020).

1. INTRODUCTION

Given the success and popularity of H^∞ design methods during the past several years for linear systems, there has recently been some research in extending these methods to the nonlinear case. The common theme of all of this work is the attempt to generalize Nevanlinna-Pick-Sarason-Nagy-Foias interpolation theory to nonlinear operators. For example, in [2], [3] an extension of the commutant lifting theorem to a local nonlinear setting was given, together with a discussion of how this result could be used to develop a design procedure for nonlinear systems. In [5] Ball and Helton approach this problem via a nonlinear version of Ball-Helton theory, which in particular gives a nonlinear Beurling-Lax type theorem. Another approach to a nonlinear Beurling-Lax theorem is explored in [4].

The present paper is based on on our previous work in [9]. Our basic approach in that paper is to apply an *iterative commutant lifting* procedure to a large class of analytic input/output operators (called *majorizable*) in designing a feedback compensator for a given physical plant. (See Section 4 for the details.) It turns out that this technique optimizes a certain "weighted sensitivity function" in the sense of germs of analytic functions, and reduces to classic H^∞ sensitivity optimization in the linear case. An attractive feature of this method is that it is constructive and physically natural, while avoiding some of the nontrivial technical difficulties of some of the approaches mentioned before.

The point of our work here is to show how the iterative commutant lifting procedure may be explicitly implemented when we begin with rational data. Namely, for a nonlinear rational plant (see Section 7 below for the definition of a rational nonlinear input/output operator), and a nonlinear rational weighting filter, the iterative commutant lifting procedure at each step leads to a rational compensator, and thus gives a promising method of doing nonlinear design. The actual construction of the compensator in this circumstance may be accomplished using the skew Toeplitz ideas of [7], [8], [10]. Indeed, with rational data, the iterative commutant lifting construction immediately falls into this framework. We should add that our techniques have already been applied to a plant with a saturation nonlinearity. The design was simulated at Honeywell with some encouraging results. We plan to describe this design in detail in a forthcoming engineering paper. In order to make this paper self-contained, we will briefly review the main results of [9] in Sections 4-6.

This research was supported in part by grants from the Research Fund of Indiana University, Department of Energy DE-FG02-86ER25020, NSF (ECS-8704047), NSF(DMS-8811084), and the Air Force Office of Scientific Research (AFOSR-88-0020).

2. MAJORIZABLE MAPPINGS

In this section we will define and discuss the class of input/output operators which will model the physical systems considered in this paper. We are following the treatments of [2], [4] here.

Let G and H denote separable complex Hilbert spaces. Let

$$B_{r_o}(G) := \{g \in G : \|g\| < r_o\}$$

(the open ball of radius r_o in G about the origin). Then we say that a mapping $\phi : B_{r_o}(G) \to H$ is *analytic* if the complex function $(z_1, \cdots, z_n) \to \langle\phi(z_1 g_1 + \cdots + z_n g_n), h\rangle$ is analytic in a neighborhood of $(1, 1, ..., 1) \in \mathbf{C}^n$ as a function of the complex variables $z_1, \cdots, z_n$ for all $g_1, \cdots, g_n \in G$ such that $\|g_1 + \cdots + g_n\| < r_o$, for all $h \in H$, and for all $n > 0$.

Assume that $\phi(0) = 0$. It is easy to see that if $\phi : B_{r_o}(G) \to H$ is analytic, then ϕ admits a convergent Taylor series expansion, i.e.

$$\phi(g) = \phi_1(g) + \phi_2(g, g) + \cdots + \phi_n(g, \cdots, g) + \cdots$$

where $\phi_n : G \times \cdots \times G \to H$ is an n-linear map. Clearly, without loss of generality we may assume that the n-linear map $(g_1, \ldots, g_n) \to \phi_n(g_1, \cdots, g_n)$ is symmetric in the arguments $g_1, \cdots, g_n$. This assumption will be made throughout this paper for the various analytic maps which we consider.

Next set

$$\hat{\phi}_n(g_1 \otimes \cdots \otimes g_n) := \phi_n(g_1, \cdots, g_n).$$

Then $\hat{\phi}_n$ extends in a unique manner to a dense subset of $G^{\otimes n} := G \otimes \cdots \otimes G$ (tensor product taken n times). Notice by $G^{\otimes n}$ we mean the Hilbert space completion of the algebraic tensor product of the G's. Clearly if $\hat{\phi}_n$ has finite norm on this dense subset, then $\hat{\phi}_n$ extends by continuity to a bounded linear operator $\hat{\phi}_n : G^{\otimes n} \to H$. By abuse of notation, we will set $\phi_n := \hat{\phi}_n$, and $\phi_n(g) := \phi_n(g \otimes \cdots \otimes g)$ (the tensor product taken n times).

It is important to note that one can determine ϕ_n quite easily from the input/output operator ϕ. Indeed, we have the following elementary:

LEMMA (2.1). Let $\phi : B_{r_0}(G) \to H$ be analytic, $\phi(0) = 0$. Suppose moreover that if

$$\phi(g) = \phi_1(g) + \cdots \phi_n(g) + \cdots,$$

then each of the ϕ_n defines a bounded linear operator $G^{\otimes n} \to H$ as above (and is symmetric in its arguments). Then for $g_j \in G$ ($j = 1, \cdots, n$) with $\|g_1\| = \cdots = \|g_n\| < r_0$, we have

$$n!\phi_n(g_1 \otimes \cdots \otimes g_n) =$$

$$\frac{1}{(2\pi)^n} \int_0^{2\pi} \cdots \int_0^{2\pi} \phi(e^{i\theta_1} g_1 + \cdots + e^{i\theta_n} g_n) e^{-i(\theta_1 + \cdots + \theta_n)} d\theta_1 \cdots d\theta_n .$$

PROOF. Expand $\phi(z_1 g_1 + \cdots + z_n g_n)$ in powers of $z_1, \cdots, z_n$. Then it is easy to see that the coefficient of $z_1 \cdots z_n$ is precisely $n!\phi_n(g_1 \otimes \cdots \otimes g_n)$. The required result then follows immediately from the Cauchy formula. □

We now conclude this section with two key definitions.

DEFINITIONS (2.2). (i) Notation as above. By a *majorizing sequence* for the holomorphic map ϕ, we mean a sequence of positive numbers α_n $n = 1,2, \cdots$ such that $\|\phi_n\| < \alpha_n$ for $n \geq 1$. Suppose that $\rho := \limsup \alpha_n^{1/n} < \infty$. Then it is completely standard ([8]) that the Taylor series expansion of ϕ converges at least on the ball $B_r(G)$ of radius $r = 1/\rho$.

(ii) If ϕ admits a majorizing sequence as in (i), then we will say that ϕ is *majorizable*.

It is precisely the class of majorizable operators which we will study in this paper.

3. NONLINEAR OPTIMAL SENSITIVITY THEORY

We will define in this section a nonlinear extension of classical linear weighted H^∞ optimal sensitivity which will be the main object of study in this paper. We will also supply some physical engineering motivation for our definition. First, we will need to consider the precise kind of input/output operator we will be considering. We are basing our discussion here on [9]. All of the operators we consider are causal and are majorizable.

Throughout this paper $H^2(\mathbf{C}^k)$ will denote the standard Hardy space of $\mathbf{C}^k$-valued functions on the unit circle (k may be infinite, i.e., in this case $\mathbf{C}^k$ is replaced by h^2, the space of one-sided square summable sequences). We now make the following definition:

DEFINITION (3.1). Let $S : H^2(\mathbf{C}^k) \to H^2(\mathbf{C}^k)$ denote the canonical unilateral right shift. Then we say an input/output operator ϕ is *locally stable* if it is causal and majorizable, $\phi(0) = 0$, and if there exists an $r > 0$ such that $\phi : B_r(H^2(\mathbf{C}^k)) \to H^2(\mathbf{C}^k)$ with $S\phi = \phi \circ S$ on $B_r(H^2(\mathbf{C}^k))$. We set

$$C_l := \{\text{space of locally stable operators}\}.$$

Since the theory we are considering is local, the notion of local stability is sufficient for all of the applications we have in mind. The interested reader can compare this notion with the more global notions of stability as for example discussed in [5].

The theory we are about to give holds for systems which admit coprime locally stable factorizations. However, for simplicity we will assume that all of our systems are locally stable. Accordingly, let P, W denote locally stable operators, with W invertible. In a typical feedback context, P represents the plant (i.e., the mathematical model of the physical device we wish to control), and W the weight or filter (which defines the disturbance model). Now we say that the feedback compensator C *locally stabilizes* the closed loop if the operators $(I + P \circ C)^{-1}$ and $C \circ (I + P \circ C)^{-1}$ are well-defined and locally stable. By a result of [1] C locally stabilizes the closed loop if and only if

$$C = \hat{q} \circ (I - P \circ \hat{q})^{-1} \tag{1}$$

for some $\hat{q} \in C_l$. For such a compensator C, the operator $(I + P \circ C)^{-1} \circ W$ is called the *weighted sensitivity,* and is a measure of disturbance rejection of the given feedback system (see [11] and [16]). Via (1), the weighted sensitivity can be expressed as $W - P \circ q$, where $q := \hat{q} \circ W$. (Since W

is invertible, the data q and $\hat{q}$ are equivalent.) In this context, we will call such a q a *compensating parameter*. From the compensating parameter q, we get a locally stabilizing compensator C from the formula (1).

The problem we would like to solve here is a version of the classical disturbance attenuation problem to which we just alluded. This of course corresponds to the "minimization" of the sensitivity $W - P \circ q$ taken over all locally stable q. In order to formulate a precise mathematical problem, we need to say in what sense we want to minimize $W - P \circ q$ in this nonlinear context. We will see that while the optimal H^∞ sensitivity is a real number in the linear case, the measure of performance which seems to be more natural in this nonlinear setting is a certain function defined in a real interval.

In order to define our notion of sensitivity, we will first have to partially order the space of germs of analytic mappings defined in a ball about the origin. All of the input/output operators here will be locally stable. We also follow here our convention that for given $\phi \in C_l$, ϕ_n will denote the bounded linear map on the tensor space $(H^2(\mathbf{C}^k))^{\otimes n}$ associated to the n-linear part of ϕ which we also denote by ϕ_n (and which we always assume without loss of generality is symmetric in its arguments). The context will always make the meaning of ϕ_n clear.

We can now state the following key definitions:

DEFINITIONS (3.2). (i) For $W, P, q \in C_l$ (W is the weight, P the plant, and q the compensating parameter), we define the *sensitivity function* $S(q)$,

$$S(q)(\rho) := \sum_{n=1} \rho^n \|(W - P \circ q)_n\|$$

for all $\rho > 0$ such that the sum converges. Notice that for fixed P and W, for each $q \in C_l$, we get an associated sensitivity function.

(ii) We write $S(q) \lesssim S(\tilde{q})$, if there exists a $\rho_o > 0$ such that $S(q)(\rho) \leq S(\tilde{q})(\rho)$ for all $\rho \in [0, \rho_o]$. If $S(q) \lesssim S(\tilde{q})$ and $S(\tilde{q}) \lesssim S(q)$, we write $S(q) \cong S(\tilde{q})$. This means that $S(q)(\rho) = S(\tilde{q})(\rho)$ for all

$\rho > 0$ sufficiently small, i.e. $S(q)$ and $S(\tilde{q})$ are equal as germs of functions.

(iii) If $S(q) \leq S(\tilde{q})$, but $S(\tilde{q}) \not\leq S(q)$, we will say that q *ameliorates* $\tilde{q}$. Note that this means $S(q)(\rho) < S(\tilde{q})(\rho)$ for all $\rho > 0$ sufficiently small.

Now with (3.2), we can define a notion of "optimality" relative to the sensitivity function:

DEFINITIONS (3.3). (i) $q_o \in C_l$ is called *optimal* if $S(q_o) \leq S(q)$ for all $q \in C_l$.

(ii) We say $q \in C_l$ is *optimal with respect to its n-th term* q_n, if for every n-linear $\hat{q}_n \in C_l$, we have

$$S(q_1 + \cdots + q_{n-1} + q_n + q_{n+1} \cdots) \leq S(q_1 + \cdots + q_{n-1} + \hat{q}_n + q_{n+1} + \cdots).$$

If $q \in C_l$ is optimal with respect to all of its terms, then we say that it is *partially optimal.*

Clearly, if q is optimal, then it is partially optimal, but the converse may not hold. Notice moreover that if ϕ is a Volterra series, then our definition of sensitivity measures in a precise sense the amplication of energy of each Volterra kernel on signals whose energy is bounded by a given ρ. For this reason, it seems that in this context, the definition (3.2) of the sensitivity function $S(q)$ is physically natural. In the next section, we will discuss a procedure for constructing partially optimal compensating parameters, and then in Sections 5-9, we will show how this procedure leads to the construction of optimal compensating parameters for SISO rational nonlinear systems. Of course, from formula (1) above, one can derive the corresponding partially optimal (resp., optimal) compensator from the partially optimal (resp., optimal) compensating parameter.

4. Iterative Commutant Lifting Method

In this section, we discuss the main construction of this paper from which we will derive both partially optimal and optimal compensators relative to the sensitivity function. As before, P will denote the plant, and W the weighting operator, both of which we assume are locally stable. As in the linear case, we always suppose that P_1 is an isometry, i.e. P_1 is **inner.** In order to state our results, we will need a few preliminary remarks and to set-up some notation.

We begin by noting the following key relationship:

$$(W - P\circ q)_k = W_k - \sum_{1\le j\le k} \sum_{i_1+....+i_j=k} P_j(q_{i_1}\otimes \cdots \otimes q_{i_j})$$

Note that once again for ϕ majorizable, ϕ_n denotes the n-linear part of ϕ, as well as the associated linear operator on the appropriate tensor space.

We are now ready to formulate the *iterative commutant lifting procedure.* Let $\Pi : H^2(\mathbf{C}^k) \to H^2(\mathbf{C}^k) \ominus P_1 H^2(\mathbf{C}^k)$ denote orthogonal projection. Using the linear commutant lifting theorem (CLT) (see [15]), we may choose q_1 such that

$$\|W_1 - P_1 q_1\| = \|\Pi W_1\|.$$

Now given this q_1, we choose (using CLT) q_2 such that

$$\|W_2 - P_2(q_1 \otimes q_1) - P_1 q_2\| = \|\Pi(W_2 - P_2(q_1\otimes q_1))\|.$$

Inductively, given $q_1, \cdots, q_{n-1}$, set

$$A_n := (W_n - \sum_{2\le j\le n} \sum_{i_1+....+i_j=n} P_j(q_{i_1}\otimes \cdots \otimes q_{i_j})) \tag{2}$$

for $n \ge 2$. Then from the CLT, we may choose q_n such that

$$\|A_n - P_1 q_n\| = \|\Pi A_n\|. \tag{2a}$$

We now come to the key point on the convergence of the iterative commutant lifting method to which we refer the reader to [9] for the proof:

PROPOSITION (4.1). With the above notation, let $q^{(1)} := q_1 + q_2 + \cdots$. Then $q^{(1)} \in C_l$.

Next note that given any $q \in C_l$, we can apply the iterative commutant lifting procedure to $W - P\circ q$. Set

$$S_\Pi(q)(\rho) := \sum_{n=1} \rho^n \|\Pi(W - P\circ q)_n\|.$$

Clearly, $S_\Pi(q) \le S(q)$ (as functions). We can now state the following result whose proof is immediate from the above discussion:

PROPOSITION (4.2). Given $q \in C_l$, there exists $\tilde{q} \in C_l$, such that $S(\tilde{q}) \equiv S_\Pi(q)$. Moreover $\tilde{q}$ may be constructed from the iterated commutant lifting procedure.

Moreover, we easily have the following result:

PROPOSITION (4.3). q is partially optimal if and only if $S(q) \doteq S_\Pi(q)$ (i.e., $S(q)(\rho) = S_\Pi(q)(\rho)$ for all $\rho > 0$ sufficiently small).

PROOF. Assume that q is partially optimal. Then q must be optimal with respect to its first term q_1. But we have seen that there exists $\hat{q}_1$ such that $\|W_1 - P_1\hat{q}_1\| = \|\Pi W_1\|$. If $\|W_1 - P_1 q_1\| > \|\Pi W_1\|$, then since we are considering germs of functions, we would have $S(q) \not\leq S(\hat{q}_1 + q_2 + \cdots)$, contradicting the partial optimality of q.

By induction, assume that we have proven

$$\|(W - P \circ q)_j\| = \|\Pi(W - P \circ q)_j\|$$

for $1 \leq j \leq n$. Then again if

$$\|(W - P \circ q)_{n+1}\| > \|\Pi(W - P \circ q)_{n+1}\|,$$

by the above construction using the commutant lifting theorem, we can find a $\hat{q}_{n+1}$ such that

$$\|\Pi(W - P \circ q)_{n+1}\| = \|(W - P \circ (q_1 + q_2 + \cdots + q_n + \hat{q}_{n+1} + \cdots))_{n+1})\|.$$

So once more, $S(q) \not\leq S(q_1 + \cdots + q_n + \hat{q}_{n+1} + q_{n+2} + \cdots)$, contradicting the partial optimality of q. Hence, we get that $S(q) \doteq S_\Pi(q)$. The proof of the converse direction is similar. □

We can now summarize the above discussion with the following:

THEOREM (4.4). For given P and W as above, any $q \in C_l$ is either partially optimal or can be ameliorated by a partially optimal compensating parameter.

PROOF. Immediate from (4.1), (4.2), and (4.3). □

It is important to emphasize that a partially optimal compensating parameter need not be optimal in the sense of (3.3)(i). Basically, what we have shown here is that using the iterated commutant lifting procedure, we can ameliorate any given design. The question of optimality will be considered in the next section.

5. OPTIMAL COMPENSATORS

In this section we will derive our main results about optimal compensators. Basically, we will show that in the single input / single output setting, the iterated commutant lifting procedure leads to an optimal design. We begin with the following:

THEOREM (5.1). There exist optimal compensators.

PROOF. The proof is not constructive and makes use of the weak compactness property of weakly closed, bounded, convex sets of operators on Hilbert space. For details see [9]. □

For the construction of the optimal compensator in Theorem (5.3) below, we will need one more technical result. Accordingly, we will need to set-up a bit more notation. First set $H^2 := H^2(\mathbf{C})$, and $H^\infty := H^\infty(\mathbf{C})$ (the space of bounded analytic complex-valued functions on the unit disc). Let $m \in H^\infty$ be a nonconstant inner function, let $\Pi_1 : H^2 \to H^2 \ominus mH^2 =: H(m)$ denote orthogonal projection, and set $T := \Pi_1 S \mid H(m)$, where S is the canonical unilateral shift on H^2. (T is the compressed shift.) For H a complex separable Hilbert space, let $S_\infty : H \to H$ denote a unilateral shift, i.e. an isometric operator with no unitary part. This means that $S_\infty^{*n} h \to 0$ for all $h \in H$ as $n \to \infty$. (See [15] and [19].) We state without proof the following result from [9] which generalizes a well-known result from [14].

LEMMA (5.2). Notation as above. Let $A : H \to H^2 \ominus mH^2$ be a bounded linear operator which attains its norm, i.e. such that there exists $h_o \in H$ with $\|Ah_o\| = \|A\| \|h_o\| \neq 0$. Suppose moreover that

$$AS_\infty = TA.$$

Then there exists a unique minimal intertwining dilation B of A, i.e. an operator $B : H \to H^2$ such that $AS_\infty = S_\infty B$, $\|A\| = \|B\|$, and $\Pi_1 B = A$.

We now come to the main result of this section:

THEOREM (5.3). Let W and P be SISO locally stable operators, with W the weight and P the plant. Suppose that ΠW_j is compact for $j \geq 1$, and that ΠP_k is compact for $k \geq 2$ ($\Pi : H^2 \to H^2 \ominus P_1 H^2$ denotes orthogonal projection.) Let q_{opt} be a partially optimal compensating parameter as constructed by the iterated commutant lifting procedure. Then q_{opt} is optimal.

PROOF. First of all, since ΠW_1 attains its norm, from (5.2) we have that the optimal q_1 constructed relative to W_1 and P_1 is unique. (Actually in this special case since we are working in H^2, this follows from [14].) From our hypotheses and equation (2) above, we have that each ΠA_k attains its norm for $k \geq 2$, and so by (5.2) each optimal q_k constructed by the iterated commutant lifting procedure is unique. Theorem (5.3) now follows immediately from (4.4). □

COROLLARY (5.4). Let P and W be locally stable and SISO, with linear parts P_1 rational. Then the partially optimal compensating parameter q_{opt} constructed by the iterated commutant lifting procedure is optimal.

PROOF. Indeed, since P_1 is rational $H^2 \ominus P_1 H^2$ is finite dimensional and hence this follows from (5.3). □

Corollary (5.4) gives a constructive procedure for finding the optimal compensator under the given hypotheses. Indeed, when P_1 is SISO rational, the iterative commutant lifting procedure can be reduced to *finite dimensional matrix calculations*. When both P and W are rational (in the non-linear sense given in Section 8), we will see how this leads to a straightforward procedure for the computation of the optimal compensator using the skew Toeplitz theory.

6. MAXIMAL VECTORS

In order to apply the iterative commutant lifting procedure, we will first need a generalization of a result due to Sarason [13] on the optimal interpolant. More precisely, for K a bounded linear operator on a Hilbert space, k_o is a *maximal vector,* if $\|Kk_o\| = \|K\| \|k_o\| \neq 0$. Then for SISO systems, Sarason [13] derives a formula for the optimal interpolant in terms of a maximal vector of the associated Hankel operator.

In order to state our result, we will first need a few preliminary remarks. Let $H := H^2(\mathbf{C}^k)$. As above, we let $m \in H^\infty$ be nonconstant inner, and let $\Pi_1 : H^2 \to H^2 \ominus mH^2 =: H(m)$ denote orthogonal projection, with T the compression of the canonical shift on H^2 to $H(m)$. Moreover S_∞ will denote the canonical shift on H, defined by multiplication by e^{it}. Now given $h \in H$, we can write h as a column vector (perhaps infinite)

$$h = \begin{bmatrix} h_1 \\ h_2 \\ \cdot \\ \cdot \\ \cdot \end{bmatrix}.$$

We then set

$$h^* := [\bar{h}_1 \ \ \bar{h}_2 \ \ \cdots].$$

Moreover, given any bounded linear operator $B : H \to H^2$ such that $BS_\infty = SB$, we have that for $z \in D$ (the unit disc),

$$(Bh)(z) = \sum_{j \geq 1} b_j(z) h_j(z).$$

That is, we can express B as the row matrix

$$[b_1 \ \ b_2 \ \ \cdots]$$

with $b_j \in H^\infty$ for $j \geq 1$. We will identify B with this row matrix. With this notation, we can now state the following result for whose proof we refer the reader to [9]:

PROPOSITION (6.1). Notation as above. Let $A : H \to H^2 \ominus mH^2$ be a bounded linear operator such that $AS_\infty = TA$. Suppose moreover that A has a maximal vector h_o. Let $B : H \to H^2$ be the minimal intertwining dilation of A, i.e. $\Pi_1 B = A$, $BS_\infty = SB$, and $\|A\| = \|B\|$. Then if we let $\lambda := \|A\|^2$, we have that

$$B = \frac{\lambda h_o^*}{\overline{Ah_o}}.$$

We will apply Proposition (6.1) to rational nonlinear systems in what follows.

7. REMARKS ON FOURIER REPRESENTATIONS AND SYMBOLS

In this section we collect some basic and well-known facts on the Fourier representations and symbols of time-invariant operators which we will need below. We will not give proofs of the assertions made here, but simply refer the reader to the books [6] and [15].

Let H be (a separable, complex) Hilbert space. Denote by $H^2(H)$, the set of all analytic functions $u : D \to H$ with a power series expansion

$$u(z) = \sum_{n=0}^{\infty} z^n a_n \text{ such that } \sum_{n=0}^{\infty} \|a_n\|^2 < \infty.$$

$H^2(H)$ is a Hilbert space under the norm

$$\|u\|_{H^2(H)}^2 := \sum_{n=0}^{\infty} \|a_n\|^2.$$

It is easy to see that on $H^2(H)$ we have the canonical shift defined by multiplication by z. Similarly, we can define $L^2(H)$ with its bilateral shift, and prove in the usual way that $H^2(H)$ may be identified as a closed subspace of $L^2(H)$.

Next for two Hilbert spaces K_1 and K_2, we let $L(K_1, K_2)$ stand for the space of linear mappings from K_1 to K_2. Now denote by $H^\infty(L(K_1, K_2))$ the Banach space of all bounded analytic functions $\Phi : D \to L(K_1, K_2)$. We can now state the following key result:

PROPOSITION (7.1). Let $X : H^2(K_1) \to H^2(K_2)$ be a bounded linear operator such that $XS_1 = S_2X$, where S_i is the canonical shift on $H^2(K_i)$ for $i = 1, 2$. Then there exists $\Theta[X] \in H^\infty(L(K_1, K_2))$ such that $X = T_{\Theta[X]}$ (where $T_{\Theta[X]}$ is the Toeplitz operator associated to $\Theta[X]$), and $\|X\| = \|\Theta[X]\|_\infty$. ($\Theta[X]$ is called the *symbol* of X.)

PROOF. We shall just briefly sketch the construction of $\Theta[X]$ here, referring the reader to [6], [15] for complete details about the proof. Indeed, for every $f \in K_1$, the function $Xf \in H^2(K_2)$ can be written as

$$(Xf)(z) = \sum_{n=0}^{\infty} z^n X_n f \ , \ z \in D.$$

Then $X_n \in L(K_1, K_2)$, and the power series

$$\Theta[X](z) := \sum_{n=0}^{\infty} z^n X_n$$

converges for every $z \in D$. This defines the required symbol. □

In system theoretic terms (7.1) means that linear time-invariant systems are defined by multiplication operators. Next, we will need to discuss the Fourier representation of a linear operator (the generalization of the "z-transform" of discrete-time input/output operators in systems theory). Let H be a Hilbert space with $S : H \to H$ unilateral shift, and let $R := \ker S^*$. Then

$$H = M(R) := \bigoplus_{n=0}^{\infty} S^n R. \tag{10}$$

The *Fourier transform* of H is given by the unitary operator $\Phi_R : H \to H^2(R)$ defined by

$$(\Phi_R \sum_{n=0}^{\infty} S^n a_n)(z) := \sum_{n=0}^{\infty} a_n z^n$$

for $z \in D$. As is standard practice in systems theory, we will identify H and $H^2(R)$, and in general make our computations in the Fourier domain.

Now let $H_i = M(R_i)$, $i = 1, 2$ be as in (10) above, and let $P : H_1 \to H_2$ be any operator such that $PS_1 = S_2P$. Then via the above, we have $\Phi_{R_2} P \Phi_{R_1}^* : H^2(R_1) \to H^2(R_2)$, and from (7.1), the operator $\Phi_{R_2} P \Phi_{R_1}^*$ admits a symbol $\Theta[P]$. We will say that P is *rational* if there exists a numerical polynomial $q(z)$ such that $q(z)\Theta[P](z)$ is an operator-valued polynomial. It is easy to check that $\Theta[P]$ is a polynomial of degree $\leq k$, if $\Theta[P](z)f$ is a polynomial of degree $\leq k$ for all f lying in a dense subset of R_1. (We will be using this fact several times in the next section.)

For systems which are rational in this sense, we will now see that the iterative commutant lifting procedure can be explicitly implemented.

8. SYSTEMS WITH RATIONAL SYMBOL

The point of this section is to show that if we start with rational data, then the iterative commutant lifting procedure produces compensating parameters which are also rational. Thus the iterative commutant lifting procedure may be carried out using the skew Toeplitz theory of [7], [10], and hence makes it very amenable to digitable implementation. What we must show of course, is that all of the operations involved in iterative commutant lifting preserve rationality.

Let $S : H^2(D) \to H^2(D)$ denote ordinary shift. Set $H^{(n)} := H^{\otimes n}$, and $S^{(n)} := S^{\otimes n}$ for all $n \geq 1$. (We put $H^{(1)} =: H$ and $S^{(1)} = S$.) Note that we have the identification $H^{(n)} = H^2(D^n)$, via the correspondence $1 \otimes \cdots \otimes z \otimes \cdots \otimes 1 \to z_j$ where z is in the jth place in the tensor product, for $j = 1, 2, \cdots, n$. We now state the following lemmas whose proofs are obvious:

LEMMA (8.1). Notation as above. Then

$$\ker S^{(n)*} = \{ \sum_{j=1}^{n} F_j(z_1, \cdots, z_{j-1}, z_{j+1}, \cdots, z_n) : F_j \in H^2(D^n) \}.$$

Notice that using (8.1) and our discussion in Section 7, any $F \in H^2(D^n)$

$$F = \sum_{i_1, \cdots, i_n} F_{i_1, \cdots, i_n} z_1^{i_1} z_2^{i_2} \cdots z_n^{i_n}$$

has the Fourier representation, denoted by $F(z)$ for $z \in D$, and given by

$$F(z) = \sum_{n=0}^{\infty} z^n \begin{bmatrix} F_{n,n,\cdots,n} \\ F_{n+1,n,\cdots,n} \\ \cdot \\ \cdot \\ F_{n,n,\cdots,n+1} \\ F_{n+1,n+1,\cdots,n} \\ \cdot \\ \cdot \\ F_{n+1,n,\cdots,n,n+1} \\ \cdot \\ \cdot \\ F_{n+1,n+1,\cdots,n+1} \\ F_{n+2,n,\cdots,n} \\ \cdot \\ \cdot \\ F_{n,n,\cdots,n+2} \\ F_{n+2,n+1,n,\dots,n} \\ \cdot \\ \cdot \\ F_{n+2,n+2,\dots,n+2} \\ \cdot \\ \cdot \end{bmatrix}.$$

In what follows we will be making the tacit identification between an element $F \in H^2(D^n) = H^{(n)}$ and its Fourier representation $F(z)$. This identification should always be clear from the context. Next we have the following elementary result:

LEMMA (8.2). Let H_i be a Hilbert space, equipped with a unilateral shift S_i, $i = 1,2$. Then $\ker (S_1 \otimes S_2)^* = \overline{\{\ker S_1^* \otimes H_2 + H_1 \otimes \ker S_2^*\}}$.

Given an operator $X : H^{(n)} \to H^{(m)}$, we say that X is *time-invariant* if $XS^{(n)} = S^{(m)}X$. Moreover, for $f \in H$, we let f_k denote the kth Fourier coefficient of f. We now check that rationality is closed under multiplication.

PROPOSITION (8.3). Let $P : H^{(n)} \to H$, $Q : H^{(m)} \to H$ be time-invariant operators. Define the operator $R : H^{(n+m)} \to H$, by $R(a \otimes b)_k := (P(a)_k \cdot Q(b)_k)$ for each $k \geq 0$. (R

corresponds to multiplication in the "time domain".) Then if P and Q are rational, so is R.

PROOF. Clearly $\ker S^{(k)*}$ is a generating subspace of $H^{(k)}$ for each $k \geq 1$. Hence from Lemma (8.2), it suffices to study R on elements $F \in H^{(n+m)}$ of the form $F = l \otimes g + f \otimes k$, for $l \in \ker S^{(n)*}$, $g \in H^{(m)}$, $f \in H^{(n)}$, $k \in \ker S^{(m)*}$. Then

$$\Theta[R](z)(l \otimes g + f \otimes k) = \frac{1}{2\pi i} \int_{|\zeta|=1} \Theta[P](\frac{z}{\zeta})l \cdot Q(g)(\zeta) + P(f)(\zeta) \cdot \Theta[Q](\frac{z}{\zeta})k \frac{d\zeta}{\zeta}. \quad (11)$$

(Note that (11) gives the symbol of R). Let $q(z)$ and $r(z)$ be polynomials such that

$$\deg q(z)\Theta[P](z) \leq N \ , \deg r(z)\Theta[Q](z) \leq M.$$

Clearly, $P(q(S^{(n)})l) = q(S)P(l)$ and $Q(r(S^{(m)})k) = r(S)Q(k)$. Then if $p := rq$

$$p(z)\Theta[R](l \otimes g + f \otimes k) = \Theta[R](p(S^{(n)})l \otimes p(S^{(m)})g + p(S^{(n)})k \otimes p(S^{(m)})g))$$
$$= \frac{1}{2\pi i} \int_{|\zeta|=1} p(\frac{z}{\zeta})\Theta[P](\frac{z}{\zeta})l \cdot Q(p(S^{(m)})g)(\zeta) + P(p(S^{(n)})f)(\zeta) \cdot p(\frac{z}{\zeta})\Theta[Q](\frac{z}{\zeta})k \frac{d\zeta}{\zeta}$$

which shows that pR is a polynomial of degree $\leq N + M$ and completes the proof. □

We now consider the tensor product of symbols.

PROPOSITION (8.4). Let H_j, H_j' be Hilbert spaces equipped with the unilateral shifts S_j, S_j' for $j = 1, 2$. Suppose moreover that the H_j, and H_j' admit generating subspaces so that they are each unitarily equivalent to the corresponding Fourier space (see Section 7). Let $P_j : H_j \to H_j'$, $P_j S_j = S_j' P_j$ for $j = 1, 2$. Then if P_1 and P_2 are rational, so is $P_1 \otimes P_2$.

PROOF. Clearly, with the above hypotheses, P_1 and P_2 admit symbols, and so the notion of rationality makes sense. Moreover, it is easy to check that $\ker (S_1 \otimes S_2)^*$ is the generating subspace of $H_1 \otimes H_2$. Hence by Lemma (8.2), it suffices to study the symbol $\Theta[P_1 \times P_2]$ on elements of the form $l \otimes g \in \ker S_1^* \otimes H_2$ and $f \otimes k \in H_1 \otimes \ker S_2^*$. Then we see that

$$\Theta[P_1 \otimes P_2](z)(l \otimes g + f \otimes k) = \Theta[P_1](z)l \otimes P_2 g + P_1 f \otimes \Theta[P_2](z)k. \quad (12)$$

(Note that (12) gives the symbol of $P_1 \otimes P_2$.) Since the P_j are rational, there exist polynomials p_j, $j = 1, 2$ such that $p_j(z)\Theta[P_j](z) =: Q_j(z)$ are polynomials. Thus

$$\begin{aligned} p_1(z)p_2(z)\Theta[P_1 \otimes P_2](z)(l \otimes g + f \otimes k) &= p_1(z)p_2(z)\{\Theta[P_1](z)l \otimes P_2 g + P_1 f \otimes \Theta[P_2](z)k\} \\ &= p_2(z)Q_1(z)l \otimes (p_1p_2)(S_2')P_2 g + (p_1p_2)(S_1')P_1 f \otimes p_1(z)Q_2(z)k \\ &= p_2(z)Q_1(z)l \otimes P_2(p_1p_2)(S_2)g + P_1(p_1p_2)(S_1)f \otimes p_1(z)Q_2(z)k \end{aligned}$$

from which the result follows. □

REMARK (8.5). If $\Theta[P_1]$ and $\Theta[P_2]$ are symbols which come from operators which can be composed and are rational, then clearly $\Theta[P_1] \cdot \Theta[P_2] = \Theta[P_1 \circ P_2]$ and is rational.

In (6.1), we saw how to compute the optimal interpolant from a maximal vector. We now want to show that again here, rationality is preserved. More precisely, we have

PROPOSITION (8.6). Notation as in (6.1). If m rational, then B is rational.

PROOF. We will be working on the unit circle ∂D in the proof, so we let $\zeta \in \partial D$. z will denote an ordinary complex variable. Set

$$k_0 := \frac{Ah_o}{\|A\|} \in H(m).$$

Since m is rational, $H(m)$ is finite dimensional, and so k_o is rational. Moreover if $m = \prod_{i=1}^{n}(\zeta - \alpha_i)/(1 - \overline{\alpha_i}\zeta)$ where $\alpha_i \in D$, then $k_o := p/q$ for

$$q := \prod_{i=1}^{n} 1/(1 - \overline{\alpha_i}\zeta),$$

and where p is a polynomial of degree $\le n - 1$. Now ($P_{H^2} : L^2 \to H^2$ is orthogonal projection)

$$A^* k_o = \begin{bmatrix} P_{H^2}(\overline{a}_1 k_o) \\ \cdot \\ \cdot \\ \cdot \\ P_{H^2}(\overline{a}_j k_o) \\ \cdot \\ \cdot \\ \cdot \end{bmatrix}$$

for some $a_j \in H^\infty$, $j \geq 1$. We claim that $P_{H^2}(\overline{a}_j k_o)$ is rational for all $j \geq 1$. In fact, we claim that for any $a \in H^\infty$, $P_{H^2}(\overline{a}\, k_o) \in H(\zeta^n m)$, and so $P_{H^2}(\overline{a}\, k_o)$ is rational. Indeed, set $f := P_{H^2}(\overline{a}\, k_o)$, and

$$b := \frac{am\zeta^n}{\prod_{j=1}^{n}(\zeta - \alpha_j)} \in H^\infty.$$

Then $\overline{a}\, k_o = \overline{b}mp$, and $f = P_{H^2}(\overline{a}\, k_o) = P_{H^2}(\overline{b}mp)$. But $P_{H^2}(\overline{m}f) = P_{H^2}(\overline{m}\,\overline{b}mp) = P_{H^2}(\overline{b}p) =: r$, where r is a polynomial of degree $\leq n-1$. Hence, $\overline{\zeta}^n P_{H^2}(\overline{m}f) = \overline{\zeta}^n r$, and so $P_{H^2}(\overline{\zeta}^n \overline{m}f) = P_{H^2}(\overline{\zeta}^n r) = 0$, which means that $f \in H(\zeta^n m)$, as claimed. In particular, we have that $f = u/q$ where u is a polynomial of degree $\leq 2n-1$. Thus

$$A^* k_o = \begin{bmatrix} \frac{u_1}{q} \\ \cdot \\ \cdot \\ \cdot \\ \frac{u_j}{q} \\ \cdot \\ \cdot \\ \cdot \end{bmatrix}$$

and is rational. Now $A^* k_o = \|A\| h_o$, and so h_o is rational. Finally since $B = \|A\|^2 h_o^* / \overline{Ah_o}$, we have that

$$B = [\ \cdots\ \frac{\overline{u_j}}{\overline{p}}\ \cdots\ ].$$

Moreover for each j, $\overline{u_j/p}$ is analytic. For $\beta(z)$ a polynomial of degree $\leq k$, set $\beta^{(k)}(z) := z^k \overline{\beta(1/\overline{z})}$. Then

$$\frac{\overline{u_j}}{\overline{p}} = \frac{\zeta^{2n-1}\overline{u_j}}{\zeta^{2n-1}\overline{p}} = \frac{\tilde{u}_j^{(2n-1)}}{\zeta^n \tilde{p}^{(n-1)}} = \frac{t_j}{\tilde{p}_{outer}}$$

where $\tilde{p}_{outer}$ denotes the outer part of $\tilde{p}^{(n-1)}$, and where t_j are polynomials of degree $\leq n-1$ for $j \geq 1$. Thus B is rational which completes our proof. □

We are now ready to formulate and prove the main result of this section.

THEOREM (8.7). Notation as in Section 4. If W and P are rational, then so is q_n.

PROOF. By induction on n. For $n = 1$, since P_1 and W_1 are rational, certainly q_1 is as well. Suppose that $q_1, \cdots, q_n$ are rational. Then (see equation (2) above)

$$A_n = W_n - \sum_{2\leq j\leq n} \sum_{i_1+\cdots+i_j=n} P_j(q_{i_1} \otimes \cdots \otimes q_{i_j})$$

is rational from (8.3), (8.4), and (8.5). But A_{n+1} is the dilation of ΠA_n, where $\Pi : H^2 \to H^2 \ominus P_1 H^2$ denotes orthogonal projection. Thus by Proposition (8.6), A_{n+1} is rational. But $A_{n+1} = A_n - P_1 q_{n+1}$, which implies that q_{n+1} is rational as required. □

In short, the set of rational systems remains invariant under the operations employed in the iterative commutant lifting procedure. We will see next that this means we can use the skew Toeplitz theory in order to compute the q_n.

9. SKEW TOEPLITZ OPERATORS

Let us very briefly recall the notion of skew Toeplitz operator from [7], [8], [10]. We will only need a special case here, and we refer the reader to [7] for the general discussion.

As above, for $m \in H^\infty(D)$ inner, let $H(m) = H^2 \ominus mH^2$, and let $T := \Pi S \mid H(m)$ denote the compressed shift (Π is the orthogonal projection $\Pi : H^2 \to H(m)$). Then a *skew Toeplitz operator* is an operator of the form

$$X = \sum_{j,k=0}^{n} T^j c_{jk} T^{*k}$$

with the complex constants enjoying the property $c_{jk} = \overline{c_{kj}}$. (Thus $X = X^*$.)

In [7], [8], [10], it is shown that these operators can be used to solve the most general kind of H^∞ optimization problems, including the so-called "4-block" problem for distributed systems. We shall not discuss this in detail here but merely refer the reader to the aforementioned papers. In this section, we would only like to show why the computation of the optimal interpolants that arise in iterative commutant lifting starting with rational data, amounts to a skew Toeplitz problem.

So assuming we start with rational P and W, from (6.1) and (8.7), we are given a rational operator $A_n : H^{(n)} \to H$ (recall $H^{(n)} = H^{\otimes n}$ and $H = H^{(1)}$) of the form

$$A_n = [\frac{p_1}{q} \; \frac{p_2}{q} \; \cdots \;]$$

(where q and p_j, $j \geq 1$, are polynomials of degree $\leq N$), and we want to find the singular values and corresponding singular vectors of $A^{(n)} := \Pi A_n$. But $\rho^2 \in \sigma(A^{(n)}A^{(n)*})$ if and only if $0 \in \sigma(I - 1/\rho^2 A^{(n)}A^{(n)*})$ if and only if $0 \in \sigma(q(T)(I - 1/\rho^2 A^{(n)}A^{(n)*})q(T)^*)$ if and only if $0 \in \sigma(q(T)q(T)^* - 1/\rho^2\{\sum p_j(T)p_j(T)^*\})$. Now clearly we have

$$q(T)q(T)^* - 1/\rho^2\{\sum p_j(T)p_j(T)^*\} = \sum_{j,k} T^j c_{jk} T^{*k}$$

where

$$c_{jk} = \sum_{j,k}\{q_j\overline{q}_k - \sum_i (1/\rho^2) p_j^{(i)}\overline{p}_k^{(i)}\}$$

for

$$q = \sum_j q_j z^j \; , p_i = \sum_j p_j^{(i)} z^j .$$

In short, determining the singular values and vectors of A, amounts to determining the invertibility of a skew Toeplitz operator, a problem solved explicitly in [7], [8], and [10] via certain easily computable determinantal formulae. Thus in the rational case we now have an explicit nonlinear synthesis procedure via iterative commutant lifting. We expect that as was the case for the algorithms of [7], [8], and [10], we should be able to develop computer software to carry out the procedure presented in this paper in the near future.

REFERENCES

[1] V. Anantharam and C. Desoer, "On the stabilization of nonlinear systems," *IEEE Trans. Automatic Control* **AC-29** (1984), 569-573.

[2] J. Ball, C. Foias, J. W. Helton, and A. Tannenbaum, "On a local nonlinear commutant lifting theorem," *Indiana J. Mathematics* **36** (1987), 693-709.

[3] J. Ball, C. Foias, J. W. Helton, and A. Tannenbaum, "Nonlinear interpolation theory in H^∞," in *Modelling, Robustness, and Sensitivity in Control Systems,* (edited by Ruth Curtain), NATO-ASI Series, Springer-Verlag, New York, 1987.

[4] J. Ball, C. Foias, J. W. Helton, and A. Tannenbaum, "A Poincare-Dulac approach to a nonlinear Beurling-Lax-Halmos theorem," to appear in *Journal of Math. Anal. and Applications.*

[5] J. Ball and J. W. Helton, "Sensitivity bandwidth optimization for nonlinear feedback systems," Technical Report, Department of Mathematics, University of California at San Diego, 1988.

[6] H. Bercovici, *Operator Theory and Arithmetic in H^∞*, AMS Publications, Providence, Rhode Island, 1988.

[7] H. Bercovici, C. Foias, and A. Tannenbaum, "On skew Toeplitz operators, I," *Operator Theory: Advances and Applications* **29** (1988), pp. 21-44.

[8] C. Foias and A. Tannenbaum, "On the four block problem II : the singular system," *Integral Equations and Operator Theory* **11** (1988), pp. 726-767.

[9] C. Foias and A. Tannenbaum, "Weighted optimization theory for nonlinear systems," to appear in *SIAM J. on Control and Optimization.*

[10] C. Foias, A. Tannenbaum, and G. Zames, "Some explicit formulae for the singular values of certain Hankel operators with factorizable symbol," *SIAM J. on Math. Analysis* **19** (1988), pp. 1081-1091.

[11] B. Francis, *A Course in H^∞ Control Theory,* Springer-Verlag, New York, 1987.

[12] M. Rosenblum and J. Rovnyak, *Hardy Classes and Operator Theory,* Oxford University Press, New York, 1985.

[13] D. Sarason, "Generalized Interpolation in H^∞," *Transactions of the AMS* **127** (1967), 179-203.

[14] D. Sarason, *Function Theory on the Unit Circle,* Lecture Notes, Virginia Polytechnic Institute, 1978.

[15] B. Sz.-Nagy and C. Foias, *Harmonic Analysis of Operators on Hilbert Space,* North-Holland Publishing Company, Amsterdam, 1970.

[16] G. Zames, "Feedback and optimal sensitivity: model reference transformations, multiplicative seminorms, and approximate inverses," *IEEE Trans. Auto. Control* **AC-26** (1981), pp. 301-320.

Ciprian Foias
Department of Mathematics
Indiana University
Bloomington, Indiana 47405

Allen Tannenbaum
Department of Electrical Engineering
University of Minnesota
Minneapolis, Minnesota 55455

Operator Theory:
Advances and Applications, Vol. 41

Dedicated to the 60-th birthday
of Professor Israel Gohberg,

ON THE REDUCTION OF COERCIVE SINGULAR PERTURBATIONS TO REGULAR PERTURBATIONS

L.S. Frank and J.J. Heijstek

It is shown that for a Coercive Singular Perturbation $\mathcal{A}^\varepsilon$ appearing in the Linear Elasticity theory, an appropriate choice of a reducing operator S^ε leads to the asymptotic relation: $S^\varepsilon \mathcal{A}^\varepsilon = \mathcal{A}^0 + \varepsilon Q^\varepsilon$, where $\mathcal{A}^0$ is the reduced operator associated with $\mathcal{A}^\varepsilon(\varepsilon = 0)$ and Q^ε is a family of continuous linear mappings, uniformly with respect to $\varepsilon \in (0,1]$, acting from the solution spaces $\mathcal{H}^\varepsilon$ into the data spaces $\mathcal{K}^\varepsilon$.

This is an improvement of the corresponding result $S^\varepsilon \mathcal{A}^\varepsilon = \mathcal{A}^0 + \varepsilon^\gamma Q^\varepsilon, 0 < \gamma < 1$, established previously by the first author and W.D. Wendt, also for general Coercive Singular Perturbations.

1. INTRODUCTION.

Coercive Singular Perturbations, introduced in 1976 by the first author (see [2], [3]) enjoy two fundamental properties. First, two-sided a priori estimates are valid (uniformly with respect to the small parameter ε) for their solutions in Sobolev-Slobodetski type spaces of vectorial order (see [2], [3], [4]) and, secondly, any such perturbation can be reduced in a constructive way to a regular perturbation (see [3], [11], [12], [15], [8]). More specifically, for any coercive singular perturbation $\mathcal{A}^\varepsilon$, which is a continuous linear mapping from the solution space into the data space (uniformly with respect to ε), a factorizing operator $\mathcal{R}^\varepsilon$ and a reducing operator S^ε exist, and can be explicitly constructed, such that

$$\mathcal{A}^\varepsilon = \mathcal{R}^\varepsilon \mathcal{A}^0 + \mathcal{O}(\varepsilon^\gamma),$$

$$S^\varepsilon \mathcal{A}^\varepsilon = \mathcal{A}^0 + \mathcal{O}(\varepsilon^\gamma),$$

as $\varepsilon \to 0$. Here $\mathcal{A}^0$ is the reduced operator for $\mathcal{A}^\varepsilon$, γ is some constant, $0 < \gamma < 1$, and the $\mathcal{O}(\varepsilon^\gamma)$ stand for continuous linear mappings in corresponding spaces whose norms are of order $O(\varepsilon^\gamma)$.

In these factorization and reduction procedures, the use of the Wiener-Hopf method and the theory of convolution operators (see [13]), as well as the herewith connected operator calculus (see [1]), plays an important role.

The main result in this paper is the improvement of the asymptotic relations above, which are, in fact, valid with the error terms $\mathcal{O}(\varepsilon)$, i.e. with $\gamma = 1$. Although this is done here only for a specific coercive singular perturbation appearing in the linear theory of elastic rods, presumably this improvement can be extended to the whole class of coercive singular perturbations.

The improvement in the estimate of the error terms (which seem to be sharp) is achieved by using in the definition of pseudodifferential singular perturbations an extension operator for functions on a finite interval onto the whole line which preserves their original regularity, uniformly with respect to ε. The extension by zero operator turns out to be less suitable for the sharp estimate of the error terms. It is also shown, that the use of any other regularity preserving extension operator brings over a modification which may be included in the error term $\mathcal{O}(\varepsilon)$.

Now, the contents of the paper will be briefly sketched. In section 2 the problem is stated and necessary previously established results are recalled. In section 3 the main results are formulated and their proofs are given in Section 4. Finally, Section 5 contains some additional remarks.

2. STATEMENT OF THE PROBLEM.

2.1. Let q be a nonnegative function which is infinitely differentiable on the closed unit interval $\overline{U} = [0,1]$. The open unit interval (0,1) will be denoted U and the notation ∂U will be used for the collection of its endpoints: $\partial U = \{0,1\}$.

The following singularly perturbed boundary problem is considered:

$$(2.1.1)\qquad \begin{cases} \varepsilon^2 D^2(q^2(x)D^2u^\varepsilon) + D^2u^\varepsilon = f^\varepsilon & \text{in } U, \\ \left.\begin{aligned} u^\varepsilon(x') &= \varphi_1^\varepsilon(x') \\ -D^2u^\varepsilon(x') &= \varphi_2^\varepsilon(x') \end{aligned}\right\} & x' \in \partial U, \end{cases}$$

with $D = -id/dx$ and with $f^\varepsilon, \varphi_1^\varepsilon, \varphi_2^\varepsilon$ being the data of the problem which may depend on the small parameter $\varepsilon \in (0, \varepsilon_0]$.

We adopt the notation used in [11], as far as the data space $\mathcal{K}^\varepsilon$ and the solution space $\mathcal{H}^\varepsilon$ are concerned; namely,

$$(2.1.2)\qquad \mathcal{K}^\varepsilon = H_{(s_1,s_2-2,s_3-2),\varepsilon}(U) \times \mathbf{C}_{s_1,\varepsilon}(\partial U) \times \mathbf{C}_{s_1+s_2-5/2,\varepsilon}(\partial U)$$

and

$$(2.1.3)\qquad \mathcal{H}^\varepsilon = H_{(s),\varepsilon}(U),$$

where $H_{(s),\varepsilon}(U)$ is the Sobolev-Slobodetski type space of vectorial order $s = (s_1, s_2, s_3) \in \mathbf{R}^3$ and $\mathbf{C}_{\sigma,\varepsilon}(\partial U)$ is the trace space of order $\sigma \in \mathbf{R}$.

2.2. The operator

$$(2.2.1)\qquad \begin{aligned} a^\varepsilon &: H_{(s),\varepsilon}(U) \to H_{(s_1,s_2-2,s_3-2),\varepsilon}(U), \\ a^\varepsilon &= \varepsilon^2 D^2(q^2 D^2) + D^2, \end{aligned}$$

is supposed to be an elliptic singular perturbation (see [2]), the ellipticity condition in the case considered being stated in the following equivalent way:

$$(2.2.2)\qquad q(x) > 0, \qquad \forall x \in \overline{U}.$$

We use the same symbol q to denote an extension of the function q above, having the following properties:

(2.2.3)
$$\begin{cases} q : \mathbf{R} \to \overline{\mathbf{R}}_+, \\ q(x) = q_\infty + q_0(x), \qquad \forall x \in \mathbf{R}, \\ q_0 \in C_0^\infty(\mathbf{R}). \end{cases}$$

and

(2.2.4)
$$q(x) \geq q_\infty > 0, \quad \forall x \in \mathbf{R},$$

for some number q_∞. It is well-known that such an extension exists (see, for instance, [14]).

2.3. One may rewrite problem (2.1.1) as follows. Introducing the operator

(2.3.1)
$$\mathcal{A}^\varepsilon = (\pi_U(\varepsilon^2 D^2 q^2 D^2 + D^2) l_0, \pi_{\partial U}, \pi_{\partial U}(-D^2))^T$$

where $\pi_U, \pi_{\partial U}$ are the operators of restriction to U, ∂U, respectively, l_0 stands for the extension by zero operator and $(\quad)^T$ denotes the transpose of the corresponding row-vector, problem (2.1.1) can be reformulated in the following way:

(2.3.2)
$$\mathcal{A}^\varepsilon u^\varepsilon = (f^\varepsilon, \varphi_1^\varepsilon, \varphi_2^\varepsilon)^T.$$

The singularly perturbed column-operator $\mathcal{A}^\varepsilon$ satisfies the coerciveness condition for linear singular perturbations ([2], [4]).
Problem (2.1.1), or (2.3.2), will be referred to as the perturbed problem.

2.4. The reduced problem ($\varepsilon = 0$)

(2.4.1)
$$\begin{cases} D^2 u^0 = f^0 \qquad \text{in } U, \\ u^0(x') = \varphi_1^0(x'), x' \in \partial U, \end{cases}$$

can also be written in the form

(2.4.2)
$$\mathcal{A}^0 u^0 = (f^0, \varphi_1^0)^T$$

where

(2.4.3)
$$\mathcal{A}^0 = (\pi_U D^2 l_0, \pi_{\partial U})^T.$$

As in [11], the data space and the space of the solutions are, respectively,

(2.4.4)
$$\mathcal{W}^\varepsilon = H_{(s_1, s_2 - 2, s_3), \varepsilon}(U) \times \mathbf{C}_{s_1, \varepsilon}(\partial U)$$

and

(2.4.5) $$\mathcal{H}^\varepsilon = H_{(s),\varepsilon}(U).$$

It has been proved (see [2],[4]) that for $s = (s_1, s_2, s_3) \in \mathbf{R}^3$ satisfying

(2.4.6) $$s_2 > 1/2, s_2 + s_3 > 1/2,$$

the reduced operator $\mathcal{A}^0 : \mathcal{H}^\varepsilon \to \mathcal{W}^\varepsilon$ is continuous, uniformly with respect to $\varepsilon \in (0, \varepsilon_0]$, it is invertible and its inverse $(\mathcal{A}^0)^{-1} : \mathcal{W}^\varepsilon \to \mathcal{H}^\varepsilon$ is also continuous, uniformly with respect to $\varepsilon \in (0, \varepsilon_0]$.

3. STATEMENT OF THE RESULT.

3.1. For $\mathcal{A}^\varepsilon$ defined by (2.3.1) to be a linear operator from $\mathcal{H}^\varepsilon$ into $\mathcal{K}^\varepsilon$ (given by (2.1.3), respectively (2.1.2)) which is continuous uniformly with respect to $\varepsilon \in (0, \varepsilon_0], \forall \varepsilon_0 < \infty$, the natural conditions upon $s = (s_1, s_2, s_3) \in \mathbf{R}^3$ are the following ones (see [2],[4]):

(3.1.1) $$1/2 < s_2 < 5/2, s_2 + s_3 > 5/2.$$

Under these conditions it was shown in [11] (also in the general multi-dimensional case) that the singularly perturbed operator $\mathcal{A}^\varepsilon$ is invertible for all $\varepsilon \in (0, \varepsilon_0]$ and that the inverse operator $(\mathcal{A}^\varepsilon)^{-1}$ is continuous, uniformly with respect to $\varepsilon \in (0, \varepsilon_0]$, provided that ε_0 is sufficiently small. This was done by constructing a factorizing and a reducing operator, $\mathcal{R}^\varepsilon : \mathcal{W}^\varepsilon \to \mathcal{K}^\varepsilon$ respectively $S^\varepsilon : \mathcal{K}^\varepsilon \to \mathcal{W}^\varepsilon$ (with $\mathcal{W}^\varepsilon$ given by (2.4.4)), such that for γ satisfying

(3.1.2) $$0 < \gamma < \min\{1, 5/2 - s_2\},$$

the operators $\varepsilon^{-\gamma}(\mathcal{A}^\varepsilon - \mathcal{R}^\varepsilon \mathcal{A}^0), \varepsilon^{-\gamma}(\mathcal{R}^\varepsilon S^\varepsilon - Id), \varepsilon^{-\gamma}(S^\varepsilon \mathcal{R}^\varepsilon - Id)$ and $\varepsilon^{-\gamma}(S^\varepsilon \mathcal{A}^\varepsilon - \mathcal{A}^0)$ are continuous (in corresponding spaces), uniformly with respect to $\varepsilon \in (0, \varepsilon_0]$; here $\mathcal{A}^0$ is given by (2.4.3) and stands Id for the identity operator (in the corresponding space). In fact, in the case considered, the factorizing respectively the reducing operator constructed in [11], takes the following form:

(3.1.3) $$\mathcal{R}^\varepsilon = \begin{pmatrix} \pi_U(1 + \varepsilon^2 q^2 D^2) l_0 & , & 0 \\ \pi_{\partial U} 0 & , & Id \\ \pi_{\partial U}(-1) & , & 0 \end{pmatrix},$$

respectively

(3.1.4) $$S^\varepsilon = \begin{pmatrix} (1 - \pi_U E^\varepsilon \pi_{\partial U}) \pi_U (1 + \varepsilon^2 q^2 D^2)^{-1} l_0 & , & 0, & -\pi_U E^\varepsilon \\ \pi_{\partial U} 0 & , & Id, & 0 \end{pmatrix};$$

here

(3.1.5) $$\begin{gathered} \pi_U E^\varepsilon : \mathbf{C}_{s_1+s_2-5/2,\varepsilon}(\partial U) \to H_{(s_1,s_2-2,s_3),\varepsilon}(U) \\ (\pi_U E^\varepsilon \varphi)(x) = \sum_{x' \in \partial U} \varphi(x') \exp(-|x - x'|/\varepsilon q(x')), \forall x \in U, \end{gathered}$$

is a linear operator which is continuous uniformly with respect to $\varepsilon \in (0, \varepsilon_0]$ and $(1 + \varepsilon^2 q^2 D^2)^{-1}$ is the pseudodifferential singular perturbation with the symbol $(1 + \varepsilon^2 q^2 \xi^2)^{-1}$, i.e. the following integral operator s^ε:

$$(s^\varepsilon f)(x) = (2\varepsilon q(x))^{-1} \int_{\mathbf{R}} f(y) \exp(-|x-y|/(\varepsilon q(x))) dy, \forall x \in \mathbf{R}, \tag{3.1.6}$$

where $f \in L_2(\mathbf{R})$ (see also [5],[6]).

3.2. In [5] the operator S^ε was constructed in a natural way by inverting $\mathcal{R}^\varepsilon$ defined by (3.1.3) with a freezed coefficient $q(x)$, i.e. by solving the following boundary value problem with constant coefficients on $\mathbf{R}_+$:

$$\begin{cases} \pi_+(1 + \varepsilon^2 q^2 D^2) v^\varepsilon = f^\varepsilon, \\ \pi_0 v^\varepsilon = -\varphi_2^\varepsilon, \end{cases} \tag{3.2.1}$$

where π_+, π_0 are the operators of restriction to $\mathbf{R}_+, \{0\}$, respectively, and where $f^\varepsilon \in H_{(s_1, s_2-2, s_3-2),\varepsilon}(\mathbf{R}_+), \varphi_2^\varepsilon \in \mathbf{C}_{s_1+s_2-5/2,\varepsilon}$. See also [11], where a formula for S^ε is indicated in the case of a general coercive singular perturbation acting on the solution space $H_{(s),\varepsilon}(U)$ (U being a bounded domain in $\mathbf{R}^n$ with a smooth boundary ∂U) with values in the corresponding data space $\mathcal{K}^\varepsilon$.

3.3. In the case considered, it turns out to be convenient to choose slightly different factorizing and reducing operators. Instead of (3.1.3) one defines

$$\mathcal{R}^\varepsilon = \begin{pmatrix} \pi_U(1 + \varepsilon^2 D^2 q^2) l_0 & , & 0 \\ \pi_{\partial U} 0 & , & Id \\ \pi_{\partial U}(-1) & , & 0 \end{pmatrix}. \tag{3.3.1}$$

for then the factorization is exact:

$$\mathcal{A}^\varepsilon = \mathcal{R}^\varepsilon \mathcal{A}^0. \tag{3.3.2}$$

Of course, the operators defined by (3.1.3) resp. (3.3.1) differ only by some small operator $\varepsilon \cdot Q^\varepsilon$, where Q^ε is a linear operator from $\mathcal{W}^\varepsilon$ into $\mathcal{K}^\varepsilon$, which is continuous uniformly with respect to $\varepsilon \in (0, \varepsilon_0]$.

3.4. Instead of (3.1.4) one defines

$$S^\varepsilon = \begin{pmatrix} (1 - \pi_U E^\varepsilon \pi_{\partial U}) \pi_U (1 + \varepsilon^2 q^2 D^2)^{-1} l & , & 0, & -\pi_U E^\varepsilon \\ \pi_{\partial U} 0 & , & Id, & 0 \end{pmatrix} \tag{3.4.1}$$

where l is a linear extension operator

$$\begin{gathered} l : H_{(s_1, s_2-2, s_3-2),\varepsilon}(U) \to H_{(s_1, s_2-2, s_3-2),\varepsilon}(\mathbf{R}), \\ \pi_U l u = u, \forall u \in H_{(s_1, s_2-2, s_3-2),\varepsilon}(U), \end{gathered} \tag{3.4.2}$$

which is continuous, uniformly with respect to $\varepsilon \in (0, 1]$.

It will turn out later on that different choices of extension operators satisfying (3.4.2), lead to minor changes in S^ε in the following sense: if $l_k, k = 1, 2$, are two such extension operators, then

$$||\pi_U(1 - E^\varepsilon \pi_{\partial U})(1 + \varepsilon^2 q^2 D^2)^{-1}(l_1 f - l_2 f)||_{H_{(s_1, s_2 - 2, s_3), \varepsilon}(U)} \leq \tag{3.4.3}$$
$$\leq C\varepsilon ||f||_{H_{(s_1, s_2-2, s_3-2), \varepsilon}(U)}, \forall f, \forall \varepsilon \in (0, \varepsilon_0],$$

with some constant C which does not depend on f and $\varepsilon \in (0, \varepsilon_0]$. (See Theorem 4.2 below.)

3.5. The main result in this paper is the following:

REDUCTION THEOREM. *Let $q : \mathbf{R} \to \overline{\mathbf{R}}_+$ be as in section 2.2, i.e.*

$$q(x) = q_\infty + q_0(x), \quad \forall x \in \mathbf{R}, \tag{3.5.1}$$

$$q_0 \in C_0^\infty(\mathbf{R}), \tag{3.5.2}$$
$$q(x) \geq q_\infty > 0, \quad \forall x \in \mathbf{R}.$$

Let $\mathcal{K}^\varepsilon, \mathcal{H}^\varepsilon$ and $\mathcal{W}^\varepsilon$ be the spaces defined previously by (2.1.2), (2.1.3) and (2.4.4), repectively.
If $s = (s_1, s_2, s_3) \in \mathbf{R}^3$ satisfies the condition (3.1.1), then the factorizing operator $\mathcal{R}^\varepsilon : \mathcal{W}^\varepsilon \to \mathcal{K}^\varepsilon$ defined by (3.3.1) and the reducing operator $S^\varepsilon : \mathcal{K}^\varepsilon \to \mathcal{W}^\varepsilon$ given by (3.4.1) are continuous, uniformly with respect to $\varepsilon \in (0, \varepsilon_0]$, and the following asymptotic formulae hold:

$$\mathcal{R}^\varepsilon S^\varepsilon - Id = \varepsilon Q_1^\varepsilon, \tag{3.5.3}$$

$$S^\varepsilon \mathcal{R}^\varepsilon - Id = \varepsilon Q_2^\varepsilon, \tag{3.5.4}$$

$$S^\varepsilon \mathcal{A}^\varepsilon - \mathcal{A}^0 = \varepsilon Q_3^\varepsilon, \tag{3.5.5}$$

where

$$Q_1^\varepsilon : \mathcal{K}^\varepsilon \to \mathcal{K}^\varepsilon, \tag{3.5.6}$$

$$Q_2^\varepsilon : \mathcal{W}^\varepsilon \to \mathcal{W}^\varepsilon, \tag{3.5.7}$$

$$Q_3^\varepsilon : \mathcal{H}^\varepsilon \to \mathcal{W}^\varepsilon \tag{3.5.8}$$

are linear operators which are continuous, uniformly with respect to $\varepsilon \in (0, \varepsilon_0]$, for each $\varepsilon_0 < \infty$.

The proof of Reduction Theorem will be given in section 4.

REMARK. It will be clear from the proof of Reduction Theorem that the condition $1/2 < s_2$ in (3.1.1) is needed only for proving (3.5.8). Thus all other assertions of Reduction Theorem hold under the condition

$$s_2 < 5/2, s_2 + s_3 > 5/2. \tag{3.5.9}$$

3.6. As it has already been mentioned before, applying the general result in [11] to the case considered yields (3.5.3)-(3.5.5) with ε replaced by ε^γ, where γ satisfies (3.1.2), i.e. $0 < \gamma < \min\{1, 5/2 - s_2\}$.
It is of importance for the applications (see [7],[8]) that, thanks to Reduction Theorem, one has the following series representation for $(\mathcal{A}^\varepsilon)^{-1}$:

$$(\mathcal{A}^\varepsilon)^{-1} = \sum_{k \geq 0} \varepsilon^k (-1)^k ((\mathcal{A}^0)^{-1} Q_3^\varepsilon)^k (\mathcal{A}^0)^{-1} S^\varepsilon, \tag{3.6.1}$$

where the Neumann series on the right-hand side of (3.6.1) converges in the Banach algebra of all continuous linear mappings from $\mathcal{H}^\varepsilon$ into itself, uniformly with respect to $\varepsilon \in (0, \varepsilon_0]$, provided that ε_0 is sufficiently small.
The general result in [11],[12] only gives (3.6.1) with $\varepsilon^{\gamma k}$, γ as above, instead of ε^k.

4. PROOF OF REDUCTION THEOREM.

4.1. For proving Reduction Theorem we use some auxiliary results.

LEMMA 4.1. *Let $s \in \mathbf{R}^3$ be given. Then there exists a nonnegative integer N and a number C such that*

$$\begin{aligned} ||\varphi(tx)u(x)||_{H_{(s),\varepsilon}(\mathbf{R})} \leq & C||u||_{H_{(s),\varepsilon}(\mathbf{R})} \sum_{0 \leq j \leq N} \sup_{x \in \mathbf{R}} <x>^2 |D^j \varphi(x)|, \\ & \forall u \in H_{(s),\varepsilon}(\mathbf{R}), \forall \varphi \in S(\mathbf{R}), \forall (\varepsilon, t) \in (0,1] \times [0,1]. \end{aligned} \tag{4.1.1}$$

PROOF OF LEMMA 4.1. The Lemma holds if $t = 0$.
If $t \in (0,1]$, then

$$\begin{aligned} & ||\varphi(tx)u(x)||_{H_{(s),\varepsilon}(\mathbf{R})} = \\ & = ||\int_{\mathbf{R}} (<\xi>/<\eta>)^{s_2} (<\varepsilon\xi>/<\varepsilon\eta>)^{s_3} t^{-1} \tilde{\varphi}((\xi - \eta)/t) \varepsilon^{-s_1} <\eta>^{s_2} <\varepsilon\eta>^{s_3} \tilde{u}(\eta) d\eta ||_{L_2(\mathbf{R})} \\ & \leq 2^{|s_2|+|s_3|} || \int_{\mathbf{R}} <\xi - \eta>^{|s_2|+|s_3|} t^{-1} |\tilde{\varphi}(t^{-1}(\xi - \eta))| \varepsilon^{-s_1} <\eta>^{s_2} <\varepsilon\eta>^{s_3} |\tilde{u}(\eta)| d\eta ||_{L_2(\mathbf{R})} \\ & \leq 2^{|s_2|+|s_3|} \int_{\mathbf{R}} <\eta>^{|s_2|+|s_3|} t^{-1} |\tilde{\varphi}(t^{-1}\eta)| d\eta ||u||_{H_{(s),\varepsilon}(\mathbf{R})} \\ & \leq 2^{|s_2|+|s_3|} \int_{\mathbf{R}} <\xi>^{|s_2|+|s_3|} |\tilde{\varphi}(\xi)| d\xi ||u||_{H_{(s),\varepsilon}(\mathbf{R})} \end{aligned} \tag{4.1.2}$$

where Peetre's inequality $<\xi>^\tau <\eta>^{-\tau} \leq 2^{|\tau|} <\xi - \eta>^{|\tau|}, \forall (\xi, \eta, \tau) \in \mathbf{R}^3$, the inequality $||f * g||_{L_2} \leq ||f||_{L_1} ||g||_{L_2}$ and the condition $(\varepsilon, t) \in (0,1] \times (0,1]$ have been used.

Let N be the least integer such that

$$N - |s_2| - |s_3| > 1. \tag{4.1.3}$$

Then

$$\int_{\mathbf{R}} <\xi>^{|s_2|+|s_3|} |\tilde{\varphi}(\xi)| d\xi \leq C \sum_{0 \leq j \leq N} \sup_{x \in \mathbf{R}} <x>^2 |D^j \varphi(x)| \tag{4.1.4}$$

where

$$C \leq \pi \, 2^N \int_{\mathbf{R}} <\xi>^{-N+|s_2|+|s_3|} d\xi, \tag{4.1.5}$$

and that ends the proof of Lemma 4.1.

4.2. As usual, $\mathring{H}_{(s),\varepsilon}(U^c)$ stands for the subspace of $H_{(s),\varepsilon}(\mathbf{R})$ whose elements have their support in $\overline{U^c} = \mathbf{R} \backslash U$.

The result stated in Theorem 4.2 (here below) is, from our point of view, of an independent interest.

THEOREM 4.2. *Let q be as in section 2.2 and let E^ε be given by (3.1.5). Then for each $k \in \mathbf{R}$ and $s = (s_1, s_2, s_3) \in \mathbf{R}^3$ such that*

$$s_2 < 1/2 < s_2 + s_3 + k \tag{4.2.1}$$

it holds that

$$(1 - \pi_U E^\varepsilon \pi_{\partial U}) \pi_U (1 + \varepsilon^2 q^2 D^2)^{-1} \tag{4.2.2}$$

is a linear operator from $\mathring{H}_{(s),\varepsilon}(U^c)$ into $H_{(s_1+1, s_2, s_3+k),\varepsilon}(U)$ which is continuous, uniformly with respect to $\varepsilon \in (0, \varepsilon_0]$.

PROOF OF THEOREM 4.2. If $u \in \mathring{H}_{(s),\varepsilon}(U^c)$ then $u = u_- + u_+$ where $u_- \in \mathring{H}_{(s),\varepsilon}(\mathbf{R}_-)$ and $u_+ \in \mathring{H}_{(s),\varepsilon}((1,\infty))$; moreover

$$||u_\pm||_{H_{(s),\varepsilon}(\mathbf{R})} \leq ||u||_{H_{(s),\varepsilon}(\mathbf{R})}. \tag{4.2.3}$$

Hence it suffices to prove for the $u_\pm$ above, that

$$||(1 - \pi_U E^\varepsilon \pi_{\partial U}) \pi_U (1 + \varepsilon^2 q^2 D^2)^{-1} u_\pm ||_{H_{(s_1+1, s_2, s_3+k),\varepsilon}(U)} \leq \tag{4.2.4}$$

$$\leq C ||u_\pm||_{H_{(s),\varepsilon}(\mathbf{R})}, \forall \varepsilon \in (0, \varepsilon_0],$$

with a constant C which may only depend on $s_2, s_3, q, \varepsilon_0, k$. Consider u_-(the argument for u_+ is exactly the same, since the substitution $1 - x \to x$ transforms u_+ into u_-).

One has

$$\begin{aligned} \|(1-\pi_U E^\varepsilon \pi_{\partial U})\pi_U(1+\varepsilon^2 q^2 D^2)^{-1}u_-\|_{H_{(s_1+1,s_2,s_3+k),\varepsilon}(U)} &\leq \\ \leq \|(1-\pi_+ E_0^\varepsilon \pi_0)\pi_+(1+\varepsilon^2 q^2 D^2)^{-1}u_-\|_{H_{(s_1+1,s_2,s_3+k),\varepsilon}(\mathbf{R}_+)} &+ \\ +\|\pi_- E_1^\varepsilon \pi_1(1+\varepsilon^2 q^2 D^2)^{-1}u_-\|_{H_{(s_1+1,s_2,s_3+k),\varepsilon}(\mathbf{R}\setminus[1,\infty))}&, \end{aligned} \tag{4.2.5}$$

where

$$\begin{aligned} E_0^\varepsilon(x) &= \exp(-x/\varepsilon q(0)), \\ E_1^\varepsilon(x) &= \exp(-(1-x)/\varepsilon q(1)), \end{aligned} \tag{4.2.6}$$

$$\begin{aligned} \pi_+ &= \text{ restriction to } \mathbf{R}_+, \\ \pi_- &= \text{ restriction to } \mathbf{R}\setminus[1,\infty), \\ \pi_0 &= \text{ restriction to } \partial\mathbf{R}_+ = \{0\}, \end{aligned} \tag{4.2.7}$$

and

$$\pi_1 = \text{ restriction to } \partial(\mathbf{R}\setminus[1,\infty)) = \{1\}. \tag{4.2.8}$$

We are going to estimate both terms on the right-hand side of (4.2.5).
Note that

$$\pi_+(1+\varepsilon^2 q^2 D^2)^{-1}u_- = (1/2)\pi_+(1+i\varepsilon qD)^{-1}u_- = (1/2)\pi_+ s_+(x,\varepsilon D)u_- \tag{4.2.9}$$

where we have denoted:

$$\begin{aligned} s_+(x,\varepsilon D) = Op\ s_+(x,\varepsilon\xi) &= \mathcal{F}^{-1}_{\xi\to x} s_+(x,\varepsilon\xi)\mathcal{F}_{x\to\xi} \\ &= \mathcal{F}^{-1}_{\xi\to x}(1+i\varepsilon q(x)\xi)^{-1}\mathcal{F}_{x\to\xi} = (1+i\varepsilon qD)^{-1}. \end{aligned} \tag{4.2.10}$$

Furthermore,

$$\begin{aligned} \pi_+ s_+(x,\varepsilon D)u_- &= \pi_+\mathcal{F}^{-1}_{\xi\to x}\int_{\mathbf{R}} \varepsilon q(x)s_+(x,\varepsilon\xi)s_+(x,\varepsilon\eta)\tilde{u}_-(\eta)d\eta \\ &= \pi_+\ Op(a(x,\varepsilon,\varepsilon\xi,\varepsilon\eta))u_- \end{aligned} \tag{4.2.11}$$

where

$$a(x,\varepsilon,\xi,\eta) = \varepsilon q(x)s_+(x,\xi)s_+(x,\eta) \tag{4.2.12}$$

and

$$Op(a(x,\varepsilon,\xi,\eta))u_- = \mathcal{F}^{-1}_{\xi\to x}\int_{\mathbf{R}} \mathcal{F}_{y\to\eta}a(x,\varepsilon,\xi,\eta)u_-(y)d\eta. \tag{4.2.13}$$

In particular,

$$\pi_+ s_+(0, \varepsilon D)u_- = \pi_+ \, Op(a(0, \varepsilon, \varepsilon\xi, \varepsilon\eta))u_- = \pi_+ E_0^\varepsilon \pi_0 s_+(0, \varepsilon D)u_-. \tag{4.2.14}$$

Thus,

$$\begin{aligned} (1-\pi_+ E_0^\varepsilon \pi_0)\pi_+(1+\varepsilon^2 q^2 D^2)^{-1} u_- &= \\ &= (1/2)\pi_+(s_+(x, \varepsilon D) - s_+(0, \varepsilon D))u_- = \\ &= (1/2)\pi_+ \, Op(a(x, \varepsilon, \varepsilon\xi, \varepsilon\eta) - a(0, \varepsilon, \varepsilon\xi, \varepsilon\eta))u_- \\ &= (1/2) \int_0^1 \pi_+ \, Op(ix \, a_{(1)}(tx, \varepsilon, \varepsilon\xi, \varepsilon\eta))u_- dt \end{aligned} \tag{4.2.15}$$

where

$$a_{(1)}(x, \varepsilon, \xi, \eta) = (-i\partial/\partial x)a(x, \varepsilon, \xi, \eta). \tag{4.2.16}$$

Using the notation

$$a_{(1)}^{(1)}(x, \varepsilon, \xi, \eta) = (-i\partial/\partial\xi)a_{(1)}(x, \varepsilon, \xi, \eta) \tag{4.2.17}$$

one has

$$Op(ix \, a_{(1)}(tx, \varepsilon, \varepsilon\xi, \varepsilon\eta)) = -i\varepsilon \, Op(a_{(1)}^{(1)}(tx, \varepsilon, \varepsilon\xi, \varepsilon\eta)) \tag{4.2.18}$$

and therefore

$$\begin{aligned} (1 - \pi_+ E_0^\varepsilon \pi_0)\pi_+(1+\varepsilon^2 q^2 D^2)^{-1} u_- &= \\ &= -(1/2)i\varepsilon \int_0^1 \pi_+ \, Op(a_{(1)}^{(1)}(tx, \varepsilon, \varepsilon\xi, \varepsilon\eta))u_- dt. \end{aligned} \tag{4.2.19}$$

We are going to show that

$$\|\pi_+ \, Op(a_{(1)}^{(1)}(tx, \varepsilon, \varepsilon\xi, \varepsilon\eta))u_-\|_{H_{(s_1, s_2, s_3+k),\varepsilon}(\mathbf{R}_+)} \le C\|u_-\|_{H_{(s),\varepsilon}(\mathbf{R})} \tag{4.2.20}$$

where the constant C does not depend on $u_- \in \mathring{H}_{(s),\varepsilon}(\mathbf{R}_-)$ and $(\varepsilon, t) \in (0, \varepsilon_0] \times [0, 1]$. One has

$$a_{(1)}^{(1)}(x, \varepsilon, \xi, \eta) = a_1(x, \varepsilon, \xi, \eta) + a_2(x, \varepsilon, \xi, \eta) + a_3(x, \varepsilon, \xi, \eta) \tag{4.2.21}$$

where

$$\begin{cases} a_1(x, \varepsilon, \xi, \eta) = \varepsilon q(x) q_{(1)}(x)(s_+(x, \xi))^2 s_+(x, \eta), \\ a_2(x, \varepsilon, \xi, \eta) = -2\varepsilon q(x) q_{(1)}(x)(s_+(x, \xi))^3 s_+(x, \eta), \\ a_3(x, \varepsilon, \xi, \eta) = -\varepsilon q(x) \dot{q}_{(1)}(x)(s_+(x, \xi))^2 (s_+(x, \eta))^2. \end{cases} \tag{4.2.22}$$

It is sufficient to show that

$$\pi_+ Op(a_j(tx,\varepsilon,\varepsilon\xi,\varepsilon\eta)) \in \operatorname{Hom}(\mathring{H}_{(s),\varepsilon}(\mathbf{R}_-), H_{(s_1,s_2,s_3+k),\varepsilon}(\mathbf{R}_+)), j=1,2,3, \tag{4.2.23}$$

uniformly with respect to $t \in [0,1]$ and $\varepsilon \in (0,\varepsilon_0]$.

Denote

$$\varphi(x) = \int_{\mathbf{R}} q_{(1)}(x) s_+(x,\varepsilon\eta)\tilde{u}_-(\eta)d\eta \tag{4.2.24}$$

and

$$\begin{aligned} w(tx, x/\varepsilon) &= \mathcal{F}^{-1}_{\xi\to x}\varepsilon q(tx) s_+^2(tx,\varepsilon\xi) \\ &= H(x)(x/\varepsilon q(tx))\exp(-x/\varepsilon q(tx)). \end{aligned} \tag{4.2.25}$$

Then Lemma 4.1 implies that

$$\begin{aligned} &\|\pi_+ Op(a_1(tx,\varepsilon,\varepsilon\xi,\varepsilon\eta))u_-\|_{H_{(s_1,s_2,s_3+k),\varepsilon}(\mathbf{R}_+)} \le \\ &\le \inf_l \|\varphi(tx) l\pi_+ w(tx,x/\varepsilon)\|_{H_{(s_1,s_2,s_3+k),\varepsilon}(\mathbf{R})} \\ &\le C \inf_l \|l\pi_+ w(tx,x/\varepsilon)\|_{H_{(s_1,s_2,s_3+k),\varepsilon}(\mathbf{R})} \sum_{0\le j\le N} \sup_{x\in\mathbf{R}} < x >^2 |D^j\varphi(x)| \\ &\le C\|\pi_+ x/(\varepsilon q(tx))\exp(-x/\varepsilon q(tx))\|_{H_{(s_1,s_2,s_3+k),\varepsilon}(\mathbf{R}_+)} \cdot \\ &\cdot \sum_{0\le j\le N} \sup_{x\in\mathbf{R}} < x >^2 \left| \int_{\mathbf{R}} D_x^j(q_{(1)}(x)s_+(x,\varepsilon\eta))\tilde{u}_-(\eta)d\eta\right|, \forall t\in[0,1], \forall\varepsilon\in(0,\varepsilon_0]. \end{aligned} \tag{4.2.26}$$

Consider

$$E(x,\rho) = (\rho/q(x))\exp(-\rho/q(x)). \tag{4.2.27}$$

Using the Taylor's formula

$$E(x,\rho) = \sum_{0\le j\le m} (ix)^j/(j!)E_{(j)}(0,\rho) + \int_0^1 (ix)^{m+1}(1-\theta)^m/(m!)E_{(m+1)}(\theta x,\rho)d\theta, m=0,1,2,..., \tag{4.2.28}$$

with

$$E_{(j)}(x,\rho) = (-i\partial/\partial x)^j E(x,\rho), \quad j=0,1,2,... \tag{4.2.29}$$

one finds

$$\begin{aligned} &\|\pi_+ E(tx,x/\varepsilon)\|_{H_{(s_1,s_2,s_3+k),\varepsilon}(\mathbf{R}_+)} \le \\ &\le \sum_{0\le j\le m} \|\pi_+(itx)^j/(j!)E_{(j)}(0,x/\varepsilon)\|_{H_{(s_1,s_2,s_3+k),\varepsilon}(\mathbf{R}_+)} + \\ &+ \|\pi_+ \int_0^1 (itx)^{m+1}/(m!)(1-\theta)^m E_{(m+1)}(\theta tx, x/\varepsilon)d\theta\|_{H_{(s_1,s_2,s_3+k),\varepsilon}(\mathbf{R}_+)}, \\ &\forall t\in[0,1], \forall\varepsilon\in(0,1]. \end{aligned} \tag{4.2.30}$$

Using the fact that $t \in [0,1]$, one easily finds that

$$\|\pi_+(itx)^j/(j!)E_{(j)}(0,x/\varepsilon)\|_{H_{(s_1,s_2,s_3+k),\varepsilon}(\mathbf{R}_+)} \le C(k,j,q,s)\varepsilon^{j+\frac{1}{2}-s_1-s_2}, \forall\varepsilon \in (0,1], \forall t \in [0,1], \tag{4.2.31}$$

where $C(k,j,q,s)$ may depend only on $k \in \mathbf{R}$, $j = 0,1,2,\ldots,$ q and s.

Denote

$$F_m(x,\varepsilon) = \int_0^1 (ix)^{m+1}/(m!)(1-\theta)^m E_{(m+1)}(\theta x, x/\varepsilon)d\theta. \tag{4.2.32}$$

Then

$$\|\pi_+F_m(tx,t\varepsilon)\|_{H_{(s_1,s_2,s_3+k),\varepsilon}(\mathbf{R}_+)} \le \|\pi_+F_m(tx,t\varepsilon)\|_{H_{(s_1,1,M),\varepsilon}(\mathbf{R}_+)} \tag{4.2.33}$$

where M is the least integer such that

$$s_2 + s_3 + k \le M + 1. \tag{4.2.34}$$

One has

$$\begin{aligned} &\|\pi_+F_m(tx,t\varepsilon)\|_{H_{(s_1,1,M),\varepsilon}(\mathbf{R}_+)} \le \\ &\le C_M\varepsilon^{-s_1}\{\|\pi_+F_m(tx,t\varepsilon)\|_{L_2(\mathbf{R}_+)} + \|\pi_+DF_m(tx,t\varepsilon)\|_{L_2(\mathbf{R}_+)} + \\ &\quad + \|\pi_+D_x(\varepsilon D_x)^M F_m(tx,t\varepsilon)\|_{L_2(\mathbf{R}_+)}\}. \end{aligned} \tag{4.2.35}$$

Now

$$\begin{aligned} \|\pi_+F_m(tx,t\varepsilon)\|_{L_2(\mathbf{R}_+)} &\le \sup_{0\le\theta\le 1} t^{m+1}(m!)^{-1}\varepsilon^{m+1}\|(x/\varepsilon)^{m+1}E_{(m+1)}(\theta tx, x/\varepsilon)\|_{L_2(\mathbf{R}_+)} \\ &\le C_1\varepsilon^{m+3/2}, \quad \forall\varepsilon \in (0,1], \forall t \in [0,1], \end{aligned} \tag{4.2.36}$$

$$\|\pi_+D_xF_m(tx,t\varepsilon)\|_{L_2(\mathbf{R}_+)} \le C_2\varepsilon^{m+1/2}, \quad \forall\varepsilon \in (0,1], \forall t \in [0,1], \tag{4.2.37}$$

and

$$\|\pi_+D_x(\varepsilon D_x)^M F_m(tx,t\varepsilon)\|_{L_2(\mathbf{R}_+)} \le C_3\varepsilon^{m+1/2}, \quad \forall\varepsilon \in (0,1], \forall t \in [0,1], \tag{4.2.38}$$

where C_1, C_2 and C_3 are constants which may depend only on s, q, M and m.

Thus, choosing for m the least integer such that

$$m \ge \max\{0, -s_2\}, \tag{4.2.39}$$

one obtains

$$\|\pi_+F_m(tx,t\varepsilon)\|_{H_{(s_1,s_2,s_3+k),\varepsilon}(\mathbf{R}_+)} \le C\varepsilon^{1/2-s_1-s_2}, \quad \forall\varepsilon \in (0,1], \forall t \in [0,1], \tag{4.2.40}$$

where $C = C(s,q,k)$, i.e. C may depend only on s, q and k.

Thus

$$\begin{aligned} ||\pi_+(x/\varepsilon q(tx))\exp(-x/\varepsilon q(tx))||_{H_{(s_1,s_2,s_3+k),\varepsilon}(\mathbf{R}_+)} &\leq \\ &\leq C\varepsilon^{1/2-s_1-s_2}, \quad \forall\varepsilon\in(0,1], \forall t\in[0,1], \end{aligned} \tag{4.2.41}$$

where $C = C(s,q,k)$.

Since

$$\begin{aligned} (1+i\varepsilon q(x)\xi)^{-1} =& 2^n(1-i\varepsilon q(x)\xi)^{-n}(1+i\varepsilon q(x)\xi)^{-1} - \\ &- \sum_{1\leq j\leq n} 2^{j-1}(1-i\varepsilon q(x)\xi)^{-j}, \qquad n=0,1,2,3,..., \end{aligned} \tag{4.2.42}$$

the trace theorem ([4]) along with condition (4.2.1) imply for $j = 0,1,...,N$ with N defined by (4.1.3):

$$\begin{aligned} &\varepsilon^{1/2-s_1-s_2}\Big|\int_{\mathbf{R}} D_x^j(q_{(1)}(x)s_+(x,\varepsilon\eta))\tilde{u}_-(\eta)d\eta\Big| \\ &= \varepsilon^{1/2-s_1-s_2}\Big|\int_{\mathbf{R}} D_x^j(q_{(1)}(x)2^n(s_+(x,-\varepsilon\eta))^n s_+(x,\varepsilon\eta))\tilde{u}_-(\eta)d\eta\Big| \\ &\leq C||\varepsilon^{-s_1}<\xi>^{s_2}<\varepsilon\xi>^{s_3+k}D_x^j(q_{(1)}(x)s_+^n(x,-\varepsilon\xi)s_+(x,\varepsilon\xi))\tilde{u}_-(\xi)||_{L_2(\mathbf{R})} \\ &\leq C\sup_{\xi\in\mathbf{R}} <\varepsilon\xi>^k|D_x^j(q_{(1)}(x)s_+^n(x,-\varepsilon\xi)s_+(x,\varepsilon\xi)|\ ||u_-||_{H_{(s),\varepsilon}(\mathbf{R})} \\ &\leq C\sum_{0\leq p\leq j}|D_x^{p+1}q(x)|\ ||u_-||_{H_{(s),\varepsilon}(\mathbf{R})},\ \forall\varepsilon\in(0,1],\ \forall x\in\mathbf{R}, \end{aligned} \tag{4.2.43}$$

where the constant C does not depend on $\varepsilon\in(0,\varepsilon_0], u_-\in\mathring{H}_{(s),\varepsilon}(\mathbf{R}_-)$, $x\in\mathbf{R}$, and where n is the least integer such that

$$n\geq\max\{0,k-1\}. \tag{4.2.44}$$

From (4.2.26), (4.2.41) and (4.2.43) follows (4.2.23) for the case $j = 1$. In exactly the same way one proves (4.2.23) for $j = 2,3$. Now (4.2.23), (4.2.19) yield:

$$||(1-\pi_+E_0^\varepsilon\pi_0)\pi_+(1+\varepsilon^2q^2D^2)^{-1}u_-||_{H_{(s_1+1,s_2,s_3+k),\varepsilon}(\mathbf{R}_+)}\leq C||u_-||_{H_{(s),\varepsilon}(\mathbf{R})}, \tag{4.2.45}$$

where C is a constant which does not depend on $u_-\in\mathring{H}_{(s),\varepsilon}(\mathbf{R}_-)$ and $\varepsilon\in(0,\varepsilon_0]$.

The second term on the right-hand side of (4.2.5) is exponentially small. Indeed, similar to (4.2.14) it holds that

$$\pi_+s_+(1,\varepsilon D)u_- = \pi_+\exp(-x/\varepsilon q(1))\pi_0 s_+(1,\varepsilon D)u_- \tag{4.2.46}$$

which along with (4.2.42) implies that

$$\begin{aligned} \pi_+(1+\varepsilon^2q^2(1)D^2)^{-1} &= \pi_+\exp(-x/\varepsilon q(1))\pi_0(1/2)s_+(1,\varepsilon D)u_- \\ &= 2^{n-1}\pi_+\exp(-x/\varepsilon q(1))\int_{\mathbf{R}} s_+^n(1,-\varepsilon\xi)s_+(1,\varepsilon\xi)\tilde{u}_-(\xi)d\xi, \\ &(n=0,1,2,...). \end{aligned} \tag{4.2.47}$$

Hence, if n is an integer satisfying (4.2.44), then

$$\begin{aligned}
&\|\pi_- E_1^\varepsilon \pi_1 (1+\varepsilon^2 q^2 D^2)^{-1} u_-\|_{H_{(s_1+1,s_2,s_3+k),\varepsilon}(\mathbf{R}\setminus[1,\infty))} = \\
&= \|\pi_- E_1^\varepsilon\|_{H_{(s_1+1,s_2,s_3+k),\varepsilon}(\mathbf{R}\setminus[1,\infty))} |\pi_1 (1+\varepsilon^2 q^2 D^2)^{-1} u_-| \\
&= \|\pi_+ \exp(-x/\varepsilon q(1))\|_{H_{(s_1+1,s_2,s_3+k),\varepsilon}(\mathbf{R}_+)} |[\pi_+(1+\varepsilon^2 q^2(1) D^2)^{-1} u_-]_{x=1}| \qquad (4.2.48) \\
&\leq C\,\varepsilon^{-1} |\exp(-1/\varepsilon q(1))| \varepsilon^{1/2-s_1-s_2} | \int_{\mathbf{R}} s_+^n(1,-\varepsilon\xi) s_+(1,\varepsilon\xi) \tilde{u}_-(\xi) d\xi| \\
&\leq C\,\varepsilon^{-1} |\exp(-1/\varepsilon q(1))| \|u_-\|_{H_{(s),\varepsilon}(\mathbf{R})},
\end{aligned}$$

according to (4.2.43), with a constant C which does not depend on $u_- \in \mathring{H}_{(s),\varepsilon}(\mathbf{R}_-)$ and $\varepsilon \in (0, \varepsilon_0]$. The estimates (4.2.5), (4.2.45) and (4.2.48) prove (4.2.4) for u_-. As it was pointed out before, for u_+ the argument is exactly the same, and that ends the proof of Theorem 4.2.

4.3. Now, Reduction Theorem will be proved.

We begin with showing (3.5.3).

Denote

$$r^\varepsilon(x, \varepsilon D) = 1 + \varepsilon^2 D^2 q^2 \qquad (4.3.1)$$

$$s^\varepsilon(x, \varepsilon D) = (1 + \varepsilon^2 q^2 D^2)^{-1}. \qquad (4.3.2)$$

The operator Q_1^ε on the right-hand side of (3.5.3) has the following form:

$$Q_1^\varepsilon = \varepsilon^{-1} \begin{pmatrix} \pi_U(p_1^\varepsilon + g_1^\varepsilon) & , & 0, & \pi_U k_1^\varepsilon \\ \pi_{\partial U} 0 & , & 0, & 0 \\ \pi_{\partial U} t_1^\varepsilon & , & 0, & q_1^\varepsilon \end{pmatrix} \qquad (4.3.3)$$

where

$$\pi_U(p_1^\varepsilon + g_1^\varepsilon) = \pi_U r^\varepsilon l_0 (1 - \pi_U E^\varepsilon \pi_{\partial U}) \pi_U s^\varepsilon l - 1, \qquad (4.3.4)$$

$$\pi_U k_1^\varepsilon = \pi_U r^\varepsilon l_0 (-\pi_U E^\varepsilon), \qquad (4.3.5)$$

$$\pi_{\partial U} t_1^\varepsilon = -\pi_{\partial U}(1 - \pi_U E^\varepsilon \pi_{\partial U}) \pi_U s^\varepsilon l, \qquad (4.3.6)$$

$$q_1^\varepsilon = \pi_{\partial U} E^\varepsilon - Id \qquad (4.3.7)$$

and where E^ε, resp. l are given by (3.1.5) resp. (3.4.2). Using Lemma 1.4.3 in [10], one finds that the operator

$$\pi_U(r^\varepsilon s^\varepsilon - 1) l : H_{(s_1, s_2-2, s_3-2),\varepsilon}(\mathbf{R}_+) \to H_{(s_1+1, s_2-2, s_3-1),\varepsilon}(\mathbf{R}_+)$$

is continuous, uniformly with respect to $\varepsilon \in (0, \varepsilon_0]$. The trace theorem in [4] shows that $\pi_{\partial U} s^\varepsilon l$ is continuous, uniformly with respect to $\varepsilon \in (0, \varepsilon_0]$, as an operator from $H_{(s_1, s_2-2, s_3-2),\varepsilon}(\mathbf{R}_+)$ into $\mathbf{C}_{s_1+s_2-5/2,\varepsilon}(\partial U)$. Since $\pi_U r^\varepsilon E^\varepsilon : \mathbf{C}_{s_1+s_2-5/2,\varepsilon}(\partial U) \rightarrow H_{(s_1+1, s_2-2, s_3-1),\varepsilon}(U)$ is also continuous, uniformly with respect to ε, one obtains (3.5.3) with $Q_1^\varepsilon : \mathcal{K}^\varepsilon \longrightarrow \mathcal{K}^\varepsilon$ continuous uniformly with respect to $\varepsilon \in (0, \varepsilon_0]$.

4.4. We continue the proof of Reduction Theorem by showing that (3.5.4) holds.
One has

(4.4.1) $$S^\varepsilon \mathcal{R}^\varepsilon - Id = \begin{pmatrix} \pi_U(p_2^\varepsilon + g_2^\varepsilon) & , & 0 \\ \pi_{\partial U} 0 & , & 0 \end{pmatrix},$$

where

(4.4.2) $$\pi_U(p_2^\varepsilon + g_2^\varepsilon) = (1 - \pi_U E^\varepsilon \pi_{\partial U}) \pi_U s^\varepsilon l \pi_U r^\varepsilon l_0 + \pi_U E^\varepsilon \pi_{\partial U} - 1$$

with, as previously,

(4.4.3) $$r^\varepsilon(x, \varepsilon D) = 1 + \varepsilon^2 D^2 q^2$$

(4.4.4) $$s^\varepsilon(x, \varepsilon D) = (1 + \varepsilon^2 q^2 D^2)^{-1}.$$

One may rewrite (4.4.2) as follows:

(4.4.5) $$\pi_U(p_2^\varepsilon + g_2^\varepsilon) = (1 - \pi_U E^\varepsilon \pi_{\partial U}) \pi_U (s^\varepsilon l \pi_U r^\varepsilon - 1) l_1$$

with l_1 an extension operator,

(4.4.6) $$\begin{aligned} &l_1 : H_{(s_1, s_2-2, s_3),\varepsilon}(U) \rightarrow H_{(s_1, s_2-2, s_3),\varepsilon}(\mathbf{R}), \\ &\pi_U l_1 u = u, \quad \forall u \in H_{(s_1, s_2-2, s_3),\varepsilon}(U). \end{aligned}$$

which is continuous, uniformly with respect to $\varepsilon \in (0, \varepsilon_0]$.
We rewrite (4.4.5) once again, as follows:

(4.4.7) $$\pi_U(p_2^\varepsilon + g_2^\varepsilon) = (1 - \pi_U E^\varepsilon \pi_{\partial U}) \pi_U (s^\varepsilon r^\varepsilon - 1) l_1 - (1 - \pi_U E^\varepsilon \pi_{\partial U}) \pi_U s^\varepsilon (1 - l \pi_U) r^\varepsilon l_1.$$

Using [10], Lemma 1.4.3, and [4], the trace theorem, one easily finds the first term on the right-hand side of (4.4.7) to be a linear operator from $H_{(s_1, s_2-2, s_3),\varepsilon}(U)$ into $H_{(s_1+1, s_2-2, s_3),\varepsilon}(U)$ which is bounded, uniformly with respect to $\varepsilon \in (0, \varepsilon_0]$. The uniform boundedness of the second term follows immediately from Theorem 4.2.
Thus (3.5.4) holds with Q_2^ε a linear operator from $\mathcal{W}^\varepsilon$ into itself which is continuous, uniformly with respect to $\varepsilon \in (0, \varepsilon_0]$.

4.5. It remains to prove (3.5.5). However, in view of (3.3.2) and (3.5.4), one immediately obtains (3.5.5) with $Q_3^\varepsilon = Q_2^\varepsilon \mathcal{A}^0$, which completes the proof of Reduction Theorem.

5. CONCLUDING REMARKS.

5.1. The continuation by zero operator $l_0 : H_{(s'),\varepsilon}(U) \to H_{(s'),\varepsilon}(\mathbf{R})$ is continuous, uniformly with respect to $\varepsilon \in (0,1]$, if $s' = (s'_1, s'_2, s'_3) \in \mathbf{R}^3$ satisfies the condition:

$$|s'_2| < 1/2, |s'_2 + s'_3| < 1/2. \tag{5.1.1}$$

(See [12], [9].)

Therefore, if in Reduction Theorem the vectorial index $s = (s_1, s_2, s_3)$ satisfies the stronger condition

$$3/2 < s_2 < 5/2, 7/2 < s_2 + s_3 < 9/2 \tag{5.1.2}$$

(which is the case if, for instance, $s = (0,2,2)$), then one may choose in (3.4.2) the extension by zero operator and the reducing operators given by (3.1.4) and (3.4.1) coincide. This is the case if the data of the perturbed problem (2.1.1) are such that $f^\varepsilon \in L_2(U), \varphi_1^\varepsilon \in C_{0,\varepsilon}(\partial U), \varphi_2^\varepsilon \in C_{-1/2,\varepsilon}(\partial U)$, $\forall \varepsilon \in (0, \varepsilon_0]$.

5.2. The reducing operator

$$S^\varepsilon = \begin{pmatrix} \pi_U(1 - E_x^\varepsilon \pi_{\partial U})(1 + \varepsilon^2 q^2 D^2)^{-1} l_0 & , & 0, & -\pi_U E_x^\varepsilon \\ \pi_{\partial U} 0 & , & Id, & 0 \end{pmatrix}, \tag{5.2.1}$$

where

$$\begin{aligned} &E_x^\varepsilon : \mathbf{C}_{s_1+s_2-5/2,\varepsilon}(\partial U) \to H_{(s_1,s_2-2,s_3),\varepsilon}(U), \\ &(E_x^\varepsilon \varphi)(x) = \sum_{x' \in \partial U} \varphi(x')(q(x')/q(x)) \exp(-|x - x'|/\varepsilon q(x)) \end{aligned} \tag{5.2.2}$$

is a linear operator which is continuous uniformly with respect to $\varepsilon \in (0, \varepsilon_0]$ provided that s_2, s_3 satisfy (3.1.1), has the properties described in Reduction Theorem, under the conditions stated there. The argument used in section 4.3 can also be used to prove (3.5.3) in the present situation. The proof of (3.5.5) remains unchanged. Thus, we need only to consider (3.5.4). Replacing l with l_0 in section 4.4, the argument used there remains the same with the exception of the very last sentence, which should read as follows: the uniform boundedness of the second term (in (4.4.7)) follows from the trace theorem in [4], the obvious identity

$$\begin{aligned} &(1 - \pi_+(q(0)/q(x)) \exp(-x/\varepsilon q(x))\pi_0)\pi_+(1 + i\varepsilon q(x)D)^{-1}(1 - l_0\pi_+)u = \\ &= \pi_+(1 - q(0)/q(x)) \exp(-x/\varepsilon q(x)) \int_{\mathbf{R}} (1 + i\varepsilon q(0)\eta)^{-1}(1 + i\varepsilon q(x)\eta)^{-1} \mathcal{F}_\eta u \, d\eta, \ \forall u \in C_0^\infty(\mathbf{R}), \end{aligned} \tag{5.2.3}$$

and Lemma 4.1.

However, this reducing operator is not a very convenient one, because of the non-localized structure of the Poisson operator E_x^ε.

5.3. One may wonder whether the condition (5.1.2) can be weakened in such a way that the reducing operator S^ε given by (3.1.4) is still continuous uniformly with respect to $\varepsilon \in (0, \varepsilon_0]$ and has the properties summed up in Reduction Theorem.

The answer is affirmative: the condition (5.1.2) can be replaced by the following one:

$$1/2 < s_2 < 5/2, s_2 + s_3 > 7/2. \tag{5.3.1}$$

For the proof of this assertion, the following theorem will be used.

THEOREM 5.3. *Let q be as in Theorem 3.5. Then for $s = (s_1, s_2, s_3) \in \mathbf{R}^3$ such that*

$$s_2 < 5/2, s_2 + s_3 > 7/2 \tag{5.3.2}$$

holds:

$$\pi_U(1 + \varepsilon^2 q^2 D^2)^{-1} l_0 : H_{(s_1,s_2-2,s_3-2),\varepsilon}(U) \to H_{(s_1,s_2-2,s_3),\varepsilon}(U) \tag{5.3.3}$$

is a linear operator which is continuous, uniformly with respect to $\varepsilon \in (0, \varepsilon_0]$.

PROOF. Let $u \in H_{(s_1,s_2-2,s_3-2),\varepsilon}(U)$. Then there exists an extension lu of u, $lu \in H_{(s_1,s_2-2,s_3-2),\varepsilon}(\mathbf{R})$, such that

$$||lu||_{H_{(s_1,s_2-2,s_3-2),\varepsilon}(\mathbf{R})} \leq 2||u||_{H_{(s_1,s_2-2,s_3-2),\varepsilon}(U)} \tag{5.3.4}$$

(There is no need for the extension operator (3.4.2) here.)

Since $l_0 u = lu - (1 - l_0\pi_U)lu$ and since the operator

$$\pi_U(1 + \varepsilon^2 q^2 D^2)^{-1} : H_{(s_1,s_2-2,s_3-2),\varepsilon}(\mathbf{R}) \to H_{(s_1,s_2-2,s_3),\varepsilon}(U) \tag{5.3.5}$$

is continuous, uniformly with respect to ε, it suffices to show that

$$\pi_U(1 + \varepsilon^2 q^2 D^2)^{-1}(1 - l_0\pi_U) : H_{(s_1,s_2-2,s_3-2),\varepsilon}(\mathbf{R}) \to H_{(s_1,s_2-2,s_3),\varepsilon}(U) \tag{5.3.6}$$

is continuous, uniformly with respect to ε.

Let $lu \in C_0^\infty(\mathbf{R})$ (dense in $H_{(s_1,s_2-2,s_3-2),\varepsilon}(\mathbf{R})$) and denote

$$(1 - l_0\pi_U)lu = u_- + u_+ \tag{5.3.7}$$

where supp $u_- \subseteq \overline{\mathbf{R}}_-$, supp $u_+ \subseteq [1, \infty)$.

Using (4.2.9) up to (4.2.13) and the fact that $\int_{\mathbf{R}}(1 + i\varepsilon q(x)\eta)^{-1}\mathcal{F}_\eta \pi_+ lu \, d\eta = 0$, one obtains

$$\begin{aligned} \pi_+(1 + \varepsilon^2 q^2 D^2)^{-1} u_- &= (1/2)\pi_+ \mathcal{F}^{-1}_{\xi\to x} \varepsilon q(x)(1 + i\varepsilon q(x)\xi)^{-1} \int_{\mathbf{R}} (1 + i\varepsilon q(x)\eta)^{-1}\mathcal{F}_\eta u_- d\eta \\ &= (1/2)\pi_+ \exp(-x/\varepsilon q(x)) \int_{\mathbf{R}} (1 + i\varepsilon q(x)\eta)^{-1}\mathcal{F}_\eta lu \, d\eta \end{aligned} \tag{5.3.8}$$

which, together with Lemma 4.1, the trace theorem in [4], (5.3.2) and the boundedness of $\pi_+ \exp(-x/\varepsilon q(x)) : \mathbf{C}_{s_1+s_2-5/2,\varepsilon} \to H_{(s_1,s_2-2,s_3),\varepsilon}(\mathbb{R}_+)$ uniformly with respect to ε, implies that

$$(5.3.9) \qquad \|\pi_+(1+\varepsilon^2q^2D^2)^{-1}u_-\|_{H_{(s_1,s_2-2,s_3),\varepsilon}(\mathbb{R}_+)} \le C\|lu\|_{H_{(s_1,s_2-2,s_3-2),\varepsilon}(\mathbb{R})}$$

with a constant C which does not depend on lu and $\varepsilon \in (0,\varepsilon_0]$.

Hence

$$(5.3.10) \qquad \begin{aligned} &\|\pi_U(1+\varepsilon^2q^2D^2)^{-1}u_-\|_{H_{(s_1,s_2-2,s_3),\varepsilon}(U)} \le \\ &\le \|\pi_+(1+\varepsilon^2q^2D^2)^{-1}u_-\|_{H_{(s_1,s_2-2,s_3),\varepsilon}(\mathbb{R}_+)} \\ &\le C\|lu\|_{H_{(s_1,s_2-2,s_3-2),\varepsilon}(\mathbb{R})}. \end{aligned}$$

Using the same argument for the function

$$(5.3.11) \qquad v_-(x) = u_+(1-x)$$

one obtains: $\|\pi_U(1+\varepsilon^2q^2D^2)^{-1}u_+\|_{H_{(s_1,s_2-2,s_3),\varepsilon}(U)} \le C\|lu\|_{H_{(s_1,s_2-2,s_3-2),\varepsilon}(\mathbb{R})}$, which, together with (5.3.10), yields (5.3.6).

◇

It follows easily from Reduction Theorem that the operator

$$(5.3.12) \qquad \pi_U(E^\varepsilon - E^\varepsilon_x)\pi_{\partial U}(1+\varepsilon^2q^2D^2)^{-1}l_0 : H_{(s_1,s_2-2,s_3-2),\varepsilon}(U) \to H_{(s_1+1,s_2-2,s_3),\varepsilon}(U)$$

with E^ε resp. E^ε_x defined by (3.1.5) resp. (5.2.2), is continuous, uniformly with respect to $\varepsilon \in (0,\varepsilon_0]$, if condition (5.3.2) is fulfilled. Hence the difference of the operators given by (3.1.4) and (5.2.1) is a linear operator from $\mathcal{K}^\varepsilon$ into $\mathcal{W}^\varepsilon$ which is continuous, uniformly with respect to ε, and has a norm which is bounded above by $C\varepsilon$ (with some constant $C > 0$), if (5.3.2) holds.

Combining this result with the fact, that the reducing operator (5.2.1) may replace the operator (3.4.1) in Reduction Theorem (see the previous section 5.2), one comes to the conclusion that the operator (3.1.4) may take the place of the operator (3.4.1) in Reduction Theorem if condition (3.1.1) is replaced by condition (5.3.1).

5.4. One may not replace the reducing operator (3.4.1) in Reduction Theorem by operator (3.1.4) if

$$s_2 < 5/2,\ 5/2 < s_2 + s_3 \le 7/2,$$

for then the operator (5.3.12) is not well-defined; namely, the trace $\pi_{\partial U}(1+\varepsilon^2q^2D^2)^{-1}l_0 f$ does not exist for all $f \in H_{(s_1,s_2-2,s_3-2),\varepsilon}(U)$, which is a consequence of the fact that $\pi_U(1+\varepsilon^2q^2D^2)^{-1}(1-l_0\pi_U)lf$, with l the extension operator (3.4.2), does not necessarily belong to $H_{(s_1,s_2-2,s_3),\varepsilon}(U)$, on the latter space $\pi_{\partial U}$ being well defined. (See (5.3.8).)

REFERENCES

[1] L. Boutet de Monvel, Boundary Problems for Pseudo-differential operators. Acta Math. 126 (1971), 11-51.

[2] L.S. Frank, Boundary Value Problems for Ordinary Differential Equations with Small Parameter, Annali di Mat. Pura Applic. (IV), 114 (1977), 27-67.

[3] L.S. Frank, Perturbazioni Singolari Ellitiche, Rendiconti Seminario Matematico e Fisico di Milano, Vol. XLVII (1977), 135-163.

[4] L.S. Frank, Coercive Singular Perturbations I: A priori estimates, Annali di Mat. Pura Applic. (IV), 119 (1979), 41-113.

[5] L.S. Frank, Inégalité de Gårding et Perturbations Singulières elliptiques aux différences finies, C.R. Acad. Sci., Paris, série I, t.296 (1983), 93-96.

[6] L.S. Frank, The factorization Method for Coercive Singularly Perturbed Wiener-Hopf Operators and Applications, Report 8720 (1987), Catholic University of Nijmegen.

[7] L.S. Frank, Perturbations Singulières Coercives: Methode de factorisation et applications, École Polytechnique, Séminaire Aux Dérivées Partielles 1986-1987, Exposé no. XVIII, 28 Avril 1987, 1-26.

[8] L.S. Frank, Coercive Singular Perturbations and Applications, North-Holland, to appear.

[9] L.S. Frank and J.J. Heijstek, On the continuation by zero in Sobolev Spaces with a small parameter, Asymptotic Analysis 1 (1988), 51-60.

[10] L.S. Frank and W.D. Wendt, Isomorphic Elliptic Singular Perturbations, Report 8003 (1980), Catholic University of Nijmegen.

[11] L.S. Frank and W.D. Wendt, Coercive Singular Perturbations II: Reduction to Regular Perturbations and Applications, Comm. P.D.E., Vol. 7, no. 5 (1982), 469-535.

[12] L.S. Frank and W.D. Wendt: Coercive Singular Perturbations III: Wiener-Hopf Operators, Journal d'Analyse Mathematique, Vol. 43 (1983/84), 88-135.

[13] I.C. Gohberg and M.G. Kein, Systems of Integral Equations on a half Line with kernel depending on the difference of the arguments, Amer. Math. Soc. Transl. (2), 14 (1960), 217-287.

[14] R.T. Seeley, Extension of C^∞ functions defined in a halfspace, Proc. Amer. Math. Soc. 15 (1964), 625-626.

[15] W.D. Wendt, Coercive Singularly Perturbed Wiener-Hopf Operators and Applications, Ph.D. Thesis, Catholic University of Nijmegen, 1983.

Mathematisch Instituut der
Katholieke Universiteit Nijmegen,
Toernooiveld
6525 ED Nijmegen
The Netherlands.

Operator Theory:
Advances and Applications, Vol. 41

AVERAGING TECHNIQUES FOR THE TRANSPORT OPERATOR AND AN EXISTENCE THEOREM FOR THE BGK EQUATION IN KINETIC THEORY

William Greenberg and Jacek Polewczak

This paper is dedicated to Israel Gohberg on the occasion of his 60th birthday. His contributions in analysis have benefitted all of the mathematical sciences.

Application of velocity averaging techniques to prove existence theorems for the solution of various kinetic equations is surveyed. This approach is exploited to obtain the first existence result for the kinetic equation of Bhatnagar–Gross–Krook.

I. Introduction

Existence theorems for the equations of kinetic theory have been a matter of interest since Carleman [1] first tackled the existence question for solutions of the Boltzmann equation in 1932. Surprisingly, for the most important of these equations many of the existence questions are still unanswered.

In the past year important progress has been made on this problem for a variety of kinetic equations. The central result about which recent progress has been constructed is the velocity averaging lemma of Golse et al. [2] In this report we would like to explain this new approach, to survey briefly its application to the Boltzmann equation by DiPerna and Lions [3] and to the Enskog equation [4,5], and then to outline some new results on the nonlinear Bhatnagar–Gross–Krook (B.G.K.) equation.

The Boltzmann equation, the Enskog equation and the B.G.K. equation are all transport equations of the form

$$\frac{\partial f}{\partial t} + v \cdot \nabla_x f = J(f), \qquad f(0,x,v) = f_0(x,v), \tag{1.1}$$

where the function $f(t,x,v)$ is the density of a gas at time t and position x with velocity v, and the nonlinear term $J(f)$ reflects specific assumptions on the collision mechanism. The Boltzmann equation was proposed by Ludwig Boltzmann [6] in 1872, based on heuristic arguments to describe the behaviour of a gas at moderate densities. The collision term J_{Bol} can be written in the form $J_{Bol}(f) = Q^+(f) - Q^-(f)$, where

$$Q^+(f) = \int\int_{\mathbb{R}^3 \times S^2_+} f(t,x,v')f(t,x,w')B(\theta, v-w)\, d\epsilon dw, \tag{1.2a}$$

$$Q^-(f) = \int\int_{\mathbb{R}^3 \times S^2_+} f(t,x,,v)f(t,x,w)B(\theta, v-w)\, d\epsilon dw. \tag{1.2b}$$

Here v' and w' are the velocities after collision of molecules initially at velocities v and w, and are related by conservation of momentum and energy as

$$v' = v - \epsilon\langle\epsilon, v-w\rangle, \qquad w' = w + \epsilon\langle\epsilon, v-w\rangle, \tag{1.3}$$

with $\epsilon \in S^2_+ = \{\epsilon \in \mathbb{R}^3 : |\epsilon| = 1, \langle v-w, \epsilon\rangle \geq 0\}$. The angle $\theta \in [0, \pi/2]$ is defined by $\cos\theta = \frac{\langle v-w,\epsilon\rangle}{|v-w|}$, and $B(\theta, v-w)$ is the scattering kernel with the usual angular cutoff. For inverse power potentials with force law $\mathcal{F}(r) = r^{-s}$, $s > 2$, one has $B(\theta, v-w) = b(\theta)|v-w|^{\frac{s-4}{s}}$.

The Enskog equation was derived in 1921 by D. Enskog [7] to describe a dense gas of hard spheres. The collision operator J_{Ens} (in a form derived by Resibois [8]) may be written $J_{Ens}(f) = E^+(f) - E^-(f)$ with

$$E^+(f) = a^2 \int\int_{\mathbb{R}^3 \times S^2_+} Y(n(t,x), n(t, x-a\epsilon))f(t,x,v')f(t,x-a\epsilon,w')\langle\epsilon, v-w\rangle\, d\epsilon dw, \tag{1.4a}$$

$$E^-(f) = a^2 \int\int_{\mathbb{R}^3 \times S^2_+} Y(n(t,x), n(t, x+a\epsilon))f(t,x,v)f(t,x+a\epsilon,w)\langle\epsilon, v-w\rangle\, d\epsilon dw. \tag{1.4b}$$

The constant a denotes the hard sphere diameter, $\langle\cdot,\cdot\rangle$ is the inner product in $\mathbb{R}^3$, the function $n(t,x) = \int_{\mathbb{R}^3} f(t,x,v)\, dv$ is the local density of the gas, and the geometric factor Y is an essentially bounded function of the local density at x and at $x \pm a\epsilon$.

The B.G.K. equation was proposed in 1954 by P. L. Bhatnagar, E. P. Gross and M. Krook [9] and independently by P. Welander [10] in the same year, to describe a gas tending toward its equilibrium maxwellian distribution. Its collision operator J_{BGK} may be written $J_{BGK}(f)(x,v) = \nu[P(f)(x,v) - f(x,v)]$ for

$$P(f)(x,v) = \frac{\rho(x)}{[2\pi RT(x)]^{3/2}} \exp\left\{\frac{-[v-V(x)]^2}{2RT(x)}\right\}, \tag{1.5}$$

where $\rho(x) = \int_{\mathbb{R}^3} f(x,v)dv$ is the macroscopic density, $V(x) = \int_{\mathbb{R}^3} vf(x,v)dv/\rho(x)$ is the macroscopic velocity, $E(x) = \int_{\mathbb{R}^3} v^2 f(x,v)dv/2\rho(x)$ is the total macroscopic energy per unit mass, and $T(x) = (2E(x) - V(x)^2)/3R$ with R the Boltzmann constant is the macroscopic temperature. The collision frequency ν is a function of the local state of the gas, i.e., $\nu = \nu(\rho, V, E)$.

In the next section we shall outline the innovative contribution of Golse et al. on the inversion of the transport operator, and its recent application to the Boltzmann equation by DiPerna and Lions and to the Enskog equation by Polewczak and, independently, by Arkeryd and Cercignani. In the last section we shall give details on properties of the B.G.K. equations and describe the proof of an existence theorem for the B.G.K. equation.

II. Averaging Lemma and Existence Results for the Boltzmann and Enskog Equations

In 1987 Golse et al. [2] proved the following lemma on the inversion of the transport operator.

LEMMA 2.1. *Suppose that* $f_n \in L^1((0,T) \times \mathbb{R}^3 \times \mathbb{R}^3)$ *and* $g_n \in L^1((0,T) \times \mathbb{R}^3 \times \mathbb{R}^3)$ *satisfy*

$$S_v f_n \stackrel{def}{=} \frac{\partial f_n}{\partial t} + v \cdot \nabla_x f_n = g_n \tag{2.1}$$

in $\mathcal{D}'((0,T) \times \mathbb{R}^3 \times \mathbb{R}^3)$, *and that the sequences* $\{f_n\}$ *and* $\{g_n\}$ *are relatively weakly compact in* $L^1((0,T)\times\mathbb{R}^3\times\mathbb{R}^3)$. *Then for all* $\varphi \in L^\infty((0,T)\times\mathbb{R}^3\times\mathbb{R}^3)$ *the set* $\{\int_{\mathbb{R}^3} \varphi f_n dv\} = \{\int_{\mathbb{R}^3} \varphi S_v^{-1} g_n dv\}$ *is relatively compact in* $L^1((0,T) \times \mathbb{R}^3)$.

Velocity averaging compensates for the lack of regularity of the operator S_v^{-1} in the characteristic direction, in a manner similar to the behaviour of the inverse of an elliptic operator. The authors in [2] then applied this lemma to a linear transport equation in radiative transport.

The Golse lemma, being a compactness argument, provides no information on the uniqueness of solutions of transport equations, nor, of course, on the smoothness of these solutions. However, inasmuch as kinetic theory is replete with fundamental equations whose existence theory is largely unknown when smallness parameters are violated (near equilibrium, small initial data, etc.), it is not surprising that this lemma has had an important impact on a number of open problems in kinetic theory.

The most immediate notable contribution of this velocity averaging technique was the proof, by DiPerna and Lions [3], that the spatially inhomogeneous Boltzmann equation has solutions for initial data arbitrarily far from equilibrium. The announcement of this result attracted great attention, inasmuch as this question has been open, with not even partial results or heuristic *hints*, for more than one hundred years. Due to the fact that Lemma 2.1 is presently known for $\mathbb{R}^3$ and for periodic boundary conditions, global existence proofs for the Boltzmann and the Enskog equations have been provided in these two cases only. In this report we will be concerned with the Boltzmann and the Enskog equations in all $\mathbb{R}^3$. The proofs outlined here can easily be extended to the case of bounded spatial domains with periodic boundary conditions. We indicate now the proof of DiPerna and Lions for the Boltzmann equation.

A function f is a mild solution to (1.1) if $Q^{\pm}(f)(t,x,v) \in L^1(0,T)$ a.e. in $(v,x) \in \mathbb{R}^3 \times \mathbb{R}^3$ and $f^{\#}(t,x,v) - f^{\#}(s,x,v) = \int_s^t J_{Bol}(f)^{\#}(\tau,x,v)d\tau$ for any $0 < s < t \leq T$ with $f^{\#}(t,x,v) =$

$f(t, x+tv, v)$. From the conservation property of the collision operator,

$$\int_{\mathbb{R}^3} \psi J_{Bol}(f) dv = \iiint_{\mathbb{R}^3 \times \mathbb{R}^3 \times S_+^2} [\psi(x,v) + \psi(x,w) - \psi(x,v') - \psi(x,w')] \times$$
$$\times [f(x,v')f(x,w') - f(x,v)f(x,w)] B(\theta, v-w)\, d\epsilon dw dv \tag{2.2}$$

for ψ measurable on $\mathbb{R}^3 \times \mathbb{R}^3$ and $f \in C_0(\mathbb{R}^3 \times \mathbb{R}^3)$, and the Boltzmann inequality,

$$\int_{\mathbb{R}^3} J_{Bol}(f) \log f dv \le 0 \tag{2.3}$$

for $0 \le f \in C_0(\mathbb{R}^3 \times \mathbb{R}^3)$, one may conclude that if $f(t,x,v)$ is a smooth and nonnegative solution to (1.1) with a nonnegative initial value f_0 satisfying

$$\iint_{\mathbb{R}^3 \times \mathbb{R}^3} (1 + v^2 + x^2 + |\log f_0|) f_0 \, dv dx = C_0 < \infty, \tag{2.4}$$

then

$$\sup_{0 \le t \le T} \iint_{\mathbb{R}^3 \times \mathbb{R}^3} (1 + v^2 + x^2 + |\log f|) f dv dx \le C_T, \tag{2.5}$$

where C_T depends only on T and C_0.

The first step in the DiPerna-Lions proof is to approximate J_{Bol} by a truncated collision operator,

$$J_n(f) = \frac{1}{(1 + \frac{1}{n}\int_{\mathbb{R}^3} f dv)} \iint_{\mathbb{R}^3 \times S_+^2} [(f(x,v')f(x,w') - f(x,v)f(x,w)] B_n(\theta, v-w)\, d\epsilon dw,$$

where $B_n(\theta, v-w) = B(\theta, v-w)$ if $v^2 + w^2 \le n$, and 0 otherwise. This truncated collision operator is Lipschitz in $L^1((0,T) \times \mathbb{R}^3 \times \mathbb{R}^3)$ with Lipschitz constant depending only on n. Therefore, the Cauchy problem (1.1) with the truncated operator can be solved uniquely on $[0,T]$ for any $T > 0$, and the solution f_n is smooth and nonnegative. Furthermore (2.5) and the Dunford-Pettis theorem [13] imply that $\{f_n\}$ is relatively weakly compact in $L^1((0,T) \times \mathbb{R}^3 \times \mathbb{R}^3)$, and so, without loss of generality, we may assume $f_n \to f$ weakly in L^1, where $0 \le f \in L^1$.

Now, the full collision operator J_{Bol} is not weakly continuous in $L^1(\mathbb{R}^3 \times \mathbb{R}^3)$, so it is not evident that Lemma 2.1 can be of use. However, since the sequence $\left\{\frac{Q^-(f_n)}{1+f_n}\right\}$ is seen rather easily to be weakly compact, it is natural to seek the weak compactness of $\left\{\frac{Q^+(f_n)}{1+f_n}\right\}$. This is not at all trivial, and, in fact, needed a new estimation of the gain term Q^+, provided by DiPerna and Lions, namely $Q^+(f_n) \le MQ^-(f_n) + \delta(f_n)/\log M$ for all $M > 1$. Here, $\delta(f)$ is given by

$$\delta(f) = \iint_{\mathbb{R}^3 \times S_+^2} [f(t,x,v')f(t,x,w') - f(t,x,v)f(t,x,w)] B(\theta, v-w) \log \frac{f(t,x,v')f(t,x,w')}{f(t,x,v)f(t,v,w)} d\epsilon dw.$$

Then the Boltzmann inequality (2.3) implies that

$$\frac{d}{dt}\iint\limits_{\mathbb{R}^3\times\mathbb{R}^3} f_n \log f_n dv dx = -\frac{1}{4}\iint\limits_{\mathbb{R}^3\times\mathbb{R}^3} \delta(f_n) dv dx, \tag{2.6}$$

and since $\delta(f_n) \geq 0$, the functions $\{\delta(f_n)\}$ are bounded in $L^1((0,T)\times\mathbb{R}^3\times\mathbb{R}^3)$ uniformly in $n \geq 1$. Thus, the set $\left\{\frac{Q^+(f_n)}{1+f_n}\right\}$ has weakly compact closure.

The relationship between J_{Bol} and $\frac{1}{1+f}J_{Bol}$ is given by the most innovative contribution of the DiPerna-Lions paper. They define $f \in C([0,T]\times L^1_+(\mathbb{R}^3\times\mathbb{R}^3))$ to be a renormalized solution to (1.1) if $\frac{1}{1+f}Q^{\pm}(f) \in L^1((0,T)\times B_R\times B_R)$ for any $R>0$ and

$$\frac{\partial}{\partial t}\log(1+f) + v\cdot\nabla_x \log(1+f) = \frac{1}{1+f}J_{Bol}(f) \tag{2.7}$$

in $\mathcal{D}'((0,\infty)\times\mathbb{R}^3\times\mathbb{R}^3)$. They then showed that if $\frac{Q^{\pm}(f)}{1+f} \in L^1_{loc}$, then f is a renormalized solution to (1.1) if and only if it is a mild solution.

From the above, $f_n \to f$ weakly in $L^1((0,T)\times\mathbb{R}^3\times\mathbb{R}^3)$ and, by the averaging lemma,

$$\int_{\mathbb{R}^3}\varphi f_n dv \to \int_{\mathbb{R}^3}\varphi f dv$$

strongly in $L^1((0,T)\times\mathbb{R}^3)$ for all $\varphi \in L^\infty((0,T)\times\mathbb{R}^3\times\mathbb{R}^3)$. This then implies that for each $\varphi \in L^\infty((0,T)\times\mathbb{R}^3\times\mathbb{R}^3)$ with compact support,

$$\int_{\mathbb{R}^3} J_{Bol}(f_n)\varphi dv \to \int_{\mathbb{R}^3} J_{Bol}(f)\varphi dv \tag{2.8}$$

in measure on $(0,T)\times B_R$. These two limits are sufficient to prove that f is a mild solution of the Boltzmann equation.

An existence theorem for the Enskog equation in three dimensions with general initial data is still an open problem, unless some simplifying assumptions are made on the collision operator. Two approaches have been successful. Arkeryd and Cercignani [4] have treated the original Enskog equation, in which the function Y depends only on the local density at the points $x \pm a\epsilon$, by extending the range of integration of ϵ in (1.4) from S^2_+ to S^2. In this case the collision operator with Y bounded enjoys all the properties of the Boltzmann collision operator, and the DiPerna-Lions proof may be followed quite closely. Unfortunately, this alteration in the range of integration has a significant effect on the dynamics of the Enskog equation. In particular, the Enskog equation with integration over S^2_+ distinguishes between forward and backward (time reversed) collisions, while the Boltzmann equation and the Enskog equation with the integration extended are both symmetric under forward and backward collisions.

Polewczak [5] has preserved the Enskog equation in the form derived by Resibois, in which the geometric factor Y is symmetric, $Y(\tau,\sigma) = Y(\sigma,\tau)$ for $\sigma,\tau \geq 0$, but has assumed (in more than one dimension) that it satisfies the bound

$$\sup_{\tau,\sigma\geq 0} \tau Y(\tau,\sigma) = M_Y < \infty. \tag{2.9}$$

The symmetry arises naturally in the Resibois formalism and is crucial in the derivation of the H-theorem and the conservation laws. However, the bound (2.9) is physically not desirable, as one would like $\lim_{\tau\to\tau_c} Y(\tau,\sigma) = \infty$, where τ_c is the close packing density for spheres of radius a. Therefore, this assumption is satisfied only by geometric factors which model the physical situation, for example, Y arbitrarily large but finite at τ_c, and $Y = 0$ for $\tau > \tau_c$.

The Enskog collision operator satisfies a conservation property analogous to (2.2) for the Boltzmann operator,

$$\iint_{\mathbb{R}^3\times\mathbb{R}^3} \psi(x,v)J_{Ens}(f)\,dvdx = \frac{a^2}{2}\iiiint_{\mathbb{R}^3\times\mathbb{R}^3\times\mathbb{R}^3\times S^2_+} [\psi(x,v') + \psi(x+a\epsilon,w') - \psi(x,v) - \psi(x+a\epsilon,w)]\times$$
$$\times f(x,v)f(x+a\epsilon,w)Y(n(x),n(x+a\epsilon))\langle\epsilon, v-w\rangle\,d\epsilon dw dv dx, \tag{2.10}$$

for ψ measurable on $\mathbb{R}^3\times\mathbb{R}^3$ and $f \in C_0(\mathbb{R}^3\times\mathbb{R}^3)$, with the expected shift in the spatial argument and the additional integration with respect to x. Resibois also showed that an H-function which is nonincreasing in time can be associated with the solution of the Enskog equation, of the form

$$H(t) = \iint f(t,x,v)\log f(t,x,v)\,dvdx + H^v(t), \tag{2.11}$$

where f is the solution to (1.1), and the potential part $H^v(t)$ is given in terms of Resibois' grand canonical formalism but, unfortunately, not explicitly in terms of f and Y. The inability to control the potential part of H has the result that one cannot say anything about a bound for $\int\int f(t,x,v)\log f(t,x,v)\,dvdx$.

Polewczak succeeded in bypassing this problem by introducing a new Liapunov functional for the Enskog equation. With the definition

$$\Gamma(t) = \iint_{\mathbb{R}^3\times\mathbb{R}^3} f(t,x,v)\log f(t,x,v)dvdx - \int_0^t I(s)ds, \tag{2.12}$$

where

$$I(t) = \frac{1}{2}a^2 \iiiint_{\mathbb{R}^3\times\mathbb{R}^3\times\mathbb{R}^3\times S^2_+} [f(t,x-a\epsilon,w)Y(n(t,x),n(t,x-a\epsilon))$$
$$-f(t,x+a\epsilon,w)Y(n(t,x),n(t,x+a\epsilon))]f(t,x,v)\langle\epsilon,v-w\rangle\,d\epsilon dw dv dx,$$

he showed

$$\frac{d\Gamma}{dt} \leq 0. \tag{2.13}$$

The functional $\Gamma(t)$ displays the irreversibility of the system governed by the modified Enskog equation, and is defined explicitly in terms of f and Y.

Given this inequality and the bound (2.9), one can reproduce the same a priori estimate obtained for the Boltzmann equation, although the derivation is not entirely straightforward. That is to say, if $f(t,x,v)$ is a smooth and nonnegative solution to the Enskog equation with a nonnegative initial value f_0 satisfying (2.4), then f satisfies (2.5) with C_T depending only on T and C_0. To proceed, one needs again an estimate for the gain term of the collision operator. Polewczak provided the estimate

$$E^+(f) \leq M \iint\limits_{\mathbb{R}^3\times S_+^2} Y(n(f,x),n(t,x-a\epsilon))f(t,x,v)f(t,x-a\epsilon,w)\langle\epsilon, v-w\rangle\, d\epsilon dw + \frac{\alpha(f)}{\log M} \tag{2.14}$$

for each $M > 1$, where

$$\alpha(f) = \iint\limits_{\mathbb{R}^3\times S_+^2} Y(n(t,x),n(t,x-a\epsilon))f(t,x,v')f(t,x-a\epsilon,w')\times$$
$$\times \left|\log \frac{f(t,x,v')f(t,x-a\epsilon,w')}{f(t,x,v)f(t,x-a\epsilon,w)}\right| \langle\epsilon, v-w\rangle d\epsilon dw$$

has a bound in $L^1((0,T)\times\mathbb{R}^3\times\mathbb{R}^3)$ depending only on C_0 and C_T.

To utilize Lemma 2.1, one considers the solutions f_n of (1.1) with an approximate Enskog collision operator defined as J_{Ens}, but with $\langle\epsilon, v-w\rangle$ replaced by $\langle\epsilon, v-w\rangle_n = \langle\epsilon, v-w\rangle$ when $v^2+w^2 \leq n$, and 0 otherwise, and with $Y(\tau,\sigma)$ replaced by $Y_n(\tau,\sigma) = (1+\frac{1}{n}\tau)^{-1}(1+\frac{1}{n}\sigma)^{-1}Y(\tau,\sigma)$. Then, approximating f_0 by convolution to obtain $\tilde{f}_0 \in C_0^\infty(\mathbb{R}^3\times\mathbb{R}^3)$ and $\tilde{f}_0 \geq 0$, we show that f_n is smooth. By (2.5), one may show that $\{f_n\}$ is weakly relatively compact in $L^1((0,T)\times\mathbb{R}^3\times\mathbb{R}^3)$. Hence, by Lemma 2.1, we obtain that $\{\int_{\mathbb{R}^3} f_n(t,x,v)\varphi(t,x,v)dv\}_{n=1}^\infty$ is relatively compact in $L^1((0,T)\times\mathbb{R}^3)$ for all $\varphi \in L^\infty((0,T)\times\mathbb{R}^3\times\mathbb{R}^3)$. Therefore, after passing to a subsequence, if necessary, we have $\int_{\mathbb{R}^3} f_n(t,x,v)dv \underset{n\to\infty}{\rightarrow} \int_{\mathbb{R}^3} f(t,x,v)dv$ a.e. in $(t,x) \in [0,T]\times\mathbb{R}^3$, and

$$Y_n\Big(\int_{\mathbb{R}^3} f_n(t,x,v)dv, \int_{\mathbb{R}^3} f_n(t,x\pm a\epsilon,v)dv\Big) \underset{n\to\infty}{\rightarrow} Y\Big(\int_{\mathbb{R}^3} f(t,x,v)dv, \int_{\mathbb{R}^3} f(t,x\pm a\epsilon,v)dv\Big)$$

a.e. in $(t,x,\epsilon) \in [0,T]\times\mathbb{R}^3\times S_+^2$. The existence of a nonnegative renormalized solution to the Enskog equation follows as for the Boltzmann equation.

III. Existence Result for the B.G.K. Equation

The B.G.K. equation has played, and still plays, an important role in kinetic theory (see, for example, [7]). However, nothing has been known about existence of solutions, neither local nor global, not even a near-equilibrium result. In 1982 the authors [11] obtained a sequence of approximate solutions to (1.1) and its convergence in the weak topology of $L^1(\Omega \times \mathbb{R}^3)$, but were unable to say in what sense the limit of the approximate solutions satisfied the original B.G.K. equation. We shall sketch here a proof, utilizing Lemma 2.1, that this limit does indeed satisfy the B.G.K. equation under an energy saturation condition on the collision frequency.

We start with a few definitions that set up the B.G.K. equation in the framework of semilinear evolution equations. Let

$$L^1_{1+v^2}(\Omega \times \mathbb{R}^3) = \{f \in L^1(\Omega \times \mathbb{R}^3) : \|f\|_2 \equiv \iint_{\Omega \times \mathbb{R}^3} (1+v^2)|f| dv dx < \infty\}.$$

Here Ω is a three-dimensional torus, i.e., $\Omega = \mathbb{R}^3/\mathbb{Z}^3$. The choice of Ω is a convenient way to express the fact that we consider the operator $Af \equiv -v \cdot \nabla_x f$ with periodic boundary conditions. Then A generates a strongly continuous semigroup $U(t)$ in $L^1(\Omega \times \mathbb{R}^3)$, and that the restriction of $U(t)$ to $L^1_{1+v^2}(\Omega \times \mathbb{R}^3)$ (also denoted by $U(t)$) is a strongly continuous semigroup in $L^1_{1+v^2}(\Omega \times \mathbb{R}^3)$. For $M > 0$ and $C \in \mathbb{R}$ we define the set D by $D = \{f \in L^1_{1+v^2}(\Omega \times \mathbb{R}^3) : f \geq 0,\ \|f\|_2 \leq M,\ H(f) \leq C\}$, where the functional H is given by $H(f) = \int\int_{\Omega \times \mathbb{R}^3} f \log f dv dx$. The B.G.K. collision operator obeys the conservation laws $\int_{\mathbb{R}^3} \psi_i J_{BGK}(f) dv = 0$ a.e. in x for $i = 0, 1, 2, 3, 4$ and $f \in D$, where ψ_i are the collision invariants $\psi_0 = 1$, $\psi_i = v_i$ for $i = 1, 2, 3$ (components of v), and $\psi_4 = v^2$. Formally, at least, we have also the Boltzmann inequality

$$\int_{\mathbb{R}^3} J_{BGK}(f) \log f dv \leq 0 \tag{3.1}$$

for fixed x.

We say that a continuous function f from $[0, T]$ into $D \subset L^1_{1+v^2}(\Omega \times \mathbb{R}^3)$ is a mild solution to the B.G.K. equation in $L_{1+v^2}(\Omega \times \mathbb{R}^3)$ if it satisfies

$$f(t) = U(f) f_0 + \int_0^t U(t-s) J_{BGK}(f(s)) ds \tag{3.2}$$

for $t \in [0, T]$. The sense of integration is taken generally in the Bochner sense. It will, in fact, turn out to be the Riemann integral in $L^1_{1+v^2}(\Omega \times \mathbb{R}^3)$ for the problem of interest here.

First we want to state a proposition and a theorem proved in [11] under the assumption that the collision frequency ν is a constant. However, the results in [11] extend in a straightforward way to the case $\nu \in L^\infty(\Gamma)$ for $\Gamma \equiv \mathbb{R}_+ \times \mathbb{R}^3 \times \mathbb{R}_+$. The proposition follows from the lower semicontinuity of the convex functional H in L^1, together with the Dunford-Pettis theorem and the conservation laws. We recall the definition of the local maxwellian $P(f)$ given in (1.5).

PROPOSITION 3.1. *D is a convex weakly compact subset of $L^1(\Omega \times \mathbb{R}^3)$, invariant under the semigroup $U(t)$, $t > 0$, and for each sequence $\{f_n\} \subset D$ there exists a subsequence $\{f_{n_i}\}$ and $f \in D$ such that $\int\int_{\Omega\times\mathbb{R}^3} \varphi f_{n_i}\, dvdx \underset{i\to\infty}{\to} \int\int_{\Omega\times\mathbb{R}^3} \varphi f\, dvdx$ for all measurable φ satisfying $|\varphi(x,v)| \leq (1+v^2)^k$, $k < 1$. Moreover, $P(D) \subseteq D$, and P is continuous as a map from $D \subset L^1_{1+v^2}(\Omega \times \mathbb{R}^3)$ into $L^1_{1+v^2}(\Omega \times \mathbb{R}^3)$.*

As a consequence of the proposition, the following theorem is the main result of [11].

THEOREM 3.2. *For each $f_0 \in D$ there exists a sequence $\{f_n\}$ of approximate solutions to (1.1). Furthermore, $\{f_n\}$ contains a subsequence $\{f_{n_i}\}$ which converges weakly in $L^1(\Omega \times \mathbb{R}^3)$ and uniformly on $[0,T]$ to a limit f, where $f : [0,T] \to D$ is weakly continuous in $L^1(\Omega \times \mathbb{R}^3)$.*

The collision frequency $\nu = \nu(\rho, V, E)$ will be taken as an essentially bounded measurable function of the normalized moments. We will need one additional assumption. We shall say that the collision frequency satisfies the *energy saturation condition* if $\nu(\rho, V, E) \underset{E\to\infty}{\to} 0$ uniformly in ρ and V. An example of energy saturation is provided by ν with compact support with respect to E. Indeed, since the energy density per unit mass is proportional to $\int v^2 f dv / \int f dv$, this corresponds to the saturation of the energy density per unit mass in the B.G.K. collision model. Now we are ready to state global existence theorem for the B.G.K. equation.

THEOREM 3.3. *Suppose the collision frequency ν satisfies $\nu \in L^\infty(\Gamma)$ and the energy saturation condition, and $f_0 \in D$. Then there exists a global mild solution to the B.G.K. equation in $L^1_{1+v^2}(\Omega \times \mathbb{R}^3)$.*

We sketch a proof. By Theorem 3.2, passing to a subsequence if necessary, $f_n(t) \underset{n\to\infty}{\to} f(t)$ weakly in $L^1(\Omega \times \mathbb{R}^3)$, uniformly in t, where $f : [0,T] \to D \subset L^1(\Omega \times \mathbb{R}^3)$ is weakly continuous. Observe that $h_n(t) = U(t)f_0 + \int_0^t U(t-s)J(f_n(\gamma_n(s)))ds$ is a solution in $\mathcal{D}'((0,T) \times \Omega \times \mathbb{R}^3)$ to (2.1) with $g_n = J_{BGK}(f_n)$, where $\gamma_n(t)$ is the nth discrete approximation of t associated with the approximate solution f_n. By Theorem 3.2 and Proposition 3.1, $\{h_n\}$ and $\{g_n\}$ are relatively weakly compact sets in $L^1((0,T) \times \Omega \times \mathbb{R}^3)$. Thus, by Lemma 2.1, for each $\psi \in L^\infty((0,T) \times \Omega \times \mathbb{R}^3)$ the set $\{\int_{\mathbb{R}^3} \psi h_n dv\}$ is relatively compact in $L^1((0,T) \times \Omega)$. Also, by properties of approximate solutions the set $\{\int_{\mathbb{R}^3} \psi f_n dv\}$ has compact closure in $L^1((0,T) \times \Omega)$. Since $\|f_n\|_2 \leq M < \infty$ for $n \geq 1$, this last assertion is also true for each measurable ψ with $|\psi(x,v)| \leq c(1+v^2)^k$, $k < 1$.

Now, the energy saturation condition, properties of approximate solutions and the Gronwall lemma lead to the limit $\int_{\mathbb{R}^3} v^2 f_{n_i}\, dv \underset{i\to\infty}{\to} \int_{\mathbb{R}^3} v^2 f\, dv$ a.e. in t and x. Then the strong continuity of $U(t)$ in $L^1(\Omega \times \mathbb{R}^3)$ implies that the set $\{k_n\}$ is relatively compact in $L^1((0,T); L^1(\Omega \times \mathbb{R}^3))$, where $k_n(t) = \int_0^t U(t-s)J(f_n(\gamma_n(s))ds$ for $t \in [0,T]$, and the set $\{f_n\}$ is relatively compact in $L^1((0,T); L^1(\Omega \times \mathbb{R}^3))$. Therefore, after passing to a subsequence if necessary, f_n converges to f in $L^1((0,T); L^1(\Omega \times \mathbb{R}^3))$ with $f(t) \in D$ a.e. in t. Finally, by the Dominated Convergence Theorem for the Bochner integral, continuity of P, and conservation of energy, we can show that f is a mild solution to the B.G.K. equation in $L^1_{1+v^2}(\Omega \times \mathbb{R}^3)$, and the integral in (3.2) can be taken as the Riemann integral in $L^1_{1+v^2}(\Omega \times \mathbb{R}^3)$. Additional details on proof of existence of solutions will be given elsewhere.

The energy saturation condition can be removed if one of the higher moments of f_n is bounded uniformly in n, i.e.,

$$\int_0^a \int_\Omega \int_{\mathbb{R}^3} |v|^k f_n dv dx dt \equiv M_n^k < c < \infty \tag{3.3}$$

for some fixed $k > 2$. Existence of solutions can also be obtained in an Orlicz space for a strict velocity cutoff model of the B.G.K. collision operator.

In closing, we note that the strong convergence of the approximate solutions $\{f_n\}$ to a mild solution contrasts with the weak sense of convergence of the sequence of approximate solutions constructed by DiPerna-Lions in the case of the Boltzmann equation. Actually, their notion of a solution is much weaker than the notion of mild solution in $L^1_{1+v^2}(\Omega \times \mathbb{R}^3)$ obtained here for the B.G.K. equation. Finally, energy conservation can be demonstrated for the solution of the B.G.K. equation, but is not yet known for the DiPerna-Lions solution of the Boltzmann equation.

NOTE ADDED IN PROOF: In a private communication, B. Perthame has provided a short proof that the moment estimate (3.3) is satisfied with $k = 3$ if the initial datum $f_0(x, v)$ satisfies such a bound. Therefore, for such initial conditions the energy saturation condition is not needed.

ACKNOWLEDGEMENT: This work was supported in part by DOE Grant No. DE FG05 87ER 25033 and NSF Grant No. DMS 8701050.

REFERENCES

[1] T. Carleman, "Sur la théorie de l'equation intégrodifférentielle de Boltzmann," Acta Mathematica **60**, 91-146 (1932).

[2] F. Golse, P.L. Lions, B. Perthame, R. Sentis, "Regularity of the moments of the solutions of a transport equation," J. Funct. Analysis **76**, 110-125 (1987).

[3] R.L. DiPerna, P.L. Lions, "On the Cauchy problem for the Boltzmann equation: global existence and weak stability," preprint (1988).

[4] L. Arkeryd, C. Cercignani, "On convergence of solutions of the Enskog equation to solutions of the Boltzmann equation," preprint (1988).

[5] J. Polewczak, "Global existence in L^1 for the modified nonlinear Enskog equation in $\mathbb{R}^3$," preprint (1988).

[6] L. Boltzmann, "Further studies on the thermal equilibrium of gas molecules," Sitz. Wien. Akad. Wiss. **66**, 275-370 (1872).

[7] D. Enskog, "Kinetische Theorie," Kungl. Svenska Vetenskaps Akademiens Handl. **63** (1921); English transl. in S. Brush, *Kinetic Theory*, vol. 3, Pergamon, New York (1972).

[8] P. Résibois, "H-theorem for the (modified) nonlinear Enskog equation," J. Stat. Phys. **19**, 593-609 (1978).

[9] P.L. Bhatnagar, E.P. Gross, M. Krook, "A model for collision processes in gases. I. Small amplitude processes in charged and neutral one-component systems," Phys. Rev. **94**, 511-525 (1954).

[10] P. Welander, "On the temperature jump in a rarefied gas," Ark. Fys. **7**, 507 (1954).

[11] J. Polewczak, W. Greenberg, "Some remarks about continuity properties of local maxwellians and an existence theorem for the B.G.K. model of the Boltzmann equation," J. Stat. Phys. **33**, 307-316 (1983).

William Greenberg

Dept. of Mathematics and Center for
Transport Theory and Math. Physics
Virginia Tech., Blacksburg, VA 24061
U.S.A.

Jacek Polewczak

Dept. of Mathematics and Center for
Transport Theory and Math. Physics
Virginia Tech., Blacksburg, VA 24061
U.S.A.

Operator Theory:
Advances and Applications, Vol. 41

FACTORIZATION OF NONLINEAR SYSTEMS *

J. William Helton

Dedicated to Israel Gohberg on his sixtieth birthday

A basic result of Bart, Gohberg, Kaashoek, and Van Dooren parameterizes factorizations of the transfer function of a linear system in terms of the state space equations of the system. This paper extends their result to nonlinear systems.

§1. INTRODUCTION

When one system S_1 has output feeding into another system S_2 this produces another system called the *product* of two systems. The reverse procedure of factoring a given system into simpler systems is a classic tool of linear circuits, filters, and now control; indeed, it is one of the most powerful tools in theoretical engineering.

Even in the nonlinear case it has been recently shown that when factorizations of various types are possible stability results [**Hm**], [**AD**], and H^∞ control type results [**BH**] are possible. These factorizations often are not *minimal* in the sense that the dimension of the state spaces of the factors of the system sum to the dimension of the state space of the system.

This paper analyzes the existence, construction, uniqueness and properties of minimal factorizations for nonlinear systems. While factorization in terms of frequency response functions goes back at least to Cauer in the 1930's, the classic state space result is due to Bart, Gohberg, Kaashoek and Van Dooren (cf. [**BGK**]). For a linear $ABCD$ system with D invertible it sets minimal factorizations into direct correspondence with

* Research on this paper was supported in part by the National Science Foundation and the Air Force Office of Scientific Research.

a pair $\mathcal{S}$, $\mathcal{S}^{\times}$ of disjoint subspaces of the state space X which actually decompose X as $X = \mathcal{S} + \mathcal{S}^{\times}$ and which have the distinctive properties that

(1.1) $\mathcal{S}$ is invariant under A.

(1.1$^{\times}$) $\mathcal{S}^{\times}$ is invariant under $A^{\times} \triangleq A - BD^{-1}C$.

In Sections 2 and 3 we generalize this to nonlinear systems. The nonlinear state space X will be a manifold. While $\mathcal{S}$, $\mathcal{S}^{\times}$ generalize to a particular set of coordinates (X_1, X_2) on the manifold $x^1 \in \mathcal{S}$, $x^2 \in \mathcal{S}^{\times}$. These must be invariant in a natural sense which we describe explicitly.

In Section 4 we show that generically invariant coordinates are uniquely determined (and very likely overdetermined) by very little information—namely the linearization at an equilibrium point. Clearly, if systems S_1, S_2 have the same equilibrium points e_1, e_2 their product has an equilibrium point e. Also the product L of linearizations L_1, L_2 about e_1, e_2 equals the linearization of the product S_1, S_2 about e_1, e_2. Our results on $\mathcal{S}$, $\mathcal{S}^{\times}$ decompositions lead us to conjecture that

> *generically at most one 'nontrivial' smooth minimal nonlinear factorization corresponds to each factorization of the linearized system, L.*

Indeed for a particular factorization of L to come from nonlinear factors of S requires that a compatibility condition be satisfied. This article actually proves this except for one gap (Conjecture 4.5). Various experts in nonlinear ordinary differential equations assured us that Conjecture 4.5 is true; but they do not know a precise reference for it in the literature. In this paper factorization typically means minimal factorization.

The author wishes to thank Chris Byrnes for helpful discussions and Asher Ben-Artzi for comments on the manuscript. The proof presented here of Lemma 2.4 was given by Ben Artzi. Neola Crimmins cheerfully typed the manuscript from a rugged original.

§2. FACTORIZATION AND $\hat{F}$, $\hat{G}$ INVARIANT COORDINATES; DISCRETE TIME

This section extends the classic result of Bart, Gohberg, Kaashoek, and Van Dooren . on state space description of factorization mentioned in the Introduction. First we treat discrete time systems; second continuous time systems.

We start with a system $\hat{F}$, $\hat{G}$ which is the product of systems f, g and F, G

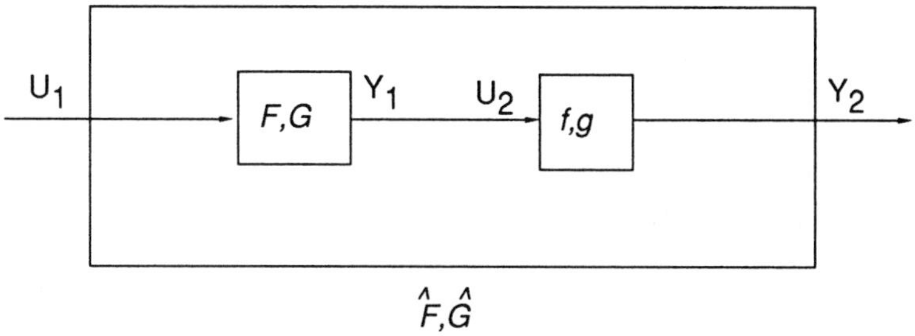

Figure 2.1.

and give a list of properties which $\hat{F}$, $\hat{G}$ must have as a result of the factorization. Later we go backwards.

Our notation is:

X_1, X_2 are finite dimensional manifolds.

U_1, U_2 are finite dimensional vector spaces.

Y_1, Y_2 are finite dimensional vector spaces.

$$F : X_1 \times U_1 \longrightarrow X_1, \qquad G : X_1 \times U_1 \longrightarrow Y_1.$$

$$f : X_2 \times U_2 \longrightarrow X_2 \qquad g : X_2 \times U_2 \longrightarrow Y_2.$$

We always assume special properties of our systems. First we assume $Y_1 = U_2$. We also assume throughout that $\dim U_1 = \dim Y_1 = \dim U_2 = \dim Y_2$, and that for each $x^2 \in X_1$ the map $G(x^1, \cdot) : U_1 \to Y_1$ is invertible with inverse denoted by $G^I(x^1, \cdot)$. Of course

$$G(x^1, G^I(x^1, y)) = y \qquad G^I(x^1, G(x^1, u)) = y.$$

Likewise for each $x^2 \in X_2$, we have $g(x^2, \cdot)$ invertible with inverse denoted $g^I(x^2, \cdot)$. Call such a system F, G an invertible system. For a system F, G denote by $F^\times : X \times Y_1 \to X_1$ the map

$$F^\times(x^1, y) = F((x^1, G^I(x^1, y))$$

for $x^1 \in X_1$, $y \in Y_1$. The key property of $F^\times$ is

LEMMA 2.1. *The system $F^\times$, G^I is the inverse of F, G.*

PROOF. By definition of G^I

$$y = G(x, u) \qquad u = G^I(x, y).$$

Therefore

$$x_{n+1} = F(x_n, u_n) \qquad y_n = G(x_n, u_n)$$

is equivalent to

$$x_{n+1} = F(x_n, G^I(x_n, y_n)) = F^\times(x_n, y_n)$$
$$u_n = G^I(x_n, y_n).$$

We say that e^1 (resp. e^2) is an equilibrium point for F, G (resp. f, g) provided that

$$F(e^1, 0) = e^1,$$

respectively,

$$f(e^2, G(e', 0)) = e^2.$$

Naturally our system dynamics are:

(2.1)
$$x^1_{n+1} = F(x^1_n, u^1_n) \quad x^n_{n+1} = f(x^2_n, u^2_n)$$
$$y^1_n = G(x^1_n, u^1_n) \qquad y^2_n = g(x^2_n, u^2_n)$$

and if one system is plugged into the other, that is $y_n^1 = u_n^2$, we have that the *product system* acts on state manifold

$$X = \{(x^1, x^2) : x^j \in X_j\} \tag{2.2}$$

and is $\hat{F}$, $\hat{G}$ given by

$$\hat{F}((x^1, x^2), u) = (F(x^1, u), f(x^2, G(x^1, u))) \tag{2.3a}$$

$$\hat{G}((x^1, x^2), u) = g(x^2, G(x^1, u)) \tag{2.3b}$$

for $(x^1, x^2) \in X$ and input $u \in U_1$. Note that invertibility of systems (2.1) implies invertibility of the system $\hat{F}$, $\hat{G}$.

We now write down properties of the system $\hat{F}$, $\hat{G}$ defined by (2.3) which arise because it is a product system, but which do not mention F, G, f, g explicitly. Indeed

$$\hat{F} : X \times U_1 \longrightarrow X \qquad\qquad \hat{G} : X \times U_2 \longrightarrow Y_2.$$

The state space X has a coordinate pair[1] X_1, X_2 so that $X = \{(x^1, x^2) : x^j \in X_j\}$ with the properties for $(x^1, x^2) \in X$, $u \in U_1$

(2.4) $\pi^1 \hat{F}((x^1, x^2), u)$ is independent of x^2.

$(2.4^\times)^2$ $\pi^2 \hat{F}^\times((x^1, x^2), y)$ is independent of x^1.

(2.5a) $\hat{G}^I((a, c), \hat{G}((b, c), u))$ is independent of c.

(2.5b) $\hat{G}((a, c), \hat{G}^I((a, d), y))$ is independent of a.

Here π^j denotes the projection $\pi^j(x^1, x^2) = x^j$.

If e_1, e_2 are equilibrium points for the factors F, G and f, g, then

(2.6) $e = (e^1, e^2)$ is an equilibrium point for $\hat{F}$, $\hat{G}$ and consequently

$$\hat{F}^\times(e, \hat{G}(e, 0)) = \hat{F}(e, 0) = e.$$

---

[1] By coordinate pair we mean direct product; there is a C^∞ diffeomorphism $\varphi : X_1 \times X_2 \to X$. The (important) case where φ is not $1 - 1$ is postponed.

[2] Actually the stronger statement: For any $a, b \in X_2$ the quantity $\pi^2 \hat{F}((x^1, a), G^I((x^1, b), y))$ is independent of x^1 holds. When $a = b$ this gives $(2.4^\times)$.

Here π^j denotes the projection $\pi^j(x^1, x^2) = x^j$.

Now we consolidate this abstraction. Suppose we are: Given a system $\hat{F}, \hat{G}$, with state manifold X, and input, output spaces U, Y. We say that (X_1, X_2) is a *product coordinate pair* for the system $\hat{F}, \hat{G}$ provided that (2.4) and (2.5) are satisfied.

THEOREM 2.1. *The invertible system $\hat{F}, \hat{G}$ has factors F, G and f, g which are invertible systems if X has a product pair. Each product pair (X_1, X_2) determines the dynamics F and $f^\times$ of the factors uniquely.*

Conversely, suppose $\hat{F}, \hat{G}$ has state manifold X of dimension $\hat{N}$. Formulas for factors are given in Recipe 2.3. The factors have an equilibrium point e^1, e^2 if $\hat{F}, \hat{G}$ has an equilibrium point e in the sense of (2.6). The converse is true in a local sense to be discussed in Theorem 2.4.

The key product pair conditions (2.4) and (2.4$^\times$) are easy to visualize. Fix $b \in X_2$, $a \in X_1$; we think of (X_1, b) and (a, X_2) as leaves of the x-coordinate foliations. Condition (2.4) says precisely that for any $u \in U_1$

(2.4a) $\quad \hat{F}((a, X_2), u) = (\tilde{a}, \tilde{X}_2) \subset (\tilde{a}, X_2)$

where for each fixed a, one gets a fixed $\tilde{a} \in X_1$ and $\tilde{X}_2$ a subset of X_2. Likewise (2.4$^\times$) is equivalent to

$$\hat{F}^\times((X_1, b), y) \subset (X_1, \tilde{b}) \tag{2.4a$^\times$}$$

for any b, y and some $\tilde{b}$. Note that when $\hat{F}, \hat{G}$ is a linear system and X_1, X_2 are linear spanning nonintersecting subspaces, then condition (2.4a) says

$$\begin{aligned} \hat{F}((a, X_2), u) &= \hat{A}(a, 0) + \hat{A}(0, X_2) + \hat{B}u \\ &\subset (\tilde{a}, 0) + (0, X_2). \end{aligned}$$

So $\hat{A}(a, 0) + \hat{B}u = (\hat{a}, \tilde{x}_2)$ and $\hat{A}(0, X_2) \subset (0, X_2)$. That is, $(0, X_2)$ is an invariant subspace for $\hat{A}$. Likewise $(X_1, 0)$ is invariant for $A^\times$ so conditions (2.4) generalize those of **[BGK]** and **[vD]**.

LEMMA 2.2. $\hat{G}^I((x^1, x^2), y) = G^I(x^1, g^I(x^2, y))$.

PROOF. $y = \hat{G}((x^1, x^2), u) \triangleq g(x^2, G(x^1, u))$.

$$g^I(x^2, \hat{G}((x^1, x^2), u) = G(x^1, u)$$

$$G^I(x^1, g^I(x^2, \hat{G}((x^1, x^2), u)) = u.$$

Since $y = \hat{G}((x^1, x^2), u)$ we have $G^I(x^1, g^I(x^2, y)) = u$ which proves the lemma.

Theorem 2.1 is proved by the following construction:

RECIPE 2.3. Given $\hat{F}, \hat{G}$ and a product coordinate pair (X_1, X_2) for it. First[3] select $c_1 \in X^1$ and φ a diffeomorphism from U^2 onto U^2. All factors of $\hat{F}, \hat{G}$ are given by

$$g(x^2, u) = \hat{G}((c^1, x^2), \varphi^{-1}(u)). \tag{2.7}$$

$$G(x^1, u) = \varphi \circ \hat{G}^I((c^1, c^2), \hat{G}((x^1, c^2), u)) \text{ independent of } c^2. \tag{2.8}$$

$$F(x^1, u) = \pi^1 \hat{F}((x^1, \beta), u) \text{ independent of } \beta. \tag{2.9}$$

$$\begin{aligned} f((x^2, u^2)) &= \pi^2\, \hat{F}((\xi, x^2), \hat{G}^I((\xi, x^2), g(x^2, u^2)) \text{ independent of } \xi \\ &\triangleq \pi^2\, \hat{F}^\times((\xi, x^2), g(x^2, u^2)) \end{aligned} \tag{2.10}$$

---

[3] There appears to be an asymmetry between c^1 and c^2 in (2.7) and (2.8). An alternative formula for g and G dispels this illusion.

First *select* $c^1 \in X^1$ $c^2 \in X^2$. Next *select* two maps

$$\delta^2 : U^2 = Y^1 \to Y^2 \qquad\qquad \delta^1 : U^2 = Y^1 \to U^1$$

which factor $\hat{G}((c^1, c^2),)$ in the sense

$$\delta^2 \circ \delta^{1^{-1}}(u) = \hat{G}((c^1, c^2), u) \text{ for all } u \in U_1$$

Define

(2.7′) $g^I(x^2, y) = \delta^{2^{-1}} \circ \hat{G}((\tau, c^2), \hat{G}^I((\tau, x^2), y))$ independent of τ.

(2.8′) $G((x^1, u)) = \delta^{1^{-1}} \circ \hat{G}^I((c^1, \rho), \hat{G}(x^1, \rho, u))$ independent of ρ.

The independences follow from (2.5a), (2.4) and (2.4$^\times$), respectively.

REMARK 2.3. A particular coordinate pair (X_1, X_2) determines F uniquely from $\hat{F}$, via equation (2.9). Likewise (X_1, X_2) determines $f^\times(x^2, y^2) = \pi^2\,\hat{F}^\times((\xi, x^2), y^2)$ uniquely.

PROOF OF THEOREM 2.1. We must prove that composition of these systems yields $\hat{F}, \hat{G}$, that is

$$\pi_1\,\hat{F}((x^1, x^2), u) = F(x^1, u), \tag{2.11}$$

$$\pi_2\,\hat{F}((x^1, x^2) = f(x^2, G(x^1, u)), \tag{2.12}$$

$$\hat{G}((x^1, x^2), u) = g(x^2, G(x^1, u)), \tag{2.13}$$

and that these are the only possibilities.

Equation (2.13) and non-uniqueness of factors is handled by

LEMMA 2.4. *The equation*

$$\hat{G}((x^1, x^2), u) = g(x^2, G(x^1, u)), \tag{2.14}$$

admits a solution (g, G) *if and only if* (2.5a) *holds, then one solution is given by*

$$g = g_1(x^2, u) = \hat{G}((c^1, x^2), u) \quad G = G_1(x^1, u) = \hat{G}^I((c^1, c^2), \hat{G}((x^1, c^2), u), \tag{2.15}$$

where $c^1 \in X^1$ *and* $c^2 \in X^2$ *are arbitrary. The most general solution of equality* (2.14) *is given by*

$$g(x^2, u) = g_1(x^2, \varphi^{-1}(u)) \quad G(x^1, u) = \varphi(G_1(x^1, u)) \tag{2.16}$$

where $\varphi : U_2 \to U_2$ *is an arbitrary diffeomorphism.*

PROOF. Assume that (2.14) holds, and denote $z = \hat{G}^I((a, c), \hat{G}((b, c), u))$. Then $\hat{G}((a, c), z) = \hat{G}((b, c), u)$. By (2.14), this leads to $g(c, G(a, z)) = g(c, G(b, u))$.

Applying $g^1(c,\cdot)$ on both sides of this equality, we obtain $G(a,z) = G(b,a)$. Thus $z = G^I(a, G(b,u))$ and therefore z is independent of c. Hence (2.5a) holds.

Assume now that (2.5a) holds. Then

$$\hat{G}^I((c^1,c^2), \hat{G}((x^1,c^2),u) = \hat{G}^I((c^1,x^2), \hat{G}((x^1,x^2),u)$$

for every $(x^1,x^2,u) \in x_1 \times x_2 \times u_1$. We now apply $\hat{G}((c^1,x^2),\cdot)$ on both sides of this equality and obtain

$$\hat{G}((c^1,x^2), \hat{G}^I((c^1,c^2), \hat{G}((x^1,c^2),u))) = \hat{G}((c^1,x^2), \hat{G}^I((c^1,x^2), \hat{G}((x^1,x^2),u))).$$

However the left hand side of this equality is equal to $g_1(x^2, G_1(x^1,u))$, while the right hand side is equal to $\hat{G}((x^1,x^2),u)$. Therefore (2.15) gives a solution of (1).

Let $\varphi : U_2 \to U_2$ be a diffeomorphism and let g, G be given by (3). Clearly, using the fact that (2) gives a solution, we have

$$g(x^2, G(x^1,u)) = g_1(x^2, \varphi^{-1}(\varphi(G_1(x^1,u)))) = g_1(x^2, G_1(x^1,u)) = \hat{G}.$$

Finally, assume that (g,G) and (g_1,G_1) are arbitrary solutions of (1).

$$g(x^2, G(x^1,u)) = g_1(x^2, G_1(x^1,u)). \tag{2.17}$$

Therefore $G(x^1,u) = g^I(x^2, g_1(x^2, G_1(x^1,u)))$. Fixing $x^2 = c^2$, and denoting $\varphi(\cdot) = g^I(c^2, g_1(c^2,\cdot))$, we obtain

$$G(x^1,u) = \varphi(G_1(x^1,u)), \tag{2.18}$$

where $\varphi : U_2 \to U_2$ is a diffeomorphism. Inserting the last equality in (2.17) we obtain

$$g(x^2, \varphi(G_1(x^1,u)) = g_1(x^2, G_1(x^1,u)).$$

Let $y \in U_2$, there is a u such that $y = \varphi(G_1(x^1,u))$, and therefore

$$g(x^2,y) = g_1(x^2, G_1(x^1,u)) = g_1(x^2, \varphi^{-1}(y)). \tag{2.19}$$

The equalities (2.16) follow from (2.18) and (2.19). ∎

We prove (2.12):

$$\begin{aligned} f(x^2, G(x^1, u)) &= \pi^2 \hat{F}((\xi, x^2), \hat{G}^I((\xi, x^2), g(x^2, G(x^1, u)) \\ &= \pi^2 \hat{F}((\xi, x^2), \hat{G}^I((\xi x^2), \hat{G}((x^1, x^2), u)) \end{aligned}$$

first by definition (2.10) and then by (2.13). Independence of ξ implies we may take $\xi = x^1$ to get

$$f(x^2, G(x^1, u)) = \pi^2 \hat{F}((x^1, x^2), u)$$

as desired.

Suppose that e is an equilibrium point of $\hat{F}$, that is, $\hat{F}(e, 0) = e$; let $e^j = \pi^j e$. Then by (2.9)

$$F(e^1, 0) = \pi^1 \hat{F}((e^1, \beta), 0) = \pi^1 \hat{F}(e, 0) = 0$$

since we may take $\beta = e^2$. By (2.10) and then (2.6),

$$\begin{aligned} f(e^2, G(e^1, 0)) &= \pi^2 \hat{F}(e, \hat{G}^I((\xi, e^2), g(e^2, G(e^1, 0))) \\ &= \pi^2 \hat{F}(\xi, e^2), \hat{G}^I(e, g(e^2, G(e^1, 0))) \\ &= \pi^2 \hat{F}(e, \hat{G}^I(e, \hat{G}(e, 0))) = \pi^2 \hat{F}^x(e, \hat{G}(e, 0)) \\ & \pi^2 e = e^2. \end{aligned}$$

Now we change directions and treat the converse of Theorem 2.1. There are generally speaking two natural notions of factoring a system $\hat{F}, \hat{G}$ and since we are operating at a greater level of detail, we will give three.

(F1) $\hat{F}, \hat{G}$ is the product of F, G and f, g as in (2.3).

(F2) The transfer operators[4] for the systems satisfy $\hat{\Phi} = \phi \circ \Phi$.

(F3) The product of systems F, G and f, g as in (2.3) which we now denote by $\tilde{F}, \tilde{G}$ is not minimal and $\hat{F}, \hat{G}$ is a restriction of it, more precisely there is a map $\psi : X_1 \times X_2 \to X$ which is smooth, onto and whose differential has constant rank, such that

---

[4]The transfer operator ϕ of a system (f, g) with equilibrium point e is a mapping from the space ℓ^+ of sequences $\{(u_0, u_1 \ldots\}$ to ℓ^+ defined by $\phi((u_0, u_1, \ldots)) = (y_0, y_1, \ldots)$ where the y's are produced by recursion (2.1) defining system f, g when $x_0 = e$ and the u_j are fed to the system. The definitions of $\hat{\Phi}$ and Φ are the same.

$$
(2.20) \qquad \begin{aligned} \hat{F}(\psi(x), u) &= \psi \circ \tilde{F}(x, u) && \forall x \in X_1 \times X_2 \\ \hat{G}(\psi(x), u) &= \tilde{G}(x, u) && \forall u \in U. \end{aligned}
$$

These definitions are related. In particular if systems $\hat{F}, \hat{G}$ and f, g are well behaved, controllable, and reachable in a strong sense one expects that (F2) and (F3) are equivalent at least locally. Since this type of equivalence is not the subject of this paper we henceforth treat only factorization of type (F1) and of type (F3). The reader may intuitively bear in mind that all three types are closely related.

Suppose that $\hat{F}, \hat{G}$ and X factors in sense (F3). We know that there is a system $\tilde{F}, \tilde{G}$ on state manifold $\tilde{X}_1 \times \tilde{X}_2$ and a smooth map $\psi : \tilde{X}_1 \times \tilde{X}_2 \to X$ satisfying (2.20). Also we know that $\tilde{F}, \tilde{G}$ is built from factors as in (F1) and so $\tilde{X}_1, \tilde{X}_2$ is a coordinate pair for $\tilde{F}, \tilde{G}$. Suppose that X and $\tilde{X}_1 \times \tilde{X}_2$ have the same dimension and ψ is nondegenerate, onto. Then the map ψ is a diffeomorphism locally on $\tilde{X}_1 \times \tilde{X}_2$, while it may not be one globally (e.g. if it is a covering map). Thus the fact that $\hat{F}, \hat{G}$ factors shows up in an awkward way on its state manifold, and we leave sorting it out as an open question.

However, locally many situations are well behaved. Intuitively if the discrete system comes from a continuous time system and the discretizing time step is small enough, then locally X must have product pairs for which (2.3) holds at least for small u.

THEOREM 2.4. Suppose that $\hat{F}, \hat{G}$ factors in sense (F3) and that the dimensions of the state manifolds of the factors sum to $\hat{N}$ the dimension of the state manifold of $\hat{F}, \hat{G}$. Given $x \in X$, there is a neighborhood $\mathcal{O}$ of x which possesses a product pair $\tilde{X}_1 \times \tilde{X}_2$. This is in the sense that $\mathcal{O} \subset \tilde{X}_1 \times \tilde{X}_2$, and if $\mathcal{O}_0$ is open inside $\mathcal{O}$ and $0 \in U_0$ (resp. $y \in Y_0$) is a neighborhood in U (resp. Y) with the property that $\hat{F}$, $\pi_1 \hat{F}$, $\pi_2 \hat{F}^{\times}$ map $\mathcal{O}_0 \times U_0$ into $\mathcal{O}$, then (2.,4), (2.4$^{\times}$) and (2.5) hold for all (x_1, x_2), (a, c), (b, c) $(a, d) \in \mathcal{O}_0$ and $u \in U_0$ and $y \in Y_0$.

PROOF OF THEOREM 2.4. First we treat the global case. Suppose that $\hat{F}, \hat{G}$ factors in the sense (F1). We prove that factors F, G and f, g produce an $\tilde{F}, \tilde{G}$ with properties (2.4), (2.5) and (2.6). Each equation takes only a few lines.

(2.4) is obvious.

$$\pi^2 \hat{F}^\times((x^1,a),y) = \pi_2 \hat{F}((x^1,a),\hat{G}^I((x^1,b),y))$$

$$(2.4^\times) \qquad = f(a,(G(x^1,G^I(x^1,g^I(b,y))) = f(a,g^I(b,y))$$

which is independent of x^1.

$$(2.5a) \qquad \hat{G}^I((a,c),\hat{G}((b,c),u))) = G^I(a,g^I(c,g(c,G(b,u)) = G^I(a,G(b,u))$$

which is independent of c.

$$(2.5b) \qquad \hat{G}((a,c),\hat{G}^I((a,d),y)) = g(c,G(a,G^I((a,g^I(d,y)) = g(c,g^I(d,y))$$

which is independent of a.

Note the equilibrium point condition (2.6) is true because

$$\hat{F}^\times(e,\hat{G}(e,0)) = \hat{F}(e,\hat{G}^I(e,\hat{G}(e,0)) = \hat{F}(e,0) = e.$$

The fact that $\hat{N} = N + n$ forces φ to be a local diffeomorphism of the global case we just treated onto the state manifold of the given system. It is straightforward to obtain the theorem from this.

§3. CONTINUOUS TIME

Our notation is

$$(3.1) \qquad \frac{dx^1}{dt} = F(x^1,u^1) \qquad y^1 = G(x^1,u^1)$$
$$\frac{dx^2}{dt} = f(x^2,u^2) \qquad y^2 = g(x^2,u^2)$$

with X_1, X_2 manifolds U_1, $U_2 = Y_1$, Y_2 subspaces having $\dim U_1 = \dim Y_2$

$$(3.2) \qquad F(x^1,u^1) \in T_{x^1}X_1 \qquad G\colon X_1 \times U_1 \to Y_1$$
$$f(x^2,u^2) \in T_{x^2}X_2 \qquad g\colon X_2 \times U_2 \to Y_2.$$

We always assume G, g are both invertible as in §2a. Equilibrium points e^1 and e^2 are defined by

$$F(e^1,0) = 0 \qquad f(e^2,G(e^1,0)) = 0.$$

The composition in Fig. 2.1 acts on state manifolds $X = X_1 \times X_2$, input space U_1, output space Y_2 and has state operators:

$$\hat{F}((x^1, x^2), u) = (F(x^1, u), f(x^2, G(x^1, u)) \tag{3.a}$$

$$\hat{G}((x^1, x^2), u) = g(x^2, G(x^1, u)). \tag{3.3b}$$

These equations are similar to (2.3). Indeed (3.3b) and (2.3b) are exactly the same while (3.3a) and (2.3a) differ only in that $A\hat{F}, F, f$ map one space into a different space.

To characterize (3.3) without reference to F, G, f, g we note that we have a coordinate pair $X = X_1 \times X_2$ of manifold X_1, X_2 for X and the corresponding tangent space $T_x X = T_{x^1} X_1 + T_{x^2} X_2$ decomposition at each $x = (x^1, x^2)$. Let π_x^j denote the projection of $T_x X$ onto $T_x X_j$ along $T_x X_k$, $k \neq j$. With this notation we get a characterization of factorizations which looks about like the discrete time characterization (2.4) and (2.5). We say that (X_1, X_2) is a *product coordinate pair* for $\hat{F}, \hat{G}$ provided that

(3.4) $\pi_x^1 \hat{F}(x, u)$ is independent of x^2; here $x = (x^1, x^2)$.

(3.4$^\times$) $\pi_x^2 \hat{F}^\times(x, y)$ is independent of x^1.

(3.5a) $\hat{G}^I((a, c), \hat{G}((b, c), u))$ is independent of c.

(3.5b) $\hat{G}((a, c), \hat{G}^I((a, d), y))$ is independent of a.

In addition one has

(3.6). If the factors have equilibrium points e_1, e_2, then $e = (e_1, e_2)$ satisfies $\hat{F}(e, 0) = 0$ and $\hat{F}^\times(e, \hat{G}(e, 0)) = 0$.

Note that not only are these equations similar to those for discrrete time, but Recipe 2.3 for factoring a given system in discrete time makes perfect sense in continuous time. Of course π_x^j replaces π^j. Indeed for continuous time one can prove

THEOREM 3.1. *A continuous time system $\hat{F}, \hat{G}$ has a factorization F, G and f, g if it has a product coordinate pair (X_1, X_2) with corresponding projections π_x^j. Recipe 2.3 again produces the factors. The factors have an equilibrium pair if (3.6) is satisfied.*

The converse to Theorem 3.1 is less encumbered in the continuous time case than it is in the discrete time case, since neighborhoods are "preserved by infinitesimal transformations."

THEOREM 3.2. *Suppose $\hat{F}, \hat{G}$ with state manifold X factors with systems where state manifolds have dimensions n and N adding to $\hat{N}$, the dimension of X. Then*

if $x \in X$ there is a neighborhood $\mathcal{O} \subset X$ of x which has a coordinate pair $\tilde{X}_1 \times \tilde{X}_2$ in the sense that $\tilde{X}_1 \times \tilde{X}_2 \supset \mathcal{O}$ and for all (x_1, x_2), (a, c), (b, c) and (a, d) in $\mathcal{O}$ equations (3.4), (3.4$^\times$) and (3.5) hold.

The proofs are so close to the discrete time proof in §2 that we omit it.

We have just seen that minimal factorization of $\hat{F}, \hat{G}$ implied the existence of a product coordinate pair (x_1, x_2) for $\hat{F}, \hat{G}$. Now we convert these conditions to infinitesimal form. This converts the problem of finding product coordinate pairs to a problem of solving a certain set of P.D.E.

Suppose we have $\hat{F}, \hat{G}$ and product pair (X_1, X_2). The invariance conditions defining product pair can be restated simply in terms of the "invariant distributions" common to nonlinear control. Namely, let Δ^1 denote the distribution defined at $x \in X$ by

$$\Delta_x^2 = T_x X_2.$$

This indeed associates a subspace of $T_x X$ to x. Since Δ^2 comes from a submanifold of X, we know that Δ^2 is involutive, that is,

(3.7) $$[\Delta^2, \Delta^2] \subset \Delta^2$$

c.f. [**I**, Ch. 1]. However, the invariance (3.3) of X^2 $\hat{F}$ says precisely that

(3.8) $$[\hat{F}\Delta^2] \subset \Delta^2.$$

This amounts to a differential equation which $\hat{F}$ and Δ^2 must satisfy. Likewise, define Δ^1 by

(3.9) $$\Delta_x^1 = T_x X_1.$$

then

(3.10) $$[\hat{F}^\times, \Delta^1] \subset \Delta^1.$$

Also

(3.11) $$[\Delta^1, \Delta^1] \subset \Delta^1.$$

We conclude (primarily from [I]) that the existence of distributions Δ^1, Δ^2 satisfying (3.7), (3.8), (3.10) and (3.11), (3.5) is necessary and sufficient for $\hat{F}, \hat{G}$ to have a product coordinate pair. Thanks are due to Chris Byrnes for for help with this.

§4. UNIQUENESS OF FACTORIZATION

Since product coordinate pairs correspond to factorization and vice versas, uniqueness of coordinate pairs is the issue here. A *leaf* of a coordinate pair $X_1 \times X_2$ for X means a set of the form (a, X_2) or (X_1, b); we call these vertical and horizontal leaves, respectively. If u is an input function, that is $u: [0,T] \to U$ define $\hat{\Phi}(\ ,u)$ to be the map on state space X which implement the dynamics of the differential equation $\frac{dx}{dt}(t) = \hat{F}(x(t), u(t))$. Namely, initialize this differential equation at x_0; then at time T the solution to the differential equation takes the value $x(T)$; define Φ by $\hat{\Phi}(x_0, u) = x(T)$. Similarly Φ_1 and Φ_2 implements the dynamics for F and f, respectively. In this article we shall say that a system is *reachable* if each state $\tilde{x}$ can be reached from any other state x in finite time when the correct input function drives the system. That is $\Phi(x,u) = \tilde{x}$. It is openly reachable if in addition $\Phi(\ ,u)$ is an invertible map real x. A key fact is the simple observation.

THEOREM 4.1. *Given a continuous time system as in Theorem 3.1 which is openly reachable, any product coordinate pair for the system is uniquely determined by one horizontal and one vertical leaf. This holds for local coordinate pairs as well under a suitably strengthened notion of reachability.*

THEOREM 4.2. *If a coordinate pair gives rise to a reachable factor F, G, then all factors F, G arising from it are reachable. Likewise for f, g.*

PROOF. Reachability is determined only by F and not influenced by G. Formula (2.9) of Recipe 2.3 for F shows that F is completely determined by the coordinate pair and $\hat{F}$. Likewise for f via formula (2.10).

PROOF OF THEOREM 4.1. We start with two leaves H and V. Since one is horizontal and one vertical, they intersect in some point p. Since $\hat{F}, \hat{G}$ is reachable any point $x \in X$ has the form $\hat{\Phi}(p, u)$ for some input function $u: [0,T] \to U$. Suppose there is another coordinate pair $\tilde{X}_1 \times \tilde{X}_2$ which contains H and V; let $\tilde{L}_x$ denote the vertical leaf through x. Now $\hat{\Phi}(\ ,u)$ maps each vertical leaf to a subset of a vertical leaf (this is the content of Theorem 3.1). Thus $\hat{\Phi}(V, u)$ is contained in $\tilde{L}_x$ and also in the vertical leaf L_x of $X_1 \times X_2$ through x. Leaves $\tilde{L}_x$ and L_x are manifolds each with dimension equal to the dimension of V. If the map $\hat{\Phi}(\ , u)$ is invertible, near p, then $L_x = \hat{\Phi}(V, u) = \tilde{L}_x$ near x.

This argument holds for any x, so the vertical leaves L_x and $\tilde{L}_x$ agree near x. An open and closed set argument shows us that all vertical leaves agree.

Horizontal leaves are treated in the same way using $F^\times$ and $(3.4^\times)$.

Now we apply Theorem 4.1 and use a Poincaré-Dulac approach to show

THEOREM 4.3. *Suppose that $\hat{F}, \hat{G}$ is an invertible system as in Theorem 3.1, that it has an equilibrium point e, is openly reachable and that the differential equations $\frac{dx}{dt} = \hat{F}(x, 0)$, $\frac{dx}{dt} = \hat{F}^\times(x, G(e, 0))$ each have no resonances*[5] *at $x = e$. Given two factorizations F, G, f, g and $\tilde{F}, \tilde{G}, \tilde{f}, \tilde{g}$ with equilibrium point $e = (e^1, e^2)$ whose state manifolds all have the same dimension. Suppose the linearized systems*

$$DF(e^1, 0) = D\tilde{F}(e^1, 0)$$
$$DG(e^1, 0) = D\tilde{G}(e^1, 0)$$
$$Df(e^2, G(e^1, 0)) = D\tilde{f}(e^2, \tilde{G}(e^1, 0)$$
$$Dg(e^2, G(e^1, 0)) = D\tilde{g}(e^2, \tilde{G}(e^1, 0))$$

are equal as shown. Then $\tilde{F} = F$ and $f^\times = \tilde{f}^\times$ in some neighborhood of e^1 and e^2.

LEMMA 4.4. (Poincaré-Dulac, see [**Har**, IX, 12.1]) *If $\hat{F}(\cdot, 0)$ has no resonances at $x = e$, then there is a diffeomorphism ψ, $\psi: \mathcal{O} \subset R^n \to X$ on a neighborhood of 0, with $\psi(0) = e$, so that the map $A: R^n \to R^n$ defined at each $s \in \mathcal{O}$ by*

$$A(s) = D_s\psi^{-1}(s)[\hat{F}(\psi(s), 0)]$$

is linear. Note that A and $D_x\hat{F}(e, 0)$ are similar.

CONJECTURE 4.5. *If $A: R^n \to R^n$ has no resonances then each smooth manifold M with $0 \in M$ which is invariant under the flow e^{At} is actually a linear invariant subspace for A.*

[5] An O.D.E. $\frac{dx}{dt}(t) = \mathcal{A}(x(t))$ with equilibrium points e is said to have no resonances provided evolution operator $D\mathcal{A}(e)$ for the linearized equations has eigenvalues $\lambda_1, \ldots, \lambda_n$ satisfying equation $m_1\lambda_1 + \cdots + m_n\lambda_n \neq \lambda_i$ for all non-negative integers m_k.

A picture is informative.

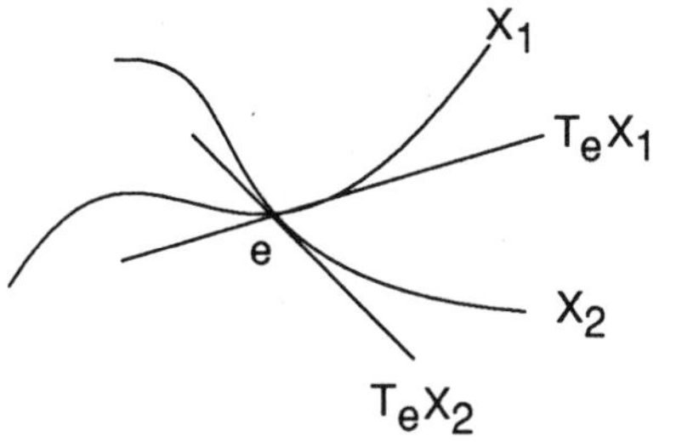

Figure 4.1.

Conjecture 4.5 and Lemma 4.4 says that generically T_eX_1 the tangent space to the $\dot{x} = F^{\times}(x,0)$ invariant submanifold X_1 through e determines X_1 uniquely. Likewise T_eX_2 determined X_2 uniquely. Theorem 4.1 and Remark 2.3 says that for reachable systems X_1 and X_2 determine the dynamical part F and $f^{\times}$ of factors uniquely.

PROOF OF THEOREM 4.3. We start with $\hat{F}, \hat{G}$ and two invariant coordinate pairs X_1, X_2 and $\tilde{X}_1, \tilde{X}_2$. These have horizontal, vertical leaves H, V and $\tilde{H}, \tilde{V}$ through e. The hypotheses are set to make the tangent spaces equal:

$$T_eH = T_e\tilde{H}$$

$$T_eV = T_e\tilde{V}.$$

Consider V and $\tilde{V}$. The map ψ of Lemma 4.4 pulls these manifolds back to

$$V_0 = \psi^{-1}(V) \qquad\qquad \tilde{V}_0 = \psi^{-1}(\tilde{V})$$

with tangent spaces

$$T_0\tilde{V}_0 = D\psi^{-1}(e)[T_e\tilde{V}]$$

etc. This implies

$$T_0V_0 = T_0\tilde{V}_0.$$

Also ψ intertwines the flow generated by $\frac{dx}{dt} = \hat{F}(x,0)$ and the flow generated by e^{At}, so V_0 and $\tilde{V}_0$ are invariant under the $\hat{F}(\ ,0)$ flow. Lemma 4.5 implies that V_0 and $\tilde{V}_0$ are linear subspaces of R^n. Consequently $V_0 = T_0V_0 = T_0\tilde{V}_0 = \tilde{V}_0$. Therefore $V = \tilde{V}$.

A similar argument applies to H and $\tilde{H}_0$. We have that they are invariant under $F^{\times}(\cdot, G(e,0))$. There is a map $\mu: R^n \to X$ which composed with these $F^{\times}$ "linearized" its action.

To this point we have shown that the horizontal leaves of X_1, X_2 and of $\tilde{X}_1$, $\tilde{X}_2$ through e are equal, so Lemma 4.5 applies directly to give these two coordinate points are actually the same. Finally Remark 2.3 shows that F and $f^\times$ are determined completely by the coordinate pair. This proves the theorem.

REFERENCES

[AD] V. Anbantharam and C. A. Desoer, On the stabilization of nonlinear systems, IEEE Trans. Automatic Control, **AC-29** (1984), 569–573.

[BGK] H. Bart, I. Gohberg, and M. A. Kaashoek, *Minimal Factorization of Matrix and Operator Functions*, Birkhäuser, 1979.

[BH1] J. A. Ball and J. W. Helton, Factorization of nonlinear systems: toward a theory for nonlinear H^∞ control, Proc. IEEE Conference on Decision and Control, December 1988.

[BH2] J. William Helton, with assistance of Joseph A. Ball, Charles R. Johnson, and John Palmer, Operator theory, analytic functions, matrices, and electrical engineering, Expository Lectures from the CBMS Regional Conference held at Lincoln, Nebraska, August 1985 (Regional conference series in mathematics, ISSN 0160-7642; no. 68, 1987).

[Har] P. Hartman, *Ordinary Differential Equations*, Birkhäuser, Boston, 1982.

[Hm] J. Hammer, Nonlinear systems: stabilization and coprimeness, International J. Control, **42**, no. 1 (1985), pp. 1–20.

[I] A. Isadori, Nonlinear control systems: an introduction, Springer Lecture Notes in Control and Information Science, no. 72, 1985.

Department of Mathematics (C-012)
University of California, San Diego
La Jolla, California 92093 U.S.A.

Operator Theory:
Advances and Applications, Vol. 41

MINIMAL LOWER SEPARABLE REPRESENTATIONS: CHARACTERIZATION AND CONSTRUCTION

M.A. Kaashoek and H.J. Woerdeman

Dedicated to Israel Gohberg, a great mathematician and a wonderful friend, on the occasion of his sixtieth birthday, with admiration and affection.

Minimality of a lower separable representation is characterized and a procedure to reduce a lower separable representation to a minimal one is described. The results yield an algorithm to construct a minimal realization for a given impulse response matrix.

Introduction

This paper concerns the problem to construct a minimal separable representation for the lower triangular part of an integral operator and to find conditions that characterize such a representation. To state the problem in a precise form, let k be an $n \times m$ matrix function in two varables defined on the square $[a,b]\times[a,b]$. Recall from [GK2] that the kernel k is said to be *lower separable*, if k admits a *lower separable representation* $\{f,g\}$, that is,

$$k(t,s) = f(t)g(s) \ , a \leqq s \leqq t \leqq b \ , \text{a.e.}, \tag{0.1}$$

with f and g matrix functions of sizes $n \times p$ and $p \times m$, respectively, of which the entries are square integrable on $[a,b]$. The number p is referred to as the *order* of the representation $\{f,g\}$. A lower separable representation $\{f,g\}$ is said to be *minimal* if among all representations of k the order of $\{f,g\}$ is as small as possible.

In this paper we answer the first question raised in [GK2], Section 6, which asked for a constructive procedure to get a minimal lower separable representation for k starting with an arbitrary one. Our construction, which may be viewed as a variation on the cutting procedure used in [GKL], is based on the following theorem which gives the conditions that characterize minimality.

THEOREM 0.1. *The lower separable representation* $\{f,g\}$ *for* k *is minimal if and only if the following conditions satisfied:*

(i) $\operatorname{Ker} F_a = (0)$, $\operatorname{Im} G_b = \mathbb{C}^p$,

(ii) $\operatorname{Ker} F_c \subset \operatorname{Im} G_c$, $a < c < b$.

Here p *is the order of the representation* $\{f,g\}$ *and for* $a \leqq \gamma \leqq b$ *the operators* F_γ *and* G_γ *are defined by*

$$F_\gamma : \mathbb{C}^p \to L_2^n[\gamma,b] \ , \ F_\gamma x = f(.)x \ ,$$

$$G_\gamma : L_2^m[a,\gamma] \to \mathbb{C}^p \ , \ G_\gamma \phi = \int_a^\gamma g(s)\phi(s)ds \ .$$

The cutting procedure, which we shall develop, reduces in a finite number of steps a lower separable representation $\{f,g\}$ for which conditions (i) and (ii) in Theorem 0.1 are not satisfied to a representation satisfying these two conditions, and hence to a minimal one.

We also obtain the discrete analogs of the above results for finite and semi-infinite matrices. In fact, first we establish the construction of a minimal lower separable representation and the conditions of minimality in the general framework of operators that are lower separable relative to chains of orthogonal projections (cf., [KW], [W]). The continuous and discrete analogs referred to above are obtained from the general framework by specifying the chains of projections.

For matrix functions a construction of a minimal lower separable representation has been given earlier in [P]. The construction in [P] is more complicated than the one given here; it involves not only a cutting procedure, but it also requires to glue together various semi-separable representations by using similarity transformations.

The paper consists of five sections. In the first section we put the problem into the general framework of operators that are lower triangular relative to chains of projections and we state the abstract version of Theorem 0.1. In Section 2 we prove the abstract version of Theorem 0.1. Section 3 describes the algorithm to construct a minimal lower separable representation starting with an arbitrary one. In Section 4 we specify the results for various concrete classes of operators and we prove Theorem 0.1. The last section concerns a problem in mathematical systems theory, and there we show that our results yield a procedure to construct a minimal realization for a given impulse response matrix.

1. General framework and a characterization of minimality

Throughout this paper Z and Y are separable Hilbert spaces over $\mathbb{C}$, and $\mathscr{P}$ and $\mathscr{Q}$ denote closed chains of orthogonal projections (see [GKr], Section I.3 for the definition) on Z and Y, respectively. Let $t \mapsto P_t$ and $t \mapsto Q_t$ be parametrizations ([GKr], Section V.1) of $\mathscr{P}$ and $\mathscr{Q}$, respectively, which we assume to be defined on the same closed subset Λ of the extended real line $\mathbb{R} \cup \{-\infty,\infty\}$. We shall refer to Λ as the *parameter set* of $\mathscr{P}$ and $\mathscr{Q}$. We do not assume that the chains $\mathscr{P}$ and $\mathscr{Q}$ contain the operators 0 and I. A finite subset $\pi = \{\alpha_0,\alpha_1, \cdots ,\alpha_n\}$ of Λ is called a *partition* of Λ if $\alpha_i < \alpha_j$ $(i < j)$, $\alpha_0 = \min \Lambda$ and $\alpha_n = \max \Lambda$. We let $\mathscr{P}_\pi$ and $\mathscr{Q}_\pi$ denote the chains

$$\mathscr{P}_\pi = \{ P_t \mid t \in \pi \}, \mathscr{Q}_\pi = \{ Q_t \mid t \in \pi \}.$$

Let $T : Z \to Y$ be a (bounded linear) operator. We recall (see [GKr], Section I.4; also [W] for the case $\mathscr{P} \neq \mathscr{Q}$) the definition of *the lower triangular part* $\mathscr{L}(T;\mathscr{P},\mathscr{Q})$ of T relative to the chains $\mathscr{P}$ and $\mathscr{Q}$. For a partition $\pi = \{\alpha_0,\alpha_1, \cdots ,\alpha_n\}$ of the parameter set Λ put

$$\mathscr{L}(T;\mathscr{P}_\pi,\mathscr{Q}_\pi) := (I-Q_{\alpha_0})TP_{\alpha_1} + \sum_{j=2}^{n} (I-Q_{\alpha_{j-1}})T(P_{\alpha_j}-P_{\alpha_{j-1}}).$$

Now define

$$\mathscr{L}(T;\mathscr{P},\mathscr{Q}) := \lim_{\pi} \mathscr{L}(T;\mathscr{P}_\pi,\mathscr{Q}_\pi), \tag{1.1}$$

provided the right hand side exists. The limit in (1.1) should be understood as follows. For every $\epsilon > 0$ there exists a partition π_ϵ of Λ such that

$$\| \mathscr{L}(T;\mathscr{P},\mathscr{Q}) - \mathscr{L}(T;\mathscr{P}_\pi,\mathscr{Q}_\pi) \| < \epsilon$$

for all partition π of Λ such that $\pi_\epsilon \subset \pi$. It is known ([GKr], Sections I.10 and III.7) that for a Hilbert-Schmidt operator T the operator $\mathscr{L}(T;\mathscr{P},\mathscr{Q})$ is well-defined when $\mathscr{P} = \mathscr{Q}$. This result also holds for $\mathscr{P} \neq \mathscr{Q}$. To see this note that $\mathscr{L}(T;\mathscr{P},\mathscr{Q})$ equals the (2,1) entry of the operator

$$\mathscr{L}\left(\begin{pmatrix} 0 & 0 \\ T & 0 \end{pmatrix} ;\mathscr{P} \oplus \mathscr{Q},\mathscr{P} \oplus \mathscr{Q}\right), \tag{1.2}$$

where $\mathscr{P}\oplus\mathscr{Q}$ is the chain on $Z\oplus Y$ given by $\mathscr{P}\oplus\mathscr{Q} = \{ P_t \oplus Q_t \mid t \in \Lambda \}$, and the operator (1.2) exists if T is Hilbert-Schmidt.

An operator $T : Z \to Y$ is called *$(\mathscr{P},\mathscr{Q})$-lower separable* if T has a lower triangular part relative to the chains $\mathscr{P}$ and $\mathscr{Q}$ and there exists a pair of operators $\{F,G\}$, with $F : X \to Y$ and $G : Z \to X$, such that $\dim X < \infty$ and $\mathscr{L}(FG;\mathscr{P},\mathscr{Q}) = \mathscr{L}(T;\mathscr{P},\mathscr{Q})$. In that case the pair $\{F,G\}$ is called a *$(\mathscr{P},\mathscr{Q})$-lower separable representation* for T and the space X is referred to as the *internal space* of the representation. A $(\mathscr{P},\mathscr{Q})$-lower separable representation of T is called *minimal* if the dimension of the internal space is as low as possible among all $(\mathscr{P},\mathscr{Q})$-lower separable representations of T. We shall prove the following theorem.

THEOREM 1.1. *Let $\{F,G\}$ be a $(\mathscr{P},\mathscr{Q})$-lower separable representation for T with internal space X. In order that $\{F,G\}$ is minimal it is necessary and sufficient that the following two conditions hold*

(i) $\operatorname{Ker}(I-Q_a)F = (0)$, $\operatorname{Im} GP_b = X$,

(ii) $\operatorname{Ker}(I-Q_c)F \subset \operatorname{Im} GP_c$, $c \in \Lambda$.

Here Λ is the parameter set of $\mathscr{P}$ and $\mathscr{Q}$, and $a = \min\Lambda$, $b = \max\Lambda$.

To check the minimality of a $(\mathscr{P},\mathscr{Q})$-lower separable representation $\{F,G\}$ it is not necessary to verify (ii) for all $c \in \Lambda$. In fact, it suffices to check (ii) for a finite number of points only. To make this more precise, we introduce the notion of a D-partition. Denote the predecessor in Λ of an element $c \in \Lambda$ by c^*, i.e.,

$$c^* = \sup\{ \alpha < c \mid \alpha \in \Lambda \}, a < c \leqq b \ , a^* := a.$$

Let $T : Z \to Y$ be $(\mathscr{P},\mathscr{Q})$-lower separable. We call a partition $\pi = \{\alpha_0, \cdots ,\alpha_n\}$ of Λ a *D-partition* associated with T (relative to the chains $\mathscr{P}$ and $\mathscr{Q}$) if for each $i = 1,\dots,n$ the operators

$$(I-Q_{\gamma^*})TP_\gamma \ , \alpha_{i-1} < \gamma \leqq \alpha_i \ , \gamma^* \neq \alpha_i \ , \tag{1.3}$$

$$(I-Q_\gamma)TP_\gamma \ , \alpha_{i-1} < \gamma < \alpha_i \ , \tag{1.4}$$

all have the same rank. Note that the definition of a D-partition associated with T only depends upon the lower triangular part of T. The existence of such a partition follows from Proposition 2.4 in the next section. Note that if π is a D-partition associated with T, then any partition π' finer than π is again a D-partition associated with T. The intersection of two D-partitions associated with T is again a D-partition associated with T. We shall show that condition (ii) in Theorem 1.1 may be replaced by:

(ii)' $\operatorname{Ker}(I-Q_{\alpha_i})F \subset \operatorname{Im} GP_{\alpha_i}$, $i = 1,\ldots,n-1$,

where $\pi = \{\alpha_0,\alpha_1, \cdots, \alpha_n\}$ *is a D-partition associated with* T.

2. Proof of Theorem 1.1.

First we do some preliminary work. Let $T : Z \to Y$ be an operator, and suppose that the lower triangular part of T (relative to $\mathscr{P}$ and $\mathscr{Q}$) exists. Recall (see [GK2], [KW], [W]) that an operator $K : Z \to Y$ is called a *finite rank extension* of the lower triangular part of T if K has finite rank and $\mathscr{L}(K;\mathscr{P},\mathscr{Q}) = \mathscr{L}(T;\mathscr{P},\mathscr{Q})$. Clearly, when T is $(\mathscr{P},\mathscr{Q})$-lower separable, then the lower triangular part of T has a finite rank extension (take $K = FG$). The converse statement also holds true. Indeed, let Π denote the orthogonal projection of Y on $\operatorname{Im} K$, and put $F = \Pi^*$, $G = \Pi K$ and $X = \operatorname{Im} K$. Then $\{F,G\}$ with internal space X is a $(\mathscr{P},\mathscr{Q})$-lower separable representation for T. A finite rank extension $K : Z \to Y$ of the lower triangular part of T is called (see [GK2], [KW], [W]) a *minimal rank extension* of the lower triangular part of T if it has the lowest possible rank among all finite rank extensions of the lower triangular part of T. It is not hard to see that if $\{F,G\}$ is a minimal $(\mathscr{P},\mathscr{Q})$-lower separable representation for T, then the operator FG is a minimal rank extension of the lower triangular part of T. Conversely, when FG is a minimal rank extension of the lower triangular part of T such that $F : X \to Y$ is injective and $G : Z \to X$ is surjective, then $\{F,G\}$ is a minimal $(\mathscr{P},\mathscr{Q})$-lower separable representation for T. Indeed, $\dim X = \operatorname{rank} FG < \infty$, and if $\{\tilde{F},\tilde{G}\}$ is another $(\mathscr{P},\mathscr{Q})$-lower separable representation for T with internal space $\tilde{X}$, then

$$\dim \tilde{X} \geqq \operatorname{rank} \tilde{F}\tilde{G} \geqq \operatorname{rank} FG = \dim X,$$

where in the second equality we use that $\tilde{F}\tilde{G}$ is a finite rank extension of the lower triangular part of T.

Next we extend the notions of reduction and dilation given in [GK2], Section 1, to arbitrary operators of finite rank. Let $K : Z \to Y$ be an operator with finite rank. A pair of operators $\{F,G\}$, where $F : X \to Y$ and $G : Z \to X$, is called a *separable representation* of K if X is a finite dimensional space and $K = FG$. The space X is called the *internal space* of the representation. Let $\{F,G\}$ and $\{F_0,G_0\}$ be separable representations of K with internal spaces X and X_0, respectively. We call the pair $\{F,G\}$ a *dilation* of $\{F_0,G_0\}$ if the space X admits a direct sum decomposition $X = X_1 \dot{+} X_0 \dot{+} X_2$ such that relative to this decomposition G and F have the following representations:

$$G = \begin{bmatrix} * \\ G_0 \\ 0 \end{bmatrix}, F = \begin{bmatrix} 0 & F_0 & * \end{bmatrix}, \tag{2.1}$$

where * denotes unspecified entries. In this case we also say that $\{F_0,G_0\}$ is a *reduction* of $\{F,G\}$. The dilation/reduction is called *proper* when $\dim X_0 < \dim X$. A separable representation $\{F,G\}$ of K is said to be *irreducible* if this pair is not a proper dilation of another separable representation $\{F_0,G_0\}$ of K.

LEMMA 2.1. *Let $K : Z \to Y$ be an operator with finite rank, and let $\{F,G\}$ be a separable representation of K. Then $\{F,G\}$ is irreducible if and only if F is injective and G is surjective.*

Proof. Suppose that F is injective and G is surjective. Then F allows a representation as in (2.1) only when $X_1 = (0)$, and G allows a representation like in (2.1) only when $X_2 = (0)$. But then $\{F,G\}$ has no proper reduction.

To prove the necessity assume that $\{F,G\}$ is irreducible. Put $X_1 = \operatorname{Ker} F$, and define X_0 to be a direct complement of $\operatorname{Ker} F \cap \operatorname{Im} G$ in $\operatorname{Im} G$. Let X_2 be a direct complement of $X_1 \dot{+} X_0$ in X. Consider the partitioning of F and G relative to this decomposition:

$$G = \begin{bmatrix} G_1 \\ G_0 \\ G_2 \end{bmatrix}, F = \begin{bmatrix} F_1 & F_0 & F_2 \end{bmatrix}.$$

Clearly, $F_1 = 0$. Further, since $\operatorname{Im} G \subset X_1 + X_0$, the operator $G_2 = 0$. But then $K = FG = F_0G_0$, and $\{F_0,G_0\}$ is a reduction of $\{F,G\}$. Since $\{F,G\}$ is irreducible, we must have that $\dim X_1 = 0 = \dim X_2$, but this implies that F is injective and G is surjective. □

We shall also need the following two lemmas.

LEMMA 2.2. *If $\{F,G\}$ is a $(\mathcal{P},\mathcal{Q})$-lower separable representation for T,* *then* $\{(I-Q_a)F, GP_b\}$ *is also a* $(\mathcal{P},\mathcal{Q})$*-lower separable representation of the lower triangular part of* T.

Proof. The lemma follows from the fact that

$$\mathcal{L}(S;\mathcal{P},\mathcal{Q}) = \mathcal{L}((I-Q_a)SP_b;\mathcal{P},\mathcal{Q})$$

for any operator $S : Z \to Y$ for which the lower triangular part relative to $\mathcal{P}$ and $\mathcal{Q}$ exists. □

LEMMA 2.3. *Let* $\{F,G\}$ *be a* $(\mathcal{P},\mathcal{Q})$*-lower separable representation for* T, *and let* X *denote its internal space. Take* $c \in \Lambda$, $a < c < b$. *Choose a subspace* X_1 *in* Ker $(I-Q_c)F$ *such that* X_1 *is a direct complement of* Im GP_c *in* Ker $(I-Q_c)F+$ Im GP_c, *and let* X_0 *be a direct complement of* X_1 *in* X *such that* Im $GP_c \subset X_0$. *Write*

$$F = \begin{pmatrix} F_1 & F_0 \end{pmatrix} : X_1 \dot{+} X_0 \to Y, G = \begin{pmatrix} G_1 \\ G_0 \end{pmatrix} : Z \to X_1 \dot{+} X_0 .$$

Then $\{F_0,G_0\}$ *is a* $(\mathcal{P},\mathcal{Q})$*-lower separable representation of* T *and*

$$\operatorname{Ker} (I-Q_c)F_0 \subset \operatorname{Im} G_0P_c .$$

Proof. Since Im $GP_c \subset X_0$, the operator $G_1P_c = 0$. Also, $X_1 \subset$ Ker $(I-Q_c)F$ yields that $(I-Q_c)F_1 = 0$. Let $\gamma \in \Lambda$. Let us compare $(I-Q_{\gamma^*})FGP_\gamma$ with $(I-Q_{\gamma^*})F_0G_0P_\gamma$. If $\gamma \leqq c$, then

$$(I-Q_{\gamma^*})F_1G_1P_\gamma = (I-Q_{\gamma^*})F_1G_1P_cP_\gamma = 0.$$

If $\gamma > c$, then $\gamma^* \geqq c$, and thus

$$(I-Q_{\gamma^*})F_1G_1P_\gamma = (I-Q_{\gamma^*})(I-Q_c)F_1G_1P_\gamma = 0.$$

Since $FG = F_1G_1+F_0G_0$, it follows that

$$(I-Q_{\gamma^*})FGP_\gamma = (I-Q_{\gamma^*})F_0G_0P_\gamma \,, \gamma \in \Lambda ,$$

which implies that

$$\mathcal{L}(F_0G_0;\mathcal{P},\mathcal{Q}) = \mathcal{L}(FG;\mathcal{P},\mathcal{Q}) = \mathcal{L}(T;\mathcal{P},\mathcal{Q}).$$

Here we use the fact that for a Hilbert-Schmidt operator S the operator $\mathscr{L}(S;\mathscr{P},\mathscr{Q}) = 0$ whenever $(I-Q_{\gamma}\cdot)SP_{\gamma} = 0$ for all $\gamma \in \Lambda$. For $\mathscr{P} = \mathscr{Q}$ this fact follows from Theorem I.10.1 and Lemma III.7.1 in [GKr]. Since $\mathscr{L}(S;\mathscr{P},\mathscr{Q})$ is the (2,1) entry of the operator in (1.2) with T replaced by S, it is also true for $\mathscr{P} \neq \mathscr{Q}$. So we may conclude that $\{F_0,G_0\}$ is a $(\mathscr{P},\mathscr{Q})$-lower separable representation of T.

Let $x_0 \in \operatorname{Ker}(I-Q_c)F_0 \ (\subset X_0)$. Then

$$x := \begin{bmatrix} 0 \\ x_0 \end{bmatrix} \in \operatorname{Ker}(I-Q_c)F \subset X_1 \dot{+} \operatorname{Im} GP_c.$$

Since $\operatorname{Im} GP_c \subset X_0$, it follows that $x \in \operatorname{Im} GP_c$. But then $x_0 \in \operatorname{Im} G_0P_c$. □

Before we prove Theorem 1.1 we establish the existence of a D-partition. Let $K : Z \to Y$ be a finite rank operator. Note that the functions

$$\gamma \mapsto \operatorname{rank}(I-Q_{\gamma})K \ , \gamma \mapsto \operatorname{rank} KP_{\gamma} \ , \tag{2.2}$$

only have a finite number of points of discontinuity. We call a partition $\pi = \{\alpha_0, \cdots, \alpha_n\}$ of the parameter set Λ a *C-partition* for $\mathscr{P},\mathscr{Q}$ and K if the functions in (2.2) are continuous on $\Lambda \setminus \pi$. Using the left continity of the first function in (2.2) and the right continuity of the second function in (2.2) it is easily checked that this definition of a C-partition coincides with the one in [W], Section 3. We have the following proposition.

PROPOSITION 2.4. *Let $T : Z \to Y$ be $(\mathscr{P},\mathscr{Q})$-lower separable. If π is a C-partition for $\mathscr{P}$, $\mathscr{Q}$ and a finite rank extension of the lower triangular part of T, then π is a D-partition associated with T. Conversely, if π is a D-partition associated with T, then π is a C-partition for $\mathscr{P}$, $\mathscr{Q}$ and any minimal rank extension of the lower triangular part of T.*

Proof. Let K be a finite rank extension of the lower triangular part of T. Let $\pi = \{\alpha_0, \cdots, \alpha_n\}$ be a C-partition for $\mathscr{P}$, $\mathscr{Q}$ and K. Fix $i \in \{1,\ldots,n\}$. Put

$$\hat{\mathscr{P}} = \{ P_t \mid t \in [\alpha_{i-1},\alpha_i] \cap \Lambda \},$$

$$\hat{\mathscr{Q}} = \{ Q_t \mid t \in [\alpha_{i-1},\alpha_i] \cap \Lambda \},$$

and consider $\hat{\mathscr{P}}$ and $\hat{\mathscr{Q}}$ as chains of orthogonal projections on $\operatorname{Im} P_{\alpha_i}$ and $\operatorname{Ker}(I-Q_{\alpha_{i-1}})$, respectively. Put $\hat{T} = (I-Q_{\alpha_{i-1}})TP_{\alpha_i} : \operatorname{Im} P_{\alpha_{i-1}} \to \operatorname{Ker}(I-Q_{\alpha_{i-1}})$. By Propositions 4.1 and 4.2 in [W] any minimal rank extension of the lower triangular part of $\hat{T}$ (relative to $\hat{\mathscr{P}}$

and $\hat{\mathscr{Q}}$) is of the form $(I-Q_{\alpha_{i-1}})HP_{\alpha_i}$, where H is a minimal rank extension of the lower triangular part of T (relative to $\mathscr{P}$ and $\mathscr{Q}$). Moreover, the last sentence of Theorem 5.1 in [W] shows that $(I-Q_{\alpha_{i-1}})HP_{\alpha_i}$ does not depend on the chosen minimal rank extension H, and thus $\hat{T}$ only has one minimal rank extension (relative to the chains $\hat{\mathscr{P}}$ and $\hat{\mathscr{Q}}$). But then Theorem 1.1 in [KW] (see also Theorem 5.2 in [W]) implies that the operators in (1.3) and (1.4) all have the same rank. Since i was arbitrary we obtain that π is a D-partition associated with T.

Conversely, suppose that $\pi = \{\alpha_0, \ldots, \alpha_n\}$ is a D-partition associated with T, and introduce $\hat{\mathscr{P}}$, $\hat{\mathscr{Q}}$ and $\hat{T}$ as in the previous paragraph. Fix $i \in \{1,\ldots,n\}$. Since the operators in (1.2) and (1.3) all have the same rank, Theorem 1.1 in [KW] yields that the lower triangular part of $\hat{T}$ relative to the chains $\hat{\mathscr{P}}$ and $\hat{\mathscr{Q}}$ only has one minimal rank extension. If K_1 and K_2 are minimal rank extensions of the lower triangular part of T (relative to $\mathscr{P}$ and $\mathscr{Q}$), then Propositions 4.1 and 4.2 in [W] yield that $(I-Q_{\alpha_{i-1}})K_1P_{\alpha_i}$ and $(I-Q_{\alpha_{i-1}})K_2P_{\alpha_i}$ viewed as operators acting $\operatorname{Im} P_{\alpha_i} \to \operatorname{Ker}(I-Q_{\alpha_{i-1}})$, are minimal rank extensions of the lower triangular part of $\hat{T}$. Hence

$$(I-Q_{\alpha_{i-1}})K_1P_{\alpha_i} = (I-Q_{\alpha_{i-1}})K_2P_{\alpha_i}.$$

Since i was arbitrary, Proposition 5.3 in [W] implies that π is a C-partition for $\mathscr{P}$, $\mathscr{Q}$ and any minimal rank extension of the lower triangular part of T. □

We are now ready to prove Theorem 1.1.

Proof of Theorem 1.1. Let $\{F,G\}$ be a $(\mathscr{P},\mathscr{Q})$-lower separable representation for T with internal space X, and let $\pi = \{\alpha_0,\alpha_1, \cdots, \alpha_n\}$ be a D-partition associated with T. Suppose that conditions (i) and (ii)′ in Section 1 hold. Note that (i) implies that F is injective and G is surjective. Since $\{F,G\}$ is a $(\mathscr{P},\mathscr{Q})$-lower separable representation for the lower triangular part of T, the operator FG is a finite rank extension of the lower triangular part of T. Conditions (i) and (ii)′ imply that for $i = 1,\ldots,n$

$$\operatorname{Ker}(I-Q_{\alpha_{i-1}})F \subset \operatorname{Im} GP_{\alpha_{i-1}} \subset \operatorname{Im} GP_{\alpha_i}.$$

But then

$$\operatorname{rank}(I-Q_{\alpha_{i-1}})FGP_{\alpha_i} = \operatorname{rank} GP_{\alpha_i} - \dim \operatorname{Ker}(I-Q_{\alpha_{i-1}})F,$$

and

$$\text{rank }(I-Q_{\alpha_{i-1}})FGP_{\alpha_{i-1}} = \text{rank } GP_{\alpha_{i-1}} - \dim \text{ Ker }(I-Q_{\alpha_{i-1}})F,$$

for $i = 1,\ldots,n$. Now

$$l := \sum_{p=1}^{n} \text{rank }(I-Q_{\alpha_{p-1}})FGP_{\alpha_p} - \sum_{p=1}^{n-1} \text{rank }(I-Q_{\alpha_p})FGP_{\alpha_p} =$$

$$= \text{rank } GP_{\alpha_n} - \dim \text{ Ker }(I-Q_{\alpha_0})F = \dim X,$$

where in the last step condition (i) is used. Proposition 2.4 implies that π is a C-partition for $\mathscr{P}$, $\mathscr{Q}$ and a minimal rank extension of the lower triangular part of T. But then Theorem 3.2 in [W] yields that $l = \dim X$ equals to the lower order of T (= the rank of a minimal rank extension of the lower triangular part of T), and hence FG is a minimal rank extension of the lower triangular part of T. Since F is injective and G is surjective, we get that $\{F,G\}$ is minimal. This proves the sufficiency of conditions (i) and (ii)$'$. Since (ii)$'$ is a weaker condition than (ii), we also showed the sufficiency of (i) and (ii).

To prove the necessity of conditions (i) and (ii) let $\{F,G\}$ be a minimal $(\mathscr{P},\mathscr{Q})$-lower separable representation for the lower triangular part of T with internal space X. By Lemma 2.2 the pair $\{(I-Q_a)F,GP_b\}$ is also a $(\mathscr{P},\mathscr{Q})$-lower separable representation for the lower triangular part of T. If condition (i) is not fulfilled, then Lemma 2.1 applied to the pair $\{(I-Q_a)F,GP_b\}$ implies that this pair has a proper reduction. Since this proper reduction is a $(\mathscr{P},\mathscr{Q})$-lower separable representation for the lower triangular part of T with a lower dimensional internal space, we get a contradiction. Hence condition (i) is fulfilled. If condition (ii) is not fulfilled, then by Lemma 2.3 there is a $(\mathscr{P},\mathscr{Q})$-lower separable representation for T with a lower dimensional internal space, and again we get a contradiction. □

3. The construction of a minimal lower separable representation

Let $\{F,G\}$ be a $(\mathscr{P},\mathscr{Q})$-lower separable representation for T with internal space X. One may produce a minimal $(\mathscr{P},\mathscr{Q})$-lower separable representation for the lower triangular part of T by appplying the following procedure.

(1) Find a partition $\pi = \{\alpha_0,\alpha_1,\ldots,\alpha_n\}$ of the parameter set Λ such that

$$\gamma \mapsto \text{rank }(I-Q_\gamma)FG \ , \gamma \mapsto \text{rank } FGP_\gamma$$

are continuous functions on $\Lambda\backslash\pi$.

(2) Check if $\{F,G\}$ satisfies

$$(i)\ \operatorname{Ker}(I-Q_a)F = (0)\ ,\ \operatorname{Im} GP_b = X\ ,$$

$$(ii)'\ \operatorname{Ker}(I-Q_{\alpha_i})F \subset \operatorname{Im} GP_{\alpha_i}\ ,\ i = 1,\ldots,n-1.$$

If so, $\{F,G\}$ is minimal. If not, go to (3).

(3a) If (i) is not satisfied, make a proper reduction $\{F_0,G_0\}$ of $\{(I-Q_a)F,GP_b\}$ as described in the proof of the necessity in Lemma 2.1, and repeat (2) with the pair $\{F_0,G_0\}$.

(3b) If (ii)′ is not satisfied for $i = i_0$, apply Lemma 2.3 (with $c = \alpha_{i_0}$) to obtain a $(\mathscr{P},\mathscr{Q})$-lower separable representation $\{F_0,G_0\}$ for T with an internal space of lower dimension. Repeat step (2) with $\{F_0,G_0\}$.

The above procedure leads to a minimal $(\mathscr{P},\mathscr{Q})$-lower separable representation for the lower triangular part of T. To see this, note that by Proposition 2.4 the partition π in step (1) is a D-partition associated with T. It is clear that the lower separable representation $\{\tilde{F},\tilde{G}\}$ obtained at the end of the procedure satisfies conditions (i) and (ii)′ . But then Theorem 1.1 yields that this pair $\{\tilde{F},\tilde{G}\}$ is a minimal $(\mathscr{P},\mathscr{Q})$-lower separable representation for T.

4. Examples

4.1. Finite operator matrices. Let

$$T = \begin{pmatrix} T_{11} & \cdots & T_{1n} \\ \vdots & & \vdots \\ T_{n1} & \cdots & T_{nn} \end{pmatrix} : Z_1 \oplus \cdots \oplus Z_n \to Y_1 \oplus \cdots \oplus Y_n \tag{4.1}$$

be a $n \times n$ operator matrix of which the (i,j)th entry is an operator acting from a separable Hilbert space Z_j into a separable Hilbert space Y_i. A pair of operator matrices $\{F,G\}$, where

$$\begin{aligned} F &= \operatorname{col}\left(F_k\right)_{k=1}^{n} = \begin{pmatrix} F_1 \\ \vdots \\ F_n \end{pmatrix} : X \to Y_1 \oplus \cdots \oplus Y_n\ , \\ G &= \operatorname{row}\left(G_k\right)_{k=1}^{n} = \begin{pmatrix} G_1 & \cdots & G_n \end{pmatrix} : Z_1 \oplus \cdots \oplus Z_n \end{aligned} \tag{4.2}$$

and X is a finite dimensional space, is called a *lower separable representation* for T if $F_i G_j = T_{ij}$, $1 \leq j \leq i \leq n$. The space X is called the *internal space* of the representation. We say that T is *lower separable* if T has a lower separable representation. Note that this is the case if and only if T_{ij} is of finite rank for $1 \leq j \leq i \leq n$. A lower separable representation of T is said to be *minimal* if among all lower separable representations for T the dimension of the internal space is as low as possible.

THEOREM 4.1. *The lower separable representation* $\{F,G\}$ *for* T, *given by* (4.2), *is minimal if and only if the following two conditions are satisfied:*

(i) F *is injective and* G *is surjective,*

(ii) $\operatorname{Ker}\operatorname{col}\left(F_k\right)_{k=j+1}^{n} \subset \operatorname{Im}\operatorname{row}\left(G_k\right)_{k=1}^{j}$, $j = 1,\ldots,n-1$.

Proof. Put $Z = Z_1 \oplus \cdots \oplus Z_n$ and $Y = Y_1 \oplus \cdots \oplus Y_n$. Let P_i be the projection of Z onto $Z_1 \oplus \cdots \oplus Z_i$ along $Z_{i+1} \oplus \cdots \oplus Z_n$ $(i = 1,..,n-1)$, and put $P_0 := 0$ and $P_n := I$. Define $Q_i : Y \to Y$ $(i = 0,..,n)$ analogously. Note that $\mathcal{P} := \{P_i \mid i = 0,..,n\}$ and $\mathcal{Q} := \{Q_i \mid i = 0,..,n\}$ are finite (closed) chains of orthogonal projections on Z and Y, respectively. The corresponding index set Λ is equal to $\{0,1,\cdots,n\}$. The definitions given in this subsection coincide with the corresponding definitions of Section 1 provided the latter are applied to the chains $\mathcal{P}$ and $\mathcal{Q}$ introduced here. Theorem 1.1 specified for the chains $\mathcal{P}$ and $\mathcal{Q}$ yields the theorem. □

4.2. Semi-infinite operator matrices. Let $T = (T_{ij})_{i,j=1}^{\infty} : l_2(Z) \to l_2(Y)$ be a (bounded linear) operator. Here $l_2(Z)$ (resp. $l_2(Y)$) stands for the space of all square summable sequences with elements in Z (resp. Y). The spaces Z and Y are given separable Hilbert spaces. A pair of operator matrices $\{F,G\}$, where

$$F = \operatorname{col}\left(F_k\right)_{k=1}^{\infty} : X \to l_2(Y)\ ,G = \operatorname{row}\left(G_k\right)_{k=1}^{\infty} : l_2(Z) \to X \tag{4.3}$$

and X is a finite dimensional space, is called a *lower separable representation* for T if $F_i G_j = T_{ij}$, $1 \leq j \leq i < \infty$. The space X is called the *internal space* of the representation. We say that T is *lower separable* if T has a lower separable representation. A lower separable representation of T is said to be *minimal* if among all lower separable representations for T the dimension of the internal space is as low as possible.

THEOREM 4.2. *The lower separable representation* $\{F,G\}$ *for* T, *given by*

(4.3), *is minimal if and only if the following two conditions are satisfied:*

(i) F is injective and G is surjective,

(ii) $\operatorname{Ker} \operatorname{col} \left[F_k \right]_{k=j+1}^{\infty} \subset \operatorname{Im} \operatorname{row} \left[G_k \right]_{k=1}^{j}, \ j = 1,2,\ldots.$

Proof. For $i = 1,2,\ldots$ let $P_i : l_2(Z) \to l_2(Z)$ be the orthogonal projection upon the first i coordinates, and put $P_0 = 0$ and $P_\infty = I$. Define $Q_i : l_2(Y) \to l_2(Y)$ $(i = 0,1,\ldots,\infty)$ analogously. Then $\mathcal{P} := \{P_0, P_1, P_2, \cdots, P_\infty\}$ and $\mathcal{Q} := \{Q_0, Q_1, Q_2, \cdots, Q_\infty\}$ are closed chains. The definitions given in this subsection coincide with the corresponding definitions of Section 1 provided the latter are applied to the chains $\mathcal{P}$ and $\mathcal{Q}$ introduced here. Theorem 1.1 specified for the chains $\mathcal{P}$ and $\mathcal{Q}$ yields the theorem. □

4.3. The continuous case. We prove Theorem 0.1.

Proof of Theorem 0.1. Let P_t be the projection in $L_2^n([a,b])$ defined by

$$(P_t\phi)(s) := \begin{cases} \phi(s) & a \leqq s \leqq t, \\ 0 & t < s \leqq b. \end{cases}$$

Then $\mathcal{P} = \{P_t \mid a \leqq t \leqq b\}$ is a closed chain. Define $\mathcal{Q}$ in $L_2^m([a,b])$ analogously. Note that in this case $\Lambda = [a,b]$ and $\gamma^* = \gamma$ for each $\gamma \in [a,b]$. Furthermore, the operators G_c and F_c in Theorem 0.1 coincide with the operators $G_b P_c$ and $(I - Q_c)F_a$ appearing in Theorem 1.1, respectively. With these remarks and the above choices of $\mathcal{P}$ and $\mathcal{Q}$ Theorem 0.1 follows directly from Theorem 1.1. □

For the continuous case the necessity of condition (ii)′ appeared earlier in a somewhat different form in [P, Lemma 2.1].

5. Minimality for time-variant systems

Consider the time variant causal system

$$\theta \begin{cases} \dot{x}(t) = A(t)x(t) + B(t)u(t), & a \leqq t \leqq b, \\ y(t) = C(t)x(t) & , a \leqq t \leqq b, \\ x(a) = 0. \end{cases} \tag{5.1}$$

Here $A(t)$, $B(t)$ and $C(t)$ are matrices of size $r \times r$, $r \times n$ and $m \times r$, respectively. We assume that $A(t)$, as a function of t, is integrable on $[a,b]$, and that $B(.)$ and $C(.)$ are square integrable. The matrix function $A(.)$ is called the *main coefficient* and the number r is called the *state space dimension* of the system. To simplify the notation we denote the system (5.1) by $\theta = (A(t),B(t),C(t))$. The *impulse response matrix* (see [K], Section 9.1) of the system θ is given by

$$h(t,s)=C(t)U(t)U(s)^{-1}B(s) \quad a \leqq s < t \leqq b,$$

where $U(t)$ is the fundamental operator of the system, i.e., $U(t)$ is the unique absolutely continuous solution of the matrix differential equation

$$\dot{U}(t) = A(t)U(t), \ a \leqq t \leqq b, \ U(a) = I_r. \tag{5.2}$$

Here I_r denotes the identity matrix of order r. Obviously, the impulse response matrix is the lower triangular part of a finite rank matrix kernel. The converse statement is also true (cf., [GK1], Section I.4).

A causal time-variant system θ is said to be a *realization* of h if the impulse response matrix of θ is equal to h, and θ is called a *minimal realization* of h if θ is a realization of h and among all realizations of h the state space dimension of θ is as small as possible. The construction of a minimal lower separable representation given in Section 3 yields an algorithm to construct a minimal realization for a given impulse response matrix h. The procedure is as follows:

(1) Choose an arbitrary realization $\theta = (A(t),B(t),C(t))$ of the given impulse response matrix θ.

(2) Put $f(t) = C(t)U(t)$ and $g(t) = U(t)^{-1}B(t)$, where $U(t)$ is given by (5.2).

(3) Construct operators G_γ and F_γ, $a \leqq \gamma \leqq b$, as follows

$$F_\gamma : \mathbb{C}^r \to L_2^n[\gamma,b] \ , \ F_\gamma x = f(.)x ,$$

$$G_\gamma : L_2^m[a,\gamma] \to \mathbb{C}^r \ , \ G_\gamma \phi = \int_a^\gamma g(s)\phi(s)ds \ ,$$

where r is the state space dimension of θ.

(4) Find a partition $\pi = \{\alpha_0,\alpha_1, \ldots ,\alpha_n\}$ of $[a,b]$ such that

$$\gamma \mapsto \operatorname{rank} F_\gamma G_b \ , \gamma \mapsto \operatorname{rank} F_a G_\gamma$$

are continuous on $[a,b]\backslash\pi$.

(5) Check if the conditions

$$(i) \ \operatorname{Ker} F_a = (0) \ , \ \operatorname{Im} G_b = \mathbb{C}^r \ ,$$

$$(ii)' \ \operatorname{Ker} F_{\alpha_i} \subset \operatorname{Im} G_{\alpha_i} \ , i = 1,\ldots,n-1,$$

are satisfied. If so, then $(0,g(t),f(t))$ is a minimal realization for h. If not, go to (6).

(6a) If (i) is not satisfied, then choose a direct complement X_0 of $\operatorname{Ker} F_a \cap \operatorname{Im} G_b$ in $\operatorname{Im} G_b$, and let P denote a projection of $\mathbb{C}^r$ onto X_0 such that $\operatorname{Ker} F_a \subset \operatorname{Ker} P$. Put

$$f_0(t) = f(t)\,|_{X_0} \ , g_0(t) = Pg(t) \ , a \leqq t \leqq b,$$

and apply steps (3) and (5) to the pair $\{f_0,g_0\}$ with the space $\mathbb{C}^r$ replaced by X_0.

(6b) If (ii)$'$ is not satisfied for $i = i_0$, say, then choose a subspace Y_1 in $\operatorname{Ker} F_{\alpha_{i_0}}$ such that Y_1 is a direct complement of $\operatorname{Im} G_{\alpha_{i_0}}$ in $\operatorname{Ker} F_{\alpha_{i_0}} + \operatorname{Im} G_{\alpha_{i_0}}$, and choose a direct complement Y_0 of Y_1 in $\mathbb{C}^r$ such that $\operatorname{Im} G_{\alpha_{i_0}} \subset Y_0$. Let Q be the projection of $\mathbb{C}^r$ onto Y_0 along Y_1, and put

$$f_0(t) = f(t)\,|_{Y_0} \ , g_0(t) = Qg(t) \ , a \leqq t \leqq b,$$

and apply steps (3) and (5) to the pair $\{f_0,g_0\}$ with $\mathbb{C}^r$ replaced by Y_0.

Since the state space dimension of a minimal realization of h is equal to the order of a minimal lower separable representation for h, the correctness of the above procedure follows from the correctness of the reduction procedure in Section 3. Here the choice of the chains is the same as in Subsection 4.3. For the construction of minimal discrete time systems a similar algorithm may be applied.

REFERENCES

[GK1] Gohberg, I. and Kaashoek, M.A.: Time varying linear systems with boundary conditions and integral operators, I. The transfer operator and its properties, *Integral Equations and Operator Theory* 7 (1984), 325-391.

[GK2] Gohberg, I. and Kaashoek, M.A.: Minimal representations of semi-separable kernels and systems with separable boundary conditions, *J. Math. Anal. Appl.* 124 (1987), 436-458.

[GKL] Gohberg I., M.A. Kaashoek, and L. Lerer: On minimality in the partial realization problem, *Systems and Control Letters* 9 (1987), 97-104.

[GKr] Gohberg, I. and Krein, M.G.: "Theory and applications of Volterra operators in Hilbert space", Transl. Math. Monographs, Vol 4, Amer. Math. Soc., Providence, R.I., 1970.

[KW] Kaashoek, M.A. and Woerdeman, H.J.: Unique minimal rank extensions of triangular operators, *J. Math. Anal. Appl.* 131 (1988), 501-516.

[K] Kailath, T.: "Linear systems", Prentice-Hall, Inc., Englewood Cliffs, N.J., 1980.

[P] Peeters, G.: Construction and classification of minimal representations of semi-separable kernels, *J. Math. Anal. Appl.* 137 (1989), 264-287.

[W] Woerdeman H.J.: The lower order of lower triangular operators and minimal rank extensions, *Integral Equations and Operator Theory* 10 (1987), 859-879.

M.A. Kaashoek and H.J. Woerdeman,
Department of Mathematics and Compter Science,
Vrije Universiteit,
De Boelelaan 1081,
1081 HV Amsterdam,
The Netherlands.

Operator Theory:
Advances and Applications, Vol. 41

ON THE INCLINATION OF HYPERINVARIANT SUBSPACES OF C_{11}–CONTRACTIONS

L. Kérchy

Dedicated to Professor I. Gohberg on the occasion of his 60th birthday

1. In this note we shall answer in the negative the following question posed by L.A. Fialkow in [3, p. 227]. Let U be a (linear, bounded) operator acting on the (complex) Hilbert space K, and let us assume that $\sigma(U) = \sigma_1 \cup \sigma_2$ is a non–trivial decomposition of the spectrum $\sigma(U)$ of U into the union of two disjoint closed sets. For $j = 1,2$, let $K(\sigma_j,U)$ denote the spectral subspace obtained by the Riesz–Dunford functional calculus. It is well–known that K splits into the direct sum of the hyperinvariant subspaces $K(\sigma_1,U)$ and $K(\sigma_2,U) : K = K(\sigma_1,U) \dot{+} K(\sigma_2,U)$. (See [10, Section 148].) Fialkow's question is whether for every operator T which is quasisimilar to U, there exist complementary T–invariant subspaces H_1 and H_2 such that $T|H_j$ is quasisimilar to $U|K(\sigma_j,U)$, for $j = 1,2$. We recall that T and U are quasisimilar, in notation: $T \sim U$, if there are quasiaffinities (i.e. injective operators with dense range) Y and Z which intertwine T and $U : YT = UY$ and $TZ = ZU$. (Cf. [11, Section II.3].)

It will be shown that the operator constructed in [1] provides a counterexample even in the special case when U is a unitary operator and T is a contraction. It turns out that the inclination of the hyperinvariant subspaces H_1 and H_2 of T, corresponding to $K(\sigma_1,U)$ and $K(\sigma_2,U)$, is of angle zero. On the other hand sufficient conditions will be given for the case when the angle between two hyperinvariant subspaces of T is positive.

2. We start by recalling some notations and results from the theory of contractions.

Let $B(H)$ denote the set of bounded, linear operators acting on the Hilbert space H, and let us consider a contraction $T \in B(H)$, i.e. $\|T\| \leq 1$. T is called of class C_{11} if for every non–zero vector $x \in H$ we have $\lim_{n\to\infty} \|T^n x\| \neq 0 \neq \lim_{n\to\infty} \|T^{*n} x\|$. With every C_{11}–contraction T we can associate a unitary operator in the following asymptotic way. Let us introduce a new inner product on H by $[x,y] := \lim_{n\to\infty} \langle T^n x, T^n y\rangle$, and denote by $H^{(a)} = H_T^{(a)}$ the completion of the inner product space $(H,[.,.])$. Then there exists a uniquely determined unitary operator $T^{(a)}$ on $H^{(a)}$ such that $T^{(a)}x = Tx$, for every $x \in H$. $T^{(a)}$ is called the unitary asymptote of T, it is—up to unitary equivalence—the unique unitary operator, which is quasisimilar to T. The natural imbedding $X = X_T$ of H into $H^{(a)}$ is an intertwining quasiaffinity: $XT = T^{(a)}X$.

Let us assume from now on that $T \in B(H)$ is a contraction of class C_{11}. We say that an invariant subspace N of T is quasireducing for T, if the restriction $T|N$ is of class C_{11}, as well. The set of all quasireducing subspaces of T is denoted by $\mathrm{Lat}_1 T$, while $\mathrm{Hyplat}_1 T$ stands for the set of all quasireducing hyperinvariant subspaces. We remark that if T is unitary, acting on a separable space, then $\mathrm{Hyplat}_1 T = \mathrm{Hyplat}\ T$ coincides with the set of all spectral subspaces. In general, $\mathrm{Hyplat}_1 T$ is isomorphic to $\mathrm{Hyplat}\ T^{(a)}$. More precisely, the mapping

$$q_X : \mathrm{Hyplat}_1 T \to \mathrm{Hyplat}\ T^{(a)},\ q_X : N \mapsto (XN)^-,$$

induced by the canonical transformation $X = X_T$, is the unique lattice isomorphism with the property that $T|N$ is quasisimilar to $T^{(a)}|(XN)^-$, for every subspace $N \in \mathrm{Hyplat}_1 T$.

For a detailed discussion of these topics we refer to [11], [6], [7] and [9].

Finally, we recall that the (minimal) angle $\varphi(M,N)$ $(0 \le \varphi \le \pi/2)$ between two non–zero subspaces M and N of H is defined by

$$\cos \varphi(M,N) := \sup\{|\langle x,y\rangle| : x \in M,\ y \in N,\ \|x\| = \|y\| = 1\}.$$

(See e.g. [4, p. 409].) It is easy to see that if $M \cap N = \{0\}$, then $M \dotplus N = M \vee N$ exactly when $\varphi(M,N) > 0$.

In the sequel, for any unitary operator U, $E(\cdot,U)$ and m_U mean the spectral measure and the spectral multiplicity function of U, respectively.

3. The following theorem illuminates the pathological behavior of hyperinvariant subspaces of contractions in the quasisimilarity orbit of a unitary operator.

THEOREM 1. *Let* $U \in B(K)$ *be an absolutely continuous unitary operator, and let us be given two disjoint, measurable sets* α_1 *and* α_2 *on the unit circle* $\partial\mathbb{D}$ *satisfying the condition* $E(\alpha_j,U) \neq 0$, *for* $j = 1,2,$. *Then there exists a* C_{11}*–contraction* $T \in B(H)$ *such that its unitary asymptote* $T^{(a)}$ *is unitarily equivalent to* $U : T^{(a)} \cong U$, *and* $\varphi(H_1,H_2) = 0$ *for the subspaces* $H_j = q_X^{-1}(E(\alpha_j,T^{(a)})H^{(a)})$, $j = 1,2$.

PROOF. Let us consider the decomposition $K = K_1 \oplus K_2 \oplus K'$, where $K_j = E(\alpha_j,U)K$ $(j = 1,2)$. For $j = 1,2$, there exists a non–invertible C_{11}–contraction $T_j \in B(H_j')$ such that $T_j^{(a)} \cong U|K_j =: U_j$. (See [2].) Moreover by [1, Lemma 3.2] there exist vectors $f \in H_1'$ and $g \in H_2'$ such that the operators $(T_1,f) : H_1' \oplus \mathbb{C} \longrightarrow H_1'$, $(T_1,f)(x \oplus \lambda) = T_1x + \lambda f$ and $(T_2^*,g) : H_2' \oplus \mathbb{C} \longrightarrow H_2'$, $(T_2^*,g)(y \oplus \lambda) = T_2^*y + \lambda g$ are injective contractions. Let us define the operator T acting on the Hilbert space $H = H_1' \oplus \mathbb{C} \oplus H_2' \oplus K'$ by the matrix

$$T = \begin{bmatrix} T_1 & f & 0 & 0 \\ 0 & 0 & g^* & 0 \\ 0 & 0 & T_2 & 0 \\ 0 & 0 & 0 & U' \end{bmatrix},$$

where $U' = U|K'$ and $g^* : H_2' \longrightarrow \mathbb{C}$, $g^*(y) = \langle y,g \rangle$. It is easy to check that T is a contraction of class C_{11}. The unitary asymptote $T^{(a)}$ of T is unitarily equivalent to the orthogonal sum $T_1^{(a)} \oplus T_2^{(a)} \oplus U' \cong U_1 \oplus U_2 \oplus U' = U$. (See [9, Corollary of Theorem 3] and [1, Theorem 1.7].)

On the other hand, it follows immediately from the definition that

$$H' := H_1' \oplus \mathbb{C} \oplus H_2' = q_X^{-1}(E(\alpha_1 \cup \alpha_2,T^{(a)})H^{(a)})$$

and

$$H_1 := q_X^{-1}(E(\alpha_1, T^{(a)})H^{(a)}) = H_1' .$$

Let us consider the subspace $H_2 = q_X^{-1}(E(\alpha_2, T^{(a)})H^{(a)})$. Since q_X is a lattice isomorphism between the lattices $\mathrm{Hyplat}_1 T$ and $\mathrm{Hyplat}\, T^{(a)}$, we infer by the results in [8] that

$$H_1 \vee H_2 = H' \text{ and } H_1 \cap H_2 = \{0\} .$$

Let us denote by Q the orthogonal projection onto the subspace $\mathbb{C} \oplus H_2'$ in the Hilbert space H' . By the preceding relations, the restriction $Z = Q|H_2$ of Q is a quasiaffinity, intertwining the operators $T|H_2$ and $T' = P_{\mathbb{C}\oplus H_2'} T|\mathbb{C} \oplus H_2' : Z(T|H_2) = T'Z$. Taking into account that $\ker T' = \mathbb{C} \oplus \{0\} \neq \{0\}$ but $T|H_2$ is one–to–one because it is of class C_{11} , it follows that the operators $T|H_2$ and T' are not similar. Therefore, the linear transformation Z is not invertible, and so $\varphi(H_1, H_2) = 0$. Q.E.D.

In contrast to the previous theorem, the next result shows that under certain circumstances quasireducing subspaces are actually reducing in the Banach space sense.

PROPOSITION 2. *Let* $T \in B(H)$ *be a* C_{11}*–contraction and let* $M \in \mathrm{Lat}_1 T$ *be a quasireducing subspace of* T . *If the restriction* $T|M$ *is similar to its unitary asymptote:* $T|M \approx (T|M)^{(a)}$, *then there exists a subspace* $N \in \mathrm{Lat}_1 T$ *such that* $M \dotplus N = H$.

PROOF. Let us consider the canonical quasiaffinity X intertwining the operators T and $T^{(a)}$. Since $T|M$ is similar to $(T|M)^{(a)}$, it follows by [9, Theorem 2] that the restriction of X to N is bounded from below. Thus the subspace $\tilde{M} = (XM)^{-} = XM$ reduces $T^{(a)}$ and $T^{(a)}|\tilde{M}$ is similar to $T|M$. Let us denote by P the orthogonal projection onto $\tilde{M}$ in $H^{(a)}$, and let Z be an arbitrary affinity intertwining $T^{(a)}|\tilde{M}$ with $T|M : Z(T^{(a)}|\tilde{M}) = (T|M)Z$. It is immediate that the operator $A \in B(H)$, defined by $Ah = ZPXh$ $(h \in H)$, commutes with T . Furthermore, in view of the relations $AH = AM = M$ and $\ker(A|M) = \{0\}$ we infer that the subspace $N = \ker A$, invariant for T , is complementary for $M : M \dotplus N = N$. Finally, observing that the orthogonal sum $(T|M) \oplus (T|N)$ is similar to the C_{11}–contraction T , we get that $N \in \mathrm{Lat}_1 T$. Q.E.D.

The following Corollary involves the answer to Fialkow's question; at the same time it shows that in many cases the pathological hyperinvariant subspaces can be considered exceptional.

COROLLARY 3. *Let* $U \in B(K)$ *be an absolutely continuous unitary operator, and suppose that its spectral multiplicity function* m_U *is finite a.e.*

(a) *For any two disjoint measurable subsets* α_1, α_2 *of* $\partial\mathbb{D}$ *with the property* $E(\alpha_j, U) \neq 0$ $(j = 1,2)$, *there exists a* C_{11}*–contraction* T *such that* $T \sim U$ *and* $\varphi(M_1, M_2) = 0$ *holds for the subspaces* M_1 *and* M_2 *satisfying the conditions* $T|M_j \sim U|E(\alpha_j, U)K$, $j = 1,2$.

(b) *On the other hand, if* T *is an arbitrary* C_{11}*–contraction, quasisimilar to* U, *then there exists an increasing sequence* $\{\beta_n\}_n$ *of measurable subsets of the unit circle* $\partial\mathbb{D}$ *such that* $\bigcup_n \beta_n = \partial\mathbb{D}$ *and if* α_1, α_2 *are disjoint sets,* $E(\alpha_j, U) \neq 0$ $(j = 1,2)$, *and* $\alpha_1 \subset \beta_n$ *for some* n, *then* $\varphi(H_1, H_2) > 0$ *is true for the hyperinvariant subspaces* $H_j = q_X^{-1}(E(\alpha_j, T^{(a)})H^{(a)})$ $j = 1,2$.

PROOF. (a) Let us consider the contraction T and the subspaces H_1, H_2 constructed in Theorem 1. Since $T|M_j$ is quasisimilar to $T|H_j$ it follows by [9, Lemma 3] that M_j is included in H_j, and then—by virtue of the finiteness of m_U —we infer that M_j coincides with H_j $(j = 1,2)$. (See [5].)

(b) Using again the assumption that m_U is finite a.e., we gain that the characteristic function Θ_T of T is boundedly invertible at a.e. point of the unit circle. For every n, let β_n denote the set of those numbers $z \in \partial\mathbb{D}$ where $\|\Theta_T(z)^{-1}\| \leq n$. Then on account of [11, Theorem VII.5.2] and Proposition 2 we can easily verify that the sequence $\{\beta_n\}_n$ possesses the required property. (See also [8].) Q.E.D.

We conclude with the following proposition dealing with the case when the spectral multiplicity function m_U of U is constant infinity.

PROPOSITION 4. *Let* U *be an absolutely continuous unitary operator such that* $m_U(z) = \infty$ *for a.e.* $z \in \partial\mathbb{D}$. *Then there exists a* C_{11}*–contraction* $T \in B(H)$ *such that* $T \sim U$ *and for every decomposition* $\alpha_1 \cup \alpha_2 = \partial\mathbb{D}$ *of the unit circle, the inclination of* *the corresponding hyperinvariant subspaces* H_1 *and* H_2 *of* T *is zero:* $\varphi(H_1,H_2) = 0$. (*Recall that* H_j *is defined by* $H_j = q_X^{-1}(E(\alpha_j,T^{(a)})H^{(a)})$, $j = 1,2$.)

PROOF. We can assume that the space of U is separable. Let A be a non–invertible, injective, positive, strict contraction acting on a separable Hilbert space, and let $T \in B(H)$ be the model–operator associated with the constant function $\Theta(z) \equiv A$. (See [11, Chapter VI].) T is a contraction of class C_{11} such that $T^{(a)} \cong U$.

Let us consider a decomposition $\partial\mathbb{D} = \alpha_1 \cup \alpha_2$ and suppose that $H_1 \dot{+} H_2 = H$ is true for the hyperinvariant subspaces $H_j = q_X^{-1}(E(\alpha_j,T^{(a)})H^{(a)})$, $j = 1,2$. Then the projection P onto the subspace H_1 and parallel to H_2 belongs to the bicommutant $\{T\}''$ of T , and P is different from O and I . On the other hand, as was pointed out to the author by K. Takahashi, the bicommutant of T consists of the functions of T in the sense of the Sz.–Nagy, Foias functional calculus. Thus every non–zero element of $\{T\}''$ must have dense range (see [11, Section III.3]), and so $\{T\}''$ does not contain any non–trivial projection. This contradicts to the existence of P , therefore $\varphi(H_1,H_2) = 0$. Q.E.D.

ACKNOWLEDGMENTS. This work was done while the author was visiting the Indiana University, Bloomington, Indiana.

After this paper was written, the author learned that D.H. Herrero has independently found a different solution to Fialkow's question ("On the essential spectra of quasisimilar operators", preprint, 1987–88).

REFERENCES

1. H. Bercovici, L. Kérchy, *Quasisimilarity and properties of the commutant of* C_{11}*–contractions*, Acta Sci. Math. (Szeged) **45** (1983), 67–74.

2. H. Bercovici, L. Kérchy, *On the spectra of* C_{11}*–contractions*, Proc. Amer. Math. Soc. **95** (1985), 412–418.

3. L.A. Fialkow, *Quasisimilarity and closures of similarity orbits of operators*, J. Operator Theory **14** (1985), 215–238.

4. I.C. Gohberg, M.G. Krein, *Introduction to the theory of linear nonselfadjoint operators*, Nauka, Moskva, 1965. (In Russian.)

5. L. Kérchy, *On the commutant of* C_{11}*–contractions*, Acta Sci. Math. (Szeged) **42** (1981), 15–26.

6. L. Kérchy, *Subspace lattices connected with* C_{11}*–contractions*, Anniversary Volume on Approximation Theory and Functional Analysis (eds. P.L. Butzer, R.L. Stens, B. Sz.–Nagy), Birkhauser Verlag, Basel (1984), 89–98.

7. L. Kérchy, *A description of invariant subspaces of* C_{11}*–contractions*, J. Operator Theory **15** (1986), 327–344.

8. L. Kérchy, *Contractions being weakly similar to unitaries*, Advances in Invariant Subspaces and Other Results of Operator Theory, OT 17, Birkhauser Verlag, Basel (1986), 187–200.

9. L. Kérchy, *Isometric Asymptotes of Power Bounded Operators*, preprint.

10. F. Riesz, B. Sz.–Nagy, *Functional Analysis*, Frederick Ungar Publ. Co., New York, 1966.

11. B. Sz.–Nagy, C. Foias, *Harmonic Analysis of Operators on Hilbert Space*, North Holland—Akadémia Kiadó, Amsterdam—Budapest, 1970.

Bolyai Institute
University Szeged
Aradi vértanuk tere 1
6720 Szeged, Hungary

Operator Theory:
Advances and Applications, Vol. 41

UNIMODULAR MÖBIUS-INVARIANT CONTRACTIVE DIVISORS FOR THE BERGMAN SPACE

Boris Korenblum

To Israel Gohberg with affection and admiration.

1. Introduction: Let B be a Banach space of analytic functions in the open unit disk **D**. For a sequence $\{\alpha_\nu\}$ ($\alpha_\nu \in \mathbf{D}$) let

$$B_{\{\alpha_\nu\}} = \{f \in B \mid f(\alpha_\nu)=0, \ \forall\nu\} \tag{1}$$

(repeated values of α_ν correspond to multiple zeros of f). $\{\alpha_\nu\}$ is called a B-zero set if $B_{\{\alpha_\nu\}} \neq \{0\}$.

A holomorphic function $d_\alpha(z)$ in **D** is called an α-divisor for B if $d_\alpha(z)$ has one and only one simple zero at α and the corresponding divisor operator $T_\alpha : B_\alpha \to B$ defined by

$$(T_\alpha f)(z) = f(z)/d_\alpha(z) \tag{2}$$

is bounded (we write B_α for $B_{\{\alpha\}}$).

A family of operators T_α ($\alpha \in \mathbf{D}$), and the corresponding family of divisors $d_\alpha(z)$, is called Möbius invariant if $d_\alpha = d_0 \circ b_\alpha$ where $b_\alpha(z) = \bar{\alpha}(\alpha - z)/|\alpha|(1-\bar{\alpha}z)$, $b_0(z)=z$; $d_0(z)$ is then called the generator of $\{T_\alpha\}$.

We call a Möbius invariant family of divisors $\{d_\alpha(z)\}$ B-efficient if it has the following properties:

(a) For every B-zero set $\{\alpha_\nu\}$ the product

$d_{\{\alpha_\nu\}}(z) = \prod_\nu d_{\alpha_\nu}(z)$ converges absolutely on **D** and uniformly on compact subsets of **D**.

(b) The divisor operator $T_{\{\alpha_\nu\}}$:

$$(T_{\{\alpha_\nu\}}f)(z) = f(z)/d_{\{\alpha_\nu\}}(z) \tag{3}$$

maps $B_{\{\alpha_\nu\}}$ into B, and $\|T_{\{\alpha_\nu\}}\| \le C$, where C is the same constant for all B-zero sets $\{\alpha_\nu\}$.

If C=1, the operators $T_{\{\alpha_\nu\}}$ are called _contractive_ divisor operators.

An analytic function g(z) in **D** is called _unimodular_ if its radial limit values $g(\zeta)$ ($\zeta\in\partial$**D**) exist a.e. on ∂**D** and $|g(\zeta)|=1$ a.e.

If B is the disk algebra A, or a Hardy space $H^p (1\le p\le\infty)$, then it is classical that the function $d_0(z)=z$ generates a Möbius invariant family of isometric divisor operators, and the divisors $d_{\{\alpha_\nu\}}$ are Blaschke products. However, for spaces whose zero sets do not necessarily satisfy the Blaschke condition, the study of divisors has begun only recently (see [1],[2],[3]). It has not been known for any such space whether efficient families of unimodular contractive divisors exist or not (isometric divisors are out of the question here). An important result of C. Horowitz [1] says that for the standard Bergman space A_2 with the norm

$$\|f\|_{A_2} = \left(\frac{1}{\pi}\iint_{\mathbf{D}} |f(z)|^2 \, r dr d\theta\right)^{1/2} \tag{4}$$

(and for a whole class of Bergman-type spaces) the 0-divisor $d_0(z)=z(2-z)$ generates an efficient Möbius-invaraint family of divisors; however, these divisors are neither contractive nor unimodular. On the other hand, some efficient unimodular divisors for A^{-n} (see [2]) have never been proved to be contractions (see also the remark at the end of the paper).

2. The main result.

Theorem. The 0-divisor $q_0(z)=z\exp\{\frac{2(1-z)}{1+z}\}$ _generates a_

<u>Möbius invariant</u> A_2-<u>efficient family of unimodular contractive divisors</u>.

<u>Outline of the Proof</u>.[1] Prove first that $q_{\{\alpha_\nu\}}(0) = \prod_\nu q_{\alpha_\nu}(0)$ converges for any A_2-zero set $\{\alpha_\nu\}$. We have

$$q_\alpha(0) = (q_0 \circ b_\alpha)(0) = q_0(|\alpha|) = |\alpha| \exp\{2(1-|\alpha|)/(1+|\alpha|)\} .$$

A straightforward computation shows that

$$q_\alpha(0) = 1 + O[(1-|\alpha|)^3] \quad (|\alpha| \to 1) , \tag{5}$$

and the same estimate holds for $q_\alpha(z)$, $z \in \mathbf{D}$, with the constant involved in "O" bounded on compact subsets of $\mathbf{D}$. Therefore $q_{\{\alpha_\nu\}}$ is well defined for the zero sets satisfying

$$\sum_\nu (1-|\alpha_\nu|)^3 < \infty , \tag{6}$$

which certainly holds for the class A_2 (see, e.g., [1] and a remark at the end of this paper).

To prove that the divisor operator corresponding to $q_{\{\alpha_\nu\}}$ is a contraction, it is enough to establish the following result:

<u>Lemma 1</u>. <u>Let</u> $f \in A_2$ <u>and</u> $f(\alpha)=0$ $(0 \le \alpha < 1$ fixed). <u>If</u>

$$g(z) = f(z) \frac{1-\alpha z}{\alpha - z} \exp\{- \frac{2(1-\alpha)(1+z)}{(1+\alpha)(1-z)}\} \tag{7}$$

<u>then</u>

$$\|g\|_{A_2} \le \|f\|_{A_2} \tag{8}$$

Proof of the Lemma. Use the substitution $\zeta = \xi + i\eta = i \frac{1+z}{1-z}$ that maps $\mathbf{D}$ onto the half-plane $\mathbb{C}_+ = (\zeta | \mathrm{Im}\zeta > 0)$; we obtain from (8):

[1] A detailed proof will appear elsewhere.

$$\iint\limits_{\mathbb{C}_+}\left|f\left(\frac{\zeta-i}{\zeta+i}\right)\frac{i\beta+\zeta}{i\beta-\zeta}e^{\frac{2i\zeta}{\beta}}\right|^2\frac{d\xi d\eta}{|i+\zeta|^4}\le\iint\limits_{\mathbb{C}_+}\left|f\left(\frac{\zeta-i}{\zeta+i}\right)\right|^2\frac{d\xi d\eta}{|i+\zeta|^4}, \tag{9}$$

where $\beta = \frac{1+\alpha}{1-\alpha}$. To get rid of β we substitute $\zeta=\beta w$ and put

$$(i+\beta w)^{-2}f\left(\frac{\beta w-i}{\beta w+i}\right) = F(w) ,$$

which reduces (9) to $(w=u+iv)$:

$$\iint\limits_{\mathbb{C}_+}\left|F(w)\,\frac{i+w}{i-w}\,e^{2iw}\right|^2 dudv \le \iint\limits_{\mathbb{C}_+}\left|F(w)\right|^2 dudv . \tag{10}$$

Now we represent $F(w)$ as a Fourier integral

$$F(w) = \int_0^\infty \phi(x)e^{ixw}dx \qquad (\operatorname{Im} w > 0) ,$$

which brings (10) to the following inequality:

$$\int_0^\infty (x+2)^{-1}\left|\phi(x) - 2\int_0^\infty \phi(x+t)e^{-t}dt\right|^2 dx \le \int_0^\infty \frac{|\phi(x)|^2}{x}\,dx , \tag{11}$$

where $\int_0^\infty \phi(x)e^{-x}dx=0$. After a chain of substitutions ($e^{-x}\phi(x) = \psi(x)$; $\int_0^x \psi(t)dt=p(x)$; $e^xp(x)=u(x)$) (11) can be reduced to the following

<u>Lemma 2</u>. <u>For differentiable functions</u> $u(x)$ <u>on</u> $(0,\infty)$ <u>such that</u>

$$\lim_{x\to 0} u(x)/x = 0,\ \lim_{x\to\infty} u(x)/\sqrt{x} = 0 , \tag{12}$$

<u>the following inequality holds</u>:

$$2\int_0^\infty \frac{|u(x)|^2 dx}{x^2(x+2)^2} \le \int_0^\infty \frac{|u'(x)|^2 dx}{x(x+2)} \tag{13}$$

Proof of Lemma 2. We can assume u(x) real. Using (12) it can be shown that to prove (13) it is enough to establish for an arbitrary interval $[a,b] \subset (0,\infty)$ the following:

$$2\int_a^b \frac{[u(x)]^2 dx}{x^2(x+2)^2} \le \int_a^b \frac{[u'(x)]^2 dx}{x(x+2)} \tag{14}$$

where $u(a)=u(b)=0$. By a standard variational argument this reduces to a Sturm-Liouville-type spectral problem

$$\left(\frac{u'}{x(x+2)}\right)' + \frac{2\lambda u}{x^2(x+2)^2} = 0 \ ; \quad u(a) = u(b) = 0 \ , \tag{15}$$

and the required result is this: there are no nontrivial solutions to (15) if $\lambda<1$. But this follows immediately from Sturm's comparison theorems and the fact that $u=x^2$ satisfies the differential equation

$$\left(\frac{u'}{x(x+2)}\right)' + \frac{2u}{x^2(x+2)^2} = 0 \ .$$

This completes the proof of Lemma 2 and the Theorem.

3. Remarks, Conjectures, Problems.

(a) An easy computation shows that $q_0(z)$ is the only contracting 0-divisor (up to a constant unimodular factor and rotation) that satisfies (5) and therefore factors out a class of zero sets as large as (6).

(b) There are reasons to conjecture that $q_0(z)$ is (essentially) a unique 0-divisor having all the properties stated in the Theorem.

(c) Condition (6) allows for a much faster growth than A_2. It is natural therefore to try and find out whether q_0 is efficient for other Bergman-type spaces (see [1]) and even for spaces of analytic functions with substantially faster rate of growth.

(d) It was shown in [2], [3] that q_0 is efficient for A^{-n}, the space of analytic functions with the norm $\|f\| = \sup_{z\in\mathbf{D}}\{(1-|z|)^n|f(z)|\}$. Is it contracting on A^{-n}? Some estimates obtained in [4] suggest that it may be so.

(e) Our theorem establishes a correspondence between z-invariant subspaces of A_2 defined by zero sets, and those defined by atomic masses of the premeasure on $\partial\mathbf{D}$ (see [3], [5]). What are the implications of this "balayage" of zeros to $\partial\mathbf{D}$ for the structure of z-invariant subspaces of A_2?

References

1. C. Horowitz. Zeros of Functions in the Bergman spaces. Duke Math. J., 41 (1974), 693-710.

2. B. Korenblum, An Extension of the Nevanlinna Theory. Acta Math., 135 (1975), 187-219.

3. B. Korenblum, A Beurling-type Theorem. Acta Math., 138 (1977), 265-293.

4. W.K. Hayman and B. Korenblum. A critical growth rate for functions regular in a disk. Michigan Math. J. 27 (1980), 21-30.

5. L. Brown and B. Korenblum. Cyclic vectors in $A^{-\infty}$. Proceedings of AMS, 102 (1988), 137-138.

Department of Mathematics
and Statistics
State University of New York
1400 Washington Avenue
Albany, NY 12222, USA

Operator Theory:
Advances and Applications, Vol. 41

TRIGONOMETRIC APPROXIMATION OF SOLUTIONS OF PERIODIC PSEUDODIFFERENTIAL EQUATIONS

William McLean and Wolfgang L. Wendland [1]

This paper is dedicated to Professor Dr. I. Gohberg on the occasion of his 60-th birthday.

Galerkin's method with trigonometric polynomials as trial and test functions is used to construct approximate solutions of periodic singular integral equations and pseudo-differential equations. We present results on pointwise rates of convergence, showing amongst other things that if the solution is in C^r, then the error is $O(n^{-r}\log n)$, where n is the degree of the trigonometric polynomials. This result is the best that can be expected, since the same rate of convergence holds for the partial sums of the Fourier series of a function in C^r.

AMS subject classifications: 42A10, 45E05, 45L10, 47G05, 65R20, 45F15, 45J05, 65E05, 65N30, 65N45.

INTRODUCTION

Here we analyse the classical Ritz-Galerkin projection method with Fourier series applied to a variety of boundary integral equations, or — more generally — to systems of pseudodifferential equations on smooth closed curves. This method of reduction with the global trigonometric polynomials on the one hand is very classical — for integral equations see E. Schmidt [17] and S. G. Mikhlin [9] — on the other hand it is most efficient in connection with numerical integration, since the fast Fourier transform provides a simple, numerically stable and fast tool to handle the discrete equations for the Fourier coefficients. With the trapezoidal rule this was pointed out by P. Henrici [5], and with composite Gaussian rules this method was analysed in [7], yielding an extremely accurate and fast numerical scheme. There one also finds numerical experiments and some applications to plate and elasticity problems and acoustic scattering. These numerical results justify the efficiency of the method.

For singular integral equations, the Fourier projection method was analysed in L_p by Gohberg and Fel'dman in [3] and by Prößdorf and Silbermann in [15] also in Sobolev spaces. An extension of this analysis to pseudodifferential equations can be found in [7]. In these works, optimal order rates of convergence were shown in terms of orders of n, the degree of the trigonometric polynomials. Based on the Sobolev imbedding theorem, the latter results also provide pointwise convergence, however, the corresponding orders are less than optimal by at least 1/2. For singular integral equations in Hölder spaces,

[1]This research was carried out while the first author was a guest professor at the Universität Stuttgart.

Prößdorf and Silbermann in [15, §8.2] show convergence of optimal order, however, the case of C^r estimates is excluded. Hence, corresponding pointwise estimates via imbedding will not be optimal, either. The analysis in [15] for Cauchy singular integral equations is based on using the canonical projections defined with the Hilbert transform, the factorization of the equation's coefficients, and corresponding commutation properties with the orthogonal projectors onto finite dimensional spaces of trigonometric polynomials. By using the same tools we extend this analysis to Hölder-Zygmund spaces which allow via imbedding *sharp pointwise error estimates* of the type

$$\|u_n - u\|_{C^s} \leq c(1/n)^{r-s}(\log n)\,\|u\|_{C^r}$$

for non-negative integers $s < r$. In the case of a scalar singular integral equation we perform all the analysis explicitly. Using Bessel potentials and the factorization of matrices, these results extend to general systems of pseudodifferential equations on closed smooth curves. Our convergence results are valid for elliptic equations with vanishing left and right indices of the principal symbol $\sigma_0(x,+1)$ and $\sigma_0(x,-1)$, respectively, and of $[\sigma_0(x,-1)]^{-1}\sigma_0(x,+1)$. A further generalization to equations with additional equilibrium conditions and new unknown parameters can be made without difficulties as in [7]; here we omit these more technical details. The above class of equations includes boundary integral equations of various types, such as Fredholm integral equations of the first and second kind, Cauchy singular integral equations, hypersingular integral equations and elliptic integro-differential equations, all of which are used to solve many different problems in applications — some are listed in [7] and [23].

Our paper is organised as follows. In §1, we formulate the well known Ritz-Galerkin reduction method with trigonometric polynomials. §2 is devoted to the Hölder-Zygmund spaces and corresponding approximation theory. In §3 we analyse singular integral equations in the Hölder-Zygmund spaces by using factorization and projection methods. In §4 we present the error analysis for the Fourier-Galerkin projection method applied to singular integral equations. These four chapters deal only with the scalar case, and can be read without further background concerning the generalizations in §§5 and 6. There we show how to extend our analysis — first to a scalar pseudodifferential equation, and then to systems — by using Bessel potentials. The necessary matrix factorization results are quoted.

There are a number of technical results in §§2–4 which are marked with the symbol †, to indicate that the proof is given at the end of the paper, in §7. The main ideas of the paper can be followed without reading these proofs.

1. GALERKIN'S METHOD WITH TRIGONOMETRIC POLYNOMIALS

Let Γ be a plane Jordan curve, given by a regular parameter representation

$$\Gamma : z = (z_1(t), z_2(t)) \simeq z_1(t) + iz_2(t),$$

where z is a 2π-periodic function of a real variable t, and $|dz/dt| \neq 0$. Via the representation we have a one-to-one correspondence between functions on Γ and 2π-periodic functions.

More generally, for a system of mutually disjoint Jordan curves $\Gamma = \cup_{j=1}^{L}\Gamma_j$, we may parametrize each and identify functions on Γ with L-vector valued 2π-periodic functions. We thus limit ourselves without loss of generality to systems of equations of the form

$$Au = f, \tag{1.1}$$

where the 2π-periodic vector valued function

$$u = (u_1(t), \cdots, u_L(t)), \qquad t \in \mathbf{R},$$

denotes the desired unknown, $f = (f_1(t), \cdots, f_L(t))$ denotes a given 2π-periodic function, and A a given linear pseudodifferential operator.

For simplicity, we consider first the special case when A is a scalar ($L = 1$) singular integral operator, satisfying certain conditions listed at the beginning of §4, which guarantee invertibilty in appropriate function spaces. We are interested in constructing approximate solutions of the equation (1.1). Denote the complex trigonometric monomials by

$$e_l(t) := \exp(ilt), \qquad l \in \mathbf{Z},$$

and the space of trigonometric polynomials of degree n by

$$\mathcal{T}_n := \operatorname{span}\{e_l : |l| \leq n\}, \qquad n \in \mathbf{N}_0.$$

Here, $\mathbf{N}_0 := \{0, 1, 2, 3, \ldots\}$ is the set of natural numbers including zero. When zero is excluded, we write $\mathbf{N} := \{1, 2, 3, \ldots\}$.

Define the normalized L_2 inner product

$$(f|g) := \frac{1}{2\pi}\int_{-\pi}^{\pi} \overline{f(t)}g(t)\,dt,$$

and the complex Fourier coefficients

$$\hat{f}(l) := (e_l|f) = \frac{1}{2\pi}\int_{-\pi}^{\pi} e^{-ilt} f(t)\,dt, \qquad l \in \mathbf{Z}.$$

For $n \in \mathbf{N}$, we seek a trigonometric polynomial $u_n \in \mathcal{T}_n$ satisfying

$$(\varphi|Au_n) = (\varphi|f) \qquad \text{for all } \varphi \in \mathcal{T}_n. \tag{1.2}$$

When it exists, u_n is said to be a *trigonometric Galerkin solution* of (1.1). The Fourier coefficients of u_n satisfy the $(2n+1) \times (2n+1)$ system of linear algebraic equations

$$\sum_{|j| \leq n} (e_l|Ae_j)\hat{u}_n(j) = \hat{f}(l), \qquad |l| \leq n. \tag{1.3}$$

We shall see later that these equations are uniquely solvable for all n sufficiently large. Moreover, the Parseval-Plancherel Theorem — which states that the L_2-norm of a function equals the l_2-norm of its sequence of Fourier coefficients — implies that the l_2 condition number of the coefficient matrix in (1.3) is bounded as $n \to \infty$, provided $A : L_2 \to L_2$ is invertible, and provided the Galerkin method is inverse stable in L_2, i.e., $\|u_n\|_{L_2} \leq c\|u\|_{L_2}$ with c independent of n. Under our assumptions on A, these properties hold by the results of [3] and [9, p.51].

2. HÖLDER-ZYGMUND SPACES AND APPROXIMATION THEORY

Let C denote the set of continuous functions $f : \mathbf{R} \to \mathbf{C}$ which are 2π-periodic, that is

$$f(x + 2\pi) = f(x), \qquad x \in \mathbf{R}.$$

Write $D = d/dx$, define

$$C^s := \{f \in C : D^j f \in C \text{ for } 0 \leq j \leq s\}, \qquad s \in \mathbf{N}_0,$$

and put $C^\infty := \bigcap_{s=0}^\infty C^s$. We think of C^s as a closed subspace of L_∞, and so use the notation

$$\|f\|_\infty := \max_{|x|\leq\pi} |f(x)|, \qquad f \in C.$$

In the usual way, C^s is made into a Banach space by putting

$$\|f\|_{C^s} := \sum_{j=0}^{s} \|D^j f\|_\infty, \qquad s \in \mathbf{N}_0.$$

For $h \in \mathbf{R}$ and $k \in \mathbf{N}$, the difference operator Δ_h^k is defined recursively by

$$\Delta_h f(x) = \Delta_h^1 f(x) := f(x+h) - f(x)$$

and

$$\Delta_h^k f(x) = \Delta_h(\Delta_h^{k-1} f), \qquad k \geq 2.$$

One finds, by induction on k, that

$$\Delta_h^k f(x) = \sum_{\nu=0}^{k} (-1)^{k-\nu} \binom{k}{\nu} f(x + \nu h).$$

Introduce the seminorm

$$[f]^{s,k} := \sup_{h>0} \frac{\|\Delta_h^k f\|_\infty}{h^s}, \qquad 0 < s \leq k,$$

and the norm

(2.1) $$\|f\|^{s,k} := \|f\|_\infty + [f]^{s,k}, \qquad 0 < s \leq k.$$

Here it is assumed $s \le k$, because if $[f]^{s,k} < \infty$ for some $s > k$, then f = constant. This fact is proved in [2, p.76]. The Hölder-Zygmund space of order $s > 0$ is defined by

$$\mathcal{H}^s := \{f \in C : [f]^{s,k} < \infty\}, \qquad 0 < s < k,$$

and is independent of the choice of k, as the following result shows.

LEMMA† 2.1. *If $l, k \in \mathbf{N}$ and $0 < s < k < l$, then*

$$\frac{1}{c_l}\|f\|^{s,l} \le \|f\|^{s,k} \le \frac{c_l}{k-s}\|f\|^{s,l}.$$

We use the notation $\|\cdot\|_{\mathcal{H}^s}$ for any equivalent norm on $\mathcal{H}^s$, but $\|\cdot\|^{s,k}$ will be needed in some places in order to keep track of how certain constants depend on s.

We shall now list a few basic properties of $\mathcal{H}^s$, and refer the reader to the standard texts on function spaces, such as Triebel [22], for more details. The space $\mathcal{H}^s$ is complete, but fails to be either reflexive or separable [6, pp.32, 47]. It is possible to construct $\mathcal{H}^s$ by real interpolation. Indeed, if s_0, $s_1 \in \mathbf{N}_0$ and

$$s = (1-\theta)s_0 + \theta s_1 \qquad \text{with } 0 < \theta < 1 \text{ and } s_0 \ne s_1, \tag{2.2}$$

then [22, p.201]

$$\mathcal{H}^s = (C^{s_0}, C^{s_1})_{\theta,\infty}. \tag{2.3}$$

Moreover, one can interpolate between Hölder-Zygmund spaces, that is

$$\mathcal{H}^s = (\mathcal{H}^{s_0}, \mathcal{H}^{s_1})_{\theta,\infty}, \tag{2.4}$$

for $s_0 > 0$ and $s_1 > 0$, with s and θ as in (2.2).

There is a close connection with the more familiar (periodic) Hölder space $C^{s,\alpha}$, defined for $s \in \mathbf{N}_0$ and $0 < \alpha \le 1$ by

$$\|f\|_{C^{s,\alpha}} := \|f\|_{C^s} + [D_s f]^{\alpha,1}.$$

In fact, it can be shown [22, p.201] that

$$C^{s,\alpha} = \mathcal{H}^{s+\alpha}, \qquad s \in N_0,\ 0 < \alpha < 1, \tag{2.5}$$

however, when $\alpha = 1$, the inclusions

$$C^{s+1} \subset C^{s,1} \subset \mathcal{H}^{s+1}, \qquad s \in \mathbf{N}_0, \tag{2.6}$$

are strict (and continuous). Thus, if $f \in \mathcal{H}^1$, then f need not be Lipschitz continuous, although [24, Vol.1, p.44] its modulus of continuity can be at worst $O(\delta|\log\delta|)$.

The space $\mathcal{H}^s$ can be characterized in terms of approximation by trigonometric polynomials, as the next result shows. Proofs can be found in the monographs of Nikol'skiĭ [11, pp.197, 201] and Timan [21, pp.260, 333].

THEOREM 2.2. *Suppose $s > 0$ and $f \in C$. The condition*

$$\inf_{v \in \mathcal{T}_n} \|v - f\|_\infty = O(n^{-s}) \qquad \text{as } n \to \infty$$

is necessary and sufficient for $f \in \mathcal{H}^s$.

Finally, let $\mathcal{P}_n$ denote the orthogonal projection of L_2 onto $\mathcal{T}_n$, then

$$\mathcal{P}_n f = \sum_{|l| \le n} \hat{f}(l) e_l, \qquad n \in \mathbf{N}_0. \tag{2.7}$$

In other words, $\mathcal{P}_n f$ is the n-th partial sum of the Fourier series of f. The following properties of $\mathcal{P}_n$ will be used later.

THEOREM 2.3. *Let $0 < s < r < \infty$.*

1. *The norm of the operator $\mathcal{P}_n : \mathcal{H}^s \to \mathcal{H}^s$ is $O(\log n)$ as $n \to \infty$.*
2. $\|(\mathcal{P}_n - I)f\|_{\mathcal{H}^s} \le c(1/n)^{r-s} \log n \|f\|_{\mathcal{H}^r}$.
3. *When $s \in \mathbf{N}_0$, $\|(\mathcal{P}_n - I)f\|_{C^s} \le c(1/n)^{r-s} \log n \|f\|_{\mathcal{H}^r}$.*

Apart from certain restrictions on the values of s and r, part 2 was first proved by Prößdorf [14], and part 3 is a classical result of D. Jackson. Recently, Prestin [13] has made an extensive study of approximation in periodic Lipschitz spaces.

3. SINGULAR INTEGRAL OPERATORS

We begin by introducing the operators

$$Pu := \sum_{l \ge 1} \hat{u}(l) e_l, \qquad Qu := \sum_{l \le -1} \hat{u}(l) e_l, \qquad Ru := \hat{u}(0) e_0,$$

each of which is a projection, that is, $P^2 = P$, $Q^2 = Q$ and $R^2 = R$. Notice also that $PQ = QP = 0$, $QR = RQ = 0$, $RP = PR = 0$ and

$$P + Q + R = I. \tag{3.1}$$

These operators are closely related to the periodic Hilbert transform, which is defined by the Cauchy principal value integral

$$Hu(x) := \frac{1}{2\pi i} \operatorname{PV} \int_{-\pi}^{\pi} \operatorname{cotan}\left(\frac{t-x}{2}\right) u(t)\, dt. \tag{3.2}$$

Indeed, since [2, p.337],

$$(Hu)\hat{\ }(l) = \operatorname{sign}(l) \hat{u}(l), \qquad l \in \mathbf{Z},$$

where

$$\operatorname{sign}(l) := \begin{cases} -1, & l \le -1 \\ 0, & l = 0 \\ 1, & l \ge 1, \end{cases}$$

it follows that $H = P - Q$ and

(3.3) $$P = \tfrac{1}{2}(I + H - R), \qquad Q = \tfrac{1}{2}(I - H - R).$$

Given $a \in C^\infty$, we use the same symbol to denote the corresponding pointwise multiplication operator, so that

$$(au)(x) = a(x)u(x), \qquad x \in \mathbf{R}.$$

The operators P and Q (or equivalently I and H), together with the compact and pointwise multiplication operators, suffice to generate the usual algebra of periodic singular integral operators [19, §1], [4], [10].

DEFINTION 3.1. *We say A is a periodic singular integral operator if*

$$A = aP + bQ + R + K,$$

where $a, b \in C^\infty$ and

$$K : \mathcal{H}^s \to \mathcal{H}^s$$

is a compact linear operator for every $s > 0$. If, in addition, for all $x \in \mathbf{R}$,

$$a(x) \neq 0 \quad \text{and} \quad b(x) \neq 0,$$

then A is said to be elliptic.

In the remainder of this section, we present some ideas and notation connected with the index theorem, which are required for the stability proofs of §4. Let

$$W(a) := \frac{1}{2\pi}[\arg a(t)]_{t=-\pi}^{\pi}$$

denote the winding number, about the origin, of the closed curve $t \mapsto a(t)$, $-\pi \leq t < \pi$, and write

$$\begin{aligned} \ker(A) &:= \{ u \in \mathcal{H}^s : Au = 0 \}, \\ \operatorname{im}(A) &:= \{ f \in \mathcal{H}^s : f = Au \text{ for some } u \in \mathcal{H}^s \}, \\ \operatorname{coker}(A) &:= \mathcal{H}^s / \overline{\operatorname{im}(A)}, \\ \operatorname{ind}(A) &:= \dim \ker(A) - \dim \operatorname{coker}(A) \end{aligned}$$

for the kernel, image, cokernel and index of A, respectively. We now state some classical results of F. Noether [12]. (For $0 < s < 1$ see [4, p.196], and for partial results [10, p.143].)

THEOREM† 3.2. *Suppose A is a periodic singular integral operator. The mapping*

(3.4) $$A : \mathcal{H}^s \to \mathcal{H}^s, \qquad s > 0,$$

is bounded. It is Fredholm if and only if A is elliptic, in which case

$$\operatorname{ind}(A) = W(b) - W(a).$$

COROLLARY 3.3. *The mapping (3.4) is invertible if and only if all of the following hold:*

1. *A is elliptic;*

2. $W(a) = W(b)$;

3. $\ker(A) = \{0\}$.

We remark that, for $K = 0$, condition 2 implies condition 3.

Next, define

$$C_+^\infty := \{ f \in C^\infty : f = (P + R)u \text{ for some } u \in C^\infty \},$$

then it is not difficult to verify that the following are equivalent:

1. $f \in C_+^\infty$.

2. The Fourier coefficients of f satisfy $\hat{f}(l) = 0$ for all $l \leq -1$ and, for all $N > 0$, $\hat{f}(l) = O(l^{-N})$ as $l \to \infty$.

3. The function $f \in C^\infty$ admits an analytic continuation into the upper half plane, which is bounded and 2π-periodic (i.e. $f(z + 2\pi) = f(z)$ for $\mathrm{Im} z \geq 0$).

There are analogous characterizations for

$$C_-^\infty := \{ f \in C^\infty : f = (Q + R)u \text{ for some } u \in C^\infty \},$$

and as a consequence one has the well known factorization property [4, p.78], [3, p.191].

THEOREM 3.4. [3, p.191] *Let* $a \in C^\infty$ satisfy $a(x) \neq 0$ *for every* $x \in \mathbf{R}$. *If* $W(a) = \kappa$, *then there exist functions* $a_\pm \in C_\pm^\infty$ *such that*

1. $a(x) = a_+(x)e^{i\kappa x}a_-(x)$ *for all* $x \in \mathbf{R}$,

2. $1/a_\pm \in C_\pm^\infty$.

For the general case of matrix functions $a(x)$, the theorem is by no means trivial. For a single function $a(x)$, however, let $\psi(x) = \log[e^{-i\kappa x}a(x)]$, then ψ is single-valued and we can take $a_+(x) = \exp[(P + R)\psi(x)]$ and $a_-(x) = \exp[(Q\psi(x)]$.

The pointwise multiplication operators associated with functions in $C_\pm^\infty$ obey the following important identities, cf. [3, p.126].

THEOREM† 3.5. *If* $a_\pm \in C_\pm^\infty$, *then*

$$\begin{aligned} Pa_+P &= a_+P, & Qa_+Q &= Qa_+, & Qa_+P &= 0, \\ Pa_-P &= Pa_-, & Qa_-Q &= a_-Q, & Pa_-Q &= 0. \end{aligned}$$

Now suppose A is elliptic, with coefficients satisfying $W(a) = 0 = W(b)$, so that by Theorem 3.4, there exist *canonical factorizations*

$$a = a_+a_-, \qquad b = b_-b_+. \tag{3.5}$$

Define the singular integral operators

$$\begin{aligned} M &:= a_+P + b_-Q + R, \\ N &:= Pa_- + Qb_+ + R, \end{aligned}$$

then it follows from Theorem 3.5 that

$$\begin{aligned} M^{-1} &= (1/a_+)P + (1/b_-)Q + R, \\ N^{-1} &= P(1/a_-) + Q(1/b_+) + R. \end{aligned} \tag{3.6}$$

Also, if $[\cdot,\cdot]$ is the usual commutator bracket, and if we define

$$K_1 := M^{-1}(K - a_+[P,a_-] - b_-[Q,b_+]), \tag{3.7}$$

then, for A as in Definition 3.1 ,

$$A = M(N + K_1). \tag{3.8}$$

This representation is the key to the error analysis which follows.

4. ERROR ESTIMATES

Let $\mathcal{P}_n$ be the projection operator defined by (2.7), and let u and u_n be as in §1, then

$$Au = f, \qquad \mathcal{P}_n A u_n = \mathcal{P}_n f, \qquad \mathcal{P}_n u_n = u_n. \tag{4.1}$$

We make three basic assumptions:

A1 The singular integral operator A is elliptic in the sense of Definition 3.1, with $W(a) = 0 = W(b)$ and $\ker(A) = \{0\}$.

A2 There is an $\epsilon > 0$ such that $K : \mathcal{H}^s \to \mathcal{H}^{s+\epsilon}$ is bounded for every $s > 0$.

A3 There is an $\epsilon > 0$ such that $K : C \to \mathcal{H}^\epsilon$ is bounded.

It can be seen from Corollary 3.3 that **A1** implies $A : \mathcal{H}^s \to \mathcal{H}^s$ is invertible. Also, Gohberg and Fel'dman [3, p.152] have shown, for the case $K = 0$, that **A1** is both necessary and sufficient for $u_n \to u$ in L_p, for every $f \in L_p$ $(1 < p < \infty)$. Moreover, if it should happen that $W(a) = W(b) = l_0 \neq 0$, then one need only multiply both sides of the first equation in (4.1) by the function e_{-l_0}, to obtain a new operator $e_{-l_0}A$, whose coefficients both have zero winding numbers. The resulting Galerkin equations are the same as (1.3), except that now $|l - l_0| \leq n$ instead of $|l| \leq n$.

Assumptions **A2** and **A3** are needed for technical reasons. Both will be satisfied if, for example, K is a periodic integral operator with a weakly singular kernel function. In applications, it often happens that K is a logarithmic convolution, or the kernel function is actually C^∞.

LEMMA† 4.1. *Let* K_1 *be the operator defined by* (3.7).

1. *If* **A2** *holds, then* $K_1 : \mathcal{H}^s \to \mathcal{H}^{s+\epsilon}$ *is bounded for every* $s > 0$.

2. If **A3** *holds, then* $K_1 : C^s \to \mathcal{H}^{s+\epsilon}$ *is bounded for every* $s \in \mathbf{N}_0$.

In the proof of the next theorem, as in the book by Prößdorf and Silbermann [15, p.99], a crucial rôle is played by the following identities due to Gohberg and Fel'dman [3, p.71].

LEMMA† 4.2. *If* $a_\pm \in C^\infty_\pm$, *then*

$$\begin{aligned} \mathcal{P}_n(a_+P + a_-Q)\mathcal{P}_n &= \mathcal{P}_n(a_+P + a_-Q), \\ \mathcal{P}_n(Pa_- + Qa_+)\mathcal{P}_n &= (Pa_- + Qa_+)\mathcal{P}_n. \end{aligned}$$

THEOREM 4.3. *Suppose* $0 < s < r < \infty$. *If* **A1–A2** *hold, then for all* n *sufficiently large, there exists a unique Galerkin solution* u_n, *and*

$$\|u_n - u\|_{\mathcal{H}^s} \le c(1/n)^{r-s} \log n \, \|u\|_{\mathcal{H}^r}.$$

PROOF: Equations (3.8) and (4.1) imply

$$M(N + K_1)u = f, \tag{4.2}$$

$$\mathcal{P}_n M(N + K_1)u_n = \mathcal{P}_n f. \tag{4.3}$$

From Lemma 4.2, it follows that $\mathcal{P}_n M \mathcal{P}_n = \mathcal{P}_n M$ and $\mathcal{P}_n M^{-1}\mathcal{P}_n = \mathcal{P}_n M^{-1}$, therefore $\mathcal{P}_n M^{-1}\mathcal{P}_n$ is the inverse of the finite dimensional operator $\mathcal{P}_n M \mathcal{P}_n : \mathcal{T}_n \to \mathcal{T}_n$. Hence, multiplying (4.3) on the left by $\mathcal{P}_n M^{-1}$, one obtains

$$\mathcal{P}_n(N + K_1)u_n = \mathcal{P}_n M^{-1} f.$$

Using Lemma 4.2 again, we see $\mathcal{P}_n N \mathcal{P}_n = N\mathcal{P}_n$, and so, by the third equation in (4.1), $\mathcal{P}_n N u_n = \mathcal{P}_n N \mathcal{P}_n u_n = N\mathcal{P}_n u_n = N u_n$. Thus,

$$(N + \mathcal{P}_n K_1)u_n = \mathcal{P}_n M^{-1} f,$$

and so, using (4.2), we find

$$(N + \mathcal{P}_n K_1)(u_n - u) = (\mathcal{P}_n - I)Nu. \tag{4.4}$$

By Lemma 4.1 and the approximation result, Theorem 2.3,

$$\|(\mathcal{P}_n - I)K_1 u\|_{\mathcal{H}^s} \le c(1/n)^\epsilon \log n \, \|u\|_{\mathcal{H}^s},$$

which shows that $\mathcal{P}_n K_1$ converges in the operator norm to $K_1 : \mathcal{H}^s \to \mathcal{H}^s$. Therefore, since $N + K_1$ is invertible, the bound

$$\|(N + \mathcal{P}_n K_1)^{-1}\|_{\mathcal{H}^s \to \mathcal{H}^s} \le c,$$

holds for n sufficiently large, and hence, by (4.4),

$$\|u_n - u\|_{\mathcal{H}^s} \le c\|(\mathcal{P}_n - I)Nu\|_{\mathcal{H}^s}.$$

The result follows now by using Theorem 2.3 and the fact that N is bounded on $\mathcal{H}^r$. □

It is not possible to replace $\mathcal{H}^s$ by C^s in the proof above, since $(N + K_1)^{-1}$ fails to be bounded on C^s (unless $a_- = b_+$). However, one does have a kind of stability estimate for the C^s-norm.

THEOREM 4.4. *Let $s \in \mathbf{N}_0$. If* **A1–A3** *hold, then for all n sufficiently large,*

$$\|u_n - u\|_{C^s} \leq c\|N^{-1}(\mathcal{P}_n - I)Nu\|_{C^s}, \tag{4.5}$$

where N is the singular integral operator defined by (3.6).

PROOF: Rewrite (4.4) as

$$(I + N^{-1}\mathcal{P}_n K_1)(u_n - u) = N^{-1}(\mathcal{P}_n - I)Nu,$$

then it suffices to show that, for n sufficiently large,

$$\|(I + N^{-1}\mathcal{P}_n K_1)^{-1}\|_{C^s \to C^s} \leq c. \tag{4.6}$$

Since $K_1 : C^s \to \mathcal{H}^{s+\epsilon}$ and $N^{-1} : \mathcal{H}^{s+\epsilon} \to \mathcal{H}^{s+\epsilon}$ are bounded, and since $\mathcal{H}^{s+\epsilon} = C^{s,\epsilon}$ is compactly imbedded in C^s, it follows that the linear operator

$$N^{-1}K_1 : C^s \to C^s \tag{4.7}$$

is compact. This implies $I + N^{-1}K_1$ is invertible on C^s, since it is easily seen to be one-one. Finally, the estimate

$$\begin{aligned} \|(N^{-1}\mathcal{P}_n K_1 - N^{-1}K_1)u\|_{C^s} &\leq c\|(\mathcal{P}_n - I)K_1 u\|_{\mathcal{H}^{s+\epsilon/2}} \\ &\leq c(1/n)^{\epsilon/2}\log n\, \|K_1 u\|_{\mathcal{H}^{s+\epsilon/2}} \\ &\leq c(1/n)^{\epsilon/2}\log n\, \|u\|_{C^s} \end{aligned}$$

shows that $N^{-1}\mathcal{P}_n K_1$ converges in the operator norm to (4.7), implying the bound (4.6). □

The next step is to estimate the right hand side of (4.5). This is a simple matter if, as is the case in some applications, the operator A has constant coefficients. Indeed, when a and b are constant, we can obviously choose $a_\pm$ and $b_\pm$ to be constant, then N will commute with $\mathcal{P}_n$, so $N^{-1}(\mathcal{P}_n - I)Nu = (\mathcal{P}_n - I)u$ can be estimated directly using part 3 of Theorem 2.3. In the general case of variable coefficients, we make use of the following identity.

LEMMA 4.5. *For all n,*

$$N^{-1}(\mathcal{P}_n - I)N = (\mathcal{P}_n - I) + P(1/a_-)[\mathcal{P}_n, a_-] + Q(1/b_+)[\mathcal{P}_n, b_+].$$

PROOF: Since $\mathcal{P}_n$ commutes with P, Q and R, it is an immediate consequence of the definition (3.6) of N, that

$$\mathcal{P}_n N = N\mathcal{P}_n + P[\mathcal{P}_n, a_-] + Q[\mathcal{P}_n, b_+]. \tag{4.8}$$

Next, Theorem 3.5 implies

$$N^{-1}P = P(1/a_-), \qquad N^{-1}Q = Q(1/b_+),$$

so, multiplying (4.8) on the left by N^{-1}, we find

$$N^{-1}\mathcal{P}_n N = \mathcal{P}_n + P(1/a_-)[\mathcal{P}_n, a_-] + Q(1/b_+)[\mathcal{P}_n, b_+].$$

The result is now obvious. □

Three technical lemmas are needed for estimating the terms involving commutators. Recall the definition (2.1) of the Hölder-Zygmund norm $\|\cdot\|^{s,k}$.

LEMMA† 4.6. *If $0 < s \leq k$, then the Hilbert transform (3.2) satisfies*

$$\|Hu\|_\infty \leq c_k s^{-1} \|u\|^{s,k}.$$

LEMMA† 4.7. *If $0 < s < k$, then*

$$\|au\|^{s,k} \leq c_k (k-s)^{-2} \|a\|^{s,k} \|u\|^{s,k}.$$

LEMMA† 4.8. *If $0 < s \leq r < k$, then*

$$\|[\mathcal{P}_n, a]u\|^{s,k} \leq c_k (k-r)^{-2} (1/n)^{r-s} \|a\|_{C^{k+1}} \|u\|^{r,k}.$$

Our final result for this section, when combined with Theorem 4.4, establishes the rate of convergence in C^s.

THEOREM 4.9. *If $s \in \mathbf{N}_0$ and $s < r < \infty$, then*

$$\|N^{-1}(\mathcal{P}_n - I)Nu\|_{C^s} \leq c_k (1/n)^{r-s} \log n \, \|u\|_{\mathcal{H}^r}.$$

PROOF: One can see from part 3 of Theorem 2.3, from (3.3), and from Lemma 4.5, that it suffices to show

$$\|Ha[\mathcal{P}_n, b]u\|_{C^s} \leq c(1/n)^{r-s} \log n \, \|u\|_{\mathcal{H}^r}, \tag{4.9}$$

where a, $b \in C^\infty$. Write $a^{(j)} = D^j a$, then, since D commutes with H,

$$D^j Ha[\mathcal{P}_n, b]u = H \sum_{\mu=0}^{j} \binom{j}{\mu} a^{(j-\mu)} D^\mu [\mathcal{P}_n, b]u.$$

Fix $k > r + 1$, then Lemma 4.6 and Lemma 4.7 imply

$$\|D^j Ha[\mathcal{P}_n, b]u\|_\infty \leq c\epsilon^{-1} \sum_{\mu=0}^{j} \|D^\mu [\mathcal{P}_n, b]u\|^{\epsilon,k}, \tag{4.10}$$

with c independent of ϵ in the range $0 < \epsilon < r - s$. Also, D commutes with $\mathcal{P}_n$, so

$$D^\mu[\mathcal{P}_n, b]u = \sum_{\nu=0}^{\mu} \binom{\mu}{\nu} [\mathcal{P}_n, b^{(\mu-\nu)}] D^\nu u,$$

and hence, by Lemma 4.8,

$$\|D^\mu[\mathcal{P}_n, b]u\|^{\epsilon,k} \le c(1/n)^{r-s-\epsilon} \sum_{\nu=0}^{\mu} \|D^\nu u\|^{r-s,k}, \tag{4.11}$$

again with c independent of ϵ. Combining (4.10) and (4.11), we obtain

$$\|Ha[\mathcal{P}_n, b]u\|_{C^s} \le c\epsilon^{-1}(1/n)^{r-s-\epsilon} \sum_{j=0}^{s} \|D^j u\|^{r-s,k},$$

then (4.9) follows by letting $\epsilon^{-1} = \log n$ and noting that the operator $D^j : \mathcal{H}^r \to \mathcal{H}^{r-j} \subset \mathcal{H}^{r-s}$ is bounded for $0 \le j \le s$. The latter fact can be proved easily using the interpolation result (2.3). □

Theorems 4.4 and 4.5, together with the imbedding (2.6), imply the error estimate quoted in the Intoduction:

$$\|u_n - u\|_{C^s} \le c(1/n)^{r-s}(\log n)\|u\|_{C^r}, \tag{4.12}$$

which does not involve the Hölder-Zygmund norm. Note that in the trivial case when $A = I$, the trigonometric Galerkin solution u_n is just $\mathcal{P}_n u$ and the estimate for $u_n - u$ given by Theorems 4.4 and 4.5 is the same as that given in part 3 of Theorem 2.3.

5. PERIODIC PSEUDODIFFERENTIAL EQUATIONS

We consider equations with operators given by

$$Bv = (aP + bQ + R + K)\Lambda^\beta v = f, \tag{5.1}$$

where the operator Λ^β denotes the periodic Bessel potential operator of order $\beta \in \mathbf{R}$, given by

$$\Lambda^\beta e_l = |l + \delta_{0l}|^\beta e_l, \qquad l \in \mathbf{Z}, \tag{5.2}$$

and corresponding continuous extensions, where δ_{0l} denotes the Kronecker symbol. Obviously, if $\beta = 0$, then B is just a singular integral operator.

From Agranovich's Theorem [1], [16], it is known that any one-dimensional (classical) pseudodifferential operator of order β acting on periodic functions (or functions defined on a closed curve) can be written in the form (5.1). In fact, suppose B_0 is the principal part of B, i.e.,

$$B_0 u(x) = \sum_{k \in \mathbf{Z}} \sigma_0(x,k)\hat{u}(k)e^{ikx},$$

where $\sigma_0(x,k)$ — the principal symbol of B — is 2π-periodic in x, and positive homogeneous of degree β in $k \ne 0$, i.e.,

$$\sigma_0(x,k) = \begin{cases} |k|^\beta \sigma_0(x, k/|k|), & k \ne 0 \\ 1, & k = 0. \end{cases}$$

A simple calculation shows that with

$$a(x) := \sigma_0(x,+1), \qquad b(x) := \sigma_0(x,-1), \tag{5.3}$$

the principal part of B can be written as $B_0 = (aP + bQ + R)\Lambda^\beta$.

The following mapping property of Λ^β allows the extension of the former results to (5.1), see [18, p.149].

THEOREM 5.1 *If $s > 0$ and $s - \beta > 0$, then $\Lambda^\beta : \mathcal{H}^s \to \mathcal{H}^{s-\beta}$ is an isomorphism.*

Hence, the definition of the spaces $\mathcal{H}^s$ can be extended to arbitrary $s \in \mathbf{R}$ in the following way. Define $\mathcal{H}^s$ to be the set of all periodic distributions f satisfying $\Lambda^{-\beta} f \in \mathcal{H}^{s+\beta}$ for some (and hence all) β with $s + \beta > 0$, and then equip $\mathcal{H}^s$with the norm

$$\|f\|_{\mathcal{H}^s} := \|\Lambda^{-\beta} f\|_{\mathcal{H}^{s+\beta}}. \tag{5.4}$$

Different choices of β lead to equivalent norms.

LEMMA 5.2. *The operator Λ^β commutes with $\mathcal{P}_n$, with P and with Q.*

These commutation properties follow immediately from (5.2). Just as for the case of a singular integral operator ($\beta = 0$), we say that B is *elliptic* if the functions (5.3) satisfy

$$a(x) \neq 0 \quad \text{and} \quad b(x) \neq 0 \quad \text{for all } x \in \mathbf{R},$$

but note that $K : \mathcal{H}^s \to \mathcal{H}^s$ is now assumed to be compact for all $s \in \mathbf{R}$. Obviously, when B is elliptic, the mapping

$$B : \mathcal{H}^s \to \mathcal{H}^{s-\beta}, \qquad s \in \mathbf{R},$$

is an isomorphism if and only if $W(a) = W(b)$ and $\ker(B) = \{0\}$.

Let us write $A = aP + bQ + R + K$, then the equation

$$Bv = A\Lambda^\beta v = f \tag{5.5}$$

is equivalent to

$$Au = f \qquad \text{with } u = \Lambda^\beta v,$$

and the trigonometric Galerkin method for (5.5), i.e., to find $v_n \in \mathcal{T}_n$ satisfying

$$(\varphi | Bv_n) = (\varphi | f) \qquad \text{for all } \varphi \in \mathcal{T}_n, \tag{5.6}$$

is equivalent to finding

$$u_n = \Lambda^\beta v_n \in \mathcal{T}_n \tag{5.7}$$

satisfying

$$(\varphi | Au_n) = (\varphi | f) \qquad \text{for all } \varphi \in \mathcal{T}_n, \tag{5.8}$$

i.e., (1.2). In the usual way, (5.6) and (5.8) lead to systems of equations for the Fourier coefficients of v_n and u_n, respectively. The two systems are related by a simple rescaling of the unknowns, as can be seen from (5.7), and in practice it is better to solve for u_n than for v_n, because the ℓ_2 condition number of the coefficient matrix arising from (5.6) is $O(n^{|\beta|})$ as $n \to \infty$. This follows at once from our earlier remarks at the end of §1, and from the fact the condition number of the diagonal matrix

$$[(e_l|\Lambda^\beta e_k)] = \begin{bmatrix} n^\beta & & & & \\ & \ddots & & & \\ & & 1 & & \\ & & & \ddots & \\ & & & & n^\beta \end{bmatrix}$$

is $n^{|\beta|}$ (exactly).

Let us formulate assumptions for B corresponding to our earlier assumptions for A.

B1 The pseudodifferential operator B, given in the form (5.1), is elliptic with $W(a) = 0 = W(b)$ and $\ker(B) = \{0\}$.

B2 There is an $\epsilon > 0$ such that K is a pseudodifferential operator of order $-\epsilon$.

Assumption **B2** implies that K satisfies **A2** and **A3** from §4, because K must be a periodic integral operator whose kernel behaves along the diagonal $x = t$ like $|x - t|^{\epsilon-1}$ if $0 < \epsilon < 1$, and like $\log|x - t|$ if $\epsilon = 1$; see [20, p.40].

THEOREM 5.3. *Suppose* $-\infty < s < r < \infty$. *If* B *satisfies* **B1** *and* **B2**, *then for all* n *sufficiently large, there exists a unique Galerkin solution* v_n *of* (5.6), *and*

$$\|v_n - v\|_{\mathcal{H}^s} \le c(1/n)^{r-s}(\log n)\|v\|_{\mathcal{H}^r}.$$

Moreover, when $s \in \mathbf{N}_0$,

$$\|v_n - v\|_{C^s} \le c(1/n)^{r-s}\|v\|_{\mathcal{H}^r}.$$

PROOF: We begin by showing that Theorem 4.3 remains valid when the restriction $s > 0$ is removed. Indeed, given $s \le 0$, choose α such that $s + \alpha > 0$, then Lemma 5.2 implies

$$\begin{aligned} (\Lambda^{-\alpha}A\Lambda^\alpha)(\Lambda^{-\alpha}u) &= \Lambda^{-\alpha}f \\ \mathcal{P}_n(\Lambda^{-\alpha}A\Lambda^\alpha)(\Lambda^{-\alpha}u_n) &= \mathcal{P}_n\Lambda^{-\alpha}f. \end{aligned}$$

The composition theorem for pseudodifferential operators implies that the principal symbol of $\Lambda^{-\alpha}A\Lambda^\alpha$ is the same as that of A, hence

$$\|\Lambda^{-\alpha}u_n - \Lambda^{-\alpha}u\|_{\mathcal{H}^{s+\alpha}} \le c(1/n)^{(r+\alpha)-(s+\alpha)}(\log n)\|\Lambda^{-\alpha}u\|_{\mathcal{H}^{r+\alpha}},$$

and the error estimate in $\mathcal{H}^s$ follows at once from (5.4).

Now consider $B = A\Lambda^\beta$. The singular integral operator $\Lambda^{-\beta}B = \Lambda^{-\beta}A\Lambda^\beta$ has the same principal symbol as A, so the relations

$$(\Lambda^{-\beta}B)v = \Lambda^{-\beta}f, \qquad \mathcal{P}_n(\Lambda^{-\beta}B)v_n = \mathcal{P}_n\Lambda^{-\beta}f,$$

imply the result. □

6. SYSTEMS OF PSEUDODIFFERENTIAL EQUATIONS

Following the arguments by Prößdorf and Silbermann [15, §§1.2 and 3], the analysis of the previous chapters can be extended without difficulties to systems of singular integral equations and, further, to elliptic systems of pseudodifferential equations on Γ. As indicated in [23, §4], and using the results of [1], [16], every system of pseudodifferential equations on Γ can be written in the form

$$Bv = \Lambda^\gamma\Big(\sigma_0(x,+1)P + \sigma_0(x,-1)Q + R + K\Big)\Lambda^\beta v = f, \tag{6.1}$$

where σ_0 is the principal symbol of B, a matrix valued function associated with the 2π-periodic parametrization, and where $\gamma = (\gamma_1, \ldots, \gamma_L)$, $\beta = (\beta_1, \ldots, \beta_L) \in \mathbf{R}^L$ are suitable vectors of orders. Λ^γ and Λ^β are defined by the diagonal matrix of operators

$$\Lambda^\mu := [\delta_{jk}\Lambda^{\mu_j}], \qquad j,k = 1, \ldots, L,$$

where δ_{jk} is the Kronecker symbol, and $\mu = \gamma$ or β, respectively. Equations of the type (6.1) include a large class of integro-differential equations [7] and boundary integral equations of various types resulting from the reduction to the boundary Γ of two-dimensional elliptic boundary value problems [23]. Now, with

$$u_n = \Lambda^\beta v_n,$$

the trigonometric Galerkin method for (6.1), i.e., to find $v_n \in \mathcal{T}_n^L$ satisfying

$$(\varphi|Bv_n) = (\varphi|f) \qquad \text{for all } \varphi \in \mathcal{T}_n^L, \tag{6.2}$$

is equivalent to

$$(\psi|Au_n) = (\psi|\Lambda^{-\gamma}f) \qquad \text{for all } \psi \in \mathcal{T}_n^L, \tag{6.3}$$

with $\psi = \Lambda^\gamma\varphi$ and with the operator

$$A := \sigma_0(x,+1)P + \sigma_0(x,-1)Q + R + K, \tag{6.4}$$

which defines a classical system of singular integral equations. For the application of the analysis of §4 to (6.3), assumption **A1** is to be replaced by corresponding properties which allow the canonical factorizations (3.5) also for the system. Following Prößdorf and Silbermann [15, p.135], we assume:

B1′ $\det\sigma_0(x,\pm1) \neq 0$ for all x; the left indices of $\sigma_0(x,+1)$ and the right indices of $\sigma_0(x,-1)$ and of $\sigma_0(x,-1)^{-1}\sigma_0(x,+1)$ are all equal to zero; and $ker(A) = \{0\}$;

along with **B2**. Assumption **B1**$'$ guarantees the existence of matrix factorizations

$$\sigma_0(x,+1) = a_+ a_-, \qquad \sigma_0(x,-1) = b_- b_+$$

with $a_\pm$, $b_\pm \in (C^\infty_\mp)^{L\times L}$, having all the desired properties needed in §3 and §4. In [7] one finds some stronger but simpler conditions for σ_0 implying **B1**$'$.

THEOREM 6.1. *Suppose* $-\infty < s < r < \infty$. *If* B *satisfies* **B1**$'$ *and* **B2**, *hen for all* n *sufficienty large, there exists a unique Galerkin solution* v_n *of* (6.2), *and*

$$\|v_n - v\|_{\mathcal{H}^s} \le c(1/n)^{r-s}(\log n)\|v\|_{\mathcal{H}^r}.$$

Moreover,

$$\|v_n - v\|_{C^s} \le c(1/n)^{r-s}(\log n)\|v\|_{\mathcal{H}^r}$$

when $s \in \mathbf{N}_0$.

As was indicated above, the proof follows as before if the function factorizations are replaced by the matrix factorizations; we omit the details.

The results in Theorem 6.1 can also be extended to systems which generalize (6.1) by incorporating additional compatibility conditions and unknown scalars, as in [7].

7. TECHNICAL PROOFS

In this section, we present proofs of those results from the main part of the paper which are marked with a †.

PROOF OF LEMMA 2.1: Define the k-th order modulus of continuity of f,

$$\omega^k(f,\delta) := \sup_{0<h<\delta} \|\triangle_h^k f\|_\infty, \qquad \delta > 0,$$

then, using the elementary inequality

$$\omega^l(f,\delta) \le 2^{l-k}\omega^k(f,\delta), \qquad \delta > 0,\, k < l,$$

we obtain $[f]^{s,l} \le 2^{l-k}[f]^{s,l}$ for $0 < s \le k$. The other estimate, $[f]^{s,k} \le c_l(k-s)^{-1}[f]^{s,l}$ $(0 < s < k)$, is a consequence of the Marchaud inequality [8, p.374] (or [21, p.105])

$$\omega^k(f,\delta) \le c_l\delta^k\Big(\|f\|_\infty + \int_\delta^1 \frac{\omega^l(f,t)}{t^{k+1}}\,dt\Big), \qquad 0 < \delta \le 1,$$

and the bound

$$\delta^k \int_\delta^1 \frac{\omega^l(f,t)}{t^{k+1}}\,dt \le \delta^k \int_\delta^1 t^{s-k-1}[f]^{s,l}dt \le \frac{\delta^s}{k-s}[f]^{s,l}. \qquad \square$$

PROOF OF THEOREM 2.3: Define the convolution of periodic functions by

$$f * g(x) := \frac{1}{2\pi}\int_{-\pi}^{\pi} f(x-t)g(t)\,dt,$$

and note that, with this normalization, the Fourier coefficients satisfy $(f * g)\hat{}(l) = \hat{f}(l)\hat{g}(l)$ for all $l \in \mathbf{Z}$. Denote the Dirichlet kernel by

$$\mathcal{D}_n(t) := \sum_{|l| \le n} e^{ilt} = \frac{\sin[(n+\frac{1}{2})t]}{\sin[\frac{1}{2}t]},$$

then

$$\mathcal{P}_n f = \mathcal{D}_n * f, \tag{7.1}$$

and recall that the Lebesgue constants satisfy [24, Vol.1, p.67]

$$\mathcal{L}_n := \frac{1}{2\pi}\int_{-\pi}^{\pi} |\mathcal{D}_n(t)|\,dt = \frac{4}{\pi^2}\log n + O(1), \qquad \text{as } n \to \infty.$$

Thus, since $\triangle_h^k(\mathcal{P}_n f) = \mathcal{D}_n * (\triangle_h^k f)$, it follows by using Young's inequality, that

$$\|\mathcal{P}_n f\|^{s,k} \le \mathcal{L}_n \|f\|^{s,k}. \tag{7.2}$$

This completes the proof of part 1.

Fix $k \in \mathbf{N}$ such that $0 < s < r < k$, and define

$$l := \begin{cases} (k+2)/2, & \text{if } k \text{ is even} \\ (k+3)/2, & \text{if } k \text{ is odd,} \end{cases}$$

then

$$2l \ge k + 2. \tag{7.3}$$

Given $n \in \mathbf{N}$, choose $\lambda \in \mathbf{N}$ such that

$$l(\lambda - 1) \le n \le l\lambda, \tag{7.4}$$

and define, cf. [11, p.87] and [21, p.258],

$$g_n(t) := \beta_n \left[\frac{\sin(\lambda t/2)}{\sin(t/2)}\right]^{2l},$$

where β_n is chosen in such a way that

$$\int_{-\pi}^{\pi} g_n(t)\,dt = 1. \tag{7.5}$$

Apart from a constant factor, the function g_n is just the l-th power of the Fejér kernel of order $\lambda - 1$, and is therefore a trigonometric polynomial of degree $l(\lambda - 1)$. Hence, (7.4) implies $g_n \in \mathcal{T}_n$. Also, it can be verified that

$$0 < \beta_n \le c_k n^{1-2l}. \tag{7.6}$$

Define

$$\mathcal{G}_n(t) := 2\pi \sum_{\nu=1}^{k} (-1)^{\nu-1} \frac{1}{\nu} \sum_{\mu=1}^{\nu} g_n\left(\frac{t + 2\pi\mu}{\nu}\right)$$

and put $\mathcal{Q}_n f := \mathcal{G}_n * f$, then it can be shown [11, p.196] that $\mathcal{G}_n \in \mathcal{T}_n$ and

$$(7.7) \qquad (\mathcal{Q}_n - I)f(x) = (-1)^{k+1}\int_{-\pi}^{\pi} g_n(t)\triangle_t^k f(x)\,dt.$$

Note that $\mathcal{Q}_n f \in \mathcal{T}_n$. One can see from (7.6), that

$$(7.8) \qquad |g_n(t)| \le c_k n^{1-2l}|t|^{-2l}, \qquad 0 < |t| \le \pi,$$

and consequently (7.2), (7.4) and (7.6) imply

$$(7.9) \qquad \|(\mathcal{Q}_n - I)f\|_\infty \le c_k(1/n)^r[f]^{r,k}.$$

(Estimate the integral for $|t| \le n^{-1}$ and for $n^{-1} < |t| \le \pi$.) Since

$$\triangle_h^k(\mathcal{Q}_n - I)f(x) = \int_{-\pi}^{\pi} g_n(t)\Psi(t,h,x)\,dt,$$

where

$$\begin{aligned}\Psi(t,h,x) &:= \sum_{\nu=0}^{k}(-1)^{\nu-1}\binom{k}{\nu}\triangle_t^k f(x+\nu h)\\ &= \sum_{\mu=0}^{k}(-1)^{\mu-1}\binom{k}{\mu}\triangle_h^k f(x+\mu t),\end{aligned}$$

it follows that

$$(7.10) \qquad \|\triangle_h^k(\mathcal{Q}_n - I)f\|_\infty \le 2^{k+1}[f]^{r,k}\Phi,$$

where $\Phi := \int_0^\pi |g_n(t)|\min\{t^r, h^r\}\,dt$; we claim

$$(7.11) \qquad \Phi \le c_k(1/n)^{r-s}h^s.$$

Indeed, when $h \le n^{-1}$, one can use (7.5) to show $\Phi \le h^r \le (1/n)^{r-s}h^s$. When $h > n^{-1}$, write

$$\Phi = \Phi_1 + \Phi_2 + \Phi_3 := \int_0^{n^{-1}} + \int_{n^{-1}}^{h} + \int_h^{\pi},$$

and note that (7.3) implies $2l - r - 1 \ge 1$. It follows from (7.5) that

$$\Phi_1 = \int_0^{n^{-1}} |g_n(t)|t^r\,dt \le (1/n)^r \le (1/n)^{r-s}h^s,$$

and using (7.8) one finds

$$\Phi_2 \le c_k\int_{n^{-1}}^{h} n^{1-2l}t^{r-2l}\,dt \le c_k(1/n)^r \le c_k(1/n)^{r-s}h^s$$

and

$$\begin{aligned}\Phi_3 &\le c_k h^r n^{1-2l}\int_h^{\pi} t^{-2l}\,dt \le c_k h^{r+1-2l}n^{1-2l}\\ &\le c_k h^s(1/n)^{r-s}(1/nh)^{2l-(r-s)-1} \le c_k h^s(1/n)^{r-s}.\end{aligned}$$

This completes the proof of (7.11), and now (7.10) implies

$$[(\mathcal{Q}_n - I)f]^{s,k} \le c_k(1/n)^{r-s}[f]^{r,k}.$$

Hence, by making use of (7.9), one obtains

$$\|(\mathcal{Q}_n - I)f\|^{s,k} \le c_k(1/n)^{r-s}[f]^{r,k}.$$

To complete the proof of part 2, it suffices to observe that, since $\mathcal{P}_n\mathcal{Q}_n = \mathcal{Q}_n$, the triangle inequality and (7.2) imply

$$\|(\mathcal{P}_n - I)f\|^{s,k} \le (1 + \mathcal{L}_n)\|(\mathcal{Q}_n - I)f\|^{s,k}.$$

Finally, for the proof of part 3, we may assume $s = 0$, since D commutes with $\mathcal{P}_n$ and is bounded from $\mathcal{H}^r$ to $\mathcal{H}^{r-1}$. The result then follows from (7.9), using the triangle inequality as with (7). □

PROOF OF THEOREM 3.2: For $0 < s < 1$ these are the well known results by F. Noether [12], since in this case $\mathcal{H}^s = C^{0,s}$; see for example [10, p.198ff.], [3] or [4]. We shall outline a simple, fairly self-contained proof for the general case $s > 0$, although we will not show that ellipticity is necessary for the Fredholm property.

The pointwise multiplication operators are bounded on $\mathcal{H}^s$, by Lemma 4.7, or alternatively, by using the interpolation result (2.3) and the fact that they are bounded on C^s for $s \in \mathbf{N}_0$. To show P and Q are bounded, it suffices, because of (3.3), to show H is bounded. When $0 < s < 1$, this is a classical result of I. Privalov. Moreover, since D commutes with H, it is obvious from (2.5) that H is bounded on $\mathcal{H}^s$ for $0 < s \ne$ integer, and hence, by interpolation (2.4), for all $s > 0$. An alternative proof of the boundednes of H on $\mathcal{H}^s$, which uses approximation theory, can be found in [21, p.163] (cf. [24, Vol.1, p.121]).

It remains to prove the Fredholm property and the formula for the index. Let $W(a) = \kappa$ and $W(b) = \rho$, then, by Theorem 3.4, there exist factorizations $a = a_+e_\kappa a_-$ and $b = b_-e_\rho b_+$. Define

$$B := e_\kappa P + e_\rho Q,$$

then, after a little algebra, one finds

$$\begin{aligned} A - MBN &= R + K - RBN - a_+e_\kappa[P, a_-] - b_-e_\rho[Q, b_+] \\ &\quad -a_+[P, e_\kappa]Pa_- - a_+[P, e_\rho]Qb_+ \\ &\quad -b_-[Q, e_\kappa]Pa_- - b_-[Q, e_\rho]Qb_+, \end{aligned}$$

where M and N are as in (3.6). From the proof of Lemma 4.1, it follows that the right hand side of the last equation is a compact operator on $\mathcal{H}^s$. Therefore, since M and N are both invertible, if B is Fredholm then so is A, and moreover $\mathrm{ind}(A) = \mathrm{ind}(B)$. We shall now compute the index of B.

Since the operator $u \mapsto e_l u$ is invertible for any $l \in \mathbf{Z}$, and since

$$B = e_\kappa(P + e_{\rho-\kappa}Q) = e_\rho(e_{\kappa-\rho}P + Q),$$

we may assume, without loss of generaltiy, that $\rho = 0$ or $\kappa = 0$.

If $\kappa \geq \rho = 0$, then

$$(Bu)^\frown(l) = \begin{cases} \hat{u}(l), & l \leq -1, \\ 0, & 0 \leq l \leq \kappa, \\ \hat{u}(l-\kappa), & \kappa + 1 \leq l, \end{cases}$$

so $\ker(B) = \operatorname{span}\{e_0\}$ and $\operatorname{im}(B) = \{ f \in \mathcal{H}^s : \hat{f}(l) = 0 \text{ for } 0 \leq l \leq \kappa \}$, which shows B is Fredholm with $\operatorname{ind}(B) = 1 - (\kappa + 1) = -\kappa$.

If $\rho > \kappa = 0$, then

$$(Bu)^\frown(l) = \begin{cases} \hat{u}(l-\rho), & l \leq 0, \\ \hat{u}(l) + \hat{u}(l-\rho), & 1 \leq l \leq \rho - 1, \\ \hat{u}(l), & \rho \leq l, \end{cases}$$

so $u \in \ker(B)$ if and only if $u = \sum_{|l| \leq \rho - 1} \xi_l e_l$ and

$$\begin{bmatrix} 1 & & & & 1 & & \\ & \ddots & & & & \ddots & \\ & & 1 & & & & 1 \\ & & & 0 & & & \\ 0 & & & & 0 & & \\ & \ddots & & & & \ddots & \\ & & 0 & & & & 0 \end{bmatrix} \begin{bmatrix} \xi_{1-\rho} \\ \vdots \\ \xi_{-1} \\ \xi_0 \\ \xi_1 \\ \vdots \\ \xi_{\rho-1} \end{bmatrix} = \begin{bmatrix} 0 \\ \vdots \\ 0 \\ 0 \\ 0 \\ \vdots \\ 0 \end{bmatrix}.$$

Thus, $\dim \ker(B) = \rho$, and since $B[P + e_{-\rho}(Q + R)] = I$, it follows $\operatorname{im}(B) = \mathcal{H}^s$, so $\operatorname{ind}(B) = \rho$. □

PROOF OF THEOREM 3.5: Since

$$(a_+Pf)^\frown(j) = \sum_{l \in \mathbf{Z}} \hat{a}_+(j-l)(Pf)^\frown(l) = \begin{cases} 0, & j \leq 0 \\ \sum_{l=1}^{j} \hat{a}_+(j-l)\hat{f}(l), & j \geq 1, \end{cases} \tag{7.12}$$

it is clear that $Qa_+P = 0$ and $Pa_+P = a_+P$. Similarly, since $(a_+Rf)^\frown(j) = \hat{a}_+(j)\hat{f}(0)$, we have $Qa_+R = 0$, and hence by (3.3),

$$Qa_+Q = Qa_+[I - P - R] = Qa_+.$$

The other three identities can be proved in a similar fashion. □

PROOF OF LEMMA 4.1: (Recall the definition (3.7) of K_1.) To prove the result, it is certainly sufficient to show that $[P, a_-]$ and $[Q, b_+]$ are periodic integral operators with C^∞ kernel functions. For this, observe that

$$[H, a]u(x) = \frac{1}{2\pi} \int_{-\pi}^{\pi} \cos\left(\frac{t-x}{2}\right) \left[\frac{a(t) - a(x)}{\sin(\frac{t-x}{2})}\right] u(t)\, dt,$$

and then use (3.3). □

PROOF OF LEMMA 4.2: From (7.1), one can see

$$(\mathcal{P}_n a_+ Pf)^\frown(j) = \begin{cases} 0, & j \le 0, \\ \sum_{1\le l\le \min\{j,n\}} \hat{a}_+(j-l)\hat{f}(l), & j \ge 1, \end{cases}$$

and hence $\mathcal{P}_n a_+ Pf = \mathcal{P}_n a_+ P\mathcal{P}_n f$. Similarly,

$$(\mathcal{P}_n a_- Qf)^\frown(j) = \begin{cases} \sum_{\max\{j,-n\}\le l\le -1} \hat{a}_-(j-l)\hat{f}(l), & j \le -1, \\ 0, & j \ge 0, \end{cases}$$

so $\mathcal{P}_n a_- Qf = \mathcal{P}_n a_- Q\mathcal{P}_n f$, which proves the first identity. The proof of the second identity is similar. □

PROOF OF LEMMA 4.6: Since

$$Hu(x) = \frac{1}{2\pi i}\int_0^\pi \operatorname{cotan}(t/2)\,[u(x+t) - u(x-t)]\,dt,$$

it follows that, for $0 < s \le 1$,

$$\|Hu\|_\infty \le c[u]^{s,1}\int_0^\pi t^{s-1}\,dt \le cs^{-1}[u]^{s,1},$$

where c is an absolute constant. This proves the result in the case $k = 1$. Now let $k \ge 2$. If $0 < s \le 1/2$, then Lemma 2.1 implies

$$[u]^{s,1} \le c_k(1-s)^{-1}\|u\|^{s,k} \le 2c_k\|u\|^{s,k},$$

and if $1/2 < s \le k$, then

$$\|Hu\|_\infty \le c[u]^{1/2,1} \le c_k[u]^{1/2,k} \le c_k[u]^{s,k}. \qquad \square$$

PROOF OF LEMMA 4.7: One can verify, by induction on k, that

$$\Delta_h^k(au)(x) = \sum_{\nu=0}^{k}\binom{k}{\nu}\Delta_h^\nu a(x)\Delta_h^{k-\nu}f(x+\nu h),$$

so

$$\frac{\|\Delta_h^k(au)\|_\infty}{h^s} \le \sum_{\nu=0}^{k}\binom{k}{\nu}\frac{\|\Delta_h^\nu a\|_\infty}{h^{s\nu/k}}\,\frac{\|\Delta_h^{k-\nu}f\|_\infty}{h^{s(k-\nu)/k}}.$$

Thus, it suffices to show that the bound

$$\|\Delta_h^\nu f\|_\infty \le c_k(k-s)^{-1}h^{s\nu/k}\|f\|^{s,k} \tag{7.13}$$

is valid for $0 \le \nu \le k$. The cases $\nu = 0$ and $\nu = k$ are obvious, so we can assume $1 \le \nu \le k-1$. Using the Marchaud inequality [8, p.374]

$$\omega^k(f,\delta) \le c_k\delta^\nu\Big(\|f\|_\infty + \int_\delta^1 \frac{\omega^k(f,t)}{t^{\nu+1}}\,dt\Big), \qquad 0 < \delta \le 1,$$

and the estimate

$$\delta^\nu \int_\delta^1 \frac{\omega^k(f,t)}{t^{\nu+1}}\,dt \le \delta^\nu \int_\delta^1 [f]^{s\nu/k,k} t^{s\nu/k-\nu-1}\,dt \le \frac{k\delta^{s\nu/k}}{\nu(k-s)}[f]^{s,k},$$

we see $\omega^\nu(f,\delta) \le c_k(k-s)^{-1}\delta^{s\nu/k}\|f\|^{s,k}$, which implies (7.13). □

PROOF OF LEMMA 4.8: It follows from (7.1) that

$$\begin{aligned}[\mathcal{P}_n, a] &= \frac{1}{2\pi}\int_{-\pi}^{\pi} \mathcal{D}_n(t)[a(x-t)-a(t)]u(x-t)\,dt \\ &= \int_{-\pi}^{\pi} \sin[(n+\tfrac{1}{2})t]\,\varphi(x,t)\,dt,\end{aligned}$$

where

$$\varphi(x,t) := \psi(x,t)u(x-t), \qquad \psi(x,t) := \frac{1}{2\pi}\frac{a(x-t)-a(t)}{\sin(t/2)}.$$

Notice ψ is C^∞ in both variables. To indicate which variable the difference operator acts upon, we introduce the notation

$$\triangle_{\eta,t}\varphi(x,t) := \varphi(x,t+\eta) - \varphi(x,t).$$

Let $\eta = \pi/(n+\frac{1}{2})$, then $\sin[(n+\frac{1}{2})(t-\nu\eta)] = (-1)^\nu \sin[(n+\frac{1}{2})t]$, and consequently

$$\int_{-\pi}^{\pi} \sin[(n+\tfrac{1}{2})t]\,\triangle_{\eta,t}^k\varphi(x,t)\,dt = (-2)^k[\mathcal{P}_n, a]u(x).$$

Since $|\triangle_{\eta,t}^k\varphi(x,t)| \le \eta^r[\varphi(x,\cdot)]^{r,k}$ and, by Lemma 4.7,

$$[\varphi(x,\cdot)]^{r,k} \le c_k(k-r)^{-2}\|\psi(x,\cdot)\|^{r,k}\|u(x-\cdot)\|^{r,k},$$

we see

(7.14) $$\|[\mathcal{P}_n, a]u\|_\infty \le c_k(k-r)^{-2}(1/n)^r\|u\|^{r,k}.$$

Put $\theta(x,t,h) := \triangle_{h,x}^k\triangle_{\eta,t}^k\varphi(x,t)$, then

(7.15) $$\triangle_h^k[\mathcal{P}_n, a]u(x) = (-2)^{-k}\int_{-\pi}^{\pi} \sin[(n+\tfrac{1}{2})t]\theta(x,t,h)\,dt$$

and, using Lemma 4.7,

$$\begin{aligned}|\theta(x,t,h)| &\le c_k \min\{\|\triangle_h^k\varphi(\cdot,t)\|_\infty, \|\triangle_\eta^k\varphi(\cdot,t)\|_\infty\} \\ &\le c_k(k-r)^{-2}\|u\|^{r,k}\min\{h^r,\eta^r\}.\end{aligned}$$

Since $\min\{h^r,\eta^r\} \le h^s\eta^{r-s} \le c_k h^s(1/n)^{r-s}$, we conclude from (7.15) that

$$\frac{\|\triangle_h^k[\mathcal{P}_n, a]u\|_\infty}{h^s} \le c_k(k-r)^{-2}(1/n)^{r-s}\|u\|^{r,k}.$$

The result now follows by combining the last inequality with (7.14). □

REFERENCES

1. Agranovich, M.S.: Spectral properties of elliptic pseudo-differential operators on a closed curve, Functional Analysis Appl. **13** (1979), 279–281.

2. Butzer, P.L. and Nessel, R.J.: Fourier Analysis and Approximation Theory, Volume 1, Birkhäuser Verlag, Basel, Stuttgart, 1971.

3. Gohberg, I. and Fel'dman, I.: Convolution Equations and Projection Methods for their Solution, AMS Translations of Mathematical Monographs 41, Amer. Math. Soc., Providence, RI, 1974.

4. Gohberg, I. and Krupnik, N.: Einführung in die Theorie der eindimensionalen singulären Integralperatoren, Birkhäuser, Basel, Boston, Stuttgart, 1979.

5. Henrici, P.: Fast Fourier methods in computational complex analysis, SIAM Rev. **21** (1979), 481–527.

6. Kufner, A. ,John, O. and Fučik, S.: Function Spaces, Noordhoff, Leyden, 1977.

7. Lamp, U., Schleicher, K.-T. and Wendland, W.L.: The fast Fourier transform and the numerical solution of one-dimensional boundary integral equations, Numer. Math. **47** (1985), 15–38.

8. Marchaud, A.: Sur les dérivées sur differences des fonctions de variable réelles, J. Math. Pures Appl. **6** (1927), 337–425.

9. Mikhlin, S.G.: The Numerical Performance of Variational Methods, Wolters-Noordhoff, Groningen, 1971.

10. Muskhelishvili, N.I.: Singular Integral Equations, Boundary Problems of Function Theory and their Application to Mathematical Physics, Noordhoff, Groningen, 1953.

11. Nikol'skiǐ, S.M.: Approximation of Functions of Several Variables and Imbedding Theorems, Springer-Verlag, Berlin, Heidelberg, New York, 1975.

12. Noether, F.: Über eine Klasse singulärer Integralgleichungen, Math. Ann. **82** (1921), 42–63.

13. Prestin, J.: Approximation in periodischen Lipschitz-Räumen, Dissertation, Wilhelm-Pieck-Univ. Rostock, 1985.

14. Prößdorf, S.: Zur Konvergenz der Fourierreihen Hölder-stetiger Funktionen, Math. Nachr. **69** (1975), 7–14.

15. Prößdorf, S. and Silbermann, B.: Projektionsverfahren und die näherungsweise Lösung singulärer Gleichungen, Teubner, Leipzig, 1977.

16. Saranen, J. and Wendland, W.L.: The Fourier series representation of pseudo-differential operators on closed curves, Complex Variables **8** (1987), 55–64.

17. Schmidt, E.: Auflösung der allgemeinen linearen Integralgleichung, Math. Ann. **64** (1907), 161–174.

18. Stein, E.M.: Singular Integrals and Differentiability Properties of Functions, Princeton University Press, Princeton, 1970.

19. Taylor, M.: Pseudo Differential Operators, Springer Lecture Notes in Math. 416, Springer-Verlag, Berlin, Heidelberg, New York, 1974.

20. Taylor, M.: Pseudodifferential Operators, Princeton University Press, Princeton, 1981.

21. Timan, A.F.: Theory of Approximation of Functions of a Real Variable, Pergamon Press, Oxford, London, New York, Paris, 1963.

22. Triebel, H.: Interpolation Theory, Function Spaces, Differential Operators, North-Holland, Amsterdam, New York, Oxford, 1978.

23. Wendland, W.L.: Strongly elliptic boundary integral equations, in The State of the Art in Numerical Analysis (A. Iserles and M. J. D. Powell, eds.) Clarendon Press, Oxford, 1987, pp. 511–562.

24. Zygmund, A.: Trigonometric Series, Volumes 1 and 2, Cambridge University Press, Cambridge, 1968.

W. McLean,
Department of Mathematics,
University of Tasmania,
Hobart 7001,
Australia.

W. L. Wendland,
Mathematisches Institut A,
Universität Stuttgart,
D-7000 Stuttgart 80,
West-Germany.

Operator Theory:
Advances and Applications, Vol. 41

WIENER-HOPF FACTORIZATION OF CERTAIN NON-RATIONAL MATRIX FUNCTIONS IN MATHEMATICAL PHYSICS

E. Meister and F.-O. Speck[1]

Dedicated to Israel Gohberg on the occasion of this 60th birthday.

Explicit factorization methods are known in the scalar case for several classes of decomposing algebras. In the systems' case, algorithms of practical use are available in particular for rational matrix functions, while the existence of a canonical or generalized factorization is proved under much more general assumptions.

Elliptic boundary value problems in diffraction theory are often governed by rational combinations of both: rational matrix functions and particular non-rational scalar functions, the squareroots of the Fourier symbols of certain elliptic partial differential operators.

This paper analyses some abstract factorization procedures, which are directly related to applications, especially to the time-harmonic scattering of waves by half-plane-shaped cracks in elastic bodies.

1. INTRODUCTION

Canonical diffraction problems in mathematical physics lead to the *factorization* of continuous matrix functions G on $\mathbb{R}$ into $G_\pm$, which are meromorphically extendable into the upper/lower complex half-plane $\mathbb{C}_\pm$ with at most a finite number of zeros and poles and algebraic behavior at infinity. While this problem is completely solved for rational matrix functions [4], there is no factorization *algorithm* known in the general case, although *existence* of appropriate factorization types has been proved for many interesting matrix function classes, see the famous work by I. GOHBERG and M.G. KREIN [5] and the recent monograph [13] for instance.

In this paper we treat matrix functions of the form

$$G = c_1 Q_1 + c_2 Q_2 \tag{1.1}$$

with rational matrix functions $Q_j \in \mathcal{R}(\mathbb{R})^{2\times 2}$ without poles on $\mathbb{R}$ and *coefficients* $c_j \in C(\mathbb{R})$ in a suitable decomposing algebra (excluding the simple case where the c_j or the Q_j are linear dependent by rational scalars). Time-harmonic linear wave propagation problems are governed by partial differential equations, which yield Fourier symbols of the form

[1]**Sponsored by the Deutsche Forschungsgemeinschaft under grant number Ko 634/32-1.**

$t(\xi) = \sqrt{\xi^2 - k^2}$, $\xi \in \mathbf{R}$, with a wave number k and $\operatorname{Im} k > 0$ for lossy media [19]. Boundary and transmission conditions of the first, second or the third kind then lead to Fourier symbol matrix functions, which are rational in ξ and $t(\xi)$ [21,22]. One realizes easily that those are of the form (1.1).

In three-dimensional elastodynamics two different squareroots $t_j = t_j(\xi_1, \xi_2) = \sqrt{\xi_1^2 + \xi_2^2 - k_j^2}$ appear where $\xi_2 \in \mathbf{R}$ plays the role of a fixed parameter [16,17]. Fortunately the symbols are also of the form (1.1) or $G = \sum_{j=1}^{N} c_j Q_j$ holds with $N \leq 6$, since only particular compositions of the non-rational entries occur. A typical example is taken from the first boundary value (traction) problem, see [11,17], namely

$$(1.2) \qquad G = \frac{1}{k_2^2 t_2^2} \begin{pmatrix} \xi_1^2 \psi - k_2^2 t_2^2 & \xi_1 \xi_2 \psi \\ \xi_1 \xi_2 \psi & \xi_2^2 \psi - k_2^2 t_2^2 \end{pmatrix}$$
$$\psi = 4\tau - 3k_2^2, \quad \tau = \xi_1^2 + \xi_2^2 - t_1 t_2,$$

where we dropped the arguments ξ_1, ξ_2 of t_j, τ, ψ, G whilst k_j are constants. This symbol matrix belongs already to the reduced Wiener-Hopf system, which is lifted from suitable Sobolev (trace) spaces onto $L^2(\mathbf{R}_+)^2$ by Bessel potential operators, i.e. one likes to solve

$$(1.3) \qquad Wu(x) = 1_+(x) F^{-1} G \cdot Fu(x) = g(x), \quad x > 0$$

for $g \in L^2(\mathbf{R}_+)^2$ where G is bounded on $\mathbf{R}$, see [15] for background.

From our point of view it is most interesting to assume coefficients in the (one-dimensional) Wiener algebra (with respect to $\xi = \xi_1$, while ξ_2 is fixed), i.e.

$$(1.4) \qquad c_j \in \mathcal{W}(\mathbf{R}) = \mathbf{C} \dot{+} FL^1(\mathbf{R})$$

since $t_j^{-1}(\xi_1, \xi_2) = \frac{i}{2} F_{x_1 \mapsto \xi_1} H_0^{(1)}(\sqrt{k_j^2 - \xi_2^2}\,|x_1|)$ and $H_0^1(k|.|) \in L^1(\mathbf{R})$ holds for $\operatorname{Im} k > 0$. This also yields $\tau, \psi \in \mathcal{W}(\mathbf{R})$ according to the Wiener-Levy theorem and $\tau \to (k_1^2 + k_2^2)/2$ for $|\xi_1| \to \infty$.

But note that (1.1) and (1.4) do not imply that $G \in C(\dot{\mathbf{R}})^{2\times 2}$ is continuous on the one-point compactification of $\mathbf{R}$, since e.g. $\xi/t(\xi) \to \pm 1$ for $\xi \to \pm\infty$, in contrast to the example $G \in \mathcal{W}(\mathbf{R})^{2\times 2}$ (for every $\xi_2 \in \mathbf{R}$) in (1.2).

Therefore it makes sense either to look for a *right canonical factorization* in the sense of [18]

$$(1.5) \qquad G(\xi) = G_-(\xi) D(\xi) G_+(\xi), \quad \xi \in \mathbf{R}$$

with $D(\xi) = \operatorname{diag}\left(\left(\frac{\xi-i}{\xi+i}\right)^{\kappa_1}, \left(\frac{\xi-i}{\xi+i}\right)^{\kappa_2}\right)$ and invertible $G_\pm \in \mathcal{W}(\mathbf{R})^{2\times 2}$, which have invertible holomorphic extensions into $\mathbf{C}_\pm : \operatorname{Im} \xi \gtrless 0$, and $\kappa_j \in \mathbf{Z}$ [5], or to look for a *generalized factorization* where only $G_\pm^{\pm 1} \in L^2(\mathbf{R}, \rho)^{2\times 2}$, $\rho(\xi) = (\xi^2 + 1)^{-1/2}$, holds in the case of piece-wise continuous matrix functions G on $\dot{\mathbf{R}}$ (jumping only at ∞) [13,18].

Another factorization type turns out to be useful as an intermediate result, say the *meromorphic factorization*

$$(1.6) \qquad G(\xi) = \tilde{G}_-(\xi) \tilde{G}_+(\xi)$$

where $\tilde{G}_\pm$ are also continuous and invertible on $\mathbf{R}$ but meromorphic with a finite number of zeros and poles in $\mathbf{C}_\pm$ and algebraic behavior at infinity:

$$\det G_\pm(\xi) = O(|\xi|^n), \quad \xi \to \pm\infty. \tag{1.7}$$

This weaker result enables us to solve a single equation (1.3) by the classical Wiener-Hopf procedure [19] instead of constructing a generalized inverse of W [14,18,20].

2. FACTORING 2×2 MATRICES OF PAIRED FORM

Consider firstly matrix functions G of the form

$$\begin{aligned} G &= b_1 R_1 + b_2 R_2 \\ b_j &\in \mathcal{W}(\mathbf{R}), \quad R_j \in \mathcal{R}(\mathbf{R})^{2\times 2} \\ R_1^2 &= R_1 = O(1) \text{ at } \infty, \quad R_2 = I - R_1 \end{aligned} \tag{2.1}$$

where I denotes the 2×2 unit matrix. These obviously form a commutative subalgebra

$$\mathcal{A}(R_1; \mathcal{W}(\mathbf{R})) \subset \mathcal{W}(\mathbf{R})^{2\times 2} \tag{2.2}$$

generated by a special rational projection matrix R_1 and bI with arbitrary $b \in \mathcal{W}(\mathbf{R})$. The representation of G in (2.1) is said to be of *paired form* (for any scalar functions b_j).

We are only interested in the (main) case $R_j \neq 0$, i.e. $\operatorname{rank} R_j = 1$, $j = 1, 2$. Further we do not investigate the exceptional cases of increasing R_1 (cancelled by R_2) or rationally linear dependent b_j.

THEOREM 2.1. *Let G be given by (2.1). A meromorphic factorization exists, iff $b_j(\xi) \neq 0$ holds for $\xi \in \dot{\mathbf{R}}$, $j = 1, 2$, and is given by*

$$G = (\tilde{b}_{1-} R_1 + \tilde{b}_{2-} R_2)(\tilde{b}_{1+} R_1 + \tilde{b}_{2+} R_2) \tag{2.3}$$

where $b_j = \tilde{b}_{j-} \tilde{b}_{j+}$ are semi-strong factorizations of b_j [20]. The factors $\tilde{G}_\pm$ of G are bounded invertible on $\mathbf{R}$ in this case.

PROOF. For regular Wiener algebra elements $b = b_j$, i.e. $b(\xi) \neq 0$ for $\xi \in \dot{\mathbf{R}}$, we have

$$\begin{aligned} \kappa &= \operatorname{ind} b := \frac{1}{2\pi} [\arg b(\xi)]_{\xi=-\infty}^{\xi=+\infty} \in \mathbf{Z} \\ \kappa_+ &= \max\{\kappa, 0\}, \quad \kappa_- = \min\{\kappa, 0\} \\ \tilde{b}_\pm(\xi) &= \left(\frac{\xi - i}{\xi + i}\right)^{\kappa_\pm} \sqrt{b(\infty)} \exp\{F_{x\mapsto\xi}\, 1_\pm(x) F^{-1}_{\xi\mapsto x} \ln[(\frac{\xi - i}{\xi + i})^{-\kappa} \frac{b(\xi)}{b(\infty)}]\} \\ &= \left(\frac{\xi - i}{\xi + i}\right)^{\kappa_\pm} b_\pm(\xi), \qquad \xi \in \mathbf{R} \end{aligned} \tag{2.4}$$

with a zero of $\tilde{b}_+$ at $\xi = i$ or of $\tilde{b}_-$ at $\xi = -i$, if $\kappa > 0$ or $\kappa < 0$ holds, respectively. By assumption the projection matrices R_j also may possess poles in both open half-planes $\mathbf{C}_\pm$. The rest follows by inspection. ■

THEOREM 2.2. *A right canonical factorization (1.5) of (2.1) exists, iff*

$$b_j(\xi) \neq 0\,,\ \xi \in \dot{\mathbf{R}}\,,\ j = 1,2 \tag{2.5}$$

holds. It can explicitly be obtained from

$$G = (b_{1-}q_1R_1 + b_{2-}R_2)(\frac{\zeta^{\kappa_1}}{q_1}R_1 + \frac{\zeta^{\kappa_2}}{q_2}R_2)(b_{1+}R_1 + b_{2+}q_2R_2) \tag{2.6}$$

by canonical factorization of the middle factor $R \in \mathcal{R}(\mathbf{R})^{2\times 2}$ *where* $\zeta(\xi) = \frac{\xi - i}{\xi + i}$, $\kappa_j = \operatorname{ind} b_j$, $b_{j\pm}$ *are defined as before, and* q_j *are suitable bounded regular* $\mathcal{R}(\mathbf{R})$ *elements.*

PROOF. The first statement is a consequence [5] of $G \in \mathcal{W}(\mathbf{R})^{2\times 2}$ and

$$\det G(\xi) = b_1(\xi)b_2(\xi)\,, \quad \xi \in \dot{\mathbf{R}}\,. \tag{2.7}$$

Theorem 2.1 yields that (2.6) is an identity where, in the case $q_j = 1$, the outer factors have all desired properties but a finite number of poles in $\mathbf{C}_\pm$ due to the polynomial denominators of R_j. Note that the inverse factors, e.g. $b_{1-}^{-1}R_1 + b_{2-}^{-1}R_2$ of $b_{1-}R_1 + b_{2-}R_2$, cause the same trouble.

The behavior of the Laurent series in these points $\xi = \xi_\ell \in \mathbf{C}_-$, $\ell = 1, \ldots, m$, depends on the corresponding first terms of the Taylor series of $b_{1-} - b_{2-}$ according to

$$b_{1-}R_1 + b_{2-}R_2 = \frac{1}{2}(b_{1-} + b_{2-})I + \frac{1}{2}(b_{1-} - b_{2-})R_0 \tag{2.8}$$

where $R_0 = R_1 - R_2$ has the same critical behavior, say a pole of order n_ℓ. Obviously we can find a rational function q_1, which is bounded invertible on $\mathbf{R}$ and has no zeros and poles in $\overline{\mathbf{C}}_-$, such that $q_1b_{1-} - b_{2-}$ has n_ℓ vanishing first terms in its Taylor series around ξ_ℓ, i.e.

$$\frac{d^j}{d\xi^j}\,[q_1(\xi)b_{1-}(\xi) - b_{2-}(\xi)] = 0 \qquad \text{in } \xi = \xi_\ell\,,\ j = 0, \ldots, n_\ell - 1 \tag{2.9}$$

for all ℓ with poles of R_0 in $\mathbf{C}_-$. The inverse function has the same property and one can repeat the argument for the plus factor. The rest follows by factoring the remaining rational matrix function R in the middle [4]. ■

EXAMPLE 2.3. The matrix in (1.2) can be written as

$$G = b_1R_1 + b_2R_2 = \frac{r}{k_2^2\,t_2^2}R_1 - R_2 \tag{2.10}$$

where

$$\begin{aligned} R_1 &= \frac{1}{\xi_1^2 + \xi_2^2}\begin{pmatrix} \xi_1^2 & \xi_1\xi_2 \\ \xi_1\xi_2 & \xi_2^2 \end{pmatrix}, \quad R_2 = \frac{1}{\xi_1^2 + \xi_2^2}\begin{pmatrix} \xi_2^2 & -\xi_1\xi_2 \\ -\xi_1\xi_2 & \xi_1^2 \end{pmatrix} \\ r &= (\xi_1^2 + \xi_2^2)\psi - k_2^2t_2^2 = (t_2^2 + \xi_1^2 + \xi_2^2)^2 - 4(\xi_1^2 + \xi_2^2)t_1t_2 \end{aligned} \tag{2.11}$$

which represents the well-known Rayleigh function [1,7]. Since the coefficients b_j are even functions, we have $\kappa_j = 0$ in (2.6) and the pole cancellation trick in (2.8), (2.9) leads to the middle factor

$$(2.12)\qquad \frac{1}{q_1}R_1+\frac{1}{q_2}R_2=\lambda^{-1}R_1+\lambda R_2$$

$$=\frac{1}{\lambda(1+\lambda^2)}\begin{pmatrix}\dfrac{\xi_1-i\lambda^2|\xi_2|}{\xi_1-i|\xi_2|} & \pm i\\ \pm i\,\dfrac{\lambda^2\xi_1-i|\xi_2|}{\xi_1-i|\xi_2|} & 1\end{pmatrix}\begin{pmatrix}\dfrac{\xi_1+i\lambda^2|\xi_2|}{\xi_1+i|\xi_2|} & \mp i\,\dfrac{\lambda^2\xi_1+i|\xi_2|}{\xi_1+i|\xi_2|}\\ \mp i\lambda^2 & \lambda^2\end{pmatrix}$$

factored strongly with $\pm i=i\cdot\operatorname{sgn}\xi_2$ and

$$(2.13)\qquad \lambda=\lambda(\xi_2)=\left.\frac{b_{2+}(\xi_1,\xi_2)}{b_{1+}(\xi_1,\xi_2)}\right|_{\xi_1=i|\xi_2|}=q_1=q_2^{-1}$$

due to symmetry, cf. [17]. Further examples can be found in [22].

3. RECOGNITION OF PAIRED FORM MATRICES

Following an idea of A.A. KHRAPKOV [10] we are interested in writing G in the form

$$(3.1)\qquad G=a_1I+a_2C\,,\quad C^2=-\det C\cdot I$$

with $C\in\mathcal{R}(\mathbf{R})^{2\times2}$ and $a_j\in C(\mathbf{R})$. If C is a polynomial matrix with $-\det C=\xi^N+\alpha_{N-1}\xi^{N-1}+\ldots+\alpha_0$ and the greatest common divisor of its elements is 1, we speak of a *Khrapkov canonical form* (K-form) $G=a_1I+a_2C$ and call C the *deviator polynomial matrix* (D-pol). (In [10], $C\in\mathcal{R}(\mathbf{R})^{2\times2}$ was denoted by "commutant of G" if normalized by $-\det C=1$. We avoid this notation, which is also used for the set of matrices commuting with G [3], and we admit $\det C=0$, see examples in [22].)

LEMMA 3.1. *The matrix function (1.1) can be written in K-form, iff there exist rational scalar functions q_{11},q_{12} with*

$$(3.2)\qquad q_{11}Q_1+q_{21}Q_2=I\,.$$

In this case we obtain (3.1) by putting

$$(3.3)\qquad \begin{pmatrix}a_1\\a_2\end{pmatrix}=Q^{-1}\begin{pmatrix}c_1\\c_2\end{pmatrix},\quad Q=(q_{j\ell})$$
$$C=q_{12}Q_1+q_{22}Q_2$$

with suitable rational functions q_{12},q_{22} (in general $q_{j\ell}$ are admitted to have poles on $\mathbf{R}$, i.e. (3.2), (3.3) may be violated in a finite number of points).

PROOF. All possible representations of G in the form (1.1) obviously have the form

$$(3.4)\qquad G=a_1(q_{11}Q_1+q_{21}Q_2)+a_2(q_{12}Q_1+q_{22}Q_2)$$

where $Q=(q_{j\ell})$ is rational and poles on $\mathbf{R}$ must cancel out. Thus (3.2) is necessary.

Conversely, let (3.2) be satisfied, then choose $q_{12}, q_{22} \in \mathcal{R}(\mathbf{R})$ such that $Q(\xi) = (q_{j\ell}(\xi))$ is invertible for a.e. $\xi \in \mathbf{R}$. So far $Q_3 = q_{12}Q_1 + q_{22}Q_2$ must not have the D-pol properties, but there holds a.e.

$$(3.5) \qquad \begin{aligned} G &= a_1 I + a_2 Q_3 \\ &= (a_1 + \frac{a_2}{2} tr\, Q_3) I + a_2 (Q_3 - \frac{1}{2} tr\, Q_3 \cdot I) \\ &= \tilde{a}_1 I + \tilde{a}_2 C \end{aligned}$$

where $tr\, Q_3$ denotes the trace of Q_3 and C might be polynomial as desired in the K-form. Modifying $q_{j\ell}$ corresponding to the last formula and dropping the tildes, we find (3.3) by inspection. ■

COROLLARY 3.2. *The K-form (3.1) is unique and directly obtained from (1.1) by putting*

$$(3.6) \qquad \begin{aligned} a_1 &= \frac{1}{2} tr\, G = \frac{1}{2} [c_1 tr\, Q_1 + c_2 tr\, Q_2] \\ a_2 C &= G - a_1 I \end{aligned}$$

where the D-pol C can be split off, if it exists, as a polynomial matrix with minimal degree.

THEOREM 3.3. *The matrix (1.1) can be written uniquely in paired form (2.1) with $b_j \in C(\mathbf{R})$, iff (3.2) holds and the D-pol C of G satisfies*

$$(3.7) \qquad -\det C = g^2$$

with $g \in \mathcal{R}(\mathbf{R})$.

PROOF. If G has paired form (2.1), then

$$(3.8) \qquad \begin{aligned} G &= \frac{b_1 + b_2}{2}(R_1 + R_2) + \frac{b_1 - b_2}{2} \cdot d(R_1 - R_2) \\ &= a_1 I + a_2 C \end{aligned}$$

represents a K-form (3.1) for suitable $d \in \mathcal{R}(\mathbf{R})$ that contains the common denominator and common factors of the numerators of the elements of $R_1 - R_2$, since

$$(3.9) \qquad [d(R_1 - R_2)]^2 = d^2 I$$

holds. Thus (3.2) is necessary according to Lemma 3.1. Further R_1 and R_2 are uniquely determined (up to interchange) by C,

$$(3.10) \qquad R_1 = \frac{1}{2}(I + \frac{1}{d}C), \; R_2 = \frac{1}{2}(I - \frac{1}{d}C)$$

and one obtains

$$(3.11) \qquad \det C = d^2 \det(2R_1 - I) = d^2 \det(R_1 - R_2) = -d^2 .$$

Conversely let (3.2) and (3.7) be satisfied. Lemma 3.1 yields a representation (3.1) of G as K-form and we obtain G in paired form by substitution of (3.10) with $d = g$. ∎

4. RATIONAL TRANSFORMATION INTO PAIRED FORM

If G is given by (1.1) and Q_1 is invertible on $\mathbf{R}$ (cf. (3.4) for this purpose), it is natural to reduce G to a K-form $\tilde{G}$ by means of any factorization $Q_1 = Q_{1-}Q_{1+}$, $Q_{1\pm} \in \mathcal{R}(\mathbf{R})^{2\times 2}$, as follows

$$(4.1) \qquad \begin{aligned} G &= c_1Q_1 + c_2Q_2 = Q_{1-}(c_1I + c_2\tilde{Q})Q_{1+} \\ &= Q_{1-}(\tilde{a}_1I + \tilde{a}_2\tilde{C})Q_{1+} \end{aligned}$$

where $\tilde{Q} = Q_{1-}^{-1}Q_2Q_{1+}^{-1}$ holds and the final representation is analogous to (3.5) with D-pol $\tilde{C} = d(\tilde{Q} - \frac{1}{2}tr\,\tilde{Q}\cdot I)$.

As known from the theorems 2.2 and 3.3 it is profitable to have $\det\tilde{C}$ as simple as possible.

THEOREM 4.1. *Let $G = a_1I + a_2C$ be an invertible K-form. Then all representations (4.1) of G are given by*

$$(4.2) \qquad \begin{aligned} Q_1 &= q_{11}I + q_{21}C \quad , \quad Q_2 = q_{12}I + q_{22}C \\ \binom{c_1}{c_2} &= Q^{-1}\binom{a_1}{a_2} \quad , \quad Q = (q_{j\ell}) \end{aligned}$$

where $Q \in \mathcal{R}(\mathbf{R})^{2\times 2}$ is any invertible rational matrix function on $\mathbf{R}$. The two D-pol determinants $-q = \det C$ and $-\tilde{q} = \det\tilde{C}$ are connected by

$$(4.3) \qquad d^2\,\tilde{q} = \frac{(q_{11}q_{22} - q_{12}q_{21})^2}{(q_{11}^2 - q_{21}^2q)^2} = \left(\frac{\det Q}{\det Q_1}\right)^2 q\,.$$

PROOF. Starting with $G = a_1I + a_2C$ we find the representations $G = c_1Q_1 + c_2Q_2$ as in Lemma 3.1. We have

$$(4.4) \qquad \begin{aligned} \det G &= \det(a_1I + a_2C) = a_1^2 - a_2^2q \\ &= (q_{11}c_1 + q_{12}c_2)^2 - (q_{21}c_1 + q_{22}c_2)^2q \end{aligned}$$

and, on the other hand, from (4.1)

$$(4.5) \qquad \begin{aligned} \det G &= \det Q_{1-}\cdot(\tilde{a}_1^2 - \tilde{a}_2^2\tilde{q})\cdot\det Q_{1+} \\ &= \det Q_1\cdot\left([c_1 + \frac{c_2}{2}tr\,\tilde{Q}]^2 - c_2^2\,\tilde{q}\right). \end{aligned}$$

Here $\det Q_1 = q_{11}^2 - q_{21}^2\,q$ holds and $tr\,\tilde{Q}$ is obtained from

$$(4.6) \qquad \tilde{C} = d(\tilde{Q} - \frac{1}{2}tr\,\tilde{Q}\cdot I)$$

which yields

$$(4.7)\qquad \det\tilde{Q} = (\frac{1}{2}\,tr\,\tilde{Q})^2 - \tilde{q}/d^2$$
$$\frac{1}{2}\,tr\,\tilde{Q} = (\det\tilde{Q} + \tilde{q}/d^2)^{1/2}\,.$$

Putting this into (4.4) and (4.5) we end up with (4.3) after some elementary calculations. ∎

COROLLARY 4.2. *G can be transformed into a paired form by rational transformation (4.1), iff it can be reduced to a K-form with* $(-\det C)^{1/2} \in \mathcal{R}(\mathbf{R})$. *The D-pol determinant is invariant modulo square factors.*

COROLLARY 4.3. *If G has the form (4.1) with strong Wiener-Hopf factors* $Q_{1\pm}$, $-\det\tilde{C} = g^2$, *and the paired form of* $\tilde{a}_1 I + \tilde{a}_2\tilde{C}$ *due to Theorem 3.3 satisfies the assumptions of Theorem 2.2, then a right canonical factorization reads*

$$(4.8)\qquad G = \{Q_{1-}(b_{1-}R_1 + b_{2-}R_2)R_-\}\,D\,\{R_+(b_{1+}R_1 + b_{2+}R_2)Q_{1+}\}$$

with $Q_1 = Q_{1-}Q_{1+} \in \mathcal{R}(\mathbf{R})^{2\times 2}$, $b_{1\pm}R_1 + b_{2\pm}R_2$ *and* $R = R_-DR_+ \in \mathcal{A}(R_1;\mathcal{W})$ *(but not* $R_\pm$*), and a diagonal factor D as in (1.5).*

So the method of paired operators centers around a factorization of *non-rational* matrix functions in a *commutative* algebra (equivalent to the factorization of two scalar coefficient functions), accompanied by two *non-commutative* factorizations of *rational* matrix functions before and behind, respectively (in order to get into $\mathcal{A}(R_1;\mathcal{W})$ and to change the meromorphic into a canonical factorization).

5. FINAL COMMENTS

In this paper we concentrated on assumptions, which are compatible with problems in elastodynamics. More work is done or will be published separately on other important and relevant exceptional situations, e.g.

(1) $-\det C \neq g^2$, see examples in [8,21,22];

(2) further coefficient algebras, in particular functions with different limits at $\pm\infty$, which leads to generalized factorization [21,22];

(3) $n\times n$ matrix functions of the form $G = \sum_{j=1}^{n} c_j Q_j$ [9], see other problems in elastodynamics [17] and thermoelasticity;

(4) increasing projection matrices R_j, appearing e.g. in oblique deriveé problems;

(5) factorization with respect to systems of closed curves in $\mathbf{C}$.

Finally we like to express our gratitude to R.A. HURD and S. PRÖSSDORF for the fruitful cooperation and discussions during their stays in Darmstadt in the period April 86 - March 87 and March 88, respectively.

After finishing this work we learned about an independent investigation of similar matrix functions by A.B. Lebre [12], which was presented at the same conference. That paper treats the question of invertibility of Wiener-Hopf operators in more detail for a class of K-form matrices (3.1) with another approach and thus supplements our results by important information.

REFERENCES

[1] Achenbach, J.D.: Wave Propagation in Elastic Solids. North-Holland, Amsterdam 1984 (First ed. 1973).

[2] Bart, H., Gohberg, I. and Kaashoek, M.A.: Minimal Factorization of Matrix and Operator Functions. Birkhäuser, Basel 1979.

[3] Baumgärtel, H.: Analytic Perturbation Theory for Matrices and Operators. Birkhäuser, Basel 1985.

[4] Clancey, K. and Gohberg, I.: Factorization of Matrix Functions and Singular Integral Operators. Birkhäuser, Basel 1981.

[5] Gohberg, I. and Krein, M.G.: Systems of integral equations on the half-line with kernels depending on the difference of the arguments. Amer. Math. Soc. Transl. 14 (1960), 217-287 (Russian ed. 1958).

[6] Gohberg, I., Lancaster, P. and Rodman, L.: Matrix Polynomials. Academic Press, New York 1982.

[7] de Hoop, A.T.: Representation Theorems for the Displacement in an Elastic Solid and Their Application to Elastodynamic Diffraction Theory. Ph.D. thesis. Delft University of Technology, Delft 1958.

[8] Hurd, R.A.: The explicit factorization of 2×2 Wiener-Hopf matrices. Preprint 1040. Fachbereich Mathematik, Technische Hochschule Darmstadt 1987.

[9] Jones, D.S.: Commutative Wiener-Hopf factorization of a matrix. Proc. R. Soc. Lond. A 393 (1984), 185-192.

[10] Khrapkov, A.A.: Certain cases of the elastic equilibrium of an infinite wedge with a non-symmetric notch at the vertex, subjected to concentrated force. Prik. Mat. Mekh. 35 (1971), 625-637.

[11] Kupradze, V.D. (Ed.) : Three-Dimensional Problems of the Mathematical Theory of Elasticity and Thermoelasticity. North-Holland, Amsterdam 1979 (Russian ed. 1976).

[12] Lebre, A.B.: Factorization in the Wiener algebra of a class of 2×2 matrix functions. Preprint 1/1988. Instituto Superior Técnico, Universidade Técnica de Lisboa 1988.

[13] Litvinchuk, G.S. and Spitkovski, I.M.: Factorization of Measurable Matrix Functions. Birkhäuser, Basel 1987.

[14] Meister, E. and Speck, F.-O.: The Moore-Penrose inverse of Wiener-Hopf operators on the half axis and the quarter plane. J. Int. Eqs. 9 (1985), 45-61.

[15] Meister, E. and Speck, F.-O.: Boundary integral equation methods for canonical problems in diffraction theory. In: Boundary Elements IX, Vol. 1 (Ed. C.A. Brebbia, W.L. Wendland and G. Kuhn). Springer, Berlin 1987, 59-77.

[16] Meister, E. and Speck, F.-O.: Elastodynamic scattering and matrix factorisation. In: Elastic Wave Propagation. Proc. IUTAM Conf. Galway (Ireland) 1988, to appear.

[17] Meister, E. and Speck, F.-O.: The explicit solution of elastodynamical diffraction problems by symbol factorization. Z. Anal. Anw., to appear.

[18] Mikhlin, S.G. and Prössdorf, S.: Singular Integral Operators. Springer, Berlin 1986 (German ed. 1980).

[19] Noble, B.: The Wiener-Hopf Technique. Pergamon, London 1958.

[20] Speck, F.-O.: General Wiener-Hopf Factorization Methods. Pitman, London 1985.

[21] Speck, F.-O.: Sommerfeld diffraction problems with first and second kind boundary conditions. SIAM J. Math. Anal., to appear.

[22] Speck, F.-O., Hurd, R.A. and Meister, E.: Sommerfeld diffraction problems with third kind boundary conditions. SIAM J. Math. Anal., to appear.

E. Meister
Fachbereich Mathematik
Technische Hochschule Darmstadt
Schlossgartenstr. 7
6100 Darmstadt
West-Germany

F.-O. Speck
Fachbereich Mathematik
Technische Hochschule Darmstadt
Schlossgartenstr. 7
6100 Darmstadt
West-Germany

Operator Theory:
Advances and Applications, Vol. 41

CLASSES OF OPERATOR MONOTONE FUNCTIONS AND STIELTJES FUNCTIONS

Yoshihiro Nakamura

Dedicated to Professor I. Gohberg on the occasion of his 60th birthday

We investigate the correspondence between operator monotone functions and Stieltjes functions. Then we shall prove that the class of nonnegative operator monotone functions is closed under certain operations, and applying those we decide the case when some binary operations become operator means.

1. DEFINITION AND CHARACTERIZATION

A function $f : (0, \infty) \to \mathbf{R}$ is called a *Stieltjes function* if there exist $a \geq 0$ and a positive measure μ on $[0, \infty)$ with $\int_0^\infty (1+t)^{-1} d\mu(t) < \infty$ such that

$$f(x) = a + \int_0^\infty \frac{d\mu(t)}{t+x}, \qquad \text{for} \quad x > 0. \tag{1}$$

For each Stieltjes function f the pair (a, μ) is uniquely determined. Typical examples of Stieltjes functions are the following:

$$x^{-\alpha} = \frac{\sin \pi\alpha}{\pi} \int_0^\infty \frac{1}{t+x} t^{-\alpha} dt \qquad (0 < \alpha < 1),$$

$$\log(1 + \frac{1}{x}) = \int_0^1 \frac{1}{t+x} dt.$$

The set of all Stieltjes functions $\mathcal{S}$ is convex and closed in the topology of pointwise convergence. It is known that a function $f : (0, \infty) \to [0, \infty)$ is a Stieltjes function if and only if f can be extended analytically to $\mathbf{C} \backslash (-\infty, 0]$ and the extension F satisfies that $\Im F(z) \leq 0$ for $\Im z > 0$ (see [1], p. 127). The Stieltjes functions play an important role in potential theory.

There is another important class of functions called operator monotone functions. A function $f : (0, \infty) \to \mathbf{R}$ is said to be *operator monotone* if $f(A) \leq f(B)$ whenever A and B are Hermitian positive definite matrices with $A \leq B$. Here $f(A)$ and $f(B)$ are defined via the usual functional calculus for Hermitian matrices. The functions

x^α $(0 < \alpha \le 1)$, $\log x$ and $-1/x$ are typical examples of operator monotone functions. It is known that f is an operator monotone function if and only if f can be extended analytically to $\mathbf{C}\backslash(-\infty, 0]$ and the extention F satisfies that $\Im F(z) \ge 0$ for $\Im z > 0$. Furthermore, an operator monotone function f can be represented in a form

$$f(x) = a + bx + \int_0^\infty \Big(\frac{t}{1+t^2} - \frac{1}{t+x}\Big) d\mu(t), \tag{2}$$

where $a \in \mathbf{R}$, $b \ge 0$ and μ is a positive measure on $[0, \infty)$ with $\int_0^\infty (1+t^2)^{-1} d\mu(t) < \infty$. Here (a, b, μ) is uniquely determined by f. Conversely, a function f given in the form (2) is operator monotone.

One more different characterization is known: a function $f : (0, \infty) \to \mathbf{R}$ is operator monotone if and only if f is continuously differentiable and the function $K(x, y)$ on $(0, \infty) \times (0, \infty)$ defined by

$$K(x, y) = \begin{cases} (f(x) - f(y))/(x - y), & \text{for } x \ne y, \\ f'(x), & \text{for } x = y, \end{cases} \tag{3}$$

is a positive definite kernel. (Equivalence of these conditions are due to Loewner. See [12], pp. 41–44 and [6].)

We denote the class of nonnegative operator monotone functions by $\mathcal{R}$. For a function $f : (0, \infty) \to [0, \infty)$, $f \in \mathcal{R}$ if and only if f has an integral representation

$$f(x) = a + bx + \int_0^\infty \frac{x}{t+x} d\mu(t), \tag{4}$$

where $a, b \ge 0$ and μ is a positive measure on $(0, \infty)$ with $\int_0^\infty (t+1)^{-1} d\mu(t) < \infty$ (not the same (a, b, μ) as in (2)). The class $\mathcal{R}$ is closely related to so-called operator means. We mention this later in §3. Note that the class $\mathcal{R}$ is also a closed convex cone and has the following stability properties:

(a) $f \in \mathcal{R} \;\Rightarrow\; xf(x^{-1}) \in \mathcal{R}$,
(b) $f \in \mathcal{R}, f \ne 0 \;\Rightarrow\; f(x^{-1})^{-1} \in \mathcal{R}$,
(c) $f \in \mathcal{R}, f \ne 0 \;\Rightarrow\; x/f(x) \in \mathcal{R}$,
(d) $f, g \in \mathcal{R} \;\Rightarrow\; f \circ g \in \mathcal{R}$.

In the sequel, we use these properties and several characterizations repeatedly.

Let $\mathcal{R}_0$ and $\mathcal{S}_0$ denote the sets $\{f \in \mathcal{R} : f \ne 0\}$ and $\{f \in \mathcal{S} : f \ne 0\}$, respectively. Characterizations of Stieltjes functions and operator monotone functions stated above yield the following.

PROPOSITION 1. $\mathcal{S}_0 = \{f(x)^{-1} : f \in \mathcal{R}_0\} = \{f(x^{-1}) : f \in \mathcal{R}_0\} = \{f(x)/x : f \in \mathcal{R}_0\}$.

The stability properties of $\mathcal{S}$ corresponding to (a)–(d) can be derived using the correspondence between $\mathcal{S}_0$ and $\mathcal{R}_0$ in the last proposition, which were known independently in potential theory (cf. [4]).

The above correspondence also implies another characterization of Stieltjes functions as follows.

COROLLARY 1. *For a function $f : (0,\infty) \to [0,\infty)$, $f \in \mathcal{S}$ if and only if f is continuously differentiable and the function $K(x,y)$ on $(0,\infty) \times (0,\infty)$ defined by*

$$K(x,y) = \begin{cases} (xf(x) - yf(y))/(x-y), & \text{for } x \neq y, \\ f(x) + xf'(x), & \text{for } x = y, \end{cases}$$

is a positive definite kernel. Moreover this is equivalent to the positive definiteness of the function

$$L(x,y) = \begin{cases} -(f(x) - f(y))/(x-y), & \text{for } x \neq y, \\ -f'(x), & \text{for } x = y. \end{cases}$$

2. SUBCLASSES OF $\mathcal{R}$ AND $\mathcal{S}$

We consider the sets of functions on $(0,\infty)$ defined by

$$\mathcal{R}_0^\alpha = \{f^\alpha : f \in \mathcal{R}_0\}, \quad \mathcal{S}_0^\alpha = \{f^\alpha : f \in \mathcal{S}_0\},$$

for $\alpha \neq 0$ and

$$\log \mathcal{R}_0 = \{\log f : f \in \mathcal{R}_0\}, \quad \log \mathcal{S}_0 = \{\log f : f \in \mathcal{S}_0\}.$$

By Proposition 1 it follows that $\mathcal{R}_0^\alpha = \mathcal{S}_0^{-\alpha}$ and $\log \mathcal{R}_0 = \{-g : g \in \log \mathcal{S}_0\}$. For $0 < \alpha \leq 1$, it follows from property (d) that $\mathcal{R}_0^\alpha \subset \mathcal{R}_0$, since x^α is an operator monotone function, while $\mathcal{R}_0^\alpha \subset \mathcal{S}_0$ for $-1 \leq \alpha < 0$.

Berg [4] proved that $\mathcal{S}$ is logarithmically convex, i.e. $f^\alpha g^{1-\alpha} \in \mathcal{S}$ if $f, g \in \mathcal{S}$ and $0 < \alpha < 1$. First part of next theorem gives a transparent proof of his result.

THEOREM 1. *(i) $\log \mathcal{R}_0$ and $\log \mathcal{S}_0$ are convex. (ii) For $-1 \leq \alpha \leq 1, \alpha \neq 0$ $\mathcal{R}_0^\alpha$ and $\mathcal{S}_0^\alpha$ are convex. In fact, $\mathcal{R}_0^\alpha = \{f \in \mathcal{R}_0 : x^{1-\alpha} f(x) \in \mathcal{R}\}$ and $\mathcal{S}_0^\alpha = \{f \in \mathcal{S}_0 : x^{\alpha-1} f(x) \in \mathcal{S}\}$ for $0 < \alpha \leq 1$.*

Proof. (i) We may consider only $\log \mathcal{R}_0$. If $f \in \mathcal{R}_0$, then there exists an analytic extension F of f to $\mathbf{C} \backslash (-\infty, 0]$ satisfying $\Im F(z) \geq 0$ for $\Im z > 0$. Since $F(z)$ takes the value in $\mathbf{C} \backslash (-\infty, 0]$, $\log F(z)$ is a well-defined analytic function on $\mathbf{C} \backslash (-\infty, 0]$ and $0 \leq \Im(\log F(z)) < \pi$ for $\Im z > 0$. Hence there exist $\gamma \in \mathbf{R}$ and a measurable function η on $[0,\infty)$ with $0 \leq \eta \leq 1$ a.e. such that

$$\log F(z) = \gamma + \int_0^\infty \Big(\frac{t}{1+t^2} - \frac{1}{t+z}\Big)\eta(t)\,dt.$$

Here the pair (γ, η) is uniquely determined by F, and so by f (see [8], p. 391). Conversely, if $\gamma \in \mathbf{R}$ and $0 \le \eta \le 1$ a.e. are given, then the function

$$f(x) = \exp\Big[\gamma + \int_0^\infty \Big(\frac{t}{1+t^2} - \frac{1}{t+x}\Big)\eta(t)\,dt\Big]$$

becomes a nonnegative operator monotone function, since it has an analytic extension to $\mathbf{C}\backslash[0,\infty)$ which maps the upper half-plane into itself. Thus we have a one-to-one correspondence between $\log \mathcal{R}_0$ and the set

$$\Gamma = \{(\gamma, \eta) : \gamma \in \mathbf{R}, \eta \text{ is a measurable function on } (0,\infty) \text{ with } 0 \le \eta \le 1 \text{ a.e.}\}.$$

Now convexity of $\log \mathcal{R}_0$ is obvious.

(ii) If $\alpha = \pm 1$, then the assertion is trivial. Let $0 < \alpha < 1$ be fixed, and define $\mathcal{Q}_\alpha = \{f \in \mathcal{R}_0 : x^{1-\alpha} f(x) \in \mathcal{R}\}$. First we prove that $\mathcal{Q}_\alpha = \mathcal{R}_0^\alpha$. If $f \in \mathcal{R}_0^\alpha$, then $g = f^{1/\alpha} \in \mathcal{R}_0$. Since the identical function x is operator monotone and $\log \mathcal{R}_0$ is convex, we have $x^{1-\alpha} f(x) = x^{1-\alpha} g(x)^\alpha \in \mathcal{R}$. Thus $\mathcal{R}_0^\alpha \subset \mathcal{Q}_\alpha$. Conversely, assume that $f \in \mathcal{Q}_\alpha$, i.e. $f \in \mathcal{R}_0$ and $h(x) = x^{1-\alpha} f(x) \in \mathcal{R}$. Then there exists a pair $(\gamma, \eta) \in \Gamma$ such that

$$\log f(x) = \gamma + \int_0^\infty \Big(\frac{t}{1+t^2} - \frac{1}{t+x}\Big)\eta(t)\,dt, \tag{5}$$

Since $\log h(x) = (1-\alpha)\log x + \log f(x)$ and

$$\log x = \int_0^\infty \Big(\frac{t}{1+t^2} - \frac{1}{t+x}\Big)dt,$$

we get an integral representation of $\log h(x)$

$$\log h(x) = \gamma + \int_0^\infty \Big(\frac{t}{1+t^2} - \frac{1}{t+x}\Big)\big(\eta(t) + 1 - \alpha\big)\,dt.$$

Since $\log h \in \log \mathcal{R}_0$, we have $\eta + 1 - \alpha \le 1$ a.e. Hence $0 \le \eta \le \alpha$ a.e., which implies $f^{1/\alpha} \in \mathcal{R}_0$. Thus $\mathcal{R}_0^\alpha = \mathcal{Q}_\alpha$. Equality $\mathcal{S}_0^\alpha = \{f \in \mathcal{S}_0 : x^{\alpha-1} f(x) \in \mathcal{S}\}$ follows immediately from Proposition 1 and the above. Then convexity of $\mathcal{R}_0^\alpha$ and $\mathcal{S}_0^\alpha$ are trivial. ∎

The above theorem means that the class of nonnegative operator monotone functions $\mathcal{R}_0$ is closed under some kind of operations:

(i) $f, g \in \mathcal{R}_0,\ 0 < \alpha < 1 \quad \Rightarrow \quad f^\alpha g^{1-\alpha} \in \mathcal{R}_0,$

(ii) $f, g \in \mathcal{R}_0,\ -1 \le \alpha \le 1,\ \alpha \ne 0 \quad \Rightarrow \quad (f^\alpha + g^\alpha)^{1/\alpha} \in \mathcal{R}_0.$

In the rest of this section we present several procedures for producing an operator monotone function from the other.

The following was stated in [2], with a somewhat different proof.

COROLLARY 2. *For any $f \in \mathcal{R}$ and $\alpha \in (0,1)$, $f(x^\alpha)^{1/\alpha}$ is an operator monotone function.*

Proof. If $f \in \mathcal{R}_0$, then $x/f(x) \in \mathcal{R}_0$ by property (c), so $x^\alpha/f(x^\alpha) \in \mathcal{R}_0$ by (d). Hence $x^{1-\alpha}f(x^\alpha) = x/\big(x^\alpha/f(x^\alpha)\big) \in \mathcal{R}_0$ again by (c). Since $f(x^\alpha) \in \mathcal{R}_0$, this is equivalent to $f(x^\alpha) \in \mathcal{R}_0^\alpha$ by Theorem 1. Thus $f(x^\alpha)^{1/\alpha} \in \mathcal{R}_0$. ∎

Kubo [9] also proved this in the case $\alpha = \frac{1}{2}$ as a corollary of his result on mean equations. We present another proof.

Alternate Proof of Corollary 2 ($\alpha = \frac{1}{2}$). Let $f \in \mathcal{R}_0$ and $\varphi = f(\sqrt{x})^2$. To prove $\varphi \in \mathcal{R}$ it suffices to show that the kernel

$$K(x,y) = \begin{cases} \big(\varphi(x) - \varphi(y)\big)/(x-y), & \text{for } x \neq y, \\ \varphi'(x), & \text{for } x = y, \end{cases} \tag{6}$$

is positive definite on $(0,\infty) \times (0,\infty)$, or equivalently so is the kernel

$$L(x,y) = \begin{cases} \big(f(x)^2 - f(y)^2\big)/(x^2-y^2), & \text{for } x \neq y, \\ f'(x)f(x)/x, & \text{for } x = y. \end{cases}$$

Since $f \in \mathcal{R}_0$, $\big(f(x) - f(y)\big)/(x-y)$ is a positive definite kernel. Since a product of two positive definite kernels is also positive definite, it suffices to show that the kernel $\big(f(x)+f(y)\big)/(x+y)$ is positive definite. Since $x/f(x) \in \mathcal{R}$ by (c), the kernel

$$\frac{x/f(x) - y/f(y)}{x-y} = \frac{xf(y) - yf(x)}{f(x)f(y)(x-y)}$$

is positive definite, and so is

$$\frac{f(x)+f(y)}{x+y} = \frac{f(x)-f(y)}{x-y} + \frac{2f(x)f(y)}{x+y} \times \frac{xf(y)-yf(x)}{f(x)f(y)(x-y)},$$

because $1/(x+y)$ is positive definite. This completes the proof. ∎

Note that in the above proof we use the fact that $(x+y)^{-1}$ is positive definite, which is easily seen by the following equation:

$$\frac{1}{x+y} = \int_0^\infty e^{-tx}e^{-ty}\,dt \qquad (x,y>0).$$

This fact is a particular case of the relation between positive definite and negative definite kernels. A function $K(x,y) : (0,\infty) \times (0,\infty) \to \mathbf{R}$ is said to be negative definite if it is symmetric, i.e. $K(x,y) = K(y,x)$, and

$$\sum_{i,j=1}^{n} c_i c_j K(x_i, x_j) \leq 0,$$

for any $x_1, x_2, \cdots, x_n \in (0,\infty)$ and $c_1, c_2, \cdots, c_n \in \mathbf{R}$ satisfying $\sum_{i=1}^n c_i = 0$. It is easily seen that the function $x+y$ is negative definite. If $K(x,y)$ is negative definite and positive, then $\exp[-tK(x,y)]$ is positive definite for all $t>0$, hence so is

$$\frac{1}{K(x,y)} = \int_0^\infty \exp[-tK(x,y)]\,dt \qquad (x,y>0).$$

(See [5], pp. 74–77 for details.)

Using this relation we can prove the next result, which was first proved by Ando [3]. Later, Donoghue [7] gave the almost same proof as ours.

PROPOSITION 2. *Let f be a non-zero operator monotone function, and φ the inverse function of the restriction of $xf(x)$ to $\Omega = \{x \in (0,\infty) : f(x) > 0\}$. Then $\varphi \in \mathcal{R}$.*

Proof. To prove $\varphi \in \mathcal{R}$ it suffices to show that the kernel (6) is positive definite on $(0,\infty)\times(0,\infty)$, or equivalently so is the kernel

$$L(x,y) = \begin{cases} (x-y)/(xf(x)-yf(y)), & \text{for } x \neq y, \\ 1/(f(x)+xf'(x)), & \text{for } x = y, \end{cases}$$

on $\Omega\times\Omega$. Note that $L(x,y)>0$ on $\Omega\times\Omega$. We may assume that f has the form (2). Then

$$\begin{aligned}\frac{1}{L(x,y)} &= \frac{xf(x)-yf(y)}{x-y} \\ &= a + b(x+y) + \int_0^\infty \Big(\frac{t}{1+t^2} - \frac{t}{(t+x)(t+y)}\Big)\,d\mu(t).\end{aligned}$$

Obviously, the first and second terms of the last expression are negative definite kernels, and so is the integrand of the third term. Hence $1/L(x,y)$ is negative definite, which implies that $L(x,y)$ is positive definite. This completes the proof. ∎

THEOREM 2. *(i) Let f be a nonconstant operator monotone function, and $x_0 \in (0,\infty)$. Then the functions $g(x) = (x-x_0)/(f(x)-f(x_0))$, $h(x) = (xf(x) - x_0f(x_0))/(x-x_0)$ and $k(x) = [(xf(x)-x_0f(x_0))/(f(x)-f(x_0))]^{1/2}$ are operator monotone. (ii) Let $f \in \mathcal{R}_0$, and $x_0 \in (0,\infty)$. Then the function $g(x) = \exp[(x\log f(x) - x_0 \log f(x_0))/(x-x_0)]$ becomes an operator monotone function.*

Proof. (i) Since f is a nonconstant operator monotone function, g, h and k are well-defined even at x_0 (we define $g(x_0) = \lim_{x\to x_0} g(x)$, etc.) and positive on $(0,\infty)$. Assume that f has the integral representation (2). Then

$$\begin{aligned} g(x)^{-1} &= \frac{f(x)-f(x_0)}{x-x_0} \\ &= b + \int_0^\infty \frac{1}{x-x_0}\Big(\frac{1}{t+x_0} - \frac{1}{t+x}\Big)d\mu(t) \\ &= b + \int_0^\infty \frac{1}{(t+x)(t+x_0)}\,d\mu(t),\end{aligned}$$

which shows that $1/g \in \mathcal{S}$. Hence $g \in \mathcal{R}$ by Proposition 1. Since

$$h(x) = x\,\frac{f(x) - f(x_0)}{x - x_0} + f(x_0) = \frac{x}{g(x)} + f(x_0)$$

and $x/g \in \mathcal{R}$ by (c), h is operator monotone. By Theorem 1 $k(x) = g(x)^{1/2}h(x)^{1/2}$ is also operator monotone.

(ii) We can represent $\log f$ in the form (5). Then

$$\begin{aligned}\log g(x) &= \frac{x \log f(x) - x_0 \log f(x_0)}{x - x_0} \\ &= \gamma + \int_0^\infty \Big(\frac{t}{1+t^2} - \frac{t}{(t+x)(t+x_0)}\Big)\eta(t)\,dt \\ &= \gamma + \int_0^\infty \frac{x_0 t\,\eta(t)}{(1+t^2)(t+x_0)}\,dt + \int_0^\infty \Big(\frac{t}{1+t^2} - \frac{1}{t+x}\Big)\frac{t\,\eta(t)}{t+x_0}\,dt.\end{aligned}$$

Since $0 \le \eta \le 1$ a.e., $0 \le t\,\eta(t)/(t+x_0) \le 1$ a.e. t. Hence the above equation shows that $\log g \in \log \mathcal{R}_0$. ∎

3. APPLICATIONS TO OPERATOR MEANS

If a binary operation $M(\cdot,\cdot) : (0,\infty) \times (0,\infty) \to (0,\infty)$ satisfies the following three conditions, then we call $M(\cdot,\cdot)$ a *mean*:

(M.1) $M(a,a) = a$,
(M.2) $M(\lambda a, \lambda b) = \lambda M(a,b), \quad \lambda > 0$,
(M.3) $a \le c,\ b \le d \quad \Rightarrow \quad M(a,b) \le M(c,d)$.

By property (M.2) a mean $M(\cdot,\cdot)$ can be determined only by the function $f(x) = M(1,x)$ on $(0,\infty)$. The arithmetic mean $A(a,b) = (a+b)/2$, the geometric mean $G(a,b) = \sqrt{ab}$ and the harmonic mean $H(a,b) = 2ab/(a+b)$ are most popular means.

There is a restrictive class of means called operator means. A binary operation σ on the class of positive selfadjoint operators on an infinite dimensional Hilbert space, $(A,B) \mapsto A\sigma B$, is called an *operator mean* if the following conditions are satisfied:

(OM.1) $A\sigma A = A$,
(OM.2) $T^*(A\sigma B)T \le (T^*AT)\sigma(T^*BT)$, for any operator T,
(OM.3) $A \le C,\ B \le D \quad \Rightarrow \quad A\sigma B \le C\sigma D$,
(OM.4) $A_n \downarrow A,\ B_n \downarrow B \quad \Rightarrow \quad A_n\sigma B_n \downarrow A\sigma B$,

where $A_n \downarrow A$ means that $A_1 \ge A_2 \ge \cdots$ and A_n converges strongly to A. Let σ be an operator mean and define a binary operation $M(\cdot,\cdot)$ on $(0,\infty)$ by $M(a,b) = (aI)\sigma(bI)$, where I denotes the identity operator. Then $M(\cdot,\cdot)$ becomes a mean, and the function $f(x) = M(1,x) = I\sigma(xI)$ is operator monotone. Conversely, every nonnegative operator monotone function f with $f(1) = 1$ determines an operator mean by

$$A\sigma B = A^{1/2} f(A^{-1/2}BA^{-1/2})A^{1/2}$$

for invertible A and B. In the sequel we use the fact that a mean $M(\cdot,\cdot)$ is an operator mean if and only if the function $f(x) = M(1,x)$ on $(0,\infty)$ is operator monotone (see [10] for further details).

When a mean $M(\cdot,\cdot)$ is an operator mean, an important consequence is the following inequality:

$$M(a,b) + M(c,d) \leq M(a+c, b+d), \tag{7}$$

which is called concavity (see [10]). Further, if M is symmetric, i.e. $M(a,b) = M(b,a)$, then it follows from (7) that

$$M(a,b) = \frac{1}{2}\big(M(a,b) + M(b,a)\big) \leq A(a,b).$$

Since $M'(a,b) = M(a^{-1},b^{-1})^{-1}$ is also a symmetric operator mean, it holds that

$$H(a,b) = A(a^{-1},b^{-1})^{-1} \leq M(a,b).$$

Among others there are three particular families of means (c.f. [11]):

Hölder's Means (Power Means).

$$H_p(a,b) = [(a^p + b^p)/2]^{1/p}, \qquad p \neq 0,$$
$$H_0(a,b) = \lim_{p\to 0} H_p(a,b) = \sqrt{ab}.$$

Lehmer's Means.

$$L_p(a,b) = (a^p + b^p)/(a^{p-1} + b^{p-1}), \qquad -\infty < p < \infty.$$

Stolarsky's Means.

$$S_p(a,b) = \left[(a^p - b^p)/\big(p(a-b)\big)\right]^{1/(p-1)}, \qquad p \neq 0,1,$$
$$S_0(a,b) = \lim_{p\to 0} S_p(a,b) = (a-b)/(\log a - \log b),$$
$$S_1(a,b) = \lim_{p\to 1} S_p(a,b) = e^{-1}[a^a/b^b]^{1/(a-b)}.$$

Note that $H_1 = L_1 = S_2 = A$, $H_0 = L_{1/2} = S_{-1} = G$ and $H_{-1} = L_0 = H$. S_0 and S_1 are called the logarithmic and identric means, respectively. The aim of this section is, by applying the results in the preceding section, to show that some of the above means are operator means.

Note that all of means H_p, L_p and S_p are symmetric and monotone increasing with respect to p. In fact, it is well-known that the Hölder's mean H_p is monotone increasing with respect to p. For $p < q$

$$L_q(a,b) - L_p(a,b) = \frac{(a-b)(a^q b^p - a^p b^q)}{ab(a^{q-1} + b^{q-1})(a^{p-1} + b^{p-1})} \geq 0.$$

Hence the Lehmer's mean L_p is monotone increasing with respect to p. Since

$$S_p(a,b) = \left[\int_0^1 \left(ta + (1-t)b\right)^{p-1} dt\right]^{1/(p-1)}, \quad p \neq 1, \tag{8}$$

the Stolarsky's mean S_p is also monotone increasing with respect to p.

THEOREM 3. *(i) Hölder's mean $H_p(\cdot,\cdot)$ is an operator mean if and only if $-1 \le p \le 1$. (ii) Lehmer's mean $L_p(\cdot,\cdot)$ is an operator mean if and only if $0 \le p \le 1$. (iii) Stolarsky's mean $S_p(\cdot,\cdot)$ is an operator mean if and only if $-2 \le p \le 2$.*

Proof. (i) As noticed above, $H_p(\cdot,\cdot)$ is an operator mean if and only if $H_p(1,x)$ is an operator monotone function. Since

$$H_p(1,x) = \begin{cases} [(x^p+1)/2]^{1/p}, & p \neq 0, \\ \sqrt{x}, & p = 0, \end{cases}$$

by Theorem 1 $H_p(1,x) \in \mathcal{R}$ for $-1 \le p \le 1$. Since $H_1 = A$, $H_{-1} = H$ and $H_p(\cdot,\cdot)$ is monotone increasing with respect to p, H_p is not an operator mean for $|p| > 1$.

(ii) Since

$$L_p(1,x) = (x^p+1)/(x^{p-1}+1) = H_p(1,x)^p H_{p-1}(1,x)^{1-p},$$

by (i) and Theorem 1 $L_p(1,x) \in \mathcal{R}$ for $0 \le p \le 1$. Since $L_1 = A$, $L_0 = H$ and $L_p(\cdot,\cdot)$ is monotone increasing with respect to p, L_p is not an operator mean for $p < 0$ or $p > 1$.

(iii) Since S_p can be written in the form (8), by Theorem 1 and the fact that $\mathcal{R}$ is closed in a pointwise convergence topology $S_p(1,x) \in \mathcal{R}$ for $0 \le p \le 2$. Since $S_2 = A$ and S_p is monotone increasing with respect to p, $S_p(1,x) \notin \mathcal{R}$ for $p > 2$. For $p > 0$

$$\begin{aligned} S_{-p}(1,x) &= \left(\frac{x^{-p}-1}{-p(x-1)}\right)^{1/(-p-1)} \\ &= \left(\frac{p(x-1)x^p}{x^p-1}\right)^{1/(p+1)} = \left(x^p S_p(1,x)^{1-p}\right)^{1/(p+1)}. \end{aligned}$$

Hence, by Theorem 1 and the above, $S_{-p}(1,x) \in \mathcal{R}$ when $0 < p \le 1$, and

$$S_{-p}(1,x) = \left(x^{1/p}\left(x/S_p(1,x)\right)^{1-1/p}\right)^{p/(p+1)} \in \mathcal{R},$$

when $1 < p \le 2$. For $p > 2$, let $f(x) = S_{-p}(1,x)$. Then the function

$$G(z) = \begin{cases} \frac{1}{p+1}\left(\log(z-1) + \frac{p}{2}\log z + \log p - \log(z^{p/2} - z^{-p/2})\right), & 0 < |\arg z| < \frac{2\pi}{p}, \\ \log f(z), & z \in (0,\infty), \end{cases}$$

is an analytic extension of $\log f$ to the sector $\{z \in \mathbf{C} : |\arg z| < 2\pi/p\}$, and the limit $\lim_{z \to e^{2\pi i/p}} G(z)$ does not exist. This means that $\log f$ does not have an analytic extension to $\mathbf{C}\backslash(-\infty, 0]$ mapping the upper half-plane into itself. Hence $f \notin \mathcal{R}$. ∎

REFERENCES

1. N. I. Akhiezer, *The classical moment problem*, Oliver and Boyd, Edinburgh, 1965.

2. T. Ando, Concavity of certain maps on positive definite matrices and applications to Hadamard products, Linear Algebra Appl. 26 (1979), 203–241.

3. T. Ando, Comparison of norms $|||f(A) - f(B)|||$ and $|||f(|A - B|)|||$, Math. Z. 197 (1988), 403–409.

4. C. Berg, The Stieltjes cone is logarithmically convex, *Complex Analysis, Joensuu 1978*, pp. 46–54, Springer, Berlin-Heidelberg-New York, 1979.

5. C. Berg, J. P. R. Christensen and P. Ressel, Harmonic analysis on semigroups, Springer-Verlag, New York-Berlin-Heidelberg-Tokyo, 1984.

6. W. F. Donoghue, *Monotone matrix functions and analytic continuation*, Springer-Verlag, Berlin-Heidelberg-New York, 1974.

7. W. F. Donoghue, Communication to T. Ando.

8. M. G. Kreĭn and A. A. Nudel'man, *The Markov moment problem and extremal problems*, American Mathematical Society, Providence, R.I., 1977.

9. F. Kubo, Equations of operator means, preprint.

10. F. Kubo and T. Ando, Means of positive linear operators, Math. Ann. 246 (1980), 205–224.

11. M. E. Mays, Functions which parametrize means, Amer. Math. Monthly 90 (1983), 679–683.

12. M. Rosenblum and J. Rovnyak, *Hardy classes and operator theory*, Oxford University Press, New York, 1985.

Division of Applied Mathematics
Research Institute of Applied Electricity
Hokkaido University
Sapporo 060, Japan

Operator Theory:
Advances and Applications, Vol. 41

A UNIFIED APPROACH TO FUNCTION MODELS, AND THE TRANSCRIPTION PROBLEM

Nikolai K. Nikolskii and Vasily I. Vasyunin

Dedicated to the 60th anniversary of Professor I. Gohberg

The aim of this paper is to develop a general approach to function models of Hilbert space contractions. This approach has been drafted (mainly by the second author) in [1], [2] and in a latent form already in [3], [4], [5]. The main idea of the method is to stop the standard construction of the function model half-way from a unitary dilation to final formulae of the model. In other words we do not fix a concrete spectral representation of the unitary dilation of a given contraction but work directly with an (abstract) dilation equipped with a special "function imbedding operator". We hope you find such model more flexible to be adapting to various problems of spectral theory because as a "free parameter" for such an adaption it contains your choice of a spectral representation of the minimal unitary dilation. To demonstrate a few possibilities we consider as partial cases of our "coordinate-free" model the well-known models due to Sz.-Nagy - Foiaş (both in the original and Pavlov's forms) and to de Branges - Rovnyak. Some other possibilities are considered.

It is clear from the program declared that our approach is based on an analysis of geometric nature of the minimal unitary dilation of a given operator. Here we follow Sz.-Nagy - Foiaş [3], Douglas [6], Pavlov [7].

1. UNITARY DILATION. This section is simply to recall in a form we need some notions and formulae from the unitary dilation theory. We follows [3], [6] and [5].

Let T be a Hilbert space contraction,

$$T : H \to H .$$

An operator $U : \mathcal{H} \to \mathcal{H}$ is said to be a silation of T if $H \subset \mathcal{H}$ and

$$T^n = P_H U^n | H , \quad n \geq 0 , \tag{1}$$

and is said to be a unitary dilation of T if in addition to (1) U is unitary on $\mathcal{H}$. By Sarason's lemma [8] U is a dilation of T iff the space $\mathcal{H}$ can be decomposed into the sum

$$\mathcal{H} = G_* \oplus H \oplus G , \tag{2}$$

where

$$U G \subset G , \qquad U^* G_* \subset G_* . \tag{3}$$

If a unitary dilation U is minimal, i.e. if

$$\mathcal{H} = span \{ U^n H : n \in \mathbb{Z} \} ,$$

the spaces G and G_* in (2) are uniquely determined. In the terminology coming from the scattering theory G_* and G are called the incoming and outgoing spaces respectively. The restrictions $U|G$ and $U^*|G_*$ are pure isometries provided U is minimal. The purity means

$$G = \sum_{n \geq 0} \oplus U^n (G \ominus UG), \quad G_* = \sum_{n \geq 0} \oplus U^{*n} (G_* \ominus U^* G_*).$$

So, writing down the matrix of U with respect to decomposition (2) as

$$U = \begin{pmatrix} P_{G_*} U | G_* & \mathbb{O} & \mathbb{O} \\ A & T & \mathbb{O} \\ C & B & U|G \end{pmatrix} \tag{5}$$

and taking into account the equalities $U^*U = UU^* = I$ one can verify (see [6]) that

$$A = D_{T^*} V_*^*, \quad B = V D_T, \quad C = -VT^*V_*^* . \tag{6}$$

Here $D_T = (I - T^*T)^{1/2}$ is the so-called defect operator of T and V, V_* are partial isometries from H to G, G_* respectively with the initial spaces $\mathcal{D}_T \overset{def}{=} clos\, D_T H$, $\mathcal{D}_{T^*} \overset{def}{=} clos\, D_{T^*} H$ and the final spaces $G \ominus UG$, $G_* \ominus U^* G_*$. In particular, any two minimal unitary dilations U_1, U_2 of a contraction are unitarily equivalent.

The latter can be observed independently if we set

$$\mathcal{W}\left(\Sigma U_1^k x_k \right) = \Sigma U_2^k x_k \tag{7}$$

for every complectly supported sequence $x_k \in H$, $k \in \mathbb{Z}$. Then $\mathcal{W}$ can be extended to a unitary operator and $\mathcal{W} U_1 = U_2 \mathcal{W}$, $\mathcal{W}x = x$ for $x \in H$.

Conversely, a matrix (5) with the choice (6) and with partial isometries V, V_* mentioned above defines a minimal unitary dilation fof the contraction T.

2. OUTLINE OF THE CONTENTS. Now, we represent the action of U on the spaces span $(U^n G : n \in \mathbb{Z}$ and span $(U^n G_* : n \in \mathbb{Z})$ as the multiplication operator $f \mapsto zf$ on $L^2(E)$, $L^2(E_*)$, respectively, $dim\, E = dim(G \ominus UG)$, $dim\, E_* = dim(G_* \ominus U^* G_*)$. This leads directly to the "unified", "coordinate-free" function model of a contraction. Then we describe the transcription problem which consists in essence in the choice of a spectral representation of U. Giving general solutions of the transcription equations (see Sec.20) we distinguish three known transcriptions, namely of Sz.-Nagy - Foias, Pavlov, de Branges - Rovnyak, and an additional one useful to discern a given invariant subspace connected with a regular factorization of the characteristic function (the so-called, "misroscope transcription").

Then we write down some formulae in coordinate-free model and rewrite them in various transcriptions. We finish with the identifying the original de Branges - Rovnyak model with a transcription of ours mentioned above.

3. FUNCTION MAPPINGS. Let us start with a spectral representation of the parts of U (the minimal unitary dilation of a contraction T) as multiplication operators. Let us pick spaces E, E_* such that

$$\dim E = \dim (G \ominus UG),$$

$$\dim E_* = \dim (G_* \ominus U^* G_*),$$

and unitary operators V, V_*

$$V : E \longrightarrow G \ominus UG,$$

$$V_* : E_* \rightarrow G_* \ominus U^* G_* .$$

Now, the dilation U produces the mapping

$$\Pi = (\pi_*, \pi) : L^2(E_*) \oplus L^2(E) \longrightarrow \mathcal{H} \tag{8}$$

by the formulae

$$\pi\Big(\sum_{k\in\mathbb{Z}} z^k e_k \Big) = \sum_{k\in\mathbb{Z}} U^k v e_k , \quad e_k \in E ; \tag{9}$$

$$\pi_*\Big(\sum_{k\in\mathbb{Z}} z^k e_{*k} \Big) = \sum_{k\in\mathbb{Z}} U^{k+1} V_* e_{*k}, \quad e_{*k} \in E_* . \tag{10}$$

The operators Π, π, π_* are said to be function mappings (or function imbeddings). Since the spaces $G \ominus UG$, $G_* \ominus U^* G_*$ are wandering subspaces of the unitary dilation U, the operators π, π_* are isometries:

(i) $\pi^* \pi = I_{L^2(E)}$, $\pi_*^* \pi_* = I_{L^2(E_*)}$

Moreover,

(ii) $\pi z = U \pi$, $\pi_* z = U \pi_*$

and hence

$$\pi_*^* \pi z = z \pi_*^* \pi . \tag{11}$$

On the other hand $\pi H^2(E) = G$, $\pi_* H_-^2(E_*) = G_*$, and hence

(iii) $\pi H^2(E) \perp \pi_* H_-^2(E_*)$

Here and throughout the paper $H^2(E)$ stands for the Hardy space

$$H^2(E) = \{ \sum_{n \geq 0} z^n e_n : e_n \in E, \quad \sum_{n \geq 0} \| e_n \|^2 < \infty \} =$$

$$= \{ f \in L^2(E) : \hat{f}(k) = 0, \ k < 0 \},$$

where $\hat{f}(k)$ is the k-th Fourier coefficient of f . Next,

$$H_-^2(E) \overset{def}{=\!=} L^2(E) \ominus H^2(E).$$

The letter z denotes the identity map of $\mathbb{T}$ or $\mathbb{D}$ onto itself ($z(\zeta) = \zeta$) as well as the operator $f \mapsto zf$ of multiplication by z on any spaces $L^2(E)$, $H^2(E)$ etc.

Now, returning to our function mappings, we can note that the orthogonality (iii) means $\pi_*^* \pi H^2(E) \subset H^2(E_*)$. This entails the following elementary and well-known fact.

4. LEMMA. Under conditions (11) and (iii) there exists a uniquely determined function θ , taking values in the space of operators from E to E_*, $\theta \in H^\infty(E \to E_*)$ such that

$$\pi_*^* \pi f = \theta f , \quad f \in L^2(E),$$

and $\| \theta \|_\infty = \| \pi_*^* \pi \| \leq 1$, $\| \theta(0) e \| < \| e \| \quad \forall e \in E$ i.e. θ is pure.

Here $H^\infty(E \to E_*)$ stands for the space of all bounded analytic operator-valued functions in the unit disc $\mathbb{D}$ equipped with the usual supnorm.

Changing the imbeddings V, V_* to some others, say V', V'_* , we arrive at a function $\theta' \in H^\infty(E' \to E'_*)$ connected with the former by the equality

$$u_* \theta = \theta' u , \tag{12}$$

where $\mathcal{U}=V'^{*}V$, $\mathcal{U}_{*}=V'^{*}_{*}V_{*}$ are unitary operator. Functions θ,θ' related to each other by (12) are said to be equivalent.

Now, one can summarize the previous considerations as follows.

5. LEMMA. A unitary operator U with properties (2), (3) (e.g. a unitary dilation) produces by formulae (9), (10), function mappings π,π_{*} having properties (i)-(iii). Moreover, there exists a unique (up to equivalence) function θ, $\theta\in H^{\infty}(E\to E_{*})$, for which $\pi_{*}^{*}\pi=\theta$, $\|\theta\|_{\infty}\leqslant 1$. ●

It is worth mentioning that properties (i)-(iii) can be rewritten in terms of Π in the following way:

$$\text{(ii)}' \qquad \Pi z = U\Pi \tag{13}$$

the operator $\Pi^{*}\Pi$ commutes with z,

$$\Pi^{*}\Pi z = z\Pi^{*}\Pi$$

and in the matrix form (as an operator on the orthogonal sum $L^{2}(E_{*})\oplus L^{2}(E)$) $\Pi^{*}\Pi$ can be written as

$$\Pi^{*}\Pi=\begin{pmatrix} I & \theta \\ \theta^{*} & I\end{pmatrix}, \qquad \theta\in H^{\infty}(E\to E_{*}),\ \theta \text{ is pure}. \tag{14}$$

In general, using the function mapping Π and formulae (13), (14) one can reconstruct the operator U (and the contraction $P_{H}U\,|\,H$) but not on the whole, just the restriction of U to $\operatorname{clos}\operatorname{Range}\Pi$. Let us compute this latter in terms of the initial contraction $T=P_{H}U\,|\,H$.

6. LEMMA. Let T be a contraction on H, U its minimal unitary dilation and Π be the corresponding function mapping. Then, among the all subspaces $H'\subset H$ invariant under T and such that the restriction $T\,|\,H'$ is unitary, there exists a largest one, say H_{u}. The subspace H_{u} reduces T and

$$H_{u}=(\operatorname{Range}\Pi)^{\perp}. \tag{15}$$

PROOF. Let us define the subspace H_u by (15). It is clear that H_u reduces the operator U and $H_u \subset H$. Since $T = P_H U | H$, H_u reduces T and $T | H_u$ is unitary.

It remains to prove the maximality property of H_u. Let $TH' \subset H' \subset H$, and $T | H'$ be unitary. If $h \in H'$ we have for $n \geqslant 0$

$$\|h\| = \|T^n h\| = \|P_H U^n h\| \leqslant \|U^n h\| = \|h\| ,$$

and hence $U^n h \in H$, $n \geqslant 0$. A similar chain can be written down for $n < 0$, too, but with the replacement T^n by $T^{*|n|}$. Hence $U^n h \in H$, $n < 0$. Together with the previous series of inclusions this implies $U^n h \perp G$, $U^n h \perp G_*$ for $n \in \mathbb{Z}$ i.e. $h \in H_u$. Therefore, $H' \subset H_u$. ●

Of course, Lemma 6 implies the well-known decomposition af an arbitrary contraction T into the orthogonal sum $T = T_u \oplus T_o$ of the unitary part $T_u = T | H_u$ and a contraction $T_o = T | H \ominus H_u$ which is completely non-unitary (unitary on no its invariant subspace). A contraction T is completely non-unitary (c.n.u.) iff

$$\text{(iv)} \quad \operatorname{clos} \operatorname{Range} \Pi = \mathcal{H} ,$$

where $\mathcal{H}$ stands for the space of the minimal unitary dilation.

The following is in a sense the converse to Lemma 5.

7. LEMMA. Any mapping $\Pi : L^2(E_*) \oplus L^2(E) \longrightarrow \mathcal{H}$ with properties (14), (iv) defines by formula (13) a unitary operator U (first on $\operatorname{Range}\Pi$ then on the whole $\mathcal{H}$) which is a minimal unitary dilation of the c.n.u. contraction

$$T = P_H U | H, \quad H \overset{def}{=\!=} \mathcal{H} \ominus [\pi_* H^2_-(E_*) \oplus \pi H^2(E)] .$$

SKETCH OF THE PROOF. Let us show, for example, that U is unitary; all the rest is essentially contained in the above discussions.

Let $f, g \in L^2(E_*) \oplus L^2(E)$. Then $(U\Pi f, U\Pi g) = $ $=(\Pi z f, \Pi z g) = (\Pi^* \Pi z f, z g) = $ (in view of $z\Pi^*\Pi = \Pi^*\Pi z$) $=$ $= (z\Pi^*\Pi f, z g) = (\Pi^*\Pi f, g) = (\Pi f, \Pi g)$. So, U is an isometry. But by (13) and (iv) $U\mathcal{H} = U\,\mathrm{clos}\,\Pi[L^2(E_*)\oplus L^2(E)] = \mathrm{clos}\,\Pi[L^2(E_*)\oplus L^2(E)] = \mathcal{H}$. ●

8. A COMPLETE UNITARY INVARIANT. The function θ is a complete unitary invariant of the objects under consideration (like T, π, π_*, Π). To see this we agree to call equivalent two mappings

$$\Pi : L^2(E_*) \oplus L^2(E) \longrightarrow \mathcal{H},$$

$$\Pi' : L^2(E'_*) \oplus L^2(E') \longrightarrow \mathcal{H}',$$

if there exist unitary operators $u : E \to E'$, $u_* : E_* \to E'_*$ and $W : \mathcal{H} \to \mathcal{H}'$ such that

$$\Pi'\mathcal{U} = W\Pi \,; \qquad \mathcal{U} \overset{def}{=\!=} \begin{pmatrix} u_* & \mathbb{O} \\ \mathbb{O} & u \end{pmatrix}.$$

9. THEOREM. <u>Let</u> T, T' <u>be c.n.u. contractions acting on the spaces</u> H, H', <u>respectively, and</u> Π, Π' <u>the corresponding function mappings constructed according to their minimal unitary dilations. Let further</u> θ, θ' <u>be operator functions defined by</u> Π, Π' <u>via formula (14) (or Lemma 4). The following are equivalent.</u>

a. T <u>and</u> T' <u>are unitarily equaivalent</u>.

b. Π <u>and</u> Π' <u>are equivalent</u>.

c. θ <u>and</u> θ' <u>are equivalent (see (12)</u>).

PROOF. a$\Rightarrow$b. Suppose that a unitary $X : H \to H'$ intertwines T and T': $XT = T'X$. Then U and U', the minimal unitary dilations of T and T', respectively, are unitary equivalent. Let W be the operator (7) intertwining U and U'. It is easy to see that $W(G \ominus UG) = G' \ominus U'G'$ and $W(G_* \ominus U^*G_*) = G'_* \ominus U'^*G'_*$. Hence

$$W\pi\Big(\sum_{n\in\mathbb{Z}} z^n e_n\Big) = W\Big(\sum_{n\in\mathbb{Z}} U^n v e_n\Big) = \sum_{n\in\mathbb{Z}} U'^n W v e_n =$$

$$= \pi'\Big(\sum_{n\in\mathbb{Z}} z^n v'^* W v e_n\Big) = \pi' u\Big(\sum_{n\in\mathbb{Z}} z^n e_n\Big),$$

where $\mathcal{U} \overset{def}{=} V'^{*}\mathcal{W}V : E \longrightarrow E'$ is unitary. Similarly, $\mathcal{W}\pi_* = \pi'_*\mathcal{U}_*$, $\mathcal{U}_* = V'^{*}_*\mathcal{W}V_*$ and so $\Pi'\mathcal{U} = \mathcal{W}\Pi$ with $\mathcal{U} = diag\{\mathcal{U}_*, \mathcal{U}\}$. Q.E.D.

b $\Rightarrow$ c. Obvious.

c $\Rightarrow$ b. From the equality $\theta'\mathcal{U} = \mathcal{U}_*\theta$ (see (12)) we have $\Pi'^{*}\Pi'\mathcal{U} = \mathcal{U}\Pi^{*}\Pi$ with $\mathcal{U} = diag\{\mathcal{U}_*, \mathcal{U}\}$. Hence the operator $\mathcal{W}$ being defined by the equalities $\mathcal{W}\Pi f = \Pi'\mathcal{U}f$, $f \in L^2(E_*) \oplus L^2(E)$ is unitary:

$$(\Pi'\mathcal{U}f, \Pi'\mathcal{U}g) = (\Pi'^{*}\Pi'\mathcal{U}f, \mathcal{U}g) = (\mathcal{U}\Pi'^{*}\Pi'f, \mathcal{U}g) = (\Pi f, \Pi g)$$

and Statement "b" follows.

b $\Rightarrow$ a. Let $\Pi'\mathcal{U} = \mathcal{W}\Pi$ by the definition of equivalence. Property (ii)' implies $\mathcal{W}\mathbb{U}\Pi = \mathcal{W}\Pi z = \Pi'\mathcal{U}z = \Pi' z\,\mathcal{U} = \mathbb{U}'\Pi'\mathcal{U} = \mathbb{U}'\mathcal{W}\Pi$, and hence (by (iv))

$$\mathcal{W}\mathbb{U} = \mathbb{U}'\mathcal{W}.$$

From $\mathcal{W}\pi H^2(E) = \pi' H^2(E')$ and $\mathcal{W}\pi_* H^2_-(E_*) = \pi'_* H^2_-(E'_*)$ we conclude that $\mathcal{W}H = H'$. Setting $X = \mathcal{W}|H$ we complete the proof:

$$T'X = P_{H'}\mathbb{U}'X = P_{H'}\mathcal{W}\mathbb{U}|H = \mathcal{W}P_H\mathbb{U}|H = XT. \quad \bullet$$

10. DEFINITION. The operator function $\theta = \pi_*^{*}\pi$ is said to be the characteristic function of the initial contraction T. Notation: $\theta = \theta_T$.

This function is uniquely determined by T (up to equivalence), $\theta_T \in H^\infty(E \to E_*)$, where $dim\, E = dim\, \mathfrak{D}_T$, $dim\, E_* = dim\, \mathfrak{D}_{T^*}$ and $\|\theta_T\|_\infty = \sup_{\zeta \in \mathbb{D}} \|\theta_T(\zeta)\| \leq 1$. The correspondence $T \mapsto \theta_T$ is one-to-one (up to unitary equivalence). There exists the following explicit and well-known formula for θ_T.

11. LEMMA. <u>Let</u> T <u>be a c.n.u. contraction on a space</u> H, $\mathfrak{D}_T$ <u>and</u> $\mathfrak{D}_{T^*}$ <u>be the defect subspaces of</u> T. <u>Then</u> θ_T <u>as a function from</u> $H^\infty(\mathfrak{D}_T \to \mathfrak{D}_{T^*})$ <u>can be computed by the formula</u>

$$\theta_T(\zeta)h = -Th + \zeta D_{T^*}(I - \zeta T^*)^{-1} D_T h, \quad h \in \mathfrak{D}_T,$$

<u>or on the whole</u> H

$$\Theta_T(\zeta)\mathcal{D}_T = \mathcal{D}_{T^*}(I-\zeta T^*)^{-1}(\zeta I - T).$$

SKETCH OF THE PROOF. Keeping in mind the previous notation it is easy to see that

$$\pi_*^* x = \sum_{n\in\mathbb{Z}} z^n P_{G_*\ominus U^*G_*} U^{*(n+1)} x, \quad x\in\mathcal{H}.$$

Up to equivalence we can assume that $G_*\ominus U^*G_* = \mathfrak{D}_{T^*}$, $G\ominus UG = \mathfrak{D}_T$ and $\pi h = h$ for $h\in G\ominus UG = \mathfrak{D}_T$. Hence,

$$\Theta_T(\zeta)h = \sum_{n\geqslant 0} \zeta^n P_{\mathfrak{D}_{T^*}} P_{G_*} U^{*(n+1)} h, \quad h\in\mathfrak{D}_T.$$

Now, from (5) and (6) one can derive the needed expressions:

$$P_{\mathfrak{D}_{T^*}} P_{G_*} U^{*(n+1)} h = \begin{cases} -Th, & n=0 \\ \mathcal{D}_{T^*} T^{*(n-1)} \mathcal{D}_T h, & n\geqslant 1 \end{cases}.$$

The first formula for Θ_T follows immediately. The second one can be proved by multiplying the first by the defect operator $\mathcal{D}_T$ and taking into account the well-known intertwining property $T\mathcal{D}_T = \mathcal{D}_{T^*}T$. ●

12. UNIFIED (COORDINATE-FREE) FUNCTION MODEL of a c.n.u. contraction T is by definition the class of equivalent function mappings (8) generated by T together with the formulae representing the action of T.

The function model is uniquely determined by the characteristic function Θ_T. To obtain the model one has to get the following:

spaces $\mathcal{H}, E, E_*$ with $\dim\mathcal{H}=\infty$, $\dim E = \operatorname{rank}\mathcal{D}_T$, $\dim E_* = \operatorname{rank}\mathcal{D}_{T^*}$;

a mapping $\Pi = (\pi_*, \pi): L^2(E_*)\oplus L^2(E) \longrightarrow \mathcal{H}$ such that

α) π, π_* are isometries,

β) $\pi_*^*\pi$ is a multiplication operator by the

function equivalent to θ;

γ) $\operatorname{clos} \operatorname{Range} \Pi = \mathcal{H}$;

a unitary operator $\mathbb{U} : \mathcal{H} \to \mathcal{H}$ such that $\Pi z = \mathbb{U}\Pi$;

the model space $\mathcal{K}_\theta$,

$$\mathcal{K}_\theta = \mathcal{H} \ominus [\pi_* H^2_-(E_*) \oplus \pi H^2(E)] ; \tag{16}$$

and, finally, the model operator M_θ,

$$M_\theta = P_\theta \mathbb{U} | \mathcal{K}_\theta .$$

Here P_θ stands for the orthogonal projection $P_{\mathcal{K}_\theta}$ onto the model space $\mathcal{K}_\theta$; it is easy to see that

$$P_\theta = I - \pi P_+ \pi^* - \pi_* P_- \pi_*^* \tag{17}$$

where $P_\pm$ denote the ortogonal projections on L^2-space onto the Hardy subspaces H^2, H^2_- respectively.

Of course, the model operator M_θ, $\theta = \theta_T$ is unitarily equivalent to the initial operator T . To simplify the treatment of the model we shall need sometimes some concrete (coordinate) transcriptions of M_θ .

13. TO CHOOSE A COORDINATE TRANSCRIPTION of the model of a c.n.u. contraction T means to write down a spectral representation of the minimal unitary dilation $\mathbb{U}$.

To begin with let us mention that the spectral measure $\mathcal{E}_{\mathbb{U}}$ of $\mathbb{U}$ is mutually absolutely continuous with respect to Lebesgue measure m on the unit circle $\mathbb{T}$: $\mathcal{E}_{\mathbb{U}}(\sigma) = \mathbb{O} \Longleftrightarrow m(\sigma) = 0$. This is an immediate consequence of properties (i), (ii) and (iv). Hence, the local spectral multiplicity of $\mathbb{U}$ is defined m-almost everywhere and, moreover, the same properties (i), (ii), (iv) imply

$$\max(\dim \mathcal{D}_T, \dim \mathcal{D}_{T^*}) \leq \operatorname{rank} \mathcal{E}_{\mathbb{U}}(\zeta) \leq \dim \mathcal{D}_T + \dim \mathcal{D}_{T^*} . \tag{18}$$

The general form of the spectral representation for such an operator $\mathbb{U}$ results in the multiplication operator

$$f \mapsto z f$$

on a weighted vector-valued space $L^2(\mathcal{E},W)$. Here $\mathcal{E}$ stands for an auxiliary Hilbert space and W for an operator weight, i.e. for a measurable operator-valued function $\zeta \mapsto W(\zeta)$, $\zeta \in \mathbb{T}$ with densely defined non-negative selfadjoint values $W(\zeta) = W(\zeta)^* \geqslant \mathbb{O}$ such that $\operatorname{rank} W(\zeta) = \operatorname{rank} \mathcal{E}_U(\zeta)$ a.e. on $\mathbb{T}$. And now

$$L^2(\mathcal{E},W) = \{ f : \zeta \mapsto f(\zeta) \in \mathcal{E} \text{ is } m\text{-measurable},$$

$$f(\zeta) \in \operatorname{Dom} W(\zeta)^{1/2} \quad \text{and} \int_{\mathbb{T}} \| W(\zeta)^{1/2} f(\zeta) \|^2 dm < \infty \} .$$

The dimension of an auxiliary space $\mathcal{E}$ depends on the spectral multiplicity $\operatorname{rank} \mathcal{E}_U(\cdot)$. But in any case we can take (see (18))

$$\mathcal{E} = E_* \oplus E ; \quad \dim E = \dim \mathcal{D}_T , \quad \dim E_* = \dim \mathcal{D}_{T^*} .$$

Then, a coordinate transcription of the model consists in the choice of a weight W and a function mapping Π

$$\Pi : L^2(E_*) \oplus L^2(E) \longrightarrow L^2(\mathcal{E},W) = \mathcal{H} \tag{22}$$

such that

$$\Pi z = z \Pi \tag{23}$$

$$\Pi^* \Pi = \begin{pmatrix} I & \theta \\ \theta^* & I \end{pmatrix} \tag{24}$$

where θ stands for an operator-valued function equivalent to θ_T . Now, (23) implies that Π is a multiplication operator $(\Pi f)(\zeta) = \Pi(\zeta) f(\zeta)$, $f \in L^2(E_* \oplus E)$, and the completeness condition

$$\text{(iv)} \quad \operatorname{clos} \operatorname{Range} \Pi = L^2(\mathcal{E},W) \tag{25}$$

turns into

$$\text{(iv)}' \quad \operatorname{clos} \operatorname{Range} \Pi(\zeta) = \operatorname{clos} \operatorname{Range} W(\zeta) \qquad \text{a.e.}$$

So, to write down a concrete (coordinate) transcription of the model we have to solve transcription equations (23), (24) under condition (iv) (or (iv)').

14. HOW TO OBTAIN THE SIMPLEST TRANSCRIPTION. Now, we outline the rest of the paper as follows: after some general remarks (this Sec.) write down (in Sec.15, 17, 18) three popular coordinate transcription (of Sz.-Nagy - Foiaş, Pavlov and de Branges - Rovnyak) and additional two (Sec.19), giving some their immediate consequences (Sec.16), then discuss the general solutions of the transcription equations (Sec.20-22) and the coordinate-free formulae for T, $\mathfrak{D}_T$, $(T - \zeta I)^{-1}, \dots,$ (Sec.24-25), after that we rewrite then in various transcriptions (Sec.26) and conclude with an explanation of an explicit relation of the original form of the Branges - Rovnyak model to our model (Sec.27-28).

This program starts with a discussion of the rules of composing a transcription of the model. The following two requirements on a coordinate transcription of the function model (22)-(24):

(a) the weight W has to be as simple as possible, and at least diagonal in the matrix representation of the operators $W(\zeta): E_* \oplus E \to E_* \oplus E$;

(b) the model space $\mathcal{K}_\theta$ (see (16)), and hence the projection P_θ (see (17)), has to be as simple as possible;

in our opinion, occur as the main requirements. These requests seem to be a bit vague ones, but the first of them can be satisfied (Sec.15). To grant request (b) is much more difficult: in all known transcriptions the projection P_θ contains some Toeplitz operators with non obvious symbols. As a substitude of (b) one can demand certain simplicity of the function mappings π_*, π or even of one of them. Also in spite of a complicacy of $\mathcal{K}_\theta$ and P_θ we will be able to point (in an explicit form) some dense subsets of $\mathcal{K}_\theta$ (at least for the three main transcriptions mentioned before).

As the most natural function mapping one can consider the identical imbedding $\Pi = id$ of $L^2(E_* \oplus E)$ to $L^2(E_* \oplus E, W)$ (see Sec.17), or even though diagonal Π 's with respect to the decomposition $E_* \oplus E$. Unfortunately, it is not hard to see that this request is not compatible with the main one (see (a) above) if we restrict ourselves of non-degenerate characteristic functions $\theta \not\equiv \mathbb{0}$. In a special case, namely, when π and π_* can be factorized in the form

$$\pi : L^2(E) \to L^2(E_* \oplus E) \xrightarrow{id} L^2(E_* \oplus E, W)$$

$$\pi_* : L^2(E_*) \to L^2(E_* \oplus E) \xrightarrow{id} L^2(E_* \oplus E, W)$$

and hence W is bounded, the assertion claimed can easily be deduced from the following. Let us denote by π^+ and π_*^+ the adjoints of π and π_* respectively as the mappings to the non-weighted space. Then $\pi^* = \pi^+ W$, $\pi_*^* = \pi_*^+ W$ and transcription equation (24) can be rewritten as

$$\pi^+ W \pi = I\ , \quad \pi_*^+ W \pi_* = I\ , \quad \pi_*^+ W \pi = \theta\ . \tag{26}$$

This form of the transcription equation is useful for other purposes too.

15. SZ.-NAGY - FOIAŞ TRANSCRIPTION: no weight, one of π, π_* is natural imbedding. To be precise, let us require that W be a range function (i.e. projection-valued function) and π_* (or π) be the natural imbedding. Namely, let

$$\pi_* = \begin{pmatrix} I \\ \mathbb{0} \end{pmatrix}, \quad \pi = \begin{pmatrix} X \\ Y \end{pmatrix}, \quad W = \begin{pmatrix} A & B \\ B^* & C \end{pmatrix}$$

be matrix representation of these operators with respect to the decomposition $\mathcal{E} = E_* \oplus E$. The condition $\pi_*^+ W \pi_* = I$ implies $A = I$, and hence $B = \mathbb{0}$, $C^2 = C$, because $W^2 = W$ is a range function. Next $\theta = \pi_*^+ W \pi = X$, and the last condition $I = \pi^+ W \pi$ yields $Y^* C Y = \Delta^2$, where $\Delta \stackrel{def}{=} (I - \theta^* \theta)^{1/2}$. This equation defines Y up to a left isometry-valued factor. A natural way is to choose $Y = \Delta$. The density condition (25) means in our case $\operatorname{clos} \Pi L^2(E_* \oplus E) = W L^2(E_* \oplus E)$, i.e. $C = \operatorname{sign} \Delta$, where

$sign\,\Delta(\zeta)$ stands for the orthogonal projection onto $clos\,Range\,\Delta(\zeta)$, $\zeta\in\mathbb{T}$.*)

So we get the original Nagy-Foiaş, transcription of the model:

$$\mathcal{H}=\begin{pmatrix} L^2(E_*) \\ clos\,\Delta L^2(E)\end{pmatrix},\quad G_*=\begin{pmatrix} H^2_-(E_*) \\ 0\end{pmatrix},\quad G=\begin{pmatrix}\theta\\ \Delta\end{pmatrix}H^2(E),$$

and

$$\mathcal{K}_\theta=\begin{pmatrix} H^2(E_*) \\ clos\,\Delta L^2(E)\end{pmatrix}\ominus\begin{pmatrix}\theta\\ \Delta\end{pmatrix}H^2(E).$$

This is so-called incoming representation. We obtain another, outgoing representation choosing π as the natural imbedding i.e.

$$\pi=\begin{pmatrix}0\\ I\end{pmatrix},\quad \pi_*=\begin{pmatrix}\Delta_*\\ \theta^*\end{pmatrix},\quad W=\begin{pmatrix} sign\,\Delta_* & \mathbb{O}\\ \mathbb{O} & I\end{pmatrix},$$

$$\mathcal{H}=\begin{pmatrix} clos\,\Delta_* L^2(E_*) \\ L^2(E)\end{pmatrix},\quad G=\begin{pmatrix} 0 \\ H^2(E)\end{pmatrix},\quad G_*=\begin{pmatrix}\Delta_*\\ \theta^*\end{pmatrix}H^2_-(E_*)\ .$$

Here Δ_* stands for $(I-\theta\theta^*)^{1/2}$.

The main advantage of these representations consist in that the weight is in fact identity, so we are working in a subspace of the usual L^2 . This makes various calculations and especialy estimate much easier. Also let us mention that in a very important partial case of an inner characteristic function θ (left inner in another terminology, i.e. $\theta(\zeta)$ are isometries a.e. and $\Delta\equiv\mathbb{O}$) the transcription discussed turns into a very nice one:

$$\mathcal{K}_\theta=H^2(E_*)\ominus\Theta H^2(E)\ .$$

---

*) More generally, for a Hilbert space operator A let us denote by $sign\,A$ the partial isometry V in the polar decomposition $A=VR$, $R\geqslant\mathbb{O}$, $Ker\,V=Ker\,R$.

16. LOCAL SPECTRAL MULTIPLICITY OF THE MINIMAL UNITARY DILATION. We had an opportunity to mention that the spectral measure $\mathcal{E}_U$ of the minimal unitary dilation U of a c.n.u. contraction is mutually absolutely continuous with respect to Lebesgue measure m. Now, on the Sz.-Nagy - Foiaş transcription of the model we can see immediately that

$$\operatorname{rank} \mathcal{E}_U(\zeta) = \dim E_* + \dim \Delta(\zeta) E = \dim \Delta_*(\zeta) E_* + \dim E .$$

By the way, this implies once more inequalities (18) (recall that $\dim \mathfrak{D}_T = \dim E, \dim \mathfrak{D}_{T^*} = \dim E_*$) and further

$$\operatorname*{ess\,sup}_{\mathbb{T}} \mu \leq 2 \operatorname*{ess\,inf}_{\mathbb{T}} \mu$$

where $\mu(\zeta) = \operatorname{rank} \mathcal{E}_U(\zeta)$, $\zeta \in \mathbb{T}$. It is known ([6]) that the last inequality is indeed sufficient for a measurable integer-valued function μ to be the multiplicity function of the unitary dilation of a c.n.u. contraction. From the Sz.-Nagy - Foiaş transcription we can see this almost immediately. In fact, put $\partial = \operatorname{ess\,inf} \mu$ and let E_* and E be of dimension ∂, and $\{\vartheta_n\}_{n=1}^{\partial}$ be arbitrary non-constant H^∞-function with the moduli

$$|\vartheta_n(\zeta)| = \begin{cases} 1, & \text{if } \mu(\zeta) < \partial + n \\ 1/2, & \text{if } \mu(\zeta) \geq \partial + n \end{cases} .$$

Let further $\Theta = \operatorname{diag}\{\vartheta_n\}_{n=1}^{\partial}$, $\mathcal{K}_\Theta$ be the corresponding Sz.-Nagy - Foiaş space and U be the shift operator $Uf = zf$ on $L^2(E_*) \oplus$ $\oplus \operatorname{clos} \Delta L^2(E)$. Then U is the minimal unitary dilation*) of M_Θ and $\operatorname{rank} \mathcal{E}_U(\zeta) = \dim E_* + \dim \Delta(\zeta) E = \partial + (\mu(\zeta) - \partial) = \mu(\zeta)$ if $\partial < \infty$ and $\operatorname{rank} \mathcal{E}_U(\zeta) \equiv \infty \equiv \mu(\zeta)$ otherwise.

---

*) In fact, here we use the solution of the inverse problem (given Θ, to construct an operator T with Θ_T being equivalent to Θ). But really, we already possess such a solution, because if we start with a pure contractive function $\Theta \in H^\infty(E \to E_*)$ (i.e. such $\|\Theta\|_\infty \leq 1$ and $\|\Theta(0)e\| < \|e\|$ for $e \in E$, $e \neq 0$) and consider the operator M_Θ then it can easily be proved (and in fact we already done this) that Θ_{M_Θ} is equivalent to Θ.

17. PAVLOV TRANSCRIPTION. In the spirit of the program of Sec.14 let us try to put $\Pi = id$. Then the weight W has to be bounded and transcription equations (26) imply

$$W = W_\theta \stackrel{def}{=\!=} \begin{pmatrix} I & \theta \\ \theta^* & I \end{pmatrix}.$$

This transcription of the model was discovered by B.S.Pavlov in [7]. The transcription is symmetric with respect to the changes T to T^* and the incoming space G_* to the outgoing space G. And both latter spaces look very simple:

$$G_* = \begin{pmatrix} H^2_-(E_*) \\ 0 \end{pmatrix}, \qquad G = \begin{pmatrix} 0 \\ H^2(E) \end{pmatrix}.$$

The payment for this simplicity is rather painful. Namely, the weight W_θ is in general degenerated and deeling with the space $L^2(E_* \oplus E, W_\theta)$ we have to form first the quotient space over the kernel of W_θ (which is not too obvious) and then the completing. Moreover, the operators $W_\theta(\zeta)$ turn the "fibres" (local spectral subspaces) of the spectral resolution of the unitary dilation (in particular they are not diagonal with respect to the decomposition $\mathcal{E} = E_* \oplus E$). These all make rather misty the structure of the model space $\mathcal{K}_\theta$ in spite of the simplicity of its orthogonal complement $G_* \oplus G$.

18. DE BRANGES - ROVNYAK TRANSCRIPTION. One can try to avoid the latter difficulties reversing the situation and choosing the operator Π^* (instead of Π) as a natural (identical)imbedding. Then the space $\mathcal{K}_\theta$ will consist of analytic (or co-analytic) functions.

So let us put $\Pi = W_\theta$. Then $\Pi^* = id$ and $W = W_\theta^{-1}$; here and in what follows under W^{-1} for a positive operator W we mean $(W|(\mathrm{Ker}\,W)^\perp)^{-1} P_{(\mathrm{Ker}\,W)^\perp}$ with densely defined inverse $(W|(\mathrm{Ker}\,W)^\perp)^{-1}$.

In such a way we obtain just the de Branges - Rovnyak transcription of the model. Referring for all details to Sec.27-28 here we try only to presuppose that our coordinate-free approach to the de Branges - Rovnyak model allows us to make more accessible numerous important works by L.de Branges and his school.

19. MICROSCROPE TRANSCRIPTION. By this we call a transcription of the model which makes transparent the given subspace invariant under a c.n.u. contraction T . It is known that such a subspace corresponds to one of regular factorizations of the characteristic function $\theta = \theta_T$. Some of the known formulae [3] representing the subspace in terms of factorization are rather complicated. Some others (e.g. [9]) seem to be easier and can be included in our transcription business.

Let $\theta = \theta_2\theta_1$ be a regular factorization of the characteristic function $\theta = \theta_T$, T being a c.n.u. contraction and $\theta_1 \in H^\infty(E \to F)$, $\theta_2 \in H^\infty(F \to E_*)$. Set $\mathcal{E} = E_* \oplus F \oplus E$, $W = diag\{I, sign\,\Delta_2, sign\,\Delta_1\}$ and

$$\Pi_* = \begin{pmatrix} I \\ \mathbb{O} \\ \mathbb{O} \end{pmatrix}, \quad \Pi = \begin{pmatrix} \theta \\ \Delta_2\theta_1 \\ \Delta_1 \end{pmatrix} \tag{27}$$

or in a dual way $W = \{sign\,\Delta_{*2}, I, sign\,\Delta_1\}$,

$$\Pi_* = \begin{pmatrix} \Delta_{*2} \\ \theta_2^* \\ \mathbb{O} \end{pmatrix}, \quad \Pi = \begin{pmatrix} \mathbb{O} \\ \theta_1 \\ \Delta_1 \end{pmatrix} \tag{28}$$

where $\Delta_i = (I - \theta_i^*\theta_i)^{1/2}$, $\Delta_{*i} = (I - \theta_i\theta_i^*)^{1/2}$, $i = 1, 2$. It is easy to see that the both choices give solution of the transcription equation (24) and so lead to two representations of the model space $\mathcal{K}_\theta$. In the first of them (via formula (27)) we have

$$\mathcal{K}_\theta = \mathcal{K}_1 \oplus \mathcal{K}_2$$

with M_θ^*-invariant subspace $\mathcal{K}_2$ (corresponding to the given factorization) just represented in the standard Sz.-Nagy - Foiaş transcription (Sec.15):

$$\mathcal{K}_2 = \begin{pmatrix} H^2(E_*) \\ clos\, \Delta_2 L^2(F) \\ \mathbb{O} \end{pmatrix} \ominus \begin{pmatrix} \theta_2 \\ \Delta_2 \\ \mathbb{O} \end{pmatrix} H^2(F) .$$

In the second transcription (via formula (28)) we get

$$\mathcal{K}_\theta = \mathcal{K}_1 \oplus \mathcal{K}_2$$

with a similar form of M_θ-invariant subspace $\mathcal{K}_1$,

$$\mathcal{K}_1 = \begin{pmatrix} \mathbb{O} \\ H^2(F) \\ clos\, \Delta_1 L^2(E) \end{pmatrix} \ominus \begin{pmatrix} \mathbb{O} \\ \theta_1 \\ \Delta_1 \end{pmatrix} H^2(E) .$$

20. GENERAL SOLUTION OF THE TRANSCRIPTION EQUATIONS. First let us recall that the unique restriction on the weight function $W, W(\zeta) : \mathcal{E} \to \mathcal{E}$, is the rank restriction

$$dim\, W(\zeta)\mathcal{E} = rank\, \mathcal{E}_U(\zeta) \tag{29}$$

a.e.on $\mathbb{T}$, where $\mathcal{E}_U$ stands for the spectral measure of the minimal unitary dilation U . Then note that the unitary equivalence of the shift operators $f \to zf$ on the spaces $L^2(\mathcal{E}, W)$ and $L^2(\mathcal{E}, sign\, W)$, $sign\, W$ being a range function is carried out by the multiplication operator $f \mapsto W^{1/2} f$. We use this remark proving the following simple proposition.

21. LEMMA. <u>Let W be a weight satisfying (29). Then the most general solution Π of equations (23), (24) and (25) is of the form</u>

$$\Pi = W^{-1/2}\, \Phi \Pi_\theta$$

<u>where Φ, $\Phi(\zeta) : E_* \oplus E \longrightarrow \mathcal{E}$, is an arbitrary measurable operator-valued function which isometrically imbeds $E_* \oplus clos\, \Delta(\zeta) E$ onto $\mathcal{E}(\zeta) \overset{def}{=} clos\, W(\zeta)\mathcal{E}$, and Π_θ stands for the Sz.-Nagy - Foiaş solution,</u>

$$\Pi_\theta = \begin{pmatrix} I & \theta \\ \mathbb{O} & \Delta \end{pmatrix} .$$

PROOF. By the previous remark, a given pair Π, W is a solution of (23), (24) iff so is the pair $W^{1/2}\Pi \overset{def}{=} \Pi_0, \operatorname{sign} W$. Now we can use (26) which means

$$\Pi_0^+(\operatorname{sign} W)\Pi_0 = W_\theta = \Pi_\theta^+\Pi_\theta .$$

The latter is equivalent to the existence of Φ with properties described in the statement and such that $\Pi_0 = \Phi\Pi_\theta$.

22. REMARK. In a dual way we can write down the general solution Π in the form

$$\Pi = W^{-1/2}\Psi\Pi_{\theta^*}$$

where $\Pi_{\theta^*} = \begin{pmatrix} \Delta_* & \mathbb{O} \\ \theta^* & I \end{pmatrix}$ stands for the dual Sz.-Nagy - Foiaş solution and Ψ has a similar meaning.

23. RESIDUAL PARTS. To use our coordinate-free model we need a formula for the orthogonal projections onto the residual and $*$-residual parts of the model space $\mathcal{H}$, i.e. onto

$$\mathcal{R} \overset{def}{=} \mathcal{H} \ominus \pi_* L^2(E_*) \quad , \quad \mathcal{R}_* = \mathcal{H} \ominus \pi L^2(E)$$

respectively. Also we need concrete spectral representations of the residual $U|\mathcal{R}$ and $*$-residual $U|\mathcal{R}_*$ parts of the unitary dilation U. To do this let as consider Φ of Lemma 21 (and Ψ of Remark 22) and write it in the coordinate form:

$$\Phi = (\varphi_*, \varphi) \quad , \quad \Psi = (\psi_*, \psi) .$$

Then for $\Pi = (\pi_*, \pi)$ of Lemma 21 we get

$$\pi_* = W^{-1/2}\varphi_* \quad , \quad \pi = W^{-1/2}(\varphi_*\theta + \varphi\Delta)$$

and for Π of Remark 22

$$\pi_* = W^{-1/2}(\psi_*\Delta_* + \psi\theta^*) \;, \qquad \pi = W^{-1/2}\psi \;.$$

This implies $\pi - \pi_*\theta = W^{-1/2}\varphi\Delta$ and since $(\pi - \pi_*\theta)^*(\pi - \pi_*\theta) = \Delta^2$, we conclude that there exists an isometry τ from $\operatorname{clos}\Delta L^2(E)$ to $\mathcal{H}$ such that

$$\pi - \pi_* \theta = \tau \Delta$$

and hence

$$\tau = W^{-1/2} \varphi \,|\, \operatorname{clos} \Delta L^2(E).$$

Similarly, $\pi_* - \pi\theta^* = \tau_* \Delta_*$ and

$$\tau_* = W^{-1/2} \psi_* \,|\, \operatorname{clos} \Delta_* L^2(E_*).$$

24. LEMMA.

$$\tau^* \pi = \Delta, \qquad \tau^* \pi_* = \mathbb{O},$$
$$\tau_*^* \pi = \mathbb{O}, \qquad \tau_*^* \pi_* = \Delta_*$$
$$\tau\tau^* = I - \pi_* \pi_*^*, \quad \tau_* \tau_*^* = I - \pi \pi^*$$

<u>and hence</u> $\tau\tau^*(\tau_*\tau_*^*)$ <u>is the projection onto</u> $\mathcal{R}$ (<u>resp</u>. $\mathcal{R}_*$) <u>and</u> $\tau\,(\tau_*)$ <u>realizes the spectral representation for</u> $\mathbb{U} \,|\, \mathcal{R}$ (<u>resp</u>. $\mathbb{U} \,|\, \mathcal{R}_*$).

PROOF. The first line of the lemma follows from an elementary calculation: $\pi^*\tau\Delta = \pi^*\pi - \pi^*\pi_*\theta = I - \theta^*\theta = \Delta^2$; $\pi_*^*\tau\Delta = \pi_*^*\pi - \pi_*^*\pi_*\theta = \theta - \theta = \mathbb{O}$. Therefore for arbitrary $x \in L^2(E)$ and $x_* \in L^2(E_*)$ we have

$$(\tau\tau^* + \pi_*\pi_*^*)(\pi x + \pi_* x_*) = (\tau\Delta + \pi_*\theta)x + \pi_* x_* = \pi x + \pi_* x_* .$$

So, $\tau\tau^* + \pi_*\pi_*^* = I$ because $\pi L^2(E) + \pi_* L^2(E_*)$ is dense in $\mathcal{H}$.

The dual formulae (for τ_*) are proved in a similar way.

25. DEFECT OPERATORS AND RESOLVENT. We give now formulae for these operators in terms of embeddings π_* and π . But first note that the expression for T (and for T^*) can be rewritten in the following form:

$$T = (I - \pi h_0^* h_0 \pi^*)\, \mathbb{U} \,|\, H,$$

$$T^* = \mathbb{U}^*(I - \pi_* h_0^* h_0 \pi_*^*) \,|\, H$$

where $h_\lambda : L^2(X) \to X$ is the operator of the point evaluation at λ: $h_\lambda f = \tilde{f}(\lambda)$, $\tilde{f}$ is the harmonic continuation of f inside the unit disc.

Let $\theta_0 \overset{def}{=} \theta(0)$ denote the operator from E into E_* as well as the operator of multiplication by θ_0 from $L^2(E)$ into $L^2(E_*)$. Since $h_0 z \pi_*^* | H = \mathbb{O}$, we have

$$D_T^2 = I - T^*T = (\pi \bar{z} h_0^* h_0 z \pi^* - \pi_* \bar{z} h_0^* \theta_0 h_0 z \pi^*) | H =$$

$$= (\pi - \pi_* \theta_0) \bar{z} h_0^* h_0 z (\pi^* - \theta_0^* \pi_*^*) | H .$$

Put $\Delta_0 = (I - \theta_0^* \theta_0)^{1/2}$, then

$$h_0 z (\pi - \pi_* \theta_0)^* (\pi - \pi_* \theta_0) \bar{z} h_0^* = \Delta_0^2$$

and hence the relation $\Omega \Delta_0 = (\pi - \pi_* \theta_0) \bar{z} h_0^*$ defines an isometric embedding

$$\Omega : E \longrightarrow \mathcal{H}$$

whose range is the defect subspace $\mathcal{D}_T$. For the defect operator D_T we have

$$D_T = \Omega \Delta_0 \Omega^* .$$

Putting $\Delta_{*0} = (I - \theta_0 \theta_0^*)^{1/2}$ we get analogously

$$D_{T^*} = \Omega_* \Delta_{*0} \Omega_*^*$$

where $\Omega_* : E_* \to \mathcal{H}$ is defined by the relation

$$\Omega_* \Delta_{*0} = (\pi_* - \pi \theta_0^*) h_0^* .$$

In terms of Ω and Ω_* formulae for T and T^* can be rewritten in the form

$$T = [U(I - \Omega \Omega^*) - \Omega_* \theta_0 \Omega^*] | H$$

$$T^* = [U^*(I - \Omega_* \Omega_*^*) - \Omega \theta_0^* \Omega_*^*] | H .$$

Finally we give coordinate formulae for resolvent of T in connection with resolvent of its minimal unitary dilation

$$(T-\lambda I)^{-1} = \begin{cases} (U-\lambda I)^{-1}(I-\pi h_0^* \theta(\lambda)^{-1} h_\lambda \pi_*^*)|H & \lambda \in \mathbb{D} \setminus \sigma_T \\ (U-\lambda I)^{-1}(I-\pi h_0^* h_{1/\bar\lambda}\, \pi^*)|H & \lambda \notin \operatorname{clos}\mathbb{D} \end{cases},$$

$$(T^*-\bar\lambda I)^{-1} = \begin{cases} (U^*-\bar\lambda I)^{-1}(I-\pi_* \bar z h_0^* \theta^*(\bar\lambda)^{-1} h_{\bar\lambda}\, z\pi^*)|H & \lambda \in \mathbb{D} \setminus \sigma_{T^*} \\ (U^*-\bar\lambda I)^{-1}(I-\pi_* \bar z h_0^* h_{1/\bar\lambda}\, z\pi_*^*)|H & \lambda \notin \operatorname{clos}\mathbb{D} \end{cases}.$$

26. SOME FORMULAE in various transcription of the model will be given in this section.

Sz.-Nagy - Foiaş transcription

$$\mathcal{H} = \begin{pmatrix} L^2(E_*) \\ \operatorname{clos}\Delta L^2(E) \end{pmatrix}$$

$$\pi_* = \begin{pmatrix} I \\ \mathbb{O} \end{pmatrix}, \quad \pi = \begin{pmatrix} \theta \\ \Delta \end{pmatrix}, \quad \tau = \begin{pmatrix} \mathbb{O} \\ I \end{pmatrix}, \quad \tau_* = \begin{pmatrix} \Delta_* \\ -\theta^* \end{pmatrix}$$

$$\mathcal{K}_\theta = \left\{ \begin{pmatrix} f \\ g \end{pmatrix} : f \in H^2(E_*),\ \ g \in \operatorname{clos}\Delta L^2(E),\ \ \theta^* f + \Delta g \in H^2_-(E) \right\}$$

$$P_\theta = \begin{pmatrix} P_+ - \theta P_+ \theta^* & -\theta P_+ \Delta \\ -\Delta P_+ \theta^* & I - \Delta P_+ \Delta \end{pmatrix}$$

$$\Omega = \begin{pmatrix} \dfrac{\theta-\theta_0}{z} \\ \bar z \Delta \end{pmatrix} \Delta_0^{-1} h_0^*, \qquad \Omega_* = \begin{pmatrix} I - \theta\theta_0^* \\ -\Delta\theta_0^* \end{pmatrix} \Delta_{*0}^{-1} h_0^*$$

$$M_\theta \begin{pmatrix} f \\ g \end{pmatrix} = \begin{pmatrix} zf - \theta e \\ zg - \Delta e \end{pmatrix}, \quad \text{where} \quad e = h_0^* h_0\, z(\theta^* f + \Delta g)$$

$$M_\theta^*\begin{pmatrix} f \\ g \end{pmatrix} = \begin{pmatrix} \frac{f-f(0)}{z} \\ \bar{z}g \end{pmatrix}$$

$$(M_\theta-\lambda I)^{-1}\begin{pmatrix} f \\ g \end{pmatrix} = \begin{pmatrix} \frac{f-\theta e}{z-\lambda} \\ \frac{g-\Delta e}{z-\lambda} \end{pmatrix}, \quad \text{where} \quad e=\begin{cases} h_0^*\theta(\lambda)^{-1}h_\lambda f, & \lambda\in\mathbb{D}\setminus\sigma_\theta \\ h_0^* h_{1/\bar\lambda}(\theta^* f+\Delta g), & \lambda\notin \operatorname{clos}\mathbb{D} \end{cases}.$$

<u>Pavlov transcription</u>

$$\mathcal{H} = L^2\left(E_*\oplus E, \begin{pmatrix} I & \theta \\ \theta^* & I \end{pmatrix}\right)$$

$$\pi_* = \begin{pmatrix} I \\ \mathbb{O} \end{pmatrix}, \quad \pi_*^* = (I, \theta), \quad \pi = \begin{pmatrix} \mathbb{O} \\ I \end{pmatrix}, \quad \pi^* = (\theta^*, I)$$

$$\mathcal{K}_\theta = \mathcal{H} \ominus \left[\begin{pmatrix} H_-^2(E_*) \\ 0 \end{pmatrix} \oplus \begin{pmatrix} 0 \\ H^2(E) \end{pmatrix}\right]$$

(NB: not all elements of $\mathcal{K}_\theta$ (as well as of $\mathcal{H}$) can be written as a pair $\begin{pmatrix} f \\ g \end{pmatrix}$ of L^2-functions because the angle between E and E_* may be equal to zero in W-norm)

$$\tau = \begin{pmatrix} -\theta \\ I \end{pmatrix}\Delta^{-1}, \quad \tau^* = (\mathbb{O}, \Delta), \quad \tau_* = \begin{pmatrix} I \\ -\theta^* \end{pmatrix}\Delta_*^{-1}, \quad \tau_*^* = (\Delta_*, \mathbb{O})$$

$$P_\theta = \begin{pmatrix} P_+ & -P_-\theta \\ -P_+\theta^* & P_- \end{pmatrix}$$

$$\Omega = \bar{z}\begin{pmatrix} -\theta_0 \\ I \end{pmatrix}\Delta_0^{-1}h_0^*, \qquad \Omega_* = \begin{pmatrix} I \\ -\theta_0^* \end{pmatrix}\Delta_{*0}^{-1}h_0^*$$

$$M_\theta\begin{pmatrix} f \\ g \end{pmatrix} = \begin{pmatrix} zf \\ zg - h_0^* h_0 z(\theta^* f+g) \end{pmatrix}, \quad M_\theta^*\begin{pmatrix} f \\ g \end{pmatrix} = \begin{pmatrix} \bar{z}f - \bar{z}h_0^* h_0(f+\theta g) \\ \bar{z}g \end{pmatrix}$$

$$(M_\theta - \lambda I)^{-1}\begin{pmatrix} f \\ g \end{pmatrix} = \begin{pmatrix} \dfrac{f}{z-\lambda} \\ \dfrac{g-e}{z-\lambda} \end{pmatrix}, \quad \textbf{where} \quad e = \begin{cases} h_0^*\theta(\lambda)^{-1}h_\lambda(f+\theta g), & \lambda \in \mathbb{D} \setminus \sigma_\theta \\ h_0^* h_{1/\bar\lambda}(\theta^* f + g), & \lambda \notin \operatorname{clos}\mathbb{D} \end{cases}$$

<u>de Branges - Rovnyak transcription</u>

$$\mathcal{H} = L^2\left(E_* \oplus E, \begin{pmatrix} I & \theta \\ \theta^* & I \end{pmatrix}^{-1}\right)$$

$$\pi_* = \begin{pmatrix} I \\ \theta^* \end{pmatrix}, \quad \pi_*^* = (I, \mathbb{O}), \quad \pi = \begin{pmatrix} \theta \\ I \end{pmatrix}, \quad \pi^* = (\mathbb{O}, I)$$

$$\tau = \begin{pmatrix} \mathbb{O} \\ \Delta \end{pmatrix}, \quad \tau^* = \Delta^{-1}(-\theta^*, I), \quad \tau_* = \begin{pmatrix} \Delta_* \\ \mathbb{O} \end{pmatrix}, \quad \tau_*^* = \Delta_*^{-1}(I, -\theta)$$

$$\mathcal{K}_\theta = \left\{ \begin{pmatrix} f \\ g \end{pmatrix} : f \in H^2(E_*),\ g \in H_-^2(E),\ g - \theta^* f \in \Delta L^2(E) \right\}$$

$$P_\theta = \begin{pmatrix} P_+ & -\theta P_+ \\ -\theta^* P_- & P_- \end{pmatrix}$$

$$\Omega = \begin{pmatrix} \dfrac{\theta - \theta_0}{z} \\ \bar z (I - \theta^* \theta_0) \end{pmatrix} \Delta_0^{-1} h_0^*, \qquad \Omega_* = \begin{pmatrix} I - \theta\theta_0^* \\ \theta^* - \theta_0^* \end{pmatrix} \Delta_{*0}^{-1} h_0^*$$

$$M_\theta \begin{pmatrix} f \\ g \end{pmatrix} = \begin{pmatrix} zf - \theta e \\ zg - e \end{pmatrix}; \quad \textbf{where} \quad e = h_0^* h_0 z g$$

$$M_\theta^* \begin{pmatrix} f \\ g \end{pmatrix} = \begin{pmatrix} \dfrac{f - f(0)}{z} \\ \bar z g - \theta^* z f(0) \end{pmatrix}$$

$$(M_\theta-\lambda I)^{-1}\begin{pmatrix} f\\ g\end{pmatrix}=\begin{pmatrix} \dfrac{f-\theta e}{z-\lambda}\\[2mm] \dfrac{g-e}{z-\lambda}\end{pmatrix}, \quad \text{where} \quad e=\begin{cases} h_0^*\theta(\lambda)^{-1}h_\lambda f, & \lambda\in\mathbb{D}\setminus\sigma_\theta\\[2mm] h_0^* h_{1/\bar\lambda}g, & \lambda\notin \operatorname{clos}\mathbb{D}\end{cases}.$$

27. A CONNECTION WITH THE ORIGINAL DE BRANGES - ROVNYAK WORDING. Here we check that our model transcription given in section 18 indeed coinsides with the original de Branges - Rovnyak model presented in [10], [11]. We would remind now the original definition of the Branges and Rovnyak (however in a slightly modifyed form).

For a function $\theta\in H^\infty(E\to E_*)$, E and E_* being auxiliary Hilbert spaces, let us denote by $\boldsymbol{\theta}$ the multiplication operator

$$\boldsymbol{\theta} f=\theta f, \quad f\in H^2(E)$$

acting from $H^2(E)$ to $H^2(E_*)$. Them $\boldsymbol{\theta}^* f=P_+\theta^* f$, $f\in H^2(E_*)$, where as everywhere before $\theta^*(\zeta)=\theta(\zeta)^*$, $\zeta\in\mathbb{T}$. Now, if $\|\theta\|_\infty\le 1$ let us denote by $\mathcal{H}(\theta)$ (following de Branges) the Hilbert space

$$\mathcal{H}(\theta)=\mathcal{D}_{\boldsymbol{\theta}^*}H^2(E_*),$$

$\mathcal{D}_{\boldsymbol{\theta}^*}$ standing for the defect operator, and endow it with the range norm

$$\|x\|_{\mathcal{H}(\theta)}=\inf\{\|f\|_{H^2}: x=\mathcal{D}_{\boldsymbol{\theta}^*}f\}.$$

Next, let $\mathcal{D}(\theta)$ be the Hilbert space of all pairs $\begin{pmatrix} f\\ g\end{pmatrix}$ such that

$$f\in\mathcal{H}(\theta), \quad g\in H^2(E)$$

$$z^n f-\theta P_+ z^n Jg\in\mathcal{H}(\theta), \qquad \forall n>0,$$

where J is a unitary involution of (any) L^2-space changing H^2 and H^2_- and acting by the rule

$$Jg=\bar z g(\bar z).$$

The space $\mathcal{D}(\theta)$ is endowed with the norm

$$\left\| \begin{pmatrix} f \\ g \end{pmatrix} \right\|^2_{\mathcal{D}(\theta)} = \lim_{n\to\infty} \left(\| z^n f - \theta P_+ z^n J g \|^2_{\mathcal{H}(\theta)} + \| P_+ z^n J g \|^2_{H^2} \right),$$

and it is known that the (finite) limit exists, see [11].

And finally, the de Branges - Rovnyak model operator BR is defined as an operator on $\mathcal{D}(\theta)$ by the formula

$$BR \begin{pmatrix} f \\ g \end{pmatrix} = \begin{pmatrix} \dfrac{f - f(0)}{z} \\ zg - \theta(\bar{z})^* f(0) \end{pmatrix}, \qquad \begin{pmatrix} f \\ g \end{pmatrix} \in \mathcal{D}(\theta).$$

28. THEOREM. Let θ be a pure contractive function from $H^\infty(E \to E_*)$ and let $\mathcal{K}_\theta$ and M_θ be the model space and the model operator in the transcription of Sec. 18. Let $\mathcal{J} = \begin{pmatrix} I & \mathbb{O} \\ \mathbb{O} & J \end{pmatrix}$. Then

$$\mathcal{K}_\theta = \mathcal{J}\mathcal{D}(\theta), \qquad \mathcal{J} M_\theta^* \mathcal{J} = BR.$$

PROOF. Note first that

$$\pi_*^* P_\theta \pi_* = P_+ - \theta P_+ \theta^* = \mathcal{D}^2_{\theta^*} P_+$$

and hence there exists a partial isometry V from $H^2(E_*)$ to $\mathcal{H}$ (the space of the unitary dilation of M_θ) with the initial subspace $\mathfrak{D}_{\theta^*} = \operatorname{clos} \mathcal{D}_{\theta^*} H^2(E_*)$ such that

$$P_\theta \pi_* = V \mathcal{D}_{\theta^*} P_+ .$$

It is not difficult to verify that the final subspace of V is $\mathcal{K}_\theta \ominus \tau [\operatorname{clos} \Delta L^2(E) \ominus \operatorname{clos} \Delta H^2(E)]$.

Let now $x = \begin{pmatrix} f \\ g \end{pmatrix} \in \mathcal{K}_\theta$ then

$$f = \pi_*^* x = \pi_*^* P_\theta x = \mathcal{D}_{\theta^*} V^* x \in \mathcal{H}(\theta),$$

$$z^n f - \theta P_+ z^n g = \pi_*^* P_\theta U^n x = \mathcal{D}_{\theta^*} V^* M_\theta^n x \in \mathcal{H}(\theta) \qquad \forall n > 0,$$

and therefore $x \in \mathcal{J}\mathcal{D}(\theta)$. Let us check that $\|x\|_{\mathcal{K}_\theta} = \|\mathcal{J}x\|_{\mathcal{D}(\theta)}$.

$$\lim \| P_+ z^n g \| = \| g \| = \| \pi^* x \| .$$

Since $s\text{-}\lim P_{(\mathrm{clos}\,\Delta H^2)^\perp} z^n | \mathrm{clos}\,\Delta L^2 = \mathbb{O}$, we have

$$\lim \| V^* M_\theta^n x \| = \lim \| M_\theta^n x \| = \| \tau_*^* x \| ,$$

because

$$s\text{-}\lim U^{*n} P_\theta U^n = s\text{-}\lim (I - \pi \bar{z}^n P_+ z^n \pi^* - \pi_* \bar{z}^n P_- z^n \pi_*^*) = I - \pi\pi^* = \tau_* \tau_*^* .$$

Hence

$$\lim \| z^n f - \theta P_+ z^n g \|_{\mathcal{H}(\theta)} = \lim \| V^* M_\theta^n x \| = \| \tau_*^* x \|$$

and $\| \mathcal{J} x \|^2_{\mathfrak{D}(\theta)} = \| \pi^* x \|^2 + \| \tau_*^* x \|^2 = \| x \|^2_{\mathcal{K}_\theta}$.

To check that $\mathcal{J}\mathfrak{D}(\theta) = \mathcal{K}_\theta$ we take $x = \binom{f}{g} \in \mathcal{J}\mathfrak{D}(\theta) \ominus \mathcal{K}_\theta$ and verify that $x = 0$. Put $y = P_\theta \pi g = \begin{pmatrix} P_+ \theta g \\ g - \theta^* P_- \theta g \end{pmatrix}$ then $(x, y)_{\mathcal{J}\mathfrak{D}(\theta)} = 0$, because $x \perp \mathcal{K}_\theta$. Since

$$z^n (P_+ \theta g) - \theta P_+ z^n (g - \theta^* P_- \theta g) = -\mathfrak{D}^2_{\theta^*} P_+ z^n P_- \theta g + P_+ \theta P_- z^n g$$

and $\lim \| P_+ \theta P_- z^n g \|_{\mathcal{H}(\theta)} = 0$ (we check this a bit later), then

$$0 = (x, y)_{\mathcal{J}\mathfrak{D}(\theta)} = \lim \big[(z^n f - \theta P_+ z^n g, -P_+ z^n P_- \theta g)_{H^2} +$$

$$+ (P_+ z^n g, P_+ z^n (g - \theta^* P_- \theta g))_{H^2} \big] = \lim \big[(f, -P_- \theta g) + \| P_+ z^n g \|^2 \big] = \| g \|^2$$

therefore $g = 0$. Put now $y = P_\theta \pi_* f = \begin{pmatrix} \mathfrak{D}^2_{\theta^*} f \\ P_- \theta^* f \end{pmatrix}$, then $0 = (x, y)_{\mathcal{J}\mathfrak{D}(\theta)} = (f, \mathfrak{D}^2_{\theta^*} f)_{\mathcal{H}(\theta)} = \| f \|^2$, i.e. $f = 0$. To complete the proof we need only to check that $\lim \| Q P_- z^n g \|_{\mathcal{H}(\theta)} = 0$, where $Q = P_+ \theta | H^2_-(E) : H^2_-(E) \to H^2(E_*)$ Since

$$QQ^* = P_+ \theta P_- \theta^* = \mathfrak{D}^2_{\theta^*} - P_+ \Delta_*^2 \leqslant \mathfrak{D}^2_{\theta^*} ,$$

there exists a contraction $C : H^2_-(E) \longrightarrow H^2(E_*)$ such that $Q = \mathfrak{D}_{\theta^*} C$

and therefore

$$\overline{\lim}\, \| QP_- z^n g \|_{\mathcal{H}(\theta)} \leq \overline{\lim}\, \| CP_- z^n g \|_{H^2} \leq \lim \| P_- z^n g \| = 0 \ .$$

Finally, formulae of Sec.26 show that $\mathcal{J} M_\theta^* \mathcal{J}$ is just BR, the de Branges - Rovnyak model operator.

To conclude we would like to metion that the operator M_θ, our transcription of the de Branges - Rovnyak model operator (Sec.18), acts only between analytic functions (does not leave the Hardy space), and we consider that as an advantage. But the difficulties of the norm computations (both in our $\mathcal{K}_\theta$ and in the original de Branges - Rovnyak space $\mathcal{D}(\theta)$ bring it almost to nothing).

REFERENCES

1. N.G.Makarov, V.I.Vasyunin. A model for noncontractions and stability of the continuous spectrum. - Lect.Notes in Math., 1981, v.864, 365-412.
2. N.K.Nikolskii, V.I.Vasyunin. Notes on two function models. - Proc.Confer.on the Occasion of the proof of the Bieberbach conjecture, 1986, 113-141.
3. B.Szőkefalvi-Nagy, C.Foiaş. Harmonic Analysis of Operators on Hilbert Space. North-Holland / Akadémiai Kiadó, Amsterdam / Budapest, 1970.
4. V.M.Adamyan, D.Z.Arov. Unitary couplings of semi-unitary operators. - Mat.Issled., 1966, v.1, N 2, 3-64 (Russian).
5. V.I.Vasyunin. Construction of the functional model of B.Sz.-Nagy and C.Foiaş. - In: Investigations on linear operators and the theory of functions. VIII, Zap.Nauch.Sem.Leningrad. Otdel.Mat.Inst.Steklov (LOMI), 1977, v.73, 16-23. (Russian).
6. R.G.Douglas. Canonical Models. - Math.Surveys, v.13, AMS, Providence, 1974, 161-218.
7. B.S.Pavlov. Conditions for separation of the spectral components of a dissipative operator. - Izv.Akad.Nauk SSSR Ser. Mat., 1975, v.39, N 1, 123-148. (Russian)
8. D.Sarason. On spectral sets having connected complement. -

Acta Sci.Math. (Szeged), 1965, t.26, 289-299.

9. B.Szőkefalvi-Nagy, C.Foiaş. Sur les contractions de l'espace de Hilbert. IX. Factorisations de la fonction caracteristique. Sous-espaces invariants. - Acta Sci.Math.(Szeged), 1964, t.25, 283-316.

10. L.de Branges, J.Rovnyak. Canonical models in quantum scattering theory. In: Perturbation Theory and its applications in Quantum Mechanics, Wiley, New York, 1966, 295-392.

11. L.de Branges. Square summable power series. Heidelberg, Springer-Verlag, to appear.

Operator Theory:
Advances and Applications, Vol. 41

QUADRATURE METHODS FOR STRONGLY ELLIPTIC CAUCHY SINGULAR INTEGRAL EQUATIONS ON AN INTERVAL

Siegfried Prössdorf and Andreas Rathsfeld

This paper is concerned with a quadrature method for singular integral equations on a finite interval. It is proved that the strong ellipticity is necessary and sufficient for this numerical procedure to be stable. The proof of this result is based on the symbol calculus on the algebra generated by Toeplitz matrices due to Gohberg and Krupnik [12]. Estimates for the speed of convergence as well as numerical examples are given.

0. INTRODUCTION

Many boundary value problems in mathematical physics and engineering can be reduced to a singular integral equation of the form

$$(0.1)\quad a(t)x(t)+\frac{b(t)}{\pi i}\int_0^1 \frac{x(\tau)}{\tau-t}\,d\tau + \int_0^1 k(t,\tau)x(\tau)\,d\tau = y(t)\ ,\quad 0\le t\le 1\ .$$

Here a, b and k are continuous functions, y is at least Riemann-integrable and x is the unknown function. For the direct numerical solution of this equation, there exist two classes of approximation methods. The first class is based on the polynomial approximation of the function $z(t)=x(t)/\rho(t)$, where ρ denotes a certain weight function depending on the asymptotic behaviour of x near the end-points of the interval. Using special invariance relations for polynomials, Galerkin, collocation and even quadrature methods can be shown to converge or to be stable (cf. [8, 3, 14, 15]). However, if the functions a and b are not constant, then the implementation of these methods becomes complicated. E.g., for the quadrature methods, orthogonal polynomials corresponding to the weight ρ are needed, and the quadrature knots are the zeros of these polynomials. However, the polynomial methods are of high order. Note that in [10] a simpler method is proposed which yields a lower rate of convergence. For this modification , no convergence analysis is known.

The second class of numerical procedures consists of spline and quadrature methods with simple quadrature rules. Using weighted spline functions and collocation, good rates of convergence have been obtained in [11,13]. The stability of spline collocation has been considered in the case of non-weighted splines (cf.e.g.[18]) and in the case of a special weight which differs essentially from the asymptotics of the solution (cf.e.g.[6]). As it seems to us, the results in e.g.[6,18] are only a first step in establishing the stability for adequate spline collocation methods. Furthermore, in order to implement collocation procedures, one has to compute the singular integrals of splines. If this cannot be done analytically, then one uses quadrature rules. Thus it is natural and easier to discretize (0.1) directly using quadrature rules.

The first convergence analysis of simple quadrature methods was done for the case of the method of discrete whirls and is due to Lifanov and Polonski [16] . A generalisation of this method can be found in [21,19] (cf.also [20]).In this paper we consider a modification of a quadrature method presented in [21] (cf.(1.15)). In order to obtain the system of equations (1.15) for this modified quadrature method, we proceed as follows (cf.Sect.1).First, we transform (0.1) substituting $\tau=\sigma^\alpha$, $t=s^\alpha$ in a neighbourhood of 0 and $\tau=1-(1-\sigma)^\alpha$, $t=1-(1-s)^\alpha$ in a neighbourhood of 1,where $\alpha\geq 1$(cf.the transformation technique for quadrature methods in the case of Fredholm integral equations of the second kind in [17]).Then we set up the usual quadrature method using subtraction technique (cf. the modified quadrature methods in [1]) and Simpson's rule.

In Sect. 3 we prove the method in consideration to be stable if and only if Equ. (0.1) is strongly elliptic. For the stable quadrature method, the rates of convergence are derived in Sect. 2. Sect. 4 contains numerical examples. Finally, we remark that all the results of this paper apply also to singular integral equatons with fixed singularities (cf. [21,19] and, for equations of the second kind, [2]). Furthermore, analogous modifications carried out in the case of other quadrature methods from [21] lead to numerical procedures which are stable for certain classes of non-strongly elliptic equations.

1. DERIVATION OF THE QUADRATURE METHOD

For the sake of uniqueness, the solution is sought in the form $x = \rho z$, where $\rho(t)=t^{\rho_0}(1-t)^{\rho_1}$, $-1/2 \le \mathrm{Re}\ \rho_0, \mathrm{Re}\ \rho_1 \le 1/2$ and $z \in L^2 := L^2(0,1)$. By ${}^0\!\log$ and by ${}^1\!\log$ we denote the branch of the logarithm which takes real values on the positive real axis and is continuous on $\mathbb{C} \setminus \{\exp(i2\pi\rho_0)t, -\infty<t\le 0\}$ and $\mathbb{C} \setminus \{\exp(i2\pi\rho_1)t, -\infty<t\le 0\}$, respectively. If we set

$$(1.1) \qquad \varkappa_j := \frac{(-1)^j}{2\pi i}\ {}^j\!\log \frac{a(j)+b(j)}{a(j)-b(j)} \quad , \quad j=0,1 \quad ,$$

then (cf. [5,4] and Sect. 2) there exist $\lambda, \mu \in \mathbb{C}$ such that the first term in the asymptotics of x is $\lambda t^{\varkappa_0}$ for $t \to 0$ and $\mu(1-t)^{\varkappa_1}$ for $t \to 1$.

In order to set up our quadrature methods, we begin with transforming Equ. (0.1). Choose $\alpha \ge 1$ and $\gamma : [0,1] \to [0,1]$ such that $\gamma(\sigma)=\sigma^\alpha$ for $0 \le \sigma \le 1/3$, $\gamma(\sigma)=1-(1-\sigma)^\alpha$ for $2/3 \le \sigma \le 1$ and γ is infinitely differentiable on $[1/4,3/4]$. Substituting $x(\tau)= \rho(\gamma(\sigma))z(\gamma(\sigma))$ in (0.1), we arrive at

$$(1.2) \quad a(\gamma(s))\rho(\gamma(s))z(\gamma(s)) + \frac{b(\gamma(s))}{\pi i} \int_0^1 \frac{\rho(\gamma(\sigma))z(\gamma(\sigma))}{\gamma(\sigma)-\gamma(s)} \gamma'(\sigma)d\sigma + \int_0^1 k(\gamma(s),\gamma(\sigma))\rho(\gamma(\sigma))z(\gamma(\sigma))\gamma'(\sigma)d\sigma = y(\gamma(s)) \ , \ 0 \le s \le 1 \ .$$

Observing that

$$(1.3)\qquad \int_0^1 |\mathfrak{z}(\tau)|^2 d\tau = \int_0^1 |\mathfrak{z}(\gamma(\sigma))\gamma'(\sigma)^{1/2}|^2 d\sigma ,$$

we introduce the new unknown function $z(\sigma):=\mathfrak{z}(\gamma(\sigma))\gamma'(\sigma)^{1/2}$. We now multiply (1.2) by $\gamma'(s)^{1/2}/\rho(\gamma(s))$ to obtain

$$(1.4)\qquad a(s)z(s) + \frac{b(s)}{\pi i}\int_0^1 \frac{r(\sigma)}{r(s)} \frac{\gamma'(s)^{1/2}\gamma'(\sigma)^{1/2}}{\gamma(\sigma)-\gamma(s)} z(\sigma)d\sigma + \int_0^1 k(s,\sigma)z(\sigma)d\sigma = y(s) , \quad 0\le s\le 1 ,$$

where $a(s):=\mathfrak{a}(\gamma(s))$, $b(s):=\mathfrak{b}(\gamma(s))$, $r(\sigma):=\rho(\gamma(\sigma))$, $y(s):=\mathfrak{y}(\gamma(s))\gamma'(s)^{1/2}/\rho(\gamma(s))$ and

$$(1.5)\qquad k(s,\sigma) := \frac{\rho(\gamma(\sigma))}{\rho(\gamma(s))} \mathfrak{k}(\gamma(s),\gamma(\sigma))\gamma'(s)^{1/2}\gamma'(\sigma)^{1/2}.$$

Note that the operator on the left-hand side of (1.4) is a pseudodifferential operator of Mellin type (cf. e.g.[4]).

We write (1.4) in the form

$$(1.6)\qquad a(s)z(s)+ \frac{b(s)}{\pi i}\int_0^1 \frac{z(\sigma)d\sigma}{\sigma-s} + \int_0^1 [\frac{b(s)}{\pi i} l(s,\sigma)+k(s,\sigma)]z(\sigma)d\sigma = y(s), \quad 0\le s\le 1 ,$$

$$l(s,\sigma):=\left\{ \frac{r(\sigma)}{r(s)} \frac{\gamma'(s)^{1/2}\gamma'(\sigma)^{1/2}}{\gamma(\sigma)-\gamma(s)} - \frac{1}{\sigma-s} \right\}$$

and observe that the second term is the only integral containing a singular function. However, for α large enough, $z(0)=z(1)=0$. Setting $z(\sigma):=0$ for $\sigma\notin[0,1]$, the subtraction method leads to

$$(1.7)\qquad \int_0^1 \frac{z(\sigma)}{\sigma-s} d\sigma = \int_{-1}^{1+2s} \frac{z(\sigma)}{\sigma-s} d\sigma = \int_{-1}^{1+2s} \frac{z(\sigma)-z(s)}{\sigma-s} d\sigma .$$

Now we discretize the integrals in (1.6),(1.7). We choose an even integer $n>0$, set $t_k:=k/n$ $(k\in\mathbb{Z})$, $\omega_j:=4/3n$ if $j=1,3,5,\dots,n-1$, $\omega_j:=1/3n$ if $j=0$ or $j=n$ and $\omega_j:=2/3n$ if $j=2,4,\dots,n-2$. Then Simpson's rule gives

$$(1.8)\qquad \int_0^1 f(t)dt \sim \sum_{j=0}^{n} f(t_j)\omega_j \,.$$

Furthermore, for fixed k, we put $\upsilon_j:=4/3n$ if j is odd, $\upsilon_j:=2/3n$ if $j=-n+2,-n+4,\dots,n+2k-2$ and $\upsilon_j:=1/3n$ if $j=-n$ or $j=n+2k$ to get

$$(1.9)\qquad \int_{-1}^{1+2t_k} f(t)dt \sim \sum_{j=-n}^{n+2k} f(t_j)\upsilon_j\,,\quad k=0,\dots,n.$$

Obviously, $\upsilon_j=4/3n$ for $j=1,3,\dots,n-1$ and $\upsilon_j=2/3n$ for $j=0,2,\dots,n-1$ independently of k. For $s=t_k$, formulae (1.7),(1.9) yield

$$\int_0^1 \frac{z(\sigma)}{\sigma-t_k}\,d\sigma \sim \sum_{j=-n}^{n+2k} \frac{z(t_j)-z(t_k)}{t_j-t_k}\,\upsilon_j$$

$$= \sum_{\substack{j=0\\ j\neq k}}^{n} \frac{z(t_j)}{t_j-t_k}\,\upsilon_j - z(t_k) \sum_{\substack{j=-n\\ j\neq k}}^{n+2k} \frac{\upsilon_j}{t_j-t_k} + z'(t_k)\upsilon_k \,.$$

Using $\sum \upsilon_j(t_j-t_k)^{-1}=0$, the difference formula $z'(t_k) \sim n[8(z(t_{k+1})-z(t_{k-1}))-(z(t_{k+2})-z(t_{k-2}))]/12$ as well as $\upsilon_j z(t_j)=\omega_j z(t_j)$, we obtain

$$(1.10)\qquad \int_0^1 \frac{z(\sigma)}{\sigma-t_k}\,d\sigma \sim$$

$$\sum_{\substack{j=0\\ j\neq k}}^{n} \frac{z(t_j)}{t_j-t_k}\,\omega_j + \upsilon_k n \left\{8\,\frac{z(t_{k+1})-z(t_{k-1})}{12} - \frac{z(t_{k+2})-z(t_{k-2})}{12}\right\},$$

where $z(t_l):=0$ for $l<0$ and $l>n$.

Applying (1.8) to the integral in the third term of (1.6), we get

$$(1.11)\quad \int_0^1 k(t_k,\sigma)z(\sigma)d\sigma \sim \sum_{j=0}^{n} k(t_k,t_j)z(t_j)\omega_j ,$$

$$(1.12)\quad \int_0^1 l(t_k,\sigma)z(\sigma)d\sigma \sim \sum_{j=0}^{n} l(t_k,t_j)z(t_j)\omega_j .$$

Here, for $0<k<n$, $n\in\mathbb{Z}$, the value $l(t_k,t_k)$ is given by

$$l(t,t):=\lim_{\sigma\to t}\left\{ \frac{\rho(\gamma(\sigma))}{\rho(\gamma(t))}\,\frac{\gamma'(\sigma)^{1/2}\gamma'(t)^{1/2}}{\gamma(\sigma)-\gamma(t)} - \frac{1}{\sigma-t}\right\} = I_1+I_2 ,$$

$$I_1:=\lim_{\sigma\to t}\left\{ \frac{\rho(\gamma(\sigma))-\rho(\gamma(t))}{\gamma(\sigma)-\gamma(t)}\,\frac{\gamma'(\sigma)^{1/2}\gamma'(t)^{1/2}}{\rho(\gamma(t))}\right\} ,$$

$$I_2:=\lim_{\sigma\to t}\left\{ \frac{\gamma'(\sigma)^{1/2}\gamma'(t)^{1/2}}{\gamma(\sigma)-\gamma(t)} - \frac{1}{\sigma-t}\right\} .$$

Obviously, $I_1=\rho'(\gamma(t))\gamma'(t)/\rho(\gamma(t))$. An easy computation shows $I_2=0$, i.e.,

$$(1.13)\quad l(t_k,t_k) = (\rho\circ\gamma)'(t_k)/\rho(\gamma(t_k)) , \quad k=1,\dots,n-1 .$$

If $k=0$ or $k=n$, then we set $l(t_k,t_k):=(\rho\circ\gamma)'(t_k)/\rho(\gamma(t_k)):=0$ in (1.12). This neglect of one term is justified by

$$\lim_{\sigma\to 0} l(\sigma,0)z(\sigma) = \lim_{\sigma\to 0} z(\sigma)/\sigma =0 , \quad \lim_{\sigma\to 1} l(\sigma,1)z(\sigma) =0$$

which holds provided $\alpha>3$. Thus (1.12) and (1.13) lead to

$$(1.14)\quad \int_0^1 l(t_k,\sigma)z(\sigma)d\sigma \sim \sum_{\substack{j=0\\ j\neq k}}^{n} \left\{ \frac{r(t_j)}{r(t_k)}\,\frac{\gamma'(t_j)^{1/2}\gamma'(t_k)^{1/2}}{\gamma(t_j)-\gamma(t_k)} - \right.$$

$$\left.\frac{1}{t_j-t_k}\right\} z(t_j)\omega_j + \begin{cases} 0 & \text{if } k=0 \text{ or } k=n\,, \\ \dfrac{(\rho\circ\gamma)'(t_k)}{\rho(\gamma(t_k))}\, z(t_k)\omega_k & \text{else} \end{cases}$$

Equ. (1.6) for $s=t_k$, $k=0,\ldots,n$ together with (1.10), (1.11) and (1.14) suggests the following quadrature method: Set $u:=0$, $V:=0$ and determine approximate values ξ_k $(k=u,\ldots,n-v)$ for $z(t_k)$ by solving the system

$$a(t_k)\xi_k + \frac{b(t_k)}{\pi i}\Bigg\{\sum_{\substack{j=u\\ j\neq k}}^{n-v} \frac{\rho(\gamma(t_j))}{\rho(\gamma(t_k))}\,\frac{\gamma'(t_k)^{1/2}\gamma'(t_j)^{1/2}}{\gamma(t_j)-\gamma(t_k)}\,\omega_j\xi_j +$$

$$\upsilon_k n\,\frac{8(\xi_{k+1}-\xi_{k-1})-(\xi_{k+2}-\xi_{k-2})}{12} + \omega_k\,\frac{(\rho\circ\gamma)'(t_k)}{\rho(\gamma(t_k))}\,\xi_k\Bigg\} + \tag{1.15}$$

$$\sum_{j=u}^{n-v} k(t_k,t_j)\omega_j\xi_j = y(t_k)\quad, \ k=u,\ldots,n-v,$$

where $\xi_k=0$ for $k<u$ and $k>n-v$. An approximation for $\mathfrak{z}(\gamma(k/n))$ and $x(\gamma(k/n))$, respectively, is given by

$$\mathfrak{z}(\gamma(k/n)) \sim \xi_k\,\gamma'(k/n)^{-1/2},\quad x(\gamma(k/n))=\mathfrak{z}(\gamma(k/n))\rho(\gamma(k/n)). \tag{1.16}$$

The number $\gamma'(k/n)^{1/2}$ is small provided k or $n-k$ is small. Hence, division by $\gamma'(k/n)^{1/2}$ leads to larger errors near the end-points of the interval. However, in Sect. 2 it will be shown that our transformation technique improves the rate of convergence not only away from the end-points, but also near them.

If α is large, then $z(t)$ tends to zero very rapidly as $t\to 0$ or $t\to 1$. For small integers $u,v\geq 0$, the neglect of the first u and the last v terms in the sums of formulae (1.10), (1.11) and (1.14) leads to small errors only. Consequently, the system (1.15) yields a natural method also in the case $u,v\neq 0$. In the subsequent sections we shall fix u and v and consider the stability of (1.15) for $n \to \infty$. It turns out that if Equ. (0.1) is strongly elliptic, then there exist u and v such that the method (1.15) is stable.

Now let us propose a slight modification of the method (1.15). For the sake of simplicity, let $\alpha=3$ and $\mathfrak{k}(t,\tau)\equiv 0$. Set $\rho_j=\varkappa_j$ $(j=0,1)$, $\rho(t):=t^{\varkappa_0}(1-t)^{\varkappa_1}$ and seek $\mathfrak{z}$ and $x=\rho\mathfrak{z}$ in the form

$$(1.17)\qquad \begin{aligned} \mathfrak{z}(t) &= \mathfrak{z}(0)(1-t)^{2-\varkappa_0-\varkappa_1} + \mathfrak{z}(1)t^{2-\varkappa_0-\varkappa_1} + \mathfrak{z}_0(t)\,, \\ x(t) &= \mathfrak{z}(0)t^{\varkappa_0}(1-t)^{2-\varkappa_0} + \mathfrak{z}(1)t^{2-\varkappa_1}(1-t)^{\varkappa_1} + \rho(t)\mathfrak{z}_0(t)\,, \end{aligned}$$

where $\mathfrak{z}_0(t)$ vanishes at 0 and 1. Formulae (cf. [9])

$$(1.18)\qquad \begin{aligned} \mathfrak{r}_0(s) &:= \int_0^1 \frac{(1-\sigma)^{2-\varkappa_0}\sigma^{\varkappa_0}}{\sigma-s}\,d\sigma = -\pi\,\mathrm{ctg}(\pi\varkappa_0)(1-s)^{2-\varkappa_0}s^{\varkappa_0} + \\ &\qquad \pi/\sin(\pi\varkappa_0)\ \{s^2+(\varkappa_0-2)s+\tfrac{1}{2}(\varkappa_0-1)(\varkappa_0-2)\} \\ \mathfrak{r}_1(s) &:= \int_0^1 \frac{(1-\sigma)^{\varkappa_1}\sigma^{2-\varkappa_1}}{\sigma-s}\,d\sigma = \pi\,\mathrm{ctg}(\pi\varkappa_1)(1-s)^{\varkappa_1}s^{2-\varkappa_1} + \\ &\qquad \pi/\sin(\pi\varkappa_1)\ \{s^2-\varkappa_1 s+\tfrac{1}{2}\varkappa_1(\varkappa_1-1)\} \end{aligned}$$

yield

$$(1.19)\qquad \begin{aligned} \int_0^1 \frac{\rho(\gamma(\sigma))\mathfrak{z}(\gamma(\sigma))}{\gamma(\sigma)-\gamma(s)}\gamma'(\sigma)d\sigma &= \mathfrak{z}(0)\mathfrak{r}_0(\gamma(s)) + \mathfrak{z}(1)\mathfrak{r}_1(\gamma(s)) \\ &\quad + \int_0^1 \frac{\rho(\gamma(\sigma))\mathfrak{z}_0(\gamma(\sigma))}{\gamma(\sigma)-\gamma(s)}\gamma'(\sigma)d\sigma\,. \end{aligned}$$

Introducing $z_0(\sigma):=\mathfrak{z}_0(\gamma(\sigma))\gamma'(\sigma)^{1/2}$ and substituting (1.19) in (1.2), we proceed analogously to the derivation of (1.15). For the approximate values η_0 of $\mathfrak{z}(0)$, η_n of $\mathfrak{z}(1)$ as well as ξ_k of

$z_o(t_k)$ (k= 1,...,n-1), we obtain

$$(1.20)\quad a(t_k)\Big[\xi_k+\gamma'(t_k)^{1/2}\Big\{(1-\gamma(t_k))^{2-\varkappa_o-\varkappa_1}\eta_0+\gamma(t_k)^{2-\varkappa_o-\varkappa_1}\eta_1\Big\}\Big]$$

$$+\frac{b(t_k)}{\pi i}\Big\{\sum_{\substack{j=1\\ j\neq k}}^{n-1}\frac{\rho(\gamma(t_j))}{\rho(\gamma(t_k))}\frac{\gamma'(t_j)^{1/2}\gamma'(t_k)^{1/2}}{\gamma(t_j)-\gamma(t_k)}\,\omega_j\xi_j+$$

$$\upsilon_k n\,\frac{8(\xi_{k+1}-\xi_{k-1})-(\xi_{k+2}-\xi_{k-2})}{12}+\omega_k\frac{(\rho\circ\gamma)'(t_k)}{\rho(\gamma(t_k))}\,\xi_k+$$

$$\frac{\gamma'(t_k)^{1/2}}{\rho(\gamma(t_k))}\,[r_o(\gamma(t_k))\eta_0+r_1(\gamma(t_k))\eta_n]\Big\}=y(t_k),k=1,\dots,n-1,$$

where $\xi_k:=0$ for k<1 and k>n-1. Furthermore, in view of $\alpha=3$ the derivative $z_o'(0)$ is equal to $\mathfrak{z}_o(0)$. Therefore, we find

$$0 = z_o'(0) \sim \frac{z_o(t_1)-z_o(0)}{t_1} = n\, z_o(t_1)\ ,$$

and analogously $n\, z_o(t_{n-1})\sim 0$. Thus the approximate values η_0, η_n, ξ_j (j=1,...,n-1) are to be determined by (1.20) and the equations

$$(1.21)\qquad n\,\xi_1 = 0\ ,\ n\,\xi_{n-1} = 0\ .$$

An approximation for $\mathfrak{z}(\gamma(k/n))$ and $x(\gamma(k/n))$, respectively, is given by

$$(1.22)\quad \mathfrak{z}(\gamma(k/n))\sim\begin{cases}\eta_0 & \text{if } k=0\ ,\\ \eta_n & \text{if } k=n\ ,\\ \eta\ (1-\gamma(k/n))^{2-\varkappa_o-\varkappa_1}+\eta\ \gamma(k/n)^{2-\varkappa_o-\varkappa_1}+\dfrac{\xi_k}{\gamma'(k/n)^{1/2}} & \text{else}\ ,\end{cases}$$

$$x(\gamma(k/n)) = \mathfrak{z}(\gamma(k/n))\rho(\gamma(k/n))\ .$$

The advantage of the method (1.20)-(1.21) is that the

singular integrals of the terms $z(0)(1-t)^{2-\varkappa_0-\varkappa_1}$ and $z(1)t^{2-\varkappa_0-\varkappa_1}$ (cf. (1.17)) are evaluated exactly. Furthermore, z_0 vanishes faster at the end-points of the interval. Hence, the functions in the integrals of (1.6),(1.7) become smoother if z is replaced by z_0. Consequently, Simpson's rule converges faster. For the method (1.20)-(1.21), we do not prove convergence or stability. However, in Sect.4 numerical results for this method are given and compared with the corresponding results of (1.15).

2. RATES OF CONVERGENCE

For the sake of simplicity, let us make some assumptions. Let H^s and $H^s(\mathbb{R})$ denote the usual Sobolev space of order s on [0,1] and on $\mathbb{R}$, respectively. Suppose $a,b,y \in H^s$ for arbitrary s>0. Furthermore, we restrict our analysis to the case $k(t,\tau)\equiv 0$ since an arbitrary $k(.,.)$ can be treated analogously to the theory of Fredholm integral equations of the second kind [1]. We set u=v=0 and define $\varkappa_0,\varkappa_1$ by (1.1), where ${}^j\log$ denotes the branch of the logarithm which is continuous on $\mathbb{C}\setminus(-\infty,0]$ and takes real values on $(0,\infty)$. Choosing $\rho_0:=\varkappa_0$, $\rho_1:=\varkappa_1$, we consider the method (1.15). If the operator on the left-hand side of (0.1) is invertible in $L^2_\rho := \{\rho f, f\in L^2\}$ and x is the solution of (0.1), then (cf. [5,4]) x takes the form

$$x(t) = g_x(t) + \sum_{m=0}^{k-2}\sum_{j=0}^{m} c_{m,j,x} t^{\varkappa_0+m} \log^j t \quad , \quad 0\le t\le 2/3 \quad ,$$

$$x(t) = h_x(t) + \sum_{m=0}^{k-2}\sum_{j=0}^{m} d_{m,j,x}(1-t)^{\varkappa_1+m} \log^j(1-t) \quad , \quad 1/3\le t\le 1$$

where $k\ge 2$ is an arbitrary integer, $c_{m,j,x}, d_{m,j,x}\in\mathbb{C}$, $g_x\in H^s$ $(0\le s<\varkappa_0+k-1/2)$, $g_x^{(l)}(0)=0$ $(l\in\mathbb{Z},\ 0\le l<\varkappa_0+k-1)$, $h_x\in H^s$ $(0\le s<\varkappa_0+k-1/2)$ and $h_x^{(l)}(1)=0 (l\in\mathbb{Z}, 0\le l<\varkappa_1+k-1)$. Consequently,

$$z(t) = \theta_z(t) + \sum_{m=0}^{k-2} \sum_{j=0}^{m} c_{m,j,z} t^{\frac{\alpha-1}{2}+m\alpha} \log^j t \ , 0 \le t \le 2/3,$$
(2.1)
$$z(t) = h_z(t) + \sum_{m=0}^{k-2} \sum_{j=0}^{m} d_{m,j,z} (1-t)^{\frac{\alpha-1}{2}+m\alpha} \log^j (1-t) \ , \quad 1/3 \le t \le 1,$$

where $k \ge 2$ is an integer, $c_{m,j,z}, d_{m,j,z} \in \mathbb{C}, \theta_z, h_z \in H^s$ ($0 \le s < \frac{\alpha-1}{2}+(k-1)\alpha+1/2$) and $\theta_z^{(l)}(0)=0$, $h_z^{(l)}(1)=0$ ($l \in \mathbb{Z}$, $0 \le l < \frac{\alpha-1}{2} + (k-1)\alpha$).

In order to deduce convergence or stability, it is useful to introduce an interpolation z_n of the approximate values ξ_j $(j=0,\ldots,n)$ satisfying $z_n(t_j)=\xi_j$. For numerical computation, one would choose piecewise polynomial interpolation, i.e., a local interpolation procedure. However, to obtain error estimates also in higher Sobolev norms let us introduce the set $S_n^{\mathbb{R}}$ of smoothest splines of order 5, i.e., $S_n^{\mathbb{R}}$ is the collection of all $\varphi \in C^4(\mathbb{R})$ such that the restrictions $\varphi|(t_j, t_{j+1})$, $j \in \mathbb{Z}$ are polynomials of degree less or equal to 5. Furthermore, let the basis $\{\varphi_j\}_{j\in\mathbb{Z}}$ be defined by $\varphi_j(t_k)=\delta_{j,k}$, $j,k \in \mathbb{Z}$, and let $L_n^{\mathbb{R}}$ denote the orthogonal projection of $L^2(\mathbb{R})$ onto $S_n^{\mathbb{R}}$. For a function f on $\mathbb{R}$, we set $K_n^{\mathbb{R}} f := \sum f(t_j)\varphi_j$, and, for a function g defined on $[0,1]$, we put $K_n g := \sum_{j=0}^{n} g(t_j)\varphi_j$. If $L_n^{\mathbb{R}} f = \sum_{j\in\mathbb{Z}} \lambda_j(f)\varphi_j$, then we define L_n by $L_n g := \sum_{j=0}^{n} \lambda_j(g)\varphi_j$. Let $\mathbf{A}_n \in \mathcal{L}(\operatorname{im} L_n)$ be the operator the matrix of which corresponding to the basis $\{\varphi_j\}_{j\in\mathbb{Z}}$ is the matrix of the system (1.15). Then (1.15) can be written in the form $\mathbf{A}_n z_n = K_n \mathbf{y}$, where $z_n := \sum_{j=0}^{n} \xi_j \varphi_j$. In view of

$$(2.2)\quad \begin{aligned} &C^{-1}\frac{(\sum_{j\in\mathbb{Z}}|\xi_j|^2)^{1/2}}{n^{1/2}} \le \|\sum_{j\in\mathbb{Z}}\xi_j\varphi_j\|_{L^2(\mathbb{R})} \le C\frac{(\sum_{j\in\mathbb{Z}}|\xi_j|^2)^{1/2}}{n^{1/2}}\ ,\\ &C^{-1}\frac{(\sum_{j=0}^{n}|\xi_j|^2)^{1/2}}{n^{1/2}} \le \|\sum_{j=0}^{n}\xi_j\varphi_j\|_{L^2} \le C\frac{(\sum_{j=0}^{n}|\xi_j|^2)^{1/2}}{n^{1/2}}\end{aligned}$$

the $\mathcal{L}(\mathrm{im}\, L_n)$-norm of A_n is equivalent to the Euclidean norm of its matrix corresponding to $\{\varphi_j\}$. (Here and in the following $C>0$ denotes a generic constant the value of which varies from instance to instance.) In this section let us suppose the method (1.15) to be stable (cf. Corollary 3.1), i.e., let there exist an integer $n_o\ge 1$ such that A_n is invertible for $n\ge n_o$ and $\sup \|A_n^{-1}\| <\infty$.

We start our error analysis by estimating $\|z_n-z\|_{L^2}$. Let $A\in\mathcal{L}(L^2)$ be defined by the left-hand side of (1.6), i.e.,

$$(Az)(t):=a(t)z(t)+b(t)(Sz)(t)+\frac{b(t)}{\pi i}(K_oz)(t)\ ,$$

$$(Sz)(t):=\frac{1}{\pi i}\int_0^1 \frac{z(\tau)}{\tau-t}\,d\tau\ ,\quad (K_oz)(t):=\int_0^1 l(t,\tau)d\tau\ .$$

Then the formula

$$z_n-z = (K_n-I)z + A_n^{-1}\{K_nAz-A_nK_nz\}$$

leads to

$$(2.3)\qquad \|z_n-z\|_{L^2}\le C\ \{\ \|(K_n-I)z\|_{L^2} + \|K_nAz-A_nK_nz\|_{L^2}\ \}\ .$$

Here the first term on the right-hand side can be estimated utilizing the approximation property of $S_n^{\mathbb{R}}$. In fact, for $0\le s\le u\le r\le 6$, $1/2<r$, $s<5.5$ and any $f\in H^r(\mathbb{R}), g\in H^u(\mathbb{R})$ we have (cf. [7])

$$(2.4)\qquad \begin{aligned} \|(I-L_n^{\mathbb{R}})\vartheta\|_{H^s(\mathbb{R})} &\le C\, n^{s-u}\, \|\vartheta\|_{H^u(\mathbb{R})}, \\ \|(I-K_n^{\mathbb{R}})f\|_{H^s(\mathbb{R})} &\le C\, n^{s-r}\, \|f\|_{H^r(\mathbb{R})}. \end{aligned}$$

In particular, if $f\in H^r(\mathbb{R})$ and f vanishes outside the interval $[0,1]$, then $K_n f = K_n^{\mathbb{R}} f$ and we get

$$(2.5)\qquad \|(I-K_n)f\|_{H^s} \le C n^{s-r} \|f\|_{H^r}.$$

For the second term on the right-hand side of (2.3), we obtain

$$(2.6)\qquad K_n\mathbf{A}z - \mathbf{A}_n K_n z =$$

$$K_n \mathbf{b}|_{\operatorname{im} L_n}\left\{[K_n Sz - S_n K_n z] + \frac{1}{\pi i}\,[K_n \mathbf{K}_0 z - \mathbf{K}_{0,n} K_n z]\right\},$$

where the operators S_n and $\mathbf{K}_{0,n}$ are defined by the matrices

$$S_n := \left(\frac{1}{\pi i}\left[\frac{(1-\delta_{j,k})}{t_j - t_k}\,\omega_j + \upsilon_k n \frac{8(\delta_{j,k+1}-\delta_{j,k-1})-(\delta_{j,k+2}-\delta_{j,k-2})}{12}\right]\right)_{k,j=0}^{n}$$

$$\mathbf{K}_{0,n} := \left(\mathbf{1}(t_k,t_j)\,\omega_j\right)_{k,j=0}^{n},$$

respectively. Since $\|K_n\mathbf{b}|\operatorname{im} L_n\| \le \sup |\mathbf{b}(t)|$, it remains to estimate $\|K_n Sz - S_n K_n z\|_{L^2}$ and $\|K_n \mathbf{K}_0 z - \mathbf{K}_{0,n} K_n z\|_{L^2}$.

The first step in estimating $\|K_n Sz - S_n K_n z\|$ will be Lemma 2.1, where the rate of convergence for a more general approximate operator on the real axis is established. Thus set $K_n^0 f := \Sigma f(t_{2j})\varphi_{2j}$, $K_n^1 f := \Sigma f(t_{2j+1})\varphi_{2j+1}$, i.e., $K_n^{\mathbb{R}} = K_n^0 + K_n^1$. If $L_n^{\mathbb{R}} f = \Sigma\, \lambda_j(f)\varphi_j$, then set $L_n^0 f := \Sigma\, \lambda_{2j}(f)\varphi_{2j}$, $L_n^1 f := \Sigma\, \lambda_{2j+1}(f)\varphi_{2j+1}$. For a 2x2-matrix function $\underline{a} = (\underline{a}^{r,s})_{r,s=0}^{1}$ on the unit circle $\mathbb{T}$, denote the k-th Fourier coefficient by $\underline{a}_k := (\underline{a}_k^{r,s})_{r,s=0}^{1}$ and the block convolution matrix $(\underline{a}_{k-j})_{k,j\in\mathbb{Z}}$ by $C(\underline{a})$. We identify this matrix with the operator $C(\underline{a})\in\mathcal{L}(\operatorname{im} L_n^{\mathbb{R}})$ defined by

$$(2.7)\qquad \begin{aligned} C(\underline{a})\varphi_{2j} &= \sum_{k\in\mathbb{Z}} \{\underline{a}^{0,0}_{k-j}\varphi_{2k}+\underline{a}^{1,0}_{k-j}\varphi_{2k+1}\}\ , \\ C(\underline{a})\varphi_{2j-1} &= \sum_{k\in\mathbb{Z}} \{\underline{a}^{0,1}_{k-j}\varphi_{2k}+\underline{a}^{1,1}_{k-j}\varphi_{2k+1}\}\ .. \end{aligned}$$

Let $S_{\mathbb{R}}$ denote the Cauchy singular operator on $\mathbb{R}$ and set $\mathbf{P}_{\mathbb{R}}:=\frac{1}{2}(I+S_{\mathbb{R}}),\mathbf{Q}_{\mathbb{R}}:=I-\mathbf{P}_{\mathbb{R}}$.

LEMMA 2.1: *Let* $\underline{a}$ *be piecewice continuous and suppose that the functions* $\underline{p}(e^{i2\pi\lambda}):=\underline{a}^{0,0}(e^{i2\pi\lambda})+\underline{a}^{0,1}(e^{i2\pi\lambda})e^{-i\pi\lambda}$, $\underline{q}(e^{i2\pi\lambda}):=\underline{a}^{1,0}(e^{i2\pi\lambda})e^{i\pi\lambda}+\underline{a}^{1,1}(e^{i2\pi\lambda})$ $(-1/2<\lambda<1/2)$ *satisfy* $\underline{p}(1\pm0)=\underline{q}(1\pm0)$ *and* $\underline{p}^{(j)}(1\pm0)=\underline{q}^{(j)}(1\pm0)=0$ *for* $j=1,2,\ldots,l-1$, *where* l *is an integer and* $0<l\leq6$. *Let* $1/2<s\leq1$. *Then there exists a constant* $C>0$ *such that, for any* $f\in H^s(\mathbb{R})$,

$$(2.8)\qquad \| C(\underline{a})K^{\mathbb{R}}_n f-K^{\mathbb{R}}_n[\underline{p}(1-0)\mathbf{P}_{\mathbb{R}}+\underline{q}(1+0)\mathbf{Q}_{\mathbb{R}}]f\|_{L^2(\mathbb{R})}\leq Cn^{-s}\|f\|_{H^s(\mathbb{R})}.$$

PROOF: It is well known that the vector $(e^{-i2\pi\lambda j})_{j\in\mathbb{Z}}$ $(-1/2<\lambda\leq1/2)$ is an eigenvector of the convolution matrix $(\underline{a}^{r,s}_{k-j})_{k,j\in\mathbb{Z}}$ corresponding to the eigenvalue $\underline{a}^{r,s}(e^{i2\pi\lambda})$. Furthermore, for $\vartheta^{\lambda}(t):=e^{-i2\pi\lambda t}$, we obtain $K^0_n\vartheta^{\lambda}=\sum e^{-i2\pi(2\lambda/n)j}\varphi_{2j}$, $K^1_n\vartheta^{\lambda}=e^{-i2\pi\lambda/n}\sum e^{-i2\pi(2\lambda/n)j}\varphi_{2j+1}$ and

$$C(\underline{a})K^0_n\vartheta^{\lambda}=\underline{a}^{0,0}(e^{i2\pi(2\lambda/n)})K^0_n\vartheta^{\lambda}+\underline{a}^{1,0}(e^{i2\pi(2\lambda/n)})e^{i2\pi\lambda/n}K^1_n\vartheta^{\lambda},$$

$$C(\underline{a})K^1_n\vartheta^{\lambda}=\underline{a}^{0,1}(e^{i2\pi(2\lambda/n)})e^{-i2\pi\lambda/n}K^0_n\vartheta^{\lambda}+\underline{a}^{1,1}(e^{i2\pi(2\lambda/n)})K^1_n\vartheta^{\lambda},$$

$$C(\underline{a})K^{\mathbb{R}}_n\vartheta^{\lambda}=\underline{p}(e^{i2\pi(2\lambda/n)})K^0_n\vartheta^{\lambda}+\underline{q}(e^{i2\pi(2\lambda/n)})K^1_n\vartheta^{\lambda}\quad .$$

Hence,

$$(2.9)\qquad C(\underline{a})K^{\mathbb{R}}_n\vartheta^{\lambda}=\underline{r}(e^{i2\pi(2\lambda/n)})K^0_n\vartheta^{\lambda}+\underline{s}(e^{i2\pi(2\lambda/n)})K^1_n\vartheta^{\lambda}$$

$$+\begin{cases}\underline{p}(1+0)K^{\mathbb{R}}_n\vartheta^{\lambda} & \text{if } \lambda>0,\\ \underline{p}(1-0)K^{\mathbb{R}}_n\vartheta^{\lambda} & \text{if } \lambda<0,\end{cases}$$

where $\underline{r}$ and $\underline{s}$ satisfy

$$(2.10)\qquad |\underline{r}(e^{i2\pi\lambda})|\le \min\{C,|\lambda|^1\}\ ,\ |\underline{s}(e^{i2\pi\lambda})|\le \min\{C,|\lambda|^1\}$$

and C>0 is constant. If $\mathcal{F}f$ denotes the Fourier transform

$$\mathcal{F}f(t):=\frac{1}{2\pi}\int_{\mathbb{R}} f(\tau)e^{i\tau t}d\tau\ ,$$

then

$$(2.11)\qquad f=\int_{\mathbb{R}}\mathcal{F}f(\lambda)\theta^{\lambda}d\lambda,\ Q_{\mathbb{R}}f=\int_0^{\infty}\mathcal{F}f(\lambda)\theta^{\lambda}d\lambda,\ P_{\mathbb{R}}f=\int_{-\infty}^{0}\mathcal{F}f(\lambda)\theta^{\lambda}d\lambda\ ,$$

and we arrive at

$$C(\underline{a})K_n^{\mathbb{R}}f=\int_{\mathbb{R}}\mathcal{F}f(\lambda)[C(\underline{a})K_n^{\mathbb{R}}\theta^{\lambda}]d\lambda\ .$$

Equs.(2.9),(2.11) lead to

$$C(\underline{a})K_n^{\mathbb{R}}f=\int_0^{\infty}\mathcal{F}f(\lambda)\underline{p}(1+0)K_n^{\mathbb{R}}\theta^{\lambda}d\lambda+\int_{-\infty}^{0}\mathcal{F}f(\lambda)\underline{p}(1-0)K_n^{\mathbb{R}}\theta^{\lambda}d\lambda+$$

$$(2.12)\qquad \int_{\mathbb{R}}\mathcal{F}f(\lambda)[\underline{r}(e^{i2\pi(2\lambda/n)})K_n^{0}\theta^{\lambda}+\underline{s}(e^{i2\pi(2\lambda/n)})K_n^{1}\theta^{\lambda}]d\lambda$$

$$=K_n^{\mathbb{R}}\{\underline{p}(1-0)P_{\mathbb{R}}+\underline{q}(1+0)Q_{\mathbb{R}}\}f+I_1+I_2\ ,$$

$$I_1:=\int_{\mathbb{R}}\mathcal{F}f(\lambda)\underline{r}(e^{i2\pi(2\lambda/n)})K_n^{0}\theta^{\lambda}d\lambda,\ I_2:=\int_{\mathbb{R}}\mathcal{F}f(\lambda)\underline{s}(e^{i2\pi(2\lambda/n)})K_n^{1}\theta^{\lambda}d\lambda.$$

Setting $I_3:=L_n^{0}\int_{\mathbb{R}}\mathcal{F}f(\lambda)\underline{r}(e^{i2\pi(2\lambda/n)})\theta^{\lambda}d\lambda$ and $I_4:=I_1-I_3$, we get

$$(2.13)\qquad I_1=I_3+I_4\ ,$$

$$\|I_3\|_{L^2(\mathbb{R})}\le C\ \left\|\int_{\mathbb{R}}\mathcal{F}f(\lambda)\underline{r}(e^{i2\pi(2\lambda/n)})\theta^{\lambda}d\lambda\right\|_{L^2(\mathbb{R})}$$

$$= C \ \| \mathcal{F}f(\lambda)\underline{r}(e^{i2\pi(2\lambda/n)})\|_{L^2(\mathbb{R})}.$$

Using (2.10), we obtain

$$(2.14) \qquad \| I_3\|_{L^2(\mathbb{R})} \le C\, n^{-1} \| \mathcal{F}f(\lambda)\lambda^{l}\|_{L^2(\mathbb{R})} \le C\, n^{-1} \| f \|_{H^{l}(\mathbb{R})}.$$

The definition of I_4 and the formulae (2.2),(2.4) yield

$$\| I_4\|_{L^2(\mathbb{R})} = \| (K_n^0 - L_n^0)\int_{\mathbb{R}} \mathcal{F}f(\lambda)\underline{r}(e^{i2\pi(2\lambda/n)})\theta^{\lambda} d\lambda \|_{L^2(\mathbb{R})}$$

$$\le \| (K_n^{\mathbb{R}} - L_n^{\mathbb{R}})\int_{\mathbb{R}} \mathcal{F}f(\lambda)\underline{r}(e^{i2\pi(2\lambda/n)})\theta^{\lambda} d\lambda \|_{L^2(\mathbb{R})}$$

$$\le Cn^{-l} \| \int_{\mathbb{R}} \mathcal{F}f(\lambda)\underline{r}(e^{i2\pi(2\lambda/n)})\theta^{\lambda} d\lambda \ \|_{H^{l}(\mathbb{R})} \le Cn^{-l} \| f \|_{H^{l}(\mathbb{R})}.$$

Formulae (2.12)-(2.14) together with analogous arguments for I_2 prove (2.8) in the case s=l is an integer. Thus Lemma 2.1 follows by interpolation. □

We now consider the operator $S_{\mathbb{R},n} \in \mathcal{L}(\text{im } L_n^{\mathbb{R}})$ given by its matrix corresponding to the basis $\{\varphi_j\}_{j\in\mathbb{Z}}$, i.e.,

$$S_{\mathbb{R},n} := \left(\frac{1}{\pi i}\left[\frac{(1-\delta_{j,k})}{t_j - t_k}\theta_j + \theta_k n \frac{8(\delta_{j,k+1} - \delta_{j,k-1}) - (\delta_{j,k+2} - \delta_{j,k-2})}{12} \right] \right)_{k,j\in\mathbb{Z}},$$

where $\theta_j := 4/3n$ if j is odd and $\theta_j := 2/3n$ if j is even. Define $z(t) := 0$ for $t \notin [0,1]$. Since $K_n S z - S_n K_n z$ is a projection of $K_n^{\mathbb{R}} S_{\mathbb{R}} z - S_{\mathbb{R},n} K_n^{\mathbb{R}} z$ into im L_n, we get (cf. (2.2))

$$(2.15) \qquad \| K_n S z - S_n K_n z \|_{L^2} \le \| K_n^{\mathbb{R}} S_{\mathbb{R}} z - S_{\mathbb{R},n} K_n^{\mathbb{R}} z \|_{L^2(\mathbb{R})}.$$

On the other hand, $S_{\mathbb{R},n} = C(\underline{a})$ holds, where

$$(\underline{a}_{k-j}^{0,0})_{k,j\in\mathbb{Z}} = \left(\frac{2}{3n}\, \frac{1}{\pi i}\left[\frac{1-\delta_{2j,2k}}{t_{2j} - t_{2k}} - n \frac{\delta_{2j,2k+2} - \delta_{2j,2k-2}}{12} \right] \right)_{k,j\in\mathbb{Z}}$$

$$=\left[\frac{1}{3}\,\frac{1}{\pi i}\,\frac{1-\delta_{j,k}}{j-k}\;-\frac{1}{18}\;\frac{1}{\pi i}(\delta_{j,k+1}-\delta_{j,k-1})\right]_{k,\,j\in\mathbb{Z}},$$

$$(\underline{a}^{1,1}_{k-j})_{k,\,j\in\mathbb{Z}}=\left[\frac{4}{3n}\,\frac{1}{\pi i}\left[\frac{1-\delta_{2j+1,2k+1}}{t_{2j+1}-t_{2k+1}}-n\frac{\delta_{2j+1,2k+3}-\delta_{2j+1,2k-1}}{12}\right]\right]_{k,\,j\in\mathbb{Z}}$$

$$=\left[\frac{2}{3}\,\frac{1}{\pi i}\,\frac{1-\delta_{j,k}}{j-k}-\frac{1}{9}\,\frac{1}{\pi i}(\delta_{j,k+1}-\delta_{j,k-1})\right]_{k,\,j\in\mathbb{Z}},$$

$$(\underline{a}^{0,1}_{k-j})_{k,\,j\in\mathbb{Z}}=\left[\frac{2}{3}\,\frac{1}{\pi i}\,\frac{1}{(j-k)+1/2}+\frac{4}{9}\,\frac{1}{\pi i}(\delta_{j,k}-\delta_{j,k-1})\right]_{k,\,j\in\mathbb{Z}},$$

$$(\underline{a}^{1,0}_{k-j})_{k,\,j\in\mathbb{Z}}=\left[\frac{1}{3}\,\frac{1}{\pi i}\,\frac{1}{(j-k)-1/2}+\frac{8}{9}\,\frac{1}{\pi i}(\delta_{j,k+1}-\delta_{j,k})\right]_{k,\,j\in\mathbb{Z}}.$$

Using (3.4) of [21], we arrive at

$$\underline{a}^{0,0}(e^{i2\pi\lambda})=\frac{1}{3}\,(2\lambda-1)-\frac{1}{18}\,\frac{1}{\pi i}\,(e^{-i2\pi\lambda}-e^{i2\pi\lambda}),$$
$$\underline{a}^{1,1}(e^{i2\pi\lambda})=\frac{2}{3}\,(2\lambda-1)-\frac{1}{9}\,\frac{1}{\pi i}\,(e^{-i2\pi\lambda}-e^{i2\pi\lambda}),$$
$$\underline{a}^{0,1}(e^{i2\pi\lambda})=\frac{2}{3}\,(-e^{i\pi\lambda})+\frac{4}{9}\,\frac{1}{\pi i}\,(1-e^{i2\pi\lambda}),$$
$$\underline{a}^{1,0}(e^{i2\pi\lambda})=\frac{1}{3}\,(-e^{-i\pi\lambda})+\frac{8}{9}\,\frac{1}{\pi i}\,(e^{-i2\pi\lambda}-1).$$

Consequently, Lemma 2.1 applies with l=5 and we find

$$(2.16)\qquad \| K^{\mathbb{R}}_n S_{\mathbb{R}} f - S_{\mathbb{R},n} K^{\mathbb{R}}_n f \|_{L^2(\mathbb{R})} \le C\, n^{-s} \| f \|_{H^s(\mathbb{R})}.$$

By (2.1) we conclude $z\in H^s(\mathbb{R})$ for $0\le s<\alpha/2$. Therefore, if $1/2<s\le 5$, $s<\alpha/2$, then formulae (2.15),(2.16) imply

$$(2.17)\qquad \| K_n S z - S_n K_n z \|_{L^2} \le C\, n^{-s}.$$

In order to simplify our considerations, we shall give coarse estimates for $\| K_n K_o z - K_{o,n} K_n z \|_{L^2}$. By (2.2) we conclude

$$\| K_n K_o z - K_{o,n} K_n z \|_{L^2} \le \sup_{k=0,\dots,n} | K_o z(t_k) - K_{o,n} K_n z(t_k) |$$

$$(2.18)\qquad \le \sup_{0\le t\le 1} \left|\int_0^1 l(t,\tau)z(\tau)d\tau - \sum_{j=0}^{n} l(t,t_k)z(t_k)\omega_k\right| \le C\, n^{-s} \sup_{0\le t\le 1} \| l(t,.)z\|_{W_1^s} ,$$

where $0\le s\le 4$ and W_1^s denotes the Sobolev-Słobodetski space of power 1 and order s (cf. [24],Sects. 2.3,2.5). It remains to examine whether the norms $\| l(t,.)z\|_{W_1^s}$ are uniformly bounded or not.

In case $\tau,t<1/3$, we have

$$l(t,\tau)=\begin{cases} -1/\tau & \text{if } t=0,\\ \dfrac{\alpha t^{\frac{\alpha-1}{2}-\alpha\varkappa_o}\tau^{\frac{\alpha-1}{2}+\alpha\varkappa_o}}{\tau^{\alpha}-t^{\alpha}} - \dfrac{1}{\tau-t} & \text{if } t\ne 0.\end{cases}$$

Setting $x=t/\tau$ $(0<x<\infty)$, we may write

$$(2.19)\qquad l(t,\tau)=\frac{1}{\tau}\begin{cases} -1 & \text{if } t=0,\\ \dfrac{\alpha x^{\frac{\alpha-1}{2}-\alpha\varkappa_o}}{1-x^{\alpha}} - \dfrac{1}{1-x} & \text{if } t\ne 0,\end{cases}$$

$$(2.20)\qquad l(t,\tau)=\frac{1}{\tau}\begin{cases} -1 & \text{if } t=0,\\ \dfrac{I_1-I_2}{\int_0^1 (x+\mu(1-x))^{\alpha-1}d\mu} & \text{if } t\ne 0,\end{cases}$$

$$I_1:=[\tfrac{\alpha-1}{2}+\alpha\varkappa_o]\times$$

$$\times\int_0^1\!\!\int_0^1 (1-\mu)[x+\mu(1-x)]^{\frac{\alpha-1}{2}-\alpha\varkappa_o}[1-(1-\nu)(1-\mu)(1-x)]^{\frac{\alpha-1}{2}+\alpha\varkappa_o-1}d\nu d\mu,$$

$$I_2:=[\tfrac{\alpha-1}{2}-\alpha\varkappa_o]\int_0^1\!\!\int_0^1 \mu[x+\mu(1-\nu)(1-x)]^{\frac{\alpha-1}{2}-\alpha\varkappa_o-1}d\nu d\mu .$$

Consequently, the function $\tau\to\tau\, l(t,\tau)$ is uniformly bounded with respect to t. Differentiating (2.19) and (2.20), we conclude that the function $\tau \longrightarrow \tau^{l+1}(\partial/\partial\tau)^{l} \mathbf{l}(t,\tau)$ is uniformly bounded, if l is a positive integer. In case $\tau>2/3$, $t<1/3$, we get

$$\mathbf{l}(t,\tau)=\begin{cases} -1/\tau & \text{if } t=0, \\ \dfrac{\alpha t^{\frac{\alpha-1}{2}-\alpha\varkappa_0}(1-\tau)^{\frac{\alpha-1}{2}+\alpha\varkappa_0}}{1-(1-\tau)^{\alpha} - t^{\alpha}} - \dfrac{1}{\tau-t} & \text{if } t\neq 0. \end{cases}$$

We now have to suppose $\frac{\alpha-1}{2}-\alpha\varkappa_0\geq 0$ in order to guarantee the uniform boundedness with respect to t. If this assumption is fulfilled, then $\tau\longrightarrow (1-\tau)^{l+1}(\partial/\partial\tau)^{l}\mathbf{l}(t,\tau)$ is uniformly bounded for $\tau\geq 2/3$, $t\leq 1/3$. The other cases can be treated analogously. Thus ,assuming $\alpha\geq 1/(1-2\varkappa_i)$ $(i=0,1)$, we obtain that the function $\tau\longrightarrow \tau^{l+1}(1-\tau)^{l+1}(\partial/\partial\tau)^{l}\mathbf{l}(t,\tau)$ is uniformly bounded with respect to t, where l is an arbitrary non-negative integer.

This fact together with (2.1) implies

$$\sup_{0\leq t\leq 1} \| \mathbf{l}(t,.)z\|_{W_1^s} < \infty \tag{2.21}$$

for any s satisfying $0\leq s<\frac{\alpha-1}{2}$. In fact, for example, consider $z(\tau)=\tau^{\frac{\alpha-1}{2}}$ (cf. (2.1)), $0\leq\tau\leq 1/3$ and $s\in\mathbb{Z}$, $s>0$. Then, for certain constants C_j, we obtain

$$(\partial/\partial\tau)^{s}\{ \mathbf{l}(t,\tau)\tau^{\frac{\alpha-1}{2}} \} = \sum_{j=0}^{s} C_j(\partial/\partial\tau)^{j}\mathbf{l}(t,\tau)\tau^{\frac{\alpha-1}{2}-(s-j)}$$
$$= \left\{ \sum_{j=0}^{s} C_j\tau^{j+1}(\partial/\partial\tau)^{j}\mathbf{l}(t,\tau) \right\} \tau^{\frac{\alpha-1}{2}-s-1}.$$

The last function is integrable if $\frac{\alpha-1}{2}-s-1>-1$, i.e., $s<\frac{\alpha-1}{2}$. If $s\notin\mathbb{Z}$, $s>0$, then another straightforward argumentation including the special definition of the norm in W_1^s leads to the same result.

THEOREM 2.1: *Suppose all the assumptions of this section are fulfilled. Then* $\|z_n - z\|_{L^2} \leq C\ n^{-s}$, *where* $1/2 < s < \frac{\alpha-1}{2}$, $s \leq 4$.

PROOF: This theorem follows from (2.3), (2.5), (2.6), (2.17), (2.18) and (2.21). □

We now define $\mathfrak{z}_n$ by $\mathfrak{z}_n(\gamma(\tau))\gamma'(\tau)^{1/2} = z_n(\tau)$, $0 \leq \tau \leq 1$, i.e., we set $\mathfrak{z}_n(t) := z_n(\gamma^{-1}(t))/\gamma'(\gamma^{-1}(t))^{1/2}$. In view of (1.3) we obtain $\|\mathfrak{z}_n - \mathfrak{z}\|_{L^2} = \|z_n - z\|_{L^2}$.

COROLLARY 2.1: *Suppose all the assumptions of this section are fulfilled,* $1/2 < s < \frac{\alpha-1}{2}$ *and* $s \leq 4$. *Then the following estimates hold.*

(i) $\|\mathfrak{z}_n - \mathfrak{z}\|_{L^2} \leq C\ n^{-s}$.

(ii) *If* $0 \leq r \leq s$, *then* $\|z_n - z\|_{H^r} \leq C\ n^{r-s}$.

(iii) *Suppose* $0 < \varepsilon < 1/2$ *is fixed and* $s < 4$.

Then

$$\sup_{\varepsilon \leq t \leq 1-\varepsilon} |\mathfrak{z}_n(t) - \mathfrak{z}(t)| \leq C\ n^{1/2-s}.$$

PROOF: Assertion (i) follows from Theorem 2.1. Furthermore, an arbitrary function φ from im $L_n^{\mathbb{R}}$ satisfies the so called inverse property (cf. [7]), i.e., $0 \leq s \leq r < 5.5$ implies

$$\|\varphi\|_{H^r(\mathbb{R})} \leq C\ n^{r-s} \|\varphi\|_{H^s(\mathbb{R})} . \tag{2.22}$$

Thus we obtain

$$\|z_n - z\|_{H^r} \leq \|z_n - L_n^{\mathbb{R}} z\|_{H^r} + \|z - L_n^{\mathbb{R}} z\|_{H^r} \leq C\ n^r \|z_n - L_n^{\mathbb{R}} z\|_{L^2(\mathbb{R})} + \|z - L_n^{\mathbb{R}} z\|_{H^r}$$
$$\leq C\ n^r \|z_n - z\|_{L^2} + C\ n^r \|z - L_n^{\mathbb{R}} z\|_{L^2(\mathbb{R})} + \|z - L_n^{\mathbb{R}} z\|_{H^r(\mathbb{R})} .$$

Together with (2.4) and Theorem 2.1, thisimplies (ii). Assertion (iii) follows by

$$\sup_{\varepsilon \leq t \leq 1-\varepsilon} |\mathfrak{z}_n(t) - \mathfrak{z}(t)| \leq C \sup_{0 \leq t \leq 1} |z_n(t) - z(t)| ,$$

the embedding $C[0.1] \supseteq H^{1/2+\delta}[0,1]$ $(\delta > 0)$ and (ii). □

REMARK 2.1: *Corollary* 2.1 *states that* $\|z_n - z\|_{H^r}$ *is of order* n^{r-s}, $s := \min\{4, \frac{\alpha-1}{2}\}$. *In view of the coarse estimate* (2.18), *we conjecture* $\|z_n - z\|_{H^r} \le Cn^{r-s}$, $s := \min\{4, \frac{\alpha}{2}\}$. *Furthermore, the use of* $C[0,1] \supseteq H^{1/2+\delta}[0,1]$ *in the proof of* (iii) *leads to a similar effect. Thus we conjecture that* $\sup_{\varepsilon \le t \le 1-\varepsilon} |\mathfrak{z}_n(t) - \mathfrak{z}(t)|$ *is of order* n^{-s}, *where* $s := \min\{4, \frac{\alpha}{2}\}$.

Now let us consider the convergence near the end-points of the interval. By Corollary 2.1 (ii) this problem is solved for z_n. However, passing from $z_n(t_k)$ to $\mathfrak{z}_n(\gamma(t_k))$, we devide by the small number $\gamma'(t_k)^{1/2}$. Multiplication by an unbounded function is a simple special case of an ill-posed problem. The corresponding theory proposes a regularisation. More exactly, we approximate $\mathfrak{z}(0)$ not by $\mathfrak{z}_n(0)$, but by $\mathfrak{z}_n(\gamma(\psi_n))$, where ψ_n is a small number which is still to be determined. By the assymptotics of $\mathfrak{z}$ (cf. (2.1)) we conclude

$$(2.23) \qquad |\mathfrak{z}(0) - \mathfrak{z}(\gamma(\psi_n))| \le C\, \gamma(\psi_n) \log \gamma(\psi_n).$$

Using the definition of γ, z and $\mathfrak{z}_n$ as well as Corollary 2.1 (ii), we get

$$(2.24) \qquad |\mathfrak{z}(\gamma(\psi_n)) - \mathfrak{z}_n(\gamma(\psi_n))| \le C \frac{|z(\psi_n) - z_n(\psi_n)|}{\gamma'(\psi_n)^{1/2}} \le C \frac{n^{-s}}{\psi_n^{\frac{\alpha-1}{2}}},$$

where $0 < s < \min\{3.5, \frac{\alpha-2}{2}\}$. The sum of the left-hand sides in (2.23),(2.24) becomes small if we choose ψ_n such that $\psi_n^{\alpha} \log \psi_n \approx n^{-s}/\psi_n^{\frac{\alpha-1}{2}}$, i.e., $\psi_n := C\, n^{-\frac{2s}{3\alpha-1}}$. By this choice and formulae (2.23), (2.24) we arrive at

$$\gamma(\psi_n) = n^{-\alpha\frac{2s}{3\alpha-1}}, \quad |\mathfrak{z}(0) - \mathfrak{z}_n(\gamma(\psi_n))| \le C\, n^{-\alpha\frac{2s}{3\alpha-1}} \log n \quad .$$

COROLLARY 2.2: *Suppose all the assumptions of this section are fulfilled,* $\alpha > 2$, $0 < s < \min\{\frac{7\alpha}{3\alpha-1}, \frac{\alpha(\alpha-2)}{3\alpha-1}\}$ *and choose*

$\gamma\ \psi_n)=C\ n^{-s}$. *Then*

$$|\mathfrak{z}(0)-\mathfrak{z}_n(\gamma(\psi_n))|\le C\ n^{-s}\log\ n\ .$$

REMARK 2.2: *If our conjectures in Remark* 2.1 *are true, then Corollary* 2.2 *holds with* $s<\min\{\frac{8\alpha}{3\alpha-1},\frac{\alpha^2}{3\alpha-1}\}$.

3. STABILITY OF THE QUADRATURE METHOD

In this section we shall give necessary and sufficient conditions for the stability of method (1.15). These conditions are of local nature (cf. Theorem 3.5). Therefore, in a first step we analyse the quadrature method for a singular integral operator with constant coefficients on the half axis or on the real axis . Analogously to the Fredholm theory of singular integral operators, we introduce the operator $\mathbf{A}^0:=a(0)I+b(0)S_{\mathbb{R}_+}+\frac{b(0)}{\pi i}\mathbf{K}_0^0\in\mathcal{L}(L^2(\mathbb{R}_+))$, where $S_{\mathbb{R}_+}$ is the Cauchy singular integral operator on the positive real half axis and $\mathbf{K}_0^0$ is defined by

$$\mathbf{K}_0^0 z(t):=\int_0^\infty \mathbf{k}_0^0(t,\tau)z(\tau)d\tau\ ,$$

$$\text{(3.1)}\qquad \mathbf{k}_0^0(t,\tau):=\left\{\alpha\ \frac{t^{\frac{\alpha-1}{2}-\alpha\rho_0}\tau^{\frac{\alpha-1}{2}+\alpha\rho_0}}{\tau^\alpha-t^\alpha}-\frac{1}{\tau-t}\right\}.$$

In a certain sense, $\mathbf{A}^0$ is locally equivalent at 0 to the operator $\mathbf{A}$ defined by the left-hand side of (1.6) (cf. also the operator $\mathbf{A}$ introduced in Sect. 2). From

$$\text{(3.2)}\qquad \mathbf{k}_0^0(t,\tau)=\frac{1}{\tau}\ \mathbf{k}^0(\tfrac{t}{\tau})\ ,\ \mathbf{k}^0(x):=\left\{\frac{\alpha x^{\frac{\alpha-1}{2}-\alpha\rho_0}}{1-x^\alpha}-\frac{1}{1-x}\right\}$$

it follows that $\mathbf{A}^0$ is a Mellin convolution operator, i.e., $\mathbf{A}^0:=\mathcal{M}^{-1}\mathrm{symb}\ \mathcal{M}$ (cf. [4]), where $\mathcal{M}$ is the Mellin transform

$$\mathcal{M}z(\zeta):=\int_0^\infty t^{\zeta-1}z(t)dt, \mathcal{M}:L^2(\mathbb{R}_+)\rightarrow L^2(\{\zeta\in\mathbb{C}, \mathrm{Re}\,\zeta=1/2\}) ,$$

and symb denotes the operator of multiplication by the symbol function

$$(3.3)\quad \mathit{symb}(\zeta):=a(0)+b(0)(-i)\mathrm{ctg}\left[\pi\left(\frac{\zeta+\frac{\alpha-1}{2}}{\alpha}-\rho_o\right)\right] , \zeta\in\mathbb{C}, \mathrm{Re}\,\zeta=1/2.$$

Obviously, $\mathbf{A}^o$ is invertible if and only if $\mathit{symb}(\zeta)$ does not vanish for all $\zeta\in\mathbb{C}$, Re $\zeta=1/2$, i.e., if and only if

$$(3.4)\quad \frac{a(0)+b(0)}{a(0)-b(0)} \notin \{ te^{i2\pi\rho_o}, -\infty<t\le 0 \} .$$

The "local representative" of the method (1.15) at the point $\tau=0$ will be the following quadrature method for the equation $\mathbf{A}^o\mathbf{z}=\mathbf{y}$: Let $\mathbf{y}\in L^2(\mathbb{R}_+)$ be Riemann integrable. Determine approximate values ξ_j for $z(t_j)$ (k=u,u+1,...) by solving the system (cf.(1.15))

$$(3.5)\quad \mathbf{y}(t_k) = a(0)\xi_k+ \frac{b(0)}{\pi i}\left\{\sum_{\substack{j=u\\ j\neq k}}^{\infty} \alpha\frac{t_k^{\frac{\alpha-1}{2}-\alpha\rho_o}\,t_j^{\frac{\alpha-1}{2}+\alpha\rho_o}}{t_j^\alpha - t_k^\alpha}\,\omega_j\xi_j + \right.$$

$$\left. \eta_k n\,\frac{8(\xi_{k+1}-\xi_{k-1})-(\xi_{k+2}-\xi_{k-2})}{12} +\omega_k\alpha\,\rho_o\,\frac{1}{t_k}\xi_k \right\} ,\quad k=u,u+1,...$$

Here we set $\xi_k:=0$ for k<u and

$$\eta_j:=\begin{cases} 4/3n & \text{if } j \text{ is odd,}\\ 2/3n & \text{if } j \text{ is even,}\end{cases} \qquad \omega_j:=\begin{cases} 4/3n & \text{if } j \text{ is odd,}\\ 2/3n & \text{if } j=2,4,...,\\ 1/3n & \text{if } j=0.\end{cases}$$

We start our analysis of method (3.5) in the special case u=0. Analogously to Sect.2 we introduce

$$K_n^{\mathbb{R}_+} f := \sum_{k=0}^{\infty} f(t_k)\varphi_k \ , \quad L_n^{\mathbb{R}_+} f := \sum_{k=0}^{\infty} \lambda_k(f)\varphi_k \ , \quad z_n := \sum_{k=0}^{\infty} \xi_k \varphi_k \ .$$

We define $A_n^o \in \mathscr{L}(\mathrm{im}\ L_n^{\mathbb{R}_+})$ to be the operator the matrix of which corresponding to the basis $\{\varphi_j\}$ is the matrix of the system (3.5). Then (3.5) holds if and only if $A_n^o z_n = K_n^{\mathbb{R}_+} y$. Since the norm of an operator in $\mathscr{L}(\mathrm{im}\ L_n^{\mathbb{R}_+})$ is equivalent to the ℓ^2-operator norm of its matrix, we have to prove the stability of $\{A_n^o\}_{n=1}^{\infty}$ with respect to this norm. However, it is easy to check that the matrix A_n^o is independent of n. Hence, $\{A_n^o\}_{n=1}^{\infty}$ (i.e., the method (3.5)) is stable if and only if $A_1^o \in \mathscr{L}(\ell^2)$ is invertible.

THEOREM 3.1: *The operator $A_1^o \in \mathscr{L}(\ell^2)$ is Fredholm with index zero if and only if (3.4) is fulfilled and*

$$(3.6) \qquad |\mathrm{Re}\ \varkappa_o| < 1/2 \ ,$$

where $\varkappa_o$ is given by (1.1) and $^0\log$ is defined as in Sect. 1 .

PROOF: To show this we need some results on Toeplitz operators which are due to Gohberg and Krupnik [12]. Let $\mathfrak{A} \subseteq \mathscr{L}(\ell^2)$ denote the smallest Banach algebra containing all Toeplitz operators $T(\underline{a}) = (\underline{a}_{k-j})_{k,j=0}^{\infty}$ with piecewise continuous $\underline{a} = \sum \underline{a}_k t^k$ on the unit circle $\mathbb{T}$. Then the set of 2x2-matrix operators $\mathfrak{A}_{2x2} \subseteq \mathscr{L}(\ell^2)_{2x2}$ is an algebra of continuous operators in $\ell_2^2 := \ell^2 + \ell^2$. There exists a multiplicative linear mapping $\mathfrak{A}_{2x2} \ni B \longrightarrow \mathscr{A}_B$ into the algebra of bounded 2x2-matrix functions over $\mathbb{T} \times [0,1]$. The symbol $\mathscr{A}_{T(\underline{a})}$ of $T(\underline{a})$ is given by $\mathscr{A}_{T(\underline{a})}(\tau,\mu) = \mu\underline{a}(\tau+0) + (1-\mu)\underline{a}(\tau-0)$, where $\tau \in \mathbb{T}$, $0 \le \mu \le 1$. Furthermore, $B \in \mathfrak{A}_{2x2}$ is a Fredholm operator if and only if $\det \mathscr{A}_B(\tau,\mu) \ne 0$ for all $(\tau,\mu) \in \mathbb{T} \times [0,1]$. If $\mathscr{A}_B(\tau,\mu) = \mathscr{A}_B(\tau,1)$ for $\tau \in \mathbb{T} \setminus \{1\}$ and $0 \le \mu \le 1$, then the index of B is the negative index of the curve $\{\det \mathscr{A}_B(e^{i2\pi\lambda},1), 0<\lambda<1\} \cup \{\det \mathscr{A}_B(1,\mu), 0 \le \mu \le 1\}$. Finally, $\mathfrak{A}_{2x2}$ contains all compact operators on ℓ_2^2, and $B \in \mathfrak{A}_{2x2}$ is compact if and only if $\mathscr{A}_B \equiv 0$.

For $C\in\mathcal{L}(\ell^2)_{2\times2}$, $C=(C^{r,s})_{r,s=0}^{1}$, $C^{r,s}=(C_{k,l}^{r,s})_{k,l=0}^{\infty}$, we define $C\in\mathcal{L}(\operatorname{im} L_n^{\mathbb{R}_+})$ by (cf. the definition of $C(\underline{a})$ in (2.7))

$$C\varphi_{2j}:=\sum_{k=0}^{\infty}\left\{C_{k,j}^{0,0}\varphi_{2k}+C_{k,j}^{1,0}\varphi_{2k+1}\right\},$$

$$C\varphi_{2j+1}:=\sum_{k=0}^{\infty}\left\{C_{k,j}^{0,1}\varphi_{2k}+C_{k,j}^{1,1}\varphi_{2k+1}\right\}.$$

Thus $\mathfrak{A}_{2\times2}\subseteq\mathcal{L}(\operatorname{im} L_n^{\mathbb{R}_+})$. On the other hand, there exists a compact $T\in\mathcal{L}(\ell^2)_{2\times2}$ such that $A_1^0=C+T$, where

$$C^{0,0}=\left[a(0)\delta_{2k,2j}+\frac{b(0)}{\pi i}\,\frac{t_{2k}^{\frac{\alpha-1}{2}-\alpha\rho_0}\,t_{2j}^{\frac{\alpha-1}{2}+\alpha\rho_0}}{t_{2j}^{\alpha}-t_{2k}^{\alpha}}\,\frac{2}{3}-\frac{2}{3}\,\frac{b(0)}{\pi i}\,\frac{\delta_{2j,2k+2}-\delta_{2j,2k-2}}{12}\right]_{k,j=0}^{\infty}$$

$$=\left[a(0)\delta_{k,j}+\frac{b(0)}{3\pi i}\,\frac{k^{\frac{\alpha-1}{2}-\alpha\rho_0}\,j^{\frac{\alpha-1}{2}+\alpha\rho_0}}{j^{\alpha}-k^{\alpha}}-\frac{b(0)}{18\pi i}\,(\delta_{j,k+1}-\delta_{j,k-1})\right]_{k,j=0}^{\infty},$$

$$C^{1,1}=\left[a(0)\delta_{2k+1,2j+1}+\frac{b(0)}{\pi i}\,\frac{t_{2k+1}^{\frac{\alpha-1}{2}-\alpha\rho_0}\,t_{2j+1}^{\frac{\alpha-1}{2}+\alpha\rho_0}}{t_{2j+1}^{\alpha}-t_{2k+1}^{\alpha}}\,\frac{4}{3}-\frac{4}{3}\,\frac{b(0)}{\pi i}\,\frac{\delta_{2j+1,2k+3}-\delta_{2j+1,2k-1}}{12}\right]_{k,j=0}^{\infty}$$

$$=\left[a(0)\delta_{k,j}+\frac{2}{3}\,\frac{b(0)}{\pi i}\,\frac{(k+1/2)^{\frac{\alpha-1}{2}-\alpha\rho_0}(j+1/2)^{\frac{\alpha-1}{2}+\alpha\rho_0}}{(j+1/2)^{\alpha}-(k+1/2)^{\alpha}}-\frac{b(0)}{9\pi i}\,(\delta_{j,k+1}-\delta_{j,k-1})\right]_{k,j=0}^{\infty},$$

$$C^{0,1}=\left[\frac{2}{3}\,\frac{b(0)}{\pi i}\,\frac{k^{\frac{\alpha-1}{2}-\alpha\rho_0}(j+1/2)^{\frac{\alpha-1}{2}+\alpha\rho_0}}{(j+1/2)^{\alpha}-k^{\alpha}}+\frac{4}{9}\,\frac{b(0)}{\pi i}\,(\delta_{j,k}-\delta_{j,k-1})\right]_{k,\,j=0}^{\infty},$$

$$C^{1,0}=\left[\frac{b(0)}{3\pi i}\,\frac{(k+1/2)^{\frac{\alpha-1}{2}-\alpha\rho_0}j^{\frac{\alpha-1}{2}+\alpha\rho_0}}{j^{\alpha}-(k+1/2)^{\alpha}}+\frac{8}{9}\,\frac{b(0)}{\pi i}\,(\delta_{j,k+1}-\delta_{j,k})\right]_{k,\,j=0}^{\infty}.$$

From [21], formula (1.10) and the proof of Theorem 3.3 in [21], we conclude $C\in\mathfrak{U}_{2\times 2}$, $A_1^0=C+T\in\mathfrak{U}_{2\times 2}$, $\mathscr{A}_{A_1^0}=\mathscr{A}_C$ and

(3.7) $$\mathscr{A}_C(\tau,\mu)=a(0)\begin{pmatrix}1&0\\0&1\end{pmatrix}+b(0)\mathscr{F}(\tau,\mu),$$

$$\mathscr{F}(\tau,\mu):=\begin{cases}(-i)\,ctg\left[\pi\left(\frac{1}{2}-\rho_0+\frac{i}{2\pi\alpha}\,log\frac{\mu}{1-\mu}\right)\right]\begin{pmatrix}1/3&2/3\\1/3&2/3\end{pmatrix} & \text{if } \tau=1,\ 0\le\mu\le 1,\\[2ex] \begin{pmatrix}\{\frac{1}{3}(2\lambda-1)+\frac{1}{9\pi}\sin(2\pi\lambda)\} & \{-e^{i\pi\lambda}[\frac{2}{3}+\frac{8}{9\pi}\sin(\pi\lambda)]\}\\ \{-e^{-i\pi\lambda}[\frac{1}{3}+\frac{16}{9\pi}\sin(\pi\lambda)]\} & \{\frac{2}{3}(2\lambda-1)+\frac{2}{9\pi}\sin(2\pi\lambda)\}\end{pmatrix} & \\ \qquad\qquad\text{if } \tau=e^{i2\pi\lambda},\ 0<\lambda<1,\ 0\le\mu\le 1.\end{cases}$$

The eigenvalues $\lambda_0(\tau,\mu)$ and $\lambda_1(\tau,\mu)$ of $\mathscr{F}(\tau,\mu)$ are given by

$$\lambda_k(\tau,\mu)=\begin{cases}(-i)\,ctg\left[\pi\left(\frac{1}{2}-\rho_0+\frac{i}{2\pi\alpha}\,log\frac{\mu}{1-\mu}\right)\right] & \text{if } \tau=1,\ 0\le\mu\le 1,\ k=0,\\ 0 & \text{if } \tau=1,\ 0\le\mu\le 1,\ k=1,\\ \frac{1}{2}\,\{(2\lambda-1)+\frac{1}{3\pi}\sin(2\pi\lambda)\}+(-1)^k\Big[\frac{1}{36}\,\{(2\lambda-1)+\frac{1}{3\pi}\sin(2\pi\lambda)\}^2+ & \\ \qquad\{\frac{1}{3}+\frac{16}{9\pi}\sin(\pi\lambda)\}\{\frac{2}{3}+\frac{8}{9\pi}\sin(\pi\lambda)\}\Big]^{1/2} & \\ \qquad\qquad\text{if } \tau=e^{i2\pi\lambda},\ 0<\lambda<1,\ 0\le\mu\le 1.\end{cases}$$

Obviously, $\lambda_0(e^{i2\pi\lambda},\mu)$ varies continuously from 0 to 1 if λ varies from 0 to 1. Furthermore, $\lambda_1(e^{i2\pi\lambda},\mu)=-\lambda_0(e^{i2\pi(1-\lambda)},\mu)$ varies from -1 to 0. We set $\mathscr{F}(\tau,\mu)=(\mathscr{F}^{r,s}(\tau,\mu))_{r,s=0}^{1}$ and introduce the matrix norm $\|\mathscr{F}(\tau,\mu)\|_{\infty}:=max\{|\mathscr{F}^{r,0}(\tau,\mu)|+|\mathscr{F}^{r,1}(\tau,\mu)|,\ r=0,1\}$.

Then an easy computation yields $\| \mathcal{F}(e^{i2\pi\lambda},\mu)\|_\infty \le 1$, $0<\lambda<1$, and we arrive at $|\lambda_k(e^{i2\pi\lambda},\mu)| \le 1$, $0<\lambda<1$, $k=0,1$. On the other hand,

$$det\ \mathcal{A}_{A_1^o}(\tau,\mu) = \begin{cases} a(0)\{a(0)+b(0)(-i)ctg[\pi(\frac{1}{2}-\rho_o+\frac{i}{2\pi\alpha}\ log\frac{\mu}{1-\mu})]\} & \text{if } \tau=1, 0\le\mu\le 1, \\ \{a(0)+b(0)\lambda_o(e^{i2\pi\lambda},\mu)\}\{a(0)+b(0)\lambda_1(e^{i2\pi\lambda},\mu)\} & \text{if } \tau=e^{i2\pi\lambda}, 0<\lambda<1, 0\le\mu\le 1. \end{cases}$$

Using the properties of $\lambda_k(\tau,\mu)$, we see that $det\ \mathcal{A}_{A^o}$ does not vanish if and only if $\mathcal{A}$ does not vanish. Moreover, $ind\ det\ \mathcal{A}_{A_1^o} = ind\ \mathcal{A}$, where

$$\mathcal{A}(\tau,\mu) := \begin{cases} a(0)+b(0)(-i)ctg[\pi(\frac{1}{2}-\rho_o+\frac{i}{2\pi\alpha}\ log\frac{\mu}{1-\mu})] & \text{if } \tau=1, 0\le\mu\le 1, \\ a(0)+(2\lambda-1)b(0) & \text{if } \tau=e^{i2\pi\lambda}, 0<\lambda<1, 0\le\mu\le 1. \end{cases}$$

However, the relations $\mathcal{A}\ne 0$ and $ind\ \mathcal{A}=0$ are equivalent to (3.4) and (3.6) (cf. [19], Sect.23 or [20]). □

Now, let $u\ge 0$. We define $A_n^{o,u}$ to be the operator in $\mathcal{L}(span\ \{\varphi_j\}_{j=u}^\infty)$ the matrix of which corresponding to the basis $\{\varphi_j\}_{j=u}^\infty$ is the matrix of (3.5). Again, method (3.5) is stable if and only if $A_1^{o,u}$ is invertible.

THEOREM 3.2: *If, for* $u\in\mathbb{Z}$, $u\ge 0$, *the operator* $A_1^{o,u}\in\mathcal{L}(\ell^2)$ *is invertible, then* (3.4) *and* (3.6) *are satisfied. On the other hand, if* (3.4) *and* (3.6) *hold, then there exists an* $u\in\mathbb{Z}$, $u\ge 0$ *such that* $A_1^{o,u}$ *is invertible.*

PROOF: Let $Q_u\in\mathcal{L}(\text{im } L_n^{\mathbb{R}^+})$ be defined by $Q_u\varphi_k:=0$ if $k<u$ and $Q_u\varphi_k:=\varphi_k$ if $k\ge u$. Thus we obtain $A_1^{o,u} = Q_u A_1^{o,o}|\text{im } Q_u$. If $A_1^{o,u}$ is invertible, then $D:=(I-Q_u)+Q_u A_1^{o,o} Q_u$ is invertible. Since $A_1^{o,o}-D$ is an operator of finite rank, $A_1^{o,o}$ is Fredholm with index zero. Hence Theorem 3.1 implies (3.4) and (3.6).

Vice versa, let (3.4) and (3.6) be satisfied. Then we shall apply a theorem which is due to S.Roch (cf.[22],Sects.2 and 3). For $C\in\mathfrak{U}_{2\times2}$, this theorem states the equivalence of the following two assertions.

(i) The operator C is Fredholm with index zero. For all $\tau\in\mathbb{T}$, the restriction of the Mellin convolution operator $\mathcal{M}^{-1}\mathcal{A}_C(\tau,\frac{1+i\,ctg(\pi.)}{2})\mathcal{M}\in\mathcal{L}(L^2(\mathbb{R}_+))_{2\times2}$ to the interval $[1,\infty)$ is invertible. The Toeplitz operator $T(\underline{c})\in\mathcal{L}(\ell^2)_{2\times2}$ is invertible , where $\underline{c}(t):=\mathcal{A}_C(t,1)$, $t\in\mathbb{T}$.

(ii) There exists $u_0\in\mathbb{Z}$, $u_0\geq0$ such that $Q_{2u}C|\operatorname{im}Q_{2u}$ is invertible for $u\geq u_0$ and $\sup_{u\geq u_0}\|(Q_{2u}C|\operatorname{im}Q_{2u})^{-1}\|<\infty$.

Setting $C=\mathbf{A}_1^{0,0}$, it remains to verify (i). By our assumption and Theorem 3.1 we get the Fredholm property of $\mathbf{A}_1^{0,0}$ and $\operatorname{ind}\mathbf{A}_1^{0,0}=0$. Furthermore, if $\tau\neq1$, then $\mathcal{M}^{-1}\mathcal{A}_{\mathbf{A}_1^{0,0}}(\tau,\frac{1+i\,ctg(\pi.)}{2})\mathcal{M}$ is a scalar multiple of the identity and, consequently, invertible. In the case $\tau=1$, we choose a 2x2-matrix G satisfying

$$\begin{pmatrix}1/3 & 2/3\\ 1/3 & 2/3\end{pmatrix}=G^{-1}\begin{pmatrix}1 & 0\\ 0 & 0\end{pmatrix}G$$

to obtain

$$\mathcal{M}^{-1}\mathcal{A}_{\mathbf{A}_1^{0,0}}(\tau,\frac{1+i\,ctg(\pi.)}{2})\mathcal{M}=G^{-1}\begin{pmatrix}\mathbf{A}^0 & 0\\ 0 & \{a(0)I\}\end{pmatrix}G\ .$$

It is well known that the restriction of the last operator is equivalent to the Wiener-Hopf integral operator on the half axis which has the symbol

$$\mathbb{R}\ni t\longrightarrow a(0)+b(0)(-i)\,ctg[\pi(\tfrac{1}{2}-\rho_0+i\,\tfrac{t}{\alpha})]\ .$$

Hence, it is invertible if and only if $\mathcal{A}(\tau,\mu)\neq0$ (for the definition of $\mathcal{A}$ we refer to the proof of Theorem 3.1) for any

$(\tau,\mu)\in\mathbb{T}\times[0,1]$ and *ind* $\mathcal{A}=0$, i.e., if and only if (3.4) and (3.6) are satisfied.

Finally, we write $\underline{c} = a(0)(I_{2\times2}+\underline{d})$, where $\underline{d}(\tau):=\frac{b(0)}{a(0)}\mathcal{F}(\tau,1)$, $\tau\in\mathbb{T}$. Then (3.4), (3.6) and $\|\mathcal{F}(\tau,1)\|_\infty\le1$ yield $|b(0)/a(0)|<1$ as well as $\|\underline{d}\|_\infty<1$. Hence, $T(\underline{c})=a(0)(I+T(\underline{d}))$ is invertible. □

In the case $0<\tau<1$ we define $A^\tau:=a(\tau)+b(\tau)S_{\mathbb{R}}\in\mathcal{L}(L^2(\mathbb{R}))$. (Remark that the restriction of $\mathbf{l}(t,\tau)$ to a neighbourhood of τ is the kernel of a compact opeator.) This operator is locally equivalent to A at the point τ. It is invertible if and only if $a(\tau)\pm b(\tau)\ne0$. We consider the equation $A^\tau z=y$, where $y\in L^2(\mathbb{R})$ and y is Riemann integrable. Analogously to (1.15), we determine approximate values ξ_j of $z(t_j)$ $(j\in\mathbb{Z})$ by solving the system

$$(3.8)\quad a(\tau)\xi_k+\frac{b(\tau)}{\pi i}\left\{\sum_{\substack{j\in\mathbb{Z}\\ j\ne k}}\frac{\eta_j}{t_j-t_k}\xi_j+\eta_k\frac{8(\xi_{k+1}-\xi_{k-1})-(\xi_{k+2}-\xi_{k-2})}{12}\right\}=$$

$$y(t_k),\ k\in\mathbb{Z},\quad \eta_j:=\begin{cases}4/3n & \text{if } j \text{ is odd,}\\ 2/3n & \text{if } j \text{ is even.}\end{cases}$$

This quadrature method is the "local representative" of (1.15) at the point τ, $0<\tau<1$.

Retaining the definitions of $S_n^{\mathbb{R}}$, $\{\varphi_j\}$, $K_n^{\mathbb{R}}$, $L_n^{\mathbb{R}}$, $C(\underline{a})$ from Sect.2, we set $z_n:=\sum\xi_j\varphi_j$. Furthermore, let A_n^τ denote the operator the matrix of which corresponding to the basis $\{\varphi_j\}$ is the matrix of the system (3.8). Hence, (3.8) is equivalent to $A_n^\tau z_n=K_n^{\mathbb{R}}y$. Again, the matrix A_n^τ is independent of n, and method (3.8) is stable if and only if A_1^τ is invertible.

THEOREM 3.3: *The operator* $A_1^\tau\in\mathcal{L}(\ell^2)$ *is invertible if and only if*

$$(3.9)\qquad \frac{a(\tau)+b(\tau)}{a(\tau)-b(\tau)}\notin(-\infty,0].$$

PROOF: Section 2 yields $A_1^\tau=a(\tau)+b(\tau)C(\underline{a})$, where $\underline{a}=(\underline{a}^{r,s})_{r,s=0}^{1}$ is the matrix function defined in Sect.2. Thus A_1^τ is a discrete convolution matrix with the symbol (cf. (3.7)) $\mathbb{T}\ni\tau$

$\rightarrow \mathscr{A}_C(\tau,1)$. It is invertible if and only if this symbol does not degenerate, i.e., $a(\tau)+b(\tau)\lambda_k(\tau,1)\neq 0$ for $\tau\in\mathbb{T}$ and $k=0,1$. Due to $\{\lambda_k(\tau,1),\tau\in\mathbb{T}\ ,\ k=0,1\} = [-1,1]$, Theorem 3.3 follows. □

The case $\tau=1$ is analogous to $\tau=0$. However, instead of considering $a(1)+b(1)S_{\mathbb{R}_-}+\frac{b(1)}{\pi i}K_0^1$ we transform this operator by setting $t'=-t$. Thus we get $A^1=a(1)-b(1)S_{\mathbb{R}_+}-\frac{b(1)}{\pi i}K_0^1$, where

$$K_0^1 z(t)=\int_0^\infty k_0^1(t,\tau)z(\tau)d\tau\ ,\ k_0^1(t,\tau):=\{\alpha\frac{t^{\frac{\alpha-1}{2}-\alpha\rho_1}\tau^{\frac{\alpha-1}{2}+\alpha\rho_1}}{\tau^\alpha-t^\alpha}-\frac{1}{\tau-t}\}.$$

We define the quadrature method for A^1 in the same way we did in (3.5) for A^0. For this method, we introduce the operators A_n^1, $A_n^{1,v}$ which correspond to A_n^0 and $A_n^{0,u}$. Hence, the quadrature method for A^1 is stable if and only if $A_1^{1,v}\in\mathscr{L}(\mathrm{span}\{\varphi\}_{j=v}^\infty)$ is invertible. Thus Theorem 3.2 applies.

THEOREM 3.4: *Let* $v\in\mathbb{Z}$, $v\geq 0$. *If* $A_1^{1,v}$ *is invertible, then*

$$\text{(3.9)}\qquad \frac{a(1)-b(1)}{a(1)+b(1)}\notin\{\ te^{i2\pi\rho_1},-\infty<t\leq 0\ \}\ ,$$

$$\text{(3.10)}\qquad |\mathrm{Re}\ \varkappa_1|<1/2\ ,$$

where $\varkappa_1$ *is given by* (1.1) *and* ${}^1\ell og$ *is defined as in Sect.* 1. *On the other hand, if* (3.9) *and* (3.10) *are satisfied, then there exists a* $v\in\mathbb{Z}$, $v\geq 0$ *such that* $A_1^{1,v}$ *is invertible.*

Now, the stability of method (1.15) can be deduced from Theorems 3.2-3.4 and the following localization principle.

THEOREM 3.5: a) *The method* (1.15) *is stable in* L^2 *if and only if the following conditions are satisfied.*

(i) A *is invertible.*

(ii) *Method* (3.5) *for* A^0 *is stable.*

(iii) *For any* τ, $0<\tau<1$, *method* (3.8) *for* A^τ *is stable.*

(iv) *The quadrature method* (3.5) *applied to the operator* A^1 *is stable.*

b) *If the quadrature method* (1.15) *is stable, then the linear system* (1.15) *is uniquely solvable for* n *large enough. Furthermore, if the right-hand side* $\mathbf{y}$ *is Riemann-integrable, then* $\mathbf{z}_n$ *converges to the exact solution* $\mathbf{z}$ *in the* L^2*-norm.*

The proof of this theorem is analogous to the proof of Theorem 4.2 in [21] or to that of Theorem 25 in [19]. (Setting $\mathbf{B}_n:=\mathbf{A}_n\Omega_n$, $\Omega_n:=(\delta_{k,j}n/\omega_j)_{k,j=0}^{n}$, the sequence $\{\mathbf{A}_n\}$ is stable if and only if $\{\mathbf{B}_n\}$ is stable. For the sequence $\{\mathbf{B}_n\}$, Theorem 26 of [19] applies. Thereby, the validity of the assumptions of Theorem 26 can be shown analogously to Sect.28 in [19].Cf.also [20],Chapter 10.)

COROLLARY 3.1:*Suppose* $\varkappa_j$, j=0,1 *is given by* (1.1), *where* $^j\ell og$ *denotes the branch of the logarithm which is continuous on* $\mathbb{C}\setminus(-\infty,0]$ *and takes real values on* $\mathbb{R}_+$. *Furthermore, set* $\rho_0:=\varkappa_0$, $\rho_1:=\varkappa_1$ *and let the operator* $\mathbf{A}$ *on the left-hand side of* (1.6) *be invertible in* L^2. *Then the stability of* (1.15) *for some* u *and* v *implies the strong ellipticity condition*

$$a(t)+\lambda b(t)\neq 0 \ , \quad 0\le t\le 1 \ , \quad -1\le\lambda\le 1 \ .$$

Vice versa, if this condition is fulfilled, then there exist $u,v\in\mathbb{Z}$, $u,v\ge 0$ *such that the quadrature method* (1.15) *is stable.*

4. NUMERICAL RESULTS

Applying method (1.15), the first question is how to choose the parameters u,v and α. The number u (v) has been introduced in order to guarantee stability in case $\mathbf{A}_1^{0,0}$ ($\mathbf{A}_1^{1,0}$) (cf.Sect.3) is Fredholm with index zero , but its null space is not trivial. Actually, it is not clear whether this situation is possible.However, if it should happen, then it is an exceptional case. (Let us consider $\mathbf{A}_1^{0,0}$ as a function of $a(0)\in\mathbb{C}$. Then it is a Fredholm operator with index zero and non-trivial null space only for $a(0)$ belonging to an at most countable subset of $\mathbb{C}$. Moreover, at the accumulation points of this subset $\mathbf{A}_1^{0,0}$ is either not Fredholm or its index differs from zero. For the proof of these assertions, we refer to [19,21].) Therefore, we propose u=v=0. If it would turn out that the resulting method is not stable, then we would choose u=v=1 or u,v≥1 (cf. also the corresponding parameters in [2], where another class of equations is considered). In all the examples presented here the choice u=v=0 is satisfactory.

Sect.2 suggests to choose α large , at least greater than 9 Then we have $\frac{\alpha-1}{2}>4$ and Theorem 2.1 insures a convergence of order n^{-4}. However, α must not be too large. Namely, to obtain (1.15) we have discretized (0.1) using the non-uniform mesh $\{\gamma(j/n), j=0,\ldots,n\}$. If n<100 and α is very large, then the mesh points are concentrated near the end-points of the interval, and in the inner part of [0,1] the mesh is coarse. Hence, though the rate of convergence may be high, a large error is to be expected. For this reason, we have set $\alpha\le5$. Moreover, we have carried out our computations using $\gamma(t):=t^\alpha/\{t^\alpha+(1-t)^\alpha\}$. This transformation of the interval behaves like t^α as $t\to0$ and like $1-(1-t)^\alpha$ as $t\to1$. It is not hard to extend the results of Sects.2 and 3 to the case of this function γ.

The first example we have considered is that of K.S.Thomas [23]. If $a(t)\equiv1$, $b(t)\equiv-i$, $k(t,\tau)\equiv0$ and $y(t)\equiv\sqrt{2}$, then the solution x of (0.1) takes the form $x(t)=t^{-1/4}(1-t)^{1/4}$. Choosing $\rho_0=\varkappa_0=-1/4$, $\rho_1=\varkappa_1=1/4$, we get $z(t)\equiv1$. The error $\sup_{0.1\le t\le0.9}|z(t)-z_n(t)|$ for α=1,3,4,5 is given in Table 1.

Table 1 (1.15)	α=1	α=3	α=4	α=5
n=12	0.08	0.0017	0.008	0.013
n=24	0.053	0.00013	0.00036	0.0010
n=48		0.00005	0.000012	0.000041
n=96		0.000005	0.0000046	0.0000014

For our special choice, equation (0.1) is strongly elliptic. In accordance with the stability theorem the condition numbers of the systems (1.15) are bounded. Table 2 gives the approximate values $z_n(t)$ for α=5.

Table 2 (1.15)	n=12	n=24	n=48	n=96
t=0.5	1.011	1.00091	1.000035	1.0000011
t=0.8432	1.013	0.99940	0.999977	0.9999993
t=0.9959	1.006	1.00023	1.000008	1.0000002
t=1	1.013	0.99978	0.999982	1.0000002

The approximate value for $\vartheta(1)$ has not been determined by the way described in Sect.2. We have set $\vartheta(1)\sim\vartheta_n(\gamma(t_{j_o}))$ and

$$|\vartheta_n(\gamma(t_{j_o+1}))-\vartheta_n(\gamma(t_{j_o}))|=\min_{j=\frac{n}{2},\frac{n}{2}+1,\dots,n-2}|\vartheta_n(\gamma(t_{j+1}))-\vartheta_n(\gamma(t_j))|.$$

This principle of choosing an approximation for $\vartheta(1)$ seems to be good if $\vartheta(1)$ does not vanish. Note that the choice $\vartheta(1)\sim\vartheta_n(\gamma(\psi_n))$ is practicable only if the constants in (2.23) and (2.24) are known explicitly.

For the case of the method (1.20)-(1.21), we have put $\alpha=3$, and the approximate values for $\vartheta(t)$ (cf.(1.22)) are given in Table 3.

Table 3 (1.20)-(1.21)	n=12	n=24	n=48	n=96
t=0.5	1.006	1.00034	1.000012	1.0000004
t=0.8889	0.993	0.99983	0.999996	0.9999999
t=0.9920	1.001	1.00006	1.000002	1.0000001
t=1	1.003	1.00023	1.000022	1.0000024

It turns out that (1.20)-(1.21) provides better approximations near the end-points of the interval. The condition numbers of the systems of equations (1.20)-(1.21) seem to tend to infinity as $n\to\infty$. However, they are still acceptable if $n<100$.

The results presented in the Tables 1-3 show that quadrature methods lead, roughly speaking, to the same order of convergence as the computations in [23], where a Galerkin method has been proposed and the solution has been approximated by continuous piecewise linear functions on non-uniform meshes.

(Moreover, the spline functions have been modified near the end-points of the interval.) In [11,13] the case $a\equiv 0$, $b\equiv 1$ has been considered only (i.e.,(0.1) is not strongly elliptic). The approximation obtained by spline collocation seems to be better. However, if the methods of [11,13] are applied to the case of variable a and b, then there occur logarithmic terms in the asymptotics of the solution (cf. Sect. 2), and the convergence will be slower. For our quadrature methods, the rate of convergence remains the same. Tables 4 and 5 give the approximate values for $z(t)$ obtained by the methods (1.15), $\alpha=5$ and (1.20)-(1.21), $\alpha=3$ in the case $a(t)=1+\sin(\pi t)$, $b(t)=-1+t^2(1-t)$, $y(t)=\sin(t)$ and $\rho_0=-1/4$, $\rho_1=1/4$.

Table 4 (1.15)	n=12	n=24	n=48	n=96
t=0.5	0.275	0.27859	0.278737	0.278739
t=0.8432	0.643	0.64436	0.643071	0.643005
t=0.9959	0.802	0.83839	0.838294	0.838282
t=1	0.802	0.83839	0.839237	0.833825

Table 5 (1.20)-(1.21)	n=12	n=24	n=48	n=96
t=0.5	0.280	0.27881	0.278742	0.278739
t=0.8889	0.720	0.71778	0.717624	0.717619
t=0.9920	0.838	0.83937	0.839382	0.839382
t=1	0.838	0.83426	0.833872	0.833819

Finally, we have tested method (1.15) in the case when the stability conditions of Sect. 3 are violated. Then the condition numbers are larger, but still acceptable for n<100. However, the errors are much larger than in the case of stable methods. For instance, if $a(t)=1-4t(1-t)$, $b(t)=-1+3t(1-t)$, $y(t)=\cos(t)$ and $\rho_0=-1/4$, $\rho_1=1/4$, then the stability condition fails to be true at t=1/2. The approximate values of $z(t)$ are given in Table 6.

Table 6 (1.15)	n=12	n=24	n=48	n=96
t=0.5	5.594	3.129	1.875	1.949
t=0.8432	-3.517	0.434	-0.406	-0.406

If a, b, y are the same as in our first example and $\rho_0=0.4$, $\rho_1=-0.4$, $\alpha=5$, then the condition numbers of (1.15) slowly turn to infinity. The approximate values $z_n(t)$ seem to converge to $f(t) := t^{-0.65}(1-t)^{0.65}$ (cf. Table 7). However, though $g=\rho f$ is a solution of (0.1), it is not the one sought in the class $\{\rho z, z\in L^2\}$.

Table 7 (1.15)	n=12	n=24	n=48	n=96	$f(t)$
t=0.5	0.676	0.99720	0.999987	0.9999996	1.0
t=0.8432	0.529	0.33415	0.335024	0.3350301	0.3350301
t=0.9959	0.228	0.02824	0.028144	0.0281421	0.0281421

REFERENCES

1. Baker, C.T.H.: The numerical treatment of integral equations. Clarendon press, Oxford, 1978.

2. Chandler, G.A. and I.G.Graham: High order methods for linear functionals of solutions of second kind integral equations. SIAM Journal Numer.Anal. (to appear).

3. Elliott, D.: The approximate solution of singular integral equations. in: Solution methods for integral equations, ed. by M.A.Golberg, Plenum Press, New-York, London, 1979, pp. 83-107.

4. Elschner, J.: Asymptotics of solutions to pseudodifferential equations of Mellin type. Math.Nachr. 130 (1987), 267-305.

5. Elschner, J.: Galerkin methods with splines for singular integral equations over (0,1). Numer.Math. 43 (1984), 256-281.

6. Elschner, J.: On spline approximation for singular integral equations on an interval. Math.Nachr. 139 (1988) (to appear).

7. Elschner, J. and G.Schmidt: On spline interpolation in periodic Sobolev spaces. Preprint P-MATH 01/83, Inst. f.Math., AdW der DDR, Berlin, 1983.

8. Erdogan,F. and G.D.Gupta:On the numerical solution of singular integral equations. Quart.Appl.Math., 30 (1972), 552-534.

9. Erdogan,F., G.D.Gupta and T.S.Cook:Numerical solution of singular integral equations. in:Mechan. fracture I.Methods of analysis and solutions of crack problems.ed.by G.C.Sih. Leyden Nordhooff Int.Publ.Co., 1973.

10. Fabrikant,V.,S.V.Hoa and T.S.Sankar: On the approximate solution of singular integral equations. Computer Methods in Applied Mechanics and Engineering 29 (1981),19-33.

11. Gerasoulis,A.:The use of piecewise quadratic polynomials for the solution of singular integral equations of Cauchy type. Comp. & Maths. with Appl. 8,No.1 (1982), 15-22.

12. Гохберг,И.Ц. и Н.Я.Крупник: Об алгебре, порожденной теплицевыми матрицами. Функциональный анализ и его прил. 3,вып.2 (1969),46-56 (english translation in Funct.Anal.Appl.3).

13. Jen,E. and R.P.Srivastav:Cubic splines and approximate solution of singular integral equations. Math.of Comp. 37, No.156 (1981), 417-423.

14. Junghanns,P. and B.Silbermann:Zur Theorie der Näherungsverfahren für singuläre Integralgleichungen auf Intervallen. Math.Nachr. 103 (1981), 199-244.

15. Krenk,S.:Polynomial solution to singular integral equations, with applications to elasticity theory. Thesis, Technical University of Denmark 1981.

16. Лифанов,И.К. и Я.Е.Полонский: Обоснование численного метода дискретных вихрей решения сингулярных интегральных уравнений. ПММ, 39, вып.4 (1975), 742-746.

17. Micke,A.:Die Auflösung schwach singulärer Fredholmscher Integralgleichungen mit angepassten Quadraturen. ZAMM ,69 (1989).

18. Prössdorf,S. and A.Rathsfeld: A spline collocation method for singular integral equations with piecewise continuous coefficients. Int. Equ.Op.Theory 7 (1984), 337-352.

19. Prössdorf,S. and A.Rathsfeld:Mellin techniques in the numerical analysis for one-dimensional singular integral equations. Report R-MATH-06/88, Akad.d. Wiss. d.DDR, Karl-Weierstrass-Inst.f.Math.,Berlin,1988.

20. Prössdorf,S. and B.Silbermann:Numerical analysis for integral and operator equations . Akademie- Verlag, Berlin and Birkhäuser Verlag, Basel,Stuttgart (to appear).

21. Rathsfeld, A.: Quadraturformelmethode für Mellinope-operatoren nullter Ordnung. Math. Nachr. 137 (1988), 321-354.

22. Roch, S.: Lokale Theorie des Reduktionsverfahrens für singuläre Integraloperatoren mit Carlemannscher Verschiebung. Dissertation, TU Karl-Marx-Stadt, 1987.

23. Thomas, K.S.: Galerkin methods for singular integral equations. Math. of Comp. 36, 153 (1981), 193-205.

24. Triebel, H.: Interpolation Theory- Function Spaces- Differential Operators. VEB Deutscher Verlag der Wissenschaften, Berlin 1978.

S. Prössdorf, A. Rathsfeld
Karl-Weierstrass-Institut
für Mathematik der AdW der DDR
Mohrenstrasse 39
Berlin - 1086, DDR

Operator Theory:
Advances and Applications, Vol. 41

GENERAL WIENER-HOPF OPERATORS AND REPRESENTATION OF THEIR GENERALIZED INVERSES

A. F. dos Santos

Dedicated to Israel Gohberg on the occasion of his 60th birthday

The relationship between the generalized invertibility of a general Wiener-Hopf operator $P_2A|_{\text{Im}\,P_1}$ acting from a Banach space X into a Banach space Y (P_1, P_2 are projections on X,Y, respectively, and $A : X \to Y$ is bounded invertible) and the existence of a decomposition of the space as a direct sum of certain subspaces of Y related to Im (AP_1) and Ker P_2 is examined. The explicit calculation of the projection associated with this decomposition is studied. An example corresponding to a singular integral equation on a finite interval is given.

1 INTRODUCTION

In this paper we study invertibility conditions for a general (or abstract) Wiener-Hopf operator i.e. an operator of the form

$$T_A = P_2A|_{\text{Im}\,P_1} \tag{1.1}$$

where A is a bounded invertible operator from a Banach space X onto a Banach space Y and P_1 and P_2 are bounded projections on X and Y, respectively. Operators of the form (1.1) appear e.g. in the study of systems of Wiener-Hopf equations and convolution equations on a finite interval both of second kind ($Y = X$) and first kind ($Y \neq X$).

Except when A is a convolution operator with a scalar-valued or rational symbol and P_1 , P_2 are projections onto subspaces of functions with support in $\mathbb{R}^+$, it is very difficult to obtain useful invertibility conditions for the operator T_A and even more difficult to obtain explicit formulas for the inverse or generalized inverse when they exist.

The aim of this paper is to study conditions for T_A to have an inverse (or generalized inverse) giving explicit formulas for the inverse in terms of the projection associated with a certain direct sum decomposition of Y. The basic results are derived in section 2 for operators acting between Banach spaces. The question of existence and calculation of the above-mentioned projection is examined in section 3 in a Hilbert space context. A formula for the projection in terms of a system of biorthogonal sequences is given. In section 4 the results of the first sections are applied to the study of a singular integral equation on a finite interval in weighted L_2 spaces.

To the author's knowledge the main results presented in this paper are new. However, some of the ideas behind these results have their roots in the works of Shinbrot and Devinatz [5], [9] and Bart, Gohberg and Kaashoek [1], [2]. Generalized invertibility of abstract Wiener-Hopf operators has been previously studied by Speck [10], [11] using a different approach.

2 INVERTIBILITY AND GENERALIZED INVERTIBILITY

In all that follows X, Y denote Banach spaces, P_1 and P_2 are continuous projections on X and Y, respectively, and $A : X \to Y$ is a bounded invertible operator.

We begin our study of the abstract Wiener-Hopf operator T_A with a theorem which proves useful in the investigation of invertibility and generalized invertibility.

THEOREM 2.1 *Let T_A be the operator defined by (1.1). Then*

(1) $\dim \operatorname{Ker} T_A = \dim(\operatorname{Im}(AP_1) \cap \operatorname{Ker} P_2)$
$\operatorname{codim} \operatorname{Im} T_A = \operatorname{codim}(\operatorname{Im}(AP_1) + \operatorname{Ker} P_2)$.

(2) T_A is invertible if and only if

$$Y = \operatorname{Im}(AP_1) \oplus \operatorname{Ker} P_2 \tag{2.1}$$

If (2.1) holds the inverse of T_A is given by

$$T_A^{-1} : \operatorname{Im} P_2 \to \operatorname{Im} P_1 \,, T_A^{-1} f = A^{-1}\Pi f \tag{2.2}$$

where Π is the projection of Y onto $\operatorname{Im}(AP_1)$ *along* $\operatorname{Ker} P_2$.

PROOF: Proposition (1) and the first part of proposition (2) are simple generalizations of known results for the case $Y = X$, $P_2 = P_1$ (see. e.g. [4], [5]); so we omit the proof. As for (2.2) we note that if (2.1) holds it suffices to show that for any $f \in \operatorname{Im} P_2$ we have $T_A(A^{-1}\Pi f) = f$. Indeed

$$P_2 A(A^{-1}\Pi f) = f - P_2(I - \Pi) f = f \,. \qquad \square$$

Suppose now that condition (2.1) is not satisfied but the spaces $\mathcal{N} = \mathrm{Im}\,(AP_1)$ $\mathrm{Ker}\,P_2$ and $\mathrm{Im}\,(AP_1) + \mathrm{Ker}\,P_2$ have closed complements in $\mathrm{Ker}\,P_2$ and Y, respectively. Let Y_2^- and $\mathcal{M}$ be these complements i.e.

$$\mathrm{Ker}\,P_2 = \mathcal{N} \oplus Y_2^- \tag{2.3}$$

$$Y = (\mathrm{Im}\,(AP_1) + \mathrm{Ker}\,P_2) \oplus \mathcal{M} \tag{2.4}$$

From (2.3) it follows that

$$\mathrm{Im}\,(AP_1) + \mathrm{Ker}\,P_2 = \mathrm{Im}\,(AP_1) \quad) \tag{2.5}$$

Combining (2.4) and (2.5) we obtain

$$Y = \mathrm{Im}\,(AP_1) \oplus Y_2^- \oplus \mathcal{M} \tag{2.6}$$

REMARK From theorem 2.1 it follows that T_A *is Fredholm* if and only if

- $\dim(\mathrm{Im}\,(AP_1) \cap \mathrm{Ker}\,P_2) < \infty$, and
- $\mathrm{codim}(\mathrm{Im}\,(AP_1) + \mathrm{Ker}\,P_2) < \infty$.

Thus, in this case the complements Y_2^- and $\mathcal{M}$ always exist.

In the next theorem we show that if the space Y admits the decomposition (2.6) the abstract Wiener-Hopf operator (1.1) has a generalized inverse.

THEOREM 2.2 *Assume that the space Y admits the decomposition (2.6) and let Π be the projection of Y onto* $\mathrm{Im}\,(AP_1)$ *along $Y_2^- \oplus \mathcal{M}$. Then the operator* $T_A^\dagger : \mathrm{Im}\,P_2 \to \mathrm{Im}\,P_1$ *defined by*

$$T_A^\dagger f = A^{-1}\Pi f \tag{2.7}$$

is a generalized inverse of T_A i.e. $T_A T_A^\dagger T_A = T_A$.

PROOF: Let us check first that the operator $T_A^\dagger$ is well defined. Noting that $\Pi f \in \mathrm{Im}\,(AP_1)$ it is clear that $A^{-1}\Pi f \in \mathrm{Im}\,P_1$.

Now let φ be any element in $\mathrm{Im}\,P_1$ and compute

$$\begin{aligned} T_A T_A^\dagger T_A \varphi &= P_2 \Pi P_2 A\varphi \\ &= P_2 \Pi A\varphi - P_2\Pi(I - P_2)A\varphi \end{aligned} \tag{2.8}$$

In this equality the first term is equal to $T_A\varphi$ since $A\varphi \in \operatorname{Im}\Pi$. As to the second term write $(I - P_2)A\varphi = \psi_2^- + \psi_N$ where $\psi_2^- \in Y_2^-$ and $\psi_N \in \mathcal{N}$, according to (2.3). Hence

$$P_2\Pi(I - P_2)A\varphi = P_2\Pi\psi_2^- + P_2\psi_N = 0$$

since $\psi_2^- \in \operatorname{Ker}\Pi$ and $\psi_N \in \operatorname{Im}\Pi \cap \operatorname{Ker} P_2$. Hence from (2.8) it follows that $T_A T_A^\dagger T_A\varphi = T_A\varphi$, thus completing the proof. □

3 THE PROJECTION Π

Henceforth we assume that Y is a separable Hilbert space to be denoted by $\mathcal{H}$. This is done with a view to simplifying the calculation of the projection Π by using the orthogonality concept. As far as applications are concerned assuming Y to be a Hilbert space does not seem to be a severe restriction; for example in connection with boundary problems for partial differential equations Sobolev spaces related to $L_2(\mathbb{R})$ are often the natural spaces and these are Hilbert spaces.

To introduce the ideas that we wish to discuss consider first the decomposition (2.1) of the previous section for the case where $Y = X = \mathcal{H}$ and $P_1 = P_2 = P$ with P an orthogonal projection. Let $P^\times$ be another projection not necessarily orthogonal and assume at first that $P^\times = APA^{-1}$. Then we are concerned with the existence of the direct-sum decomposition: $\mathcal{H} = \operatorname{Im} P^\times \oplus \operatorname{Ker} P$. If this relation holds then, for any $f \in \mathcal{H}$ we have the unique decomposition

$$f = f^\times + f^- \tag{3.1}$$

where $f^- \in \operatorname{Ker} P$ and $f^\times = A\varphi$ for some $\varphi \in \operatorname{Im} P$. Assume initially that A is a positive operator; then since A is invertible it has a positive square root. Hence

$$A^{-1/2}f = A^{1/2}\varphi + A^{-1/2}f^-$$

But the subspace $\operatorname{Im}(A^{1/2}P)$ and $\operatorname{Im}\left(A^{-1/2}(I - P)\right)$ are orthogonal as can easily be checked. Letting Π_0 be the orthogonal projection onto $\operatorname{Im}(A^{1/2}P)$ we obtain

$$\varphi = A^{-1/2}\Pi_0 A^{-1/2}f$$

and thus (cf.(3.1))

$$f^\times = A^{1/2}\Pi_0 A^{-1/2}f = \Pi f$$

Now let (u_n) be an orthonormal sequence total in $\operatorname{Im}\Pi_0 = \operatorname{Im}(A^{1/2}P)$.Then, denoting by $(\cdot|\cdot)$ the inner product in $\mathcal{H}$, we have

$$\begin{aligned} \Pi f &= \sum_n (A^{-1/2} f | u_n) A^{1/2} u_n \\ &= \sum_n (f|\chi_n) A\chi_n \end{aligned} \tag{3.2}$$

where $\chi_n = A^{-1/2}u_n \in \operatorname{Im} P$. Note that $(A\chi_n|\chi_m) = \delta_{n,m}$ since the sequence u_n is orthonormal. Closely related to (3.2) is the formula for the solution of the equation $PA\varphi = f$ when A is positive,

$$\varphi = \sum_n (f|\chi_n)\chi_n$$

which was first derived by Shinbrot [9].

The preceding analysis gives a formula, (3.2), for the projection Π when A is a positive operator. The following theorem shows that by an appropriate modification of (3.2) one obtains a formula valid under much more general conditions.

THEOREM 3.1 *Let $P^\times$ and P be bounded projections on a separable Hilbert space $\mathcal{H}$ and assume that P is orthogonal. Then*

$$\mathcal{H} = \operatorname{Im} P^\times \oplus \operatorname{Ker} P \tag{3.3}$$

if and only if there exist sequences (ξ_n) and (χ_n) total in $\operatorname{Im} P^\times$ *and* $\operatorname{Im} P$, *respectively, constituting a biorthogonal system, i.e., satisfying*

$$(\xi_n|\chi_m) = \delta_{n,m} \tag{3.4}$$

and such that the series

$$\sum_{n=1}^{\infty} (f|\chi_n)\xi_n \tag{3.5}$$

is convergent for any $f \in \mathcal{H}$. If (3.3) holds, the projection Π of $\mathcal{H}$ onto $\operatorname{Im} P^\times$ *along* $\operatorname{Ker} P$ *is given by*

$$\Pi f = \sum_{n=1}^{\infty} (f|\chi_n)\xi_n \tag{3.6}$$

PROOF: (1) Assume first that the sequences (χ_n) and (ξ_n) satisfy the conditions of the theorem; we wish to prove that (3.6) defines a bounded projection on $\mathcal{H}$ and that this projection is associated with the decomposition (3.3). To begin with we note that the series (3.5) defines a linear operator (Π) on the whole space $\mathcal{H}$. Furthermore, by the principle of uniform boundedness (Cf. [7], corollary X.4.5) the operator Π is bounded. Let us now check that Π is a projection:

$$\Pi^2 f = \sum_{n=1}^{\infty}(f|\chi_n)\sum_{k=1}^{\infty}(\xi_n|\chi_k)\xi_n$$

and in view of (3.4) it follows that

$$\Pi^2 f = \sum_{n=1}^{\infty}(f|\chi_n)\xi_n = \Pi f$$

From (3.6), taking account that the sequence (ξ_n) is total in $\operatorname{Im} P^\times$, we have

$$\operatorname{Im}\Pi = \operatorname{Im} P^\times$$

On the other hand since (χ_n) is total in $\operatorname{Im} P$ and P is orthogonal, the condition $(f|\chi_n) = 0$ for all n is equivalent to $f \in \operatorname{Ker} P$. Hence

$$\operatorname{Ker}\Pi = \operatorname{Ker} P$$

and consequently

$$\mathcal{H} = \operatorname{Im}\Pi \oplus \operatorname{Ker}\Pi = \operatorname{Im} P^\times \oplus \operatorname{Ker} P$$

(2) Assume now that

$$\mathcal{H} = \operatorname{Im} P^\times \oplus \operatorname{Ker} P$$

and let Π be the projection of $\mathcal{H}$ onto $\operatorname{Im} P^\times$ along $\operatorname{Ker} P$. Let (χ_n) be an orthonormal basis for $\operatorname{Im} P$ and (χ_n^-) an orthonormal basis for $\operatorname{Ker} P$. Then for any f in $\mathcal{H}$ we have

$$f = \sum_n (f|\chi_n)\chi_n + \sum_n (f|\chi_n^-)\chi_n^-$$

and, applying Π to both sides of this equality, we obtain

$$\Pi f = \sum_n (f|\chi_n)\Pi\chi_n \tag{3.7}$$

i.e. all the elements in $\operatorname{Im}\Pi$ are given by (3.10). Hence

$$\overline{\operatorname{Span}(\Pi\chi_n)} = \operatorname{Im}\Pi$$

On the other hand

$$(\chi_n|\Pi\chi_m) = (\chi_n|\chi_m) + (\chi_n|(I-\Pi)\chi_m) = \delta_{n,m}$$

Thus the sequence (ξ_n) with $\xi_n = \Pi\chi_n$ is total in $\operatorname{Im} P^\times$ and biorthogonal to the sequence (χ_n) which is total in $\operatorname{Im} P$. Moreover the series (3.10) is convergent for any $f \in \mathcal{H}$. The proof is complete. □

REMARK The relation (3.4) shows that the sequences (χ_n) and (ξ_n) constitute a system of biorthogonal sequences. A number of properties of biorthogonal systems that may be useful in applications of theorem 3.1 can be found in [8].

In the next section we illustrate the application of this theorem to the study of a special convolution equation on a finite interval: the integral equation with Cauchy kernel on [-1,1].

4 APPLICATION: SINGULAR INTEGRAL EQUATION ON A FINITE INTERVAL

In this section we wish to apply the results of the preceding sections to the study of the singular integral equation with Cauchy kernel on a finite interval,

$$\frac{1}{\pi i}\int_{-1}^{1} \frac{\varphi(\xi)}{\xi - x} d\xi = f(x) \ , \quad -1 < x < 1 \tag{4.1}$$

in which the integral is understood in the sense of Cauchy principal value. To formulate completely our problem it remains to state the space in which equation (4.1) is to be considered. It is well known (cf. [12]) that the singular integral operator defined by the left-hand side of (4.1) is not Fredholm in $L_2(-1,1)$ so we rule out this space and consider instead a weighted L_2 space. More precisely we choose the space of Lebesgue measurable functions φ such that

$$\int_{-1}^{1} \left(1 - x^2\right)^{\alpha} |\varphi(x)|^2 dx < \infty \qquad (0 < \alpha \leq 1/2)$$

However to state our problem in the context of the theory of the preceding sections we have to consider the space of Lebesgue measurable functions defined on $\mathbb{R}$, such that

$$\int_{\mathbb{R}} w(x)|\varphi(x)|^2 dx < \infty \tag{4.2}$$

where $w(x) = |1 - x^2|^{\alpha}$ $(0 < \alpha \leq 1/2)$. By introducing the inner product

$$(\varphi|\psi) = \int_{\mathbb{R}} w(x)\varphi(x)\overline{\psi}(x)dx \tag{4.3}$$

this space becomes a Hilbert space which we shall denote by L_2^α. In this space equation (4.1) can be written in the form

$$PS\varphi = f \tag{4.4}$$

where P denotes the orthogonal projection onto to the subspace of L_2^α of functions that have support in [-1,1] and φ is sought in this subspace; S is the basic singular integral operator given by

$$(S\varphi)(x) = \frac{1}{\pi i}\int_{\mathbb{R}} \frac{\varphi(\xi)}{\xi - x} d\xi \tag{4.5}$$

which we have to prove to be invertible in L_2^α. To this end we note that every element in L_2^α can be expressed in the form

$$\varphi(x) = (x^2-1)^{-\alpha/2} g(x) \quad , \quad x \in \mathbb{R} \tag{4.6}$$

where $g \in L_2(\mathbb{R})$ and $(x^2-1)^{\alpha/2}$ is defined by

$$\begin{aligned} (x^2-1)^{\alpha/2} &= |x^2-1|^{\alpha/2} \quad , \quad x \notin [-1,1] \\ &= |x^2-1|^{\alpha/2} e^{i\pi\alpha/2} \quad , \quad x \in [-1,1] \end{aligned}$$

With this definition $(x^2-1)^{\alpha/2}$ becomes the restriction to $\mathbb{R}$ of a complex variable function analytic in $\mathbb{C} \setminus [-1,1]$.

The representation (4.6) makes it clear that every element in L_2^α can be decomposed as follows:

$$\varphi(x) = \varphi_1(x) + \varphi_2(x) \tag{4.7}$$

where

$$\varphi_k(x) = (x^2-1)^{-\alpha/2} g_k(x) \quad , \quad k = 1,2$$

with the g_k being the restriction to $\mathbb{R}$ of elements in one of the Hardy spaces $H_2^\pm$ of functions analytic in the upper or lower half planes with L_2 boundary limits ($g_1 \in H_2^+$, $g_2 \in H_2^-$). From (4.7) it is easy to verify that

$$S\varphi_1 = \varphi_1 \quad , \quad S\varphi_2 = -\varphi_2 \tag{4.8}$$

from which it follows that S is a bounded invertible operator on L_2^α. A further consequence of (4.8) is that $S^2 = I$ as in L_2.

Let us now go back to equation (4.4). To this equation we associate the general Wiener-Hopf operator defined by

$$T_S : \ P(L_2^\alpha) \to P(L_2^\alpha) \quad , \quad T_S\varphi = PS\varphi \tag{4.9}$$

where P is the above-defined projection.

To apply theorem (3.1) to the study of the operator T_S we proceed to examine the existence of sequences (ξ_n) and (χ_n) with the properties stated in the theorem.

To this end we have at our disposal, known results for the Hilbert transform of functions related to the Tchebyshef polynomials of first and second kinds [6]. Consider the functions

$$\begin{aligned} v_n(x) &= (1-x^2)^{-1/2}T_n(x) \quad , \quad x \in [-1,1] \\ &= 0 \quad , \quad x \notin [-1,1] \end{aligned} \tag{4.10}$$

where T_n is the Tchebyshef polynomial of first kind and order n $(n \geq 0)$. These functions are known to constitute a basis for $P(L_2^{1/2})$ [3], and since $L_2^{\alpha}(-1,1) \subset L_2^{1/2}(-1,1)$ for $\alpha \leq 1/2$ it follows that the sequence (v_n) is total in $L_2^{\alpha}(-1,1)$. Hence Sv_n is total in $\mathrm{Im}\,(SP)$; we put $\xi_n = Sv_n$. On the other hand, for $n \geq 1$, we have

$$(Sv_n)(x) = U_{n-1}(x) \quad , \quad x \in [-1,1] \tag{4.11}$$

where U_n is the Tchebyshef polynomial of the second kind and order n. Now, take for χ_n the functions given by

$$\begin{aligned} \chi_n(x) &= (1-x^2)^{\frac{1}{2}-\alpha}U_n(x) \quad x \in [-1,1] \\ &= 0 \quad x \notin [-1,1] \end{aligned}$$

Hence (cf.(4.3))

$$\begin{aligned} (\xi_n|\chi_m) &= \int_{-1}^{1}(1-x^2)^{1/2}U_{n-1}(x)U_m(x)dx \\ &= \delta_{n-1,m} \end{aligned} \tag{4.12}$$

where use was made of the orthogonality of the polynomials U_n with respect to the weight function $(1-x^2)^{1/2}$.

Relation (4.12) shows that the sequence (ξ_n), $n \geq 1$. is biorthogonal to the sequence (U_n) , $n \geq 0$, which is total in $\mathrm{Im}\,P$. On the other hand the series (3.5) converges for any $f \in L_2^{\alpha}$ since the sequence $(\xi)_{n=1}^{\infty}$ is biorthogonal to the sequence (ξ_n) which is a basis for L_2^{α} (Cf. theorem 1.1, Ch.VI of [8]).

For $n = 0$ in (4.11) we obtain $(Sv_0)(x) = 0$ for $x \in [-1,1]$ and $(Sv_0(x) \neq 0$ for $x \notin [-1.1]$. Hence $Sv_0 \in \mathrm{Ker}\,P$; thus from this and (4.12) we have

$$\overline{\mathrm{Span}(Sv_n)_{n=0}^{\infty}} \cap \mathrm{Ker}\,P = \mathrm{Span}(Sv_0)$$

from which it follows that T_S is not invertible (cf. theorem 2.1). Indeed from theorem 2.1 it follows that $\dim \mathrm{Ker}\,T_S = 1$ i.e. equation (4.4) is not uniquely solvable.

Let

$$X^- = (\mathrm{Im}\,P \oplus \mathrm{Span}(Sv_0))^{\perp}$$

and denote by $I - P_1$ the orthogonal projection onto X^-. Recall that the sequence $(\chi_n)_{n=0}^{\infty}$ is biorthogonal to the sequence $(Sv_n)_{n=1}^{\infty}$; to obtain a sequence total in $\operatorname{Im} P_1$ biorthogonal to $(Sv_n)_{n=0}^{\infty}$ we adjoin to (χ_n) a new element which we shall denote by χ_{-1} defined by

$$\chi_{-1} = \beta S v_0 + \sum_{n=0}^{\infty} \gamma_n \chi_n$$

The condition $(Sv_n | \chi_{-1}) = 0$ for $n \geq 1$ gives γ_n, explicitly; the normalization condition $(Sv_0 | \chi_{-1}) = 1$ gives β.

The existence of the sequences $(\xi_n)_{n=0}^{\infty}$ and $(\chi_n)_{n=-1}^{\infty}$ of theorem 3.1 has, thus, been established. Hence L_2^{α} admits the decomposition

$$L_2^{\alpha} = \operatorname{Im}(SP) \oplus X^-$$

which corresponds to (2.6) for $Y = L_2^{\alpha}$, $Y_2^- = X^-$ and $\mathcal{M} = (0)$ ($\mathcal{M} = (0)$ is equivalent to saying that (4.4) has a solution for all $f \in L_2^{\alpha}$).

Let Π be the projection of L_2^{α} onto $\operatorname{Im}(SP)$ along X^-; then

$$\Pi f = \sum_{n=0}^{\infty} (f | \chi_{n-1}) S v_n$$

Substitution of this result in formula (2.7) of theorem 2.2 gives us an expression for a generalized inverse of T_s,

$$T_S^{\dagger} : \operatorname{Im} P \longrightarrow \operatorname{Im} P\,, \quad T_S^{\dagger} f = \sum_{n=0}^{\infty} (f | \chi_{n-1}) v_n \tag{4.13}$$

The above results concerning the solvability of equation (4.1) in $L_2^{\alpha}(-1,1)$ are condensed in the following proposition.

PROPOSITION 4.1 *(1) Equation (4.1) has a solution* $\varphi \in L_2^{\alpha}(-1,1)$ $(0 < \alpha \leq \frac{1}{2})$ *for every* $f \in L_2^{\alpha}(-1,1)$.
(2) Every solution of (4.1) in $L_2^{\alpha}(-1,1)$ *has the representation*

$$\varphi = \gamma v_0 + T_S^{\dagger} f$$

where $\gamma \in \mathbb{C}$ *is arbitrary,* v_0 *is defined in (4.10) and* $T_s^{\dagger}$ *is given by (4.13).*

REFERENCES

[1] Bart,H. ,Gohberg,I., Kaashoek,M.A.: Fredholm theory of Wiener-Hopf equations in terms of realization of their symbols, Integral Equations and Operator Theory, 8(1985), 560-613.

[2] Bart,H. ,Gohberg,I. ,Kaashoek,M.A.: Wiener-Hopf equations with symbols analytic in a strip, Operator Theory: Advances and Applications Vol.21, 39-74, Birkhäuser Verlag, 1986.

[3] Blum, E.: Numerical Analysis and Computation. Addison-Wesley.1972,p.294.

[4] Bastos,M.A. ,dos Santos,A.F. ,Lebre,A.B.: On the Fredholm theory of Wiener-Hopf equations and the coupling method, Integral Equations and Operator Theory 11(1988),297-309.

[5] Devinatz,A. ,Shinbrot,H.: General Wiener-Hopf operators, Trans. A.M.S. 145(1969),467-494.

[6] Erdélyi,A.: Tables of integral transforms, Vol.II. McGraw-Hill, 1954.

[7] Gohberg, I., Goldberg, S.: Basic operator theory. Birkhäuser, 1980.

[8] Gohberg,I. ,Krein,M.G.: Theory of linear non-selfadjoint operators in Hilbert spaces, A.M.S. Monographs Vol. 18, 1969.

[9] Shinbrot,M.: On singular integral operators, J.Math.Mech. 13(1964), 395-406.

[10] Speck, F.-O.: On the generalized invertibility of Wiener-Hopf operators in Bannach spaces, Integral Equations and Operator Theory 6 (1983), 458-465.

[11] Speck, F.-O.: General Wiener-Hopf factorization methods, Pitman 1985.

[12] Widom,H.:Singular integral equations in L_p, Trans.A.M.S., 97(1960),131-160.

Departamento de Matemática
Instituto Superior Técnico, U.T.L.
Av. Rovisco Pais
1096 Lisboa Codex
Portugal
or

Centro de Análise e Processamento de Sinais
Av. Rovisco Pais
1000 Lisboa
Portugal

Operator Theory:
Advances and Applications, Vol. 41

EXPOSED POINTS IN H^1, I

Donald Sarason

To Israel Gohberg, in admiration of his pioneering and lasting contributions to operator theory.

The spaces $\mathcal{H}(b)$ of L. de Branges are used to study the exposed points of the unit ball of H^1.

1. INTRODUCTION

The space H^1 in the paper's title is the classical Hardy space in the unit disk, D. Although the exposed points of the unit ball of H^1 turn up in several connections, we still lack a structural description of them. The purpose of this paper is to introduce a new approach to their study which sheds some light on their structure and shows promise of leading further. The approach is based on the spaces $\mathcal{H}(b)$ of L. de Branges.

The paper is organized as follows. The next section is preliminary; it contains a collection of known facts about exposed points in H^1. Section 3 contains the definition of the spaces $\mathcal{H}(b)$, a review of their needed properties, and some lemmas to be used in the following section, where the spaces are applied to the study of exposed points. Section 5 contains the statement of a plausible conjecture that is partly established in Section 4. Another part will be established in a sequel to this paper; the full conjecture, however, is at present open. The concluding Section 6 explains the relation between exposed points and the Nehari interpolation problem.

The author thanks Eric Hayashi for helpful discussions.

2. EXPOSED POINTS IN H^1

A point in the unit ball of a Banach space is called an exposed point of the unit ball if there is a linear functional on the space that takes the value 1 at the point but has absolute value less than 1 elsewhere in the unit ball. Thus, an exposed point of the unit ball must be an extreme point. A theorem of K. de Leeuw and W. Rudin [8] states the extreme points of the unit ball of H^1 are the outer functions of unit norm. Each such function is thus the square of an outer function of unit norm in H^2. Throughout this paper, we let f denote an outer function in H^2 with $\|f\|_2 = 1$. In case f^2 is an exposed point of the unit ball of H^1, we shall say simply that f^2 is exposed.

A function in H^1 will be called rigid if no other functions in H^1, except for positive multiples of it, have the same argument as it almost everywhere on ∂D. A function in H^1 is rigid, for example, if its reciprocal also is in H^1; this follows from a theorem of J. Neuwirth and D. J. Newman [11] that says the only functions in $H^{1/2}$ that are positive almost everywhere on ∂D are constant. On the other hand, a nonconstant inner function is nonrigid, for if u is inner then u has the same argument on ∂D as $(1+u)^2$. More generally, then, a function in H^1 fails to be rigid if it is divisible in H^1 by $(1+u)^2$ for some nonconstant inner function u.

The following characterizations of the exposed points in H^1 can be found in [4].

PROPOSITION. *The following conditions are equivalent.*

(i) f^2 *is exposed.*

(ii) f^2 *is rigid.*

(iii) *The Toeplitz operator* $T_{\bar{f}/f}$ *has a trivial kernel.*

(iv) *The space* $H^2(|f|^2 d\theta)$ *contains no nonconstant real-valued functions.*

The notations appearing in conditions (iii) and (iv) are standard. Namely, if h is a function in L^∞, then T_h, the Toeplitz operator with symbol h, is by definition the operator on H^2 of multiplication by h followed by orthogonal projection onto H^2. If μ is a positive measure on ∂D, then $H^2(\mu)$ is, by definition, the closure of the polynomials in $L^2(\mu)$. Condition (iv) says that the stationary Gaussian process whose spectral measure is $|f|^2 d\theta$ is completely nondeterministic, that is, its past and future have a trivial intersection [4]. From this condition one sees that if f^2 is exposed then so is every outer function in H^1 that is bounded in modulus from below by some positive multiple of $|f|^2$. This suggests that a structural characterization of when f^2 is exposed should involve the affinity of $|f|^2$

to 0.

3. THE SPACES $\mathcal{H}(b)$

Let b be a nonconstant function in the unit ball of H^∞. The Hilbert space $\mathcal{H}(b)$ is, as a vector space, the range of the operator $(1 - T_bT_b^*)^{1/2}$, but with the norm that makes the preceding operator into a coisometry. The norm in $\mathcal{H}(b)$ will be denoted by $\|\cdot\|_b$. Thus, for example, if the function h of H^2 is orthogonal to $\ker(1 - T_bT_b^*)$, then $\|(1 - T_bT_b^*)^{1/2}h\|_b = \|h\|_2$. A companion space, $\mathcal{M}(b)$, is as a vector space the range of the operator T_b, with the norm that makes T_b an isometry. In the special case where b is an inner function, $\mathcal{M}(b)$ and $\mathcal{H}(b)$ are ordinary subspaces of H^2, orthogonal complements of each other.

The book [**5**] and the paper [**12**] contain the basic properties of the spaces $\mathcal{H}(b)$, and any property mentioned below without a specific reference can be found in one or the other or (most likely) both of these places. For w a point of D, we let k_w and k_w^b denote the kernel functions in H^2 and $\mathcal{H}(b)$, respectively, for evaluation at w. It is well known that $k_w(z) = (1 - \overline{w}z)^{-1}$ and easily shown that $k_w^b = (1 - \overline{b(w)}b)k_w$.

With b we associate the function $F_b = \frac{1+b}{1-b}$, which has a positive real part in D and thus has a Herglotz integral representation:

$$F_b(z) = \int \frac{e^{i\theta} + z}{e^{i\theta} - z} d\mu_b(e^{i\theta}) + ic,$$

where c is a real constant and μ_b is a positive Borel measure on ∂D, uniquely determined by F_b. The first lemma establishes a useful correspondence between $H^2(\mu_b)$ and $\mathcal{H}(b)$.

For g in $H^2(\mu_b)$, we define the function V_bg in D by

$$(V_bg)(z) = (1 - b(z))\langle g, k_z\rangle_{\mu_b},$$

where $\langle .,.\rangle_{\mu_b}$ denotes the inner product in $L^2(\mu_b)$.

LEMMA 1. *The map V_b is an isometry of $H^2(\mu_b)$ onto $\mathcal{H}(b)$.*

For the case where b is an inner function, the lemma is due to D. N. Clark [**3**]. A vector-valued version containing the one above can be found, essentially, in a paper of D. Alpay and H. Dym [**2**, Lemma 6.4]. The lemma is a simple consequence of the formula

$$(1) \qquad \langle k_w, k_z\rangle_{\mu_b} = (1 - b(z))^{-1}(1 - \overline{b(w)})^{-1}k_w^b(z).$$

In fact, (1) says that the image of k_w under V_b is $(1-\overline{b(w)})^{-1}k_w^b$, and it also guarantees that the inner product of k_w and k_z in $L^2(\mu_b)$ equals the inner product of their images in $\mathcal{H}(b)$. Thus V_b maps the linear hull in $L^2(\mu_b)$ of the functions k_w isometrically onto the linear hull in $\mathcal{H}(b)$ of the functions k_w^b. Since the functions k_w span $H^2(\mu_b)$ and the functions k_w^b span $\mathcal{H}(b)$, a straightforward limit argument completes the proof.

To establish (1) it suffices to rewrite the left side as

$$\frac{1}{1-\overline{w}z}\int \frac{1}{2}\left[\frac{\overline{e^{i\theta}}+\overline{w}}{\overline{e^{i\theta}}-\overline{w}}+\frac{e^{i\theta}+z}{e^{i\theta}-z}\right]d\mu_b(e^{i\theta}),$$

which equals

$$\frac{F_b(z)+\overline{F_b(w)}}{2(1-\overline{w}z)},$$

an expression that one can easily reduce to the right side of (1).

The properties of $\mathcal{H}(b)$ are crucially influenced by whether b is or is not an extreme point of the unit ball of H^∞, that is, by whether $\log(1-|b|^2)$ is not or is integrable on ∂D [**7**, p. 138]. This state of affairs is hinted by Lemma 1, for, due to a well-known theorem of G. Szegö, A. Kolmogorov and M. G. Krein [**7**, p. 49], the equality $H^2(\mu_b)=L^2(\mu_b)$ holds if and only if b is an extreme point. Of paramount interest below will be the case where b is not an extreme point, and where also $b(0)=0$. In that case we let a denote the outer function whose modulus on ∂D equals $(1-|b|^2)^{1/2}$ and whose value at 0 is positive. For λ in ∂D we let f_λ denote the outer function $a/(1-\lambda b)$.

LEMMA 2 (cf. [**9**]). *Under the stated conditions on b, the functions f_λ are in the unit ball of H^2, and $\|f_\lambda\|_2=1$ for almost all λ.*

In fact, inside D one has $\operatorname{Re} F_{\lambda b}=(1-|b|^2)/|1-\lambda b|^2$. Taking radial limits, one concludes that the Radon–Nikodym derivative of $\mu_{\lambda b}$ with respect to normalized Lebesgue measure on ∂D equals $|f_\lambda|^2$. The hypothesis $b(0)=0$ implies that $\mu_{\lambda b}$ is a probability measure, so

$$\frac{1}{2\pi}\int_{-\pi}^{\pi}|f_\lambda(e^{i\theta})|^2d\theta\le\|\mu_{\lambda b}\|=1,$$

showing that f_λ is in the unit ball of H^2. To prove that $\|f_\lambda\|_2=1$ for almost all λ, it suffices to show that

$$\frac{1}{2\pi}\int_{-\pi}^{\pi}\|f_{e^{it}}\|_2^2dt=1.$$

One can do that by writing out the integral expression for $\|f_{e^{it}}\|_2^2$ and applying Fubini's theorem to the resulting double integral. After the order of integration has been changed, almost all the inside integrals have the value 1.

LEMMA 3. *If b is not an extreme point and $b(0)=0$ then, for each λ in ∂D, the operator $T_{1-\lambda b}T_{\overline{f_\lambda}}$ is an isometry of H^2 into $\mathcal{H}(b)$. The range of this isometry is all of $\mathcal{H}(b)$ if and only if $\|f_\lambda\|_2 = 1$.*

Since f_λ is not necessarily bounded, the sense in which unbounded Toeplitz operators are to be interpreted needs clarification. For any function h in L^2 of the circle, we understand by T_h the operator on H^2 that maps the function g to the Cauchy integral of the function hg:

$$(T_h g)(z) = \frac{1}{2\pi}\int_{-\pi}^{\pi} h(e^{i\theta})g(e^{i\theta})\overline{k_z(e^{i\theta})}d\theta \qquad (|z|<1).$$

If h is bounded, this reduces to the usual definition. Note that, in the unbounded case, it still makes perfectly good sense to multiply T_h from the right by any bounded operator on H^2 and from the left by any analytic Toeplitz operator.

To prove Lemma 3 we note that the operator of multiplication by $1/f_\lambda$ is an isometry of H^2 onto $H^2\left(\frac{1}{2\pi}|f_\lambda|^2 d\theta\right)$. As mentioned in the proof of Lemma 2, the measure $\frac{1}{2\pi}|f_\lambda|^2 d\theta$ is the absolutely continuous component of $\mu_{\lambda b}$. By the theorem of Szegö–Kolmogorov–Krein [**7**, p. 49], the space $H^2(\mu_{\lambda b})$ is the direct sum of $H^2\left(\frac{1}{2\pi}|f_\lambda|^2 d\theta\right)$ with L^2 of the singular component of $\mu_{\lambda b}$; the second summand is trivial if and only if $\|f_\lambda\|_2 = 1$. Hence, multiplication by $1/f_\lambda$ defines an isometry of H^2 into $H^2(\mu_{\lambda b})$, and the isometry is onto if and only if $\|f_\lambda\|_2 = 1$. Following the last isometry by the isometry $V_{\lambda b}$ of Lemma 1, one obtains an isometry of H^2 into $\mathcal{H}(b)$; it is onto if and only if $\|f_\lambda\|_2 = 1$. The latter map is easily seen to coincide with $T_{1-\lambda b}T_{\overline{f_\lambda}}$.

Notice that Lemma 3 implies the operator $T_{1-\lambda b}T_{\overline{f_\lambda}}$ is a contraction of H^2 into itself, even when f_λ is unbounded.

The unilateral shift operator on H^2 will be denoted by $S : ((Sg)(z) = zg(z))$. It is known that $\mathcal{H}(b)$ is invariant under S^* and even under $T_{\overline{h}}$ for any h in H^∞. The restriction $S^* \mid \mathcal{H}(b)$ will be denoted by X.

LEMMA 4. *If b is not an extreme point, then $T_{\overline{u}}\mathcal{H}(b)$ is dense in $\mathcal{H}(b)$ for any inner function u.*

In fact, assume $T_{\overline{u}}\mathcal{H}(b)$ is not dense in $\mathcal{H}(b)$. The closure of $T_{\overline{u}}\mathcal{H}(b)$ is then a proper X-invariant subspace of $\mathcal{H}(b)$, so, by a result in [**13**], it equals $\mathcal{H}(b) \cap \mathcal{H}(v)$ for some inner function v. Then $\overline{u}\mathcal{H}(b)$ is orthogonal, relative to the inner product in L^2, to vH^2, so that $\mathcal{H}(b)$ itself is orthogonal to uvH^2. This is a contradiction because $\mathcal{H}(b)$ is the range of $(1 - T_b T_b^*)^{1/2}$, a positive operator with a trivial kernel and hence a dense range.

Finally, a technical result on Toeplitz operators is needed.

LEMMA 5. *If $\overline{h}_1$ is in H^2 and h_2 is in L^∞, then $T_{h_1}T_{h_2} = T_{h_1 h_2}$.*

This is a standard property of Toeplitz operators in case h_1 is bounded. In our case it is enough to show that $T_{h_1}T_{h_2}p = T_{h_1 h_2}p$ for all polynomials p; once that is done, the full equality can be obtained from a straightforward limit argument. Thus, we only need to verify that $\langle T_{h_1}T_{h_2}p, q\rangle = \langle T_{h_1 h_2}p, q\rangle$ for all polynomials p and q. Letting P denote the orthogonal projection fo L^2 onto H^2, we can rewrite the left side here as $\langle P(h_2 p), \overline{h}_1 q\rangle$ and the right side as $\langle h_2 p, \overline{h}_1 q\rangle$. The desired equality is now obvious.

4. RESULTS ON EXPOSED POINTS

We return to our outer function f having unit norm in H^2. To align the discussion to follow with the notational conventions of the preceding section, we assume also that f takes a positive value at the origin, an obviously inconsequential restriction. Let F be the Herglotz integral of $|f|^2$:

$$F(z) = \frac{1}{2\pi}\int_{-\pi}^{\pi} \frac{e^{i\theta} + z}{e^{i\theta} - z}|f(e^{i\theta})|^2 d\theta \qquad (|z| < 1).$$

Then, by defining $b = \frac{F-1}{F+1}$ and $a = \frac{2f}{F+1}$, we re-enter the setting of Section 3, namely, b and a are in the unit ball of H^∞, a is an outer function, $|a|^2 + |b|^2 = 1$ almost everywhere on ∂D (implying that b is not an extreme point), and $b(0) = 0$. Also $F = \frac{1+b}{1-b}$ and $f = a/(1-b)$, or, in the notation of Section 3, $F = F_b$ and $f = f_1$.

The basic connection between $\mathcal{H}(b)$ and exposed points is given by the first theorem.

THEOREM 1. *f^2 is exposed if and only if $\mathcal{M}(a)$ is dense in $\mathcal{H}(b)$.*

The statement of the theorem presupposes the inclusion $\mathcal{M}(a) \subset \mathcal{H}(b)$, which can be found in [**13**] but will also emerge from the following discussion. The theorem is a simple consequence of the preparatory work in Section 3. In view of condition (iii) in the proposition of Section 2, we need only to show that the denseness of $\mathcal{M}(a)$ in $\mathcal{H}(b)$ is equivalent to the denseness of the range of $T_{f/\bar{f}}$ in H^2. By Lemma 3, the operator $T_{1-b}T_{\bar{f}}$ is an isometry of H^2 onto $\mathcal{H}(b)$. But the product $T_{1-b}T_{\bar{f}}T_{f/\bar{f}}$ is exactly T_a; in fact, the equality $T_{\bar{f}}T_{f/\bar{f}} = T_f$ follows by Lemma 5, and the equality $T_{1-b}T_f = T_a$ is obvious. Thus $T_{1-b}T_{\bar{f}}$ maps the range of $T_{f/\bar{f}}$ onto $\mathcal{M}(a)$, and the theorem is established.

COROLLARY. *If f^2 is exposed then, for every λ in ∂D, $\|f_\lambda\|_2 = 1$ and f_λ^2 is exposed.*

In fact, suppose $\|f_\lambda\|_2 < 1$. Then, by Lemma 3, the operator $T_{1-\lambda b}T_{\bar{f}_\lambda}$ is an isometry of H^2 onto a proper subspace of $\mathcal{H}(b)$. But the range of $T_{1-\lambda b}T_{\bar{f}_\lambda}$ contains $\mathcal{M}(a)$, due to the equality $T_{1-\lambda b}T_{\bar{f}_\lambda}T_{f_\lambda/\bar{f}_\lambda} = T_a$ (explained in the proof of Theorem 1), and, in virtue of Theorem 1, the nondenseness of $\mathcal{M}(a)$ in $\mathcal{H}(b)$ implies f^2 is not exposed. Hence, if f^2 is exposed, we must have $\|f_\lambda\|_2 = 1$ for all λ in ∂D, and then it follows from Theorem 1 that f_λ^2 is exposed.

The second theorem extends the preceding corollary. For v an inner function we let f_v denote the outer function $a/(1 - vb)$. By Lemma 2, f_v is in the unit ball of H^2.

THEOREM 2. *If f^2 is exposed and v is any inner function, then $\|f_v\|_2 = 1$ and f_v^2 is exposed.*

In fact, from the equality

$$1 - T_bT_b^* = \left[T_{\bar{v}}(1 - T_{vb}T_{vb}^*)^{1/2}\right]\left[T_{\bar{v}}(1 - T_{vb}T_{vb}^*)^{1/2}\right]^*$$

one sees that $\mathcal{H}(b) = T_{\bar{v}}\mathcal{H}(vb)$ and, because $T_{\bar{v}}$ acts as a contraction in $\mathcal{H}(vb)$ [**13**], one sees also that $\mathcal{H}(b)$ is contained contractively in $\mathcal{H}(vb)$. By Lemma 4, $\mathcal{H}(b)$ is dense in $\mathcal{H}(vb)$. If f^2 is exposed then $\mathcal{M}(a)$ is dense in $\mathcal{H}(b)$, which, in view of the preceding remarks, implies that $\mathcal{M}(a)$ is dense in $\mathcal{H}(vb)$. If $\|f_v\|_2 = 1$ it then follows by Theorem 1 that f_v^2 is exposed. By Lemma 2 there is a λ in ∂D such that $\|f_{\lambda v}\|_2 = 1$, and the desired conclusions follow by the corollary to Theorem 1.

AN EXAMPLE. The example will show that the converse of Theorem 2 fails: it can happen that f^2 is not exposed, yet there is an inner function v such that $\|f_v\|_2 = 1$ and f_v^2 is exposed.

We let $b(z) = \frac{z(1-z)}{2}$, which gives $a(z) = \frac{1+z}{2}$. The corresponding function $f = a/(1 - b)$ equals $(1 + z)/(2 - z + z^2)$; it has unit norm in H^2 because the function

$F_b = \frac{1+b}{1-b}$ is bounded, guaranteeing that the measure μ_b has no singular component. Because f is divisible in H^2 by the function $1+z$, we see (as explained in Section 2) that f^2 is not exposed. (Incidentally, the function $f_{-1} = a/(1+b)$ turns out to equal $1/(2-z)$, which has H^2 norm $3^{-1/2}$.)

Let v be a Blaschke product with zero sequence $(r_n)_1^\infty$ lying in $(-1,0)$ and tending to -1. It is asserted that $\|f_v\|_2 = 1$ and that f_v^2 is exposed. In fact, the functions $(1+z)k_{r_n}$ are in $\mathcal{H}(zv) \cap \mathcal{M}(a)$, and they converge to the constant function 1 in H^2-norm as $n \to \infty$. Since $\mathcal{H}(zv)$ is contained isometrically in H^2, these functions also converge to 1 in $\mathcal{H}(zv)$-norm. But also $\mathcal{H}(zv)$ is contained isometrically in $\mathcal{H}(vb)$ (because the operator $(1 - T_{vb}T_{vb}^*)^{1/2}$ reduces to the identity on $\mathcal{H}(zv)$), so they converge to 1 in $\mathcal{H}(vb)$-norm, showing that 1 is in the $\mathcal{H}(vb)$-closure of $\mathcal{M}(a)$. Now the polynomials multiply $\mathcal{H}(vb)$ into itself and act as bounded operators in $\mathcal{H}(vb)$ [**13**]. Since the polynomials also multiply $\mathcal{M}(a)$ into itself, it follows that the $\mathcal{H}(vb)$-closure of $\mathcal{M}(a)$ contains the polynomials, and hence $\mathcal{M}(a)$ is dense in $\mathcal{H}(vb)$ [**13**]. Also, the range of the operator $T_{1-vb}T_{\bar{f}_v}$ contains $\mathcal{M}(a)$ (for reasons explained in the proof of Theorem 1), so that operator must map H^2 onto $\mathcal{H}(vb)$. We can now conclude by Lemma 3 and Theorem 1 that $\|f_{vb}\|_2 = 1$ and that f_{vb}^2 is exposed.

From condition (ii) in the proposition of Section 2 one sees that f^2 will be exposed if $a^2/\|a\|_2^2$ is. The example shows that the reverse implication fails. It raises the following question: If f^2 is not exposed, does there nevertheless always exist an inner function v such that $\|f_v\|_2 = 1$ and f_v^2 is exposed? An affirmative answer would say that one can infer nothing special about a from the knowledge that f^2 is exposed. The reasoning used in the example seems very special to the case at hand and inapplicable to the general question.

At the end of Section 2 it is suggested that whether or not f^2 is exposed depends in some way on how close $|f|^2$ gets to 0. In this connection it is natural to ask whether there is any H^p class to which $1/f$ must belong whenever f^2 is exposed. The example above shows that H^1 is not such a class, for in the example $|f_v(r_n)| = (1-|r_n|)/2$, and functions in H^1 are $o((1-|z|)^{-1})$ as $|z| \to 1$.

A modification of the example shows that there is no H^p class with the suggested property. Fix a positive integer m and set $a(z) = \left(\frac{1+z}{2}\right)^m$. Let b equal z times an outer function with modulus $(1 - |a|^2)^{1/2}$ on ∂D. Due to the smoothness of $|a|^2$ the function b is continuous in $\overline{D}$. Its modulus is less than 1 except at the point -1, and we can arrange to have $b(-1) \neq 1$. Then the function $f = a/(1 - b)$ has unit norm in H^2 by the same reasoning as was used in the original example. For v we take a Blaschke product with zeros r_n as above, but such that each zero is of multiplicity at least m. The previous reasoning goes through without change, provided we replace the functions $(1 + z)k_{r_n}$ used earlier by their m-th powers. The function f_v^2 is thus exposed, and as $|f_v(r_n)| = 2^{-m}(1 - |r_n|)^m$, the function $1/f_v$ is not in $H^{1/m}$.

In Section 6 a connection between exposed points and interpolation problems will be discussed. In the context of the Nevanlinna–Pick problem, A. Nicolau [private communication] has constructed an example that leads to an exposed point in H^1 whose reciprocal is in no H^p class. (It will appear in the Proceedings of the American Mathematical Society under the title "The coefficients of Nevanlinna's parameterization are not in H^p".)

5. A CONJECTURE

The notations of Section 4 are retained. It is conjectured that the following conditions are equivalent:

(i) *f^2 is exposed.*

(ii) *There is no nonconstant inner function u such that $(1 + u)^{-1} f$ is in H^2.*

(iii) *$\|f_\lambda\|_2 = 1$ for all λ in ∂D.*

The status of the conjecture is as follows. The implication (i) $\Rightarrow$ (ii) is true and near the surface, as was explained in Section 2. The implication (i) $\Rightarrow$ (iii) is given by the corollary to Theorem 1. The author has been able to establish the implication (ii) $\Rightarrow$ (iii); the proof will be given in a sequel to this paper. In addition, substantial progress has been made on the missing implication (iii) $\Rightarrow$ (i).

6. CONNECTION WITH THE NEHARI INTERPOLATION PROBLEM

In the Nehari problem one is given a function h_0 in L^∞ of the circle and one wants to find all functions h in the unit ball of L^∞ such that $h - h_0$ is in H^∞. The problem is named after Z. Nehari because he discovered that a solution exists if and only if the Hankel operator associated with h_0 has norm at most 1 [**10**]. A linear-fractional parameterization of the set of all solutions, in case more than one exist, was found by V. M. Adamjan, D. Z. Arov and M. G. Krein [**1**]. Their result can be stated as follows: If the Nehari problem for h_0 has a solution but is indeterminate, then the solution set is of the form

$$\left\{ \frac{a}{\overline{a}} \left(\frac{u - \overline{b}}{1 - bu} \right) : u \in \mathrm{ball}(H^\infty) \right\}, \tag{2}$$

where (i) a and b are functions in the unit ball of H^∞; (ii) a is outer and $a(0) > 0$; (iii) $b(0) = 0$; (iv) $|a|^2 + |b|^2 = 1$ almost everywhere on ∂D. The functions a and b are uniquely determined by the problem. The question now arises: Given a pair of functions a and b satisfying (i)–(iv), how can one recognize whether the set of functions (2) is the solution set for an indeterminate Nehari problem? The pair (a, b) will be called a Nehari pair if the set (2) is such a solution set.

The methods of Adamjan, Arov and Krein are operator theoretic. J. B. Garnett [**6**, pp. 157 ff.] has given a function-theoretic derivation of their parameterization from which one sees that the problem of characterizing Nehari pairs is equivalent to the problem of characterizing the exposed points in H^1. Garnett's analysis shows that if (a, b) is a Nehari pair, then the function $f = a/(1 - b)$ has unit norm in H^2 and f^2 is exposed. The converse can be deduced from the Adamjan–Arov–Krein parameterization. Namely, if a and b satisfy (i)–(iv) but (a, b) is not a Nehari pair, the parameterization yields a Nehari pair (a_0, b_0) and an inner function u such that

$$\frac{f}{\overline{f}} = \frac{a_0}{\overline{a}_0} \left(\frac{u - \overline{b}_0}{1 - b_0 u} \right).$$

The last equality can be rewritten as $f/\overline{f} = u f_0/\overline{f}_0$, where $f_0 = a_0/(1 - b_0)$. If u is a constant, λ, then, since $(a_0, \lambda b_0)$ is along with (a_0, b_0) a Nehari pair, it follows from the result of Garnett mentioned above that $f = c\eta a_0/(1 - \lambda b_0)$, where c is a positive constant and $\eta^2 = \lambda$. Since $f(0)$ and $a_0(0)$ are both positive, in fact $\eta = \lambda = 1$, so that $f = c f_0$. The equality $\|f\|_2 = 1$ (i.e., $c = 1$) implies $a = a_0$ and $b = b_0$, so necessarily $\|f\|_2 < 1$ in case u is constant. If u is nonconstant, on the other hand, the equality $\arg f^2 = \arg u f_0^2$ implies that f^2 is not exposed.

The spaces $\mathcal{H}(b)$ thus appear linked to interpolation problems, but it would be desirable to see the connection revealed at a deeper level. In the same vein, one can ask if there is a natural generalization of the Nehari problem whose solution sets encompass all sets of the form (2).

REFERENCES

1. V. M. Adamjan, D. Z. Arov and M. G. Krein, Infinite Hankel matrices and generalized problems of Carathéodory–Fejer and I. Schur. Funkcional. Anal. i Priložen 2 (1968), vyp. 4, 1–17.
2. D. Alpay and H. Dym, Hilbert spaces of analytic functions, inverse scattering, and operator models, I. Integral Equations and Operator Theory 7 (1984), 589–641.
3. D. N. Clark, One dimensional perturbations of restricted shifts. J. Analyse Math. 25 (1972), 169–191.
4. P. Bloomfield, N. P. Jewell and E. Hayashi, Characterizations of completely nondeterministic processes. Pacific J. Math. 107 (1983), 307–317.
5. L. de Branges and J. Rovnyak, "Square Summable Power Series." Holt, Rinehart and Winston, New York, 1966.
6. J. B. Garnett, "Bounded Analytic Functions." Academic Press, New York, 1981.
7. K. Hoffman, "Banach Spaces of Analytic Functions." Prentice–Hall, Englewood Cliffs, New Jersey, 1962.
8. K. de Leeuw and W. Rudin, Extreme points and extremum problems in H_1. Pacific J. Math. 8 (1958), 467–485.
9. Y. Nakamura, One-dimensional perturbations of isometries. Integral Equations and Operator Theory 9 (1986), 286–294.
10. Z. Nehari, On bounded bilinear forms. Ann. of Math. 65 (1957), 153–162.
11. J. Neuwirth and D. J. Newman, Positive $H^{1/2}$ functions are constants. Proc. Amer. Math. Soc. 18 (1967), 958.
12. D. Sarason, Shift-invariant spaces from the Brangesian point of view. The Bierberbach Conjecture – Proceedings of the Symposium on the Occasion

of the Proof, Mathematical Surveys, Amer. Math. Soc., Providence, 1986, pp. 153–166.

13. D. Sarason, Doubly shift-invariant spaces in H^2. J. Operator Theory 16 (1986), 75–97.

D. Sarason
Department of Mathematics
University of California
Berkeley, CA 94720

Operator Theory:
Advances and Applications, Vol. 41

GEOMETRICAL PROPERTIES OF A UNIT SPHERE OF THE OPERATOR SPACES IN L_p

Dedicated to professor Israel Gohberg on the occasion of his 60 th birthday

By E.M.Semenov and I.Ya.Shneiberg

Let S be an operator of orthogonal projecting on the set of constant functions on $[0,1]$:

$$Sx(t) = \int_0^1 x(s)\,ds\,\mathbb{1}$$

Denote by $B(L_q, L_p)$ the space of linear continuous operators acting from $L_q[0,1]$ into $L_p[0,1]$ and by $\mathcal{U}$ the set of $T \in B(L_\infty, L_1)$ such that $T\mathbb{1} = \mathbb{1}$ and $TH \subset H$ where

$$H = \left\{ x : \ x \in L_1, \ \int_0^1 x(s)\,ds = 0 \right\}$$

Instead of $\|T\|_{L_q \to L_p}$ we shall write $\|T\|_{q,p}$. The following theorem was announced in $[1]$.

THEOREM I . For any $1 < q \leq p < \infty$ there exists $\varepsilon = \varepsilon(q, p) > 0$ such that from the assumptions $T \in \mathcal{U}$ and

$$\max(\|T\|_{q,p}, \|T\|_{2,2}) \leq \varepsilon$$

the equality $\|S + T\|_{q,p} = 1$ follows.

The full proof of theorem 1 will appear in [2] . The paper of Ch.Fefferman - H.Shapiro [3] was a start point for this work. Actually it contained the theorem for the particular case $q = p$. In the present work theorem 1 is strengthened and developed in different directions.

For simplicity we shall consider a real spaces, although some results can be transfered on complex spaces without difficulties.

Denote by V_{qp} the set of such $T \in B(L_q, L_p)$ that $\|S + \varepsilon T\|_{q,p} = 1$ for all sufficienty small ε . It is easy to show that for each $T \in V_{qp}$ the equality $T\mathbb{1} = T^*\mathbb{1} = 0$ holds. It means that $V_{qp} \subset \mathcal{U}$ for any $q, p \in]1, \infty[$.

The next theorem shows that the assumption of theorem 1 about the boundedness of an operator T is not only sufficient but also necessary.

THEOREM 2. Suppose that $1 < q, p < \infty$. If $T \in B(L_p, L_q) \cap \mathcal{U}$ and $\|S + T\|_{q,p} = 1$ then $T \in B(L_2, L_2)$ and

$$\|T\|_{2,2} \leq \sqrt{\frac{q-1}{p-1}} \qquad (1)$$

PROOF. Denote by E_n the set of step-functions of the form

$$z(t) = \sum_{k=1}^{n} z_k \, \varkappa_{\left(\frac{k-1}{n}, \frac{k}{n}\right)}(t)$$

In [4], p 167 the following statement was proved. If $z \in E_n \cap H$ then

$$\| \mathbb{1} + z \|_q = 1 + \frac{q-1}{2} \|z\|_2^2 + o\left(\|z\|_2^2\right)$$

In particular,

$$\| \mathbb{1} + \lambda z \|_q = 1 + \frac{q-1}{2} \|z\|_2^2 |\lambda|^2 + o(\lambda^2) \qquad (2)$$

Denote by P the operator of orthogonal projecting on E_n :

$$P_n x(t) = n \sum_{k=1}^{n} \int_{\frac{k-1}{n}}^{\frac{k}{n}} x(s)\,ds \; \varkappa_{\left(\frac{k-1}{n}, \frac{k}{n}\right)}(t)$$

It is well known that $\|P_n\|_{r,r} = 1$ for each $r \in [1, \infty]$ Therefore $\|S + P_n T\|_{q,p} = 1$. Consequently, if $z \in E_n \cap H$ then

$$\| \mathbb{1} + \lambda P_n T z \|_p = \| (S + P_n T)(\mathbb{1} + \lambda z) \|_p \leq \| \mathbb{1} + \lambda z \|_q$$

Applying to both parts of this inequality the relation (2) we obtain

$$\frac{p-1}{2}\|P_n Tz\|_2^2 \leqslant \frac{q-1}{2}\|z\|_2^2$$

Hence,

$$\|P_n Tz\|_2 \leqslant \sqrt{\frac{q-1}{p-1}}\,\|z\|_2$$

and

$$\|P_n T(z+c)\|_2 = \|P_n Tz\|_2 \leqslant \sqrt{\frac{q-1}{p-1}}\,\|z+c\|_2$$

for each $c \in R^1$. As E_n is dense in L_2 and the operators P_n converge to the indentical operator, the last inequality implies the continuity of T in L_2 and the estimate

$$\|T\|_{2,2} \leqslant \sqrt{\frac{q-1}{p-1}}$$

COROLLARY. If $q \in \,]1,\infty[$, then $V_{q\infty} = \{0\}$ Indeed, let $T \in B(L_q, L_\infty)$ and

$$\|S+\varepsilon T\|_{q\infty} = 1$$

is true for some $\varepsilon \neq 0$. Then

$$\|S+\varepsilon T\|_{qp} = 1$$

for any $p \in [q,\infty[$ and according to (1)

$$\|\varepsilon T\|_{2,2} \leq \sqrt{\frac{q-1}{p-1}}$$

Passing to the limit we obtain $T=0$. By analogy we can show that $V_{1p}=\{0\}$ for $1<p<\infty$. Hence the assumptions in theorem 1 $q\neq 1$ and $p\neq\infty$ are essential.

The first part of theorem 2 and theorem 1 solves the problem when a continuous operator from L_q into L_p $(1<q\leq p<\infty)$ belonge to V_{qp}.

THEOREM 3. Suppose that $1<q\leq p<\infty$ and $T\in B(L_q, L_p)$. Then $T\in V_{qp}$ iff $T\in B(L_2,L_2)\cap$ $\cap\, \mathcal{U}$.

The Lorents and Marcinkiewicz spaces possess important extremal properties in the class of Banach functional spaces [5]. They naturally arise while studying some properties of L_p. Let us introduce the necessary definition. Suppose that $1<p<\infty$. By $L_{p,1}$ we denote the Lorentz space with norm

$$\|x\|_{L_{p,1}} = \frac{1}{p}\int_0^1 x^*(t)\, t^{\frac{1}{p}-1}\, dt$$

and by $L_{p,\infty}$ we denote the Marcinkiewicz space with the norm

$$\|x\|_{L_{p,\infty}} = \sup_{0<\tau\le 1} \frac{\int_0^\tau x^*(t)\,dt}{\tau^{1-\frac{1}{p}}}$$

where $x^*(t)$ is a decreasing rearrangement of $|x(t)|$.

LEMMA 1. Suppose that $1 < q_1 < p < q_2$ and $\frac{1}{p} = \frac{\alpha}{q_1} + \frac{1-\alpha}{q_2}$. A linear operator T is continuous from $L_{p,1}$ into L_p iff for some constant $C > 0$ and for each $x \in L_{q_2}$ the following inequality

$$\|Tx\|_p \le C \|x\|_{q_1}^{\alpha} \|x\|_{q_2}^{1-\alpha} \tag{3}$$

is valid.

PROOF. Necessity. Show that for some constant $C_1 = C_1(p, q_1, q_2)$ the following inequality

$$\|Z\|_{L_{p,1}} \le C_1 \|Z\|_{q_1}^{\alpha} \|Z\|_{q_2}^{1-\alpha} \tag{3'}$$

takes place for each $Z \in L_{q_2}$. To prove this inequality we use the method of real interpolation.

If (E_0, E_1) is a pair of Banach spaces then for each element $x \in E_0 + E_1$ and $t > 0$ we construct the functional

$$K(t,x) = K(t,x,E_0,E_1) = \inf_{\substack{u\in E_0,\, v\in E_1 \\ u+v=x}} \left(\|u\|_{E_0} + t\|v\|_{E_1}\right)$$

Then we construct the space $(E_0, E_1)_{\theta,p}$ with the norm

$$\|x\|_{(E_0,E_1)_{\theta,p}} = \left(\int_0^\infty \left(K(t,x)\,t^{-\theta} \right)^p \frac{dt}{t} \right)^{\frac{1}{p}}$$

where $0<\theta<1,\ 1\leq p\leq\infty$. Applying the theorems 3.4.2 and 5.3 from [6] we obtain

$$\|x\|_{(E_0,E_1)_{\theta,1}} \leq c_1 \|x\|^{1-\theta}_{E_0} \|x\|^{\theta}_{E_1}$$

and $(L_\infty, L_1)_{\theta,1} = L_{1/\theta,1}$. Hence we have (3'). So the first part of the lemma is established.

Sufficiency. Suppose that $e\subset[0,1],\ me>0$ and

$$y(t) = \begin{cases} 1, & t\in e \\ 0, & t\overline{\in} e \end{cases}$$

Then (3) may by rewritten in the following form

$$\|Ty\|_p \leq C(me)^{\frac{1}{p}}$$

By the corollary 1 of lemma 5.2 of chapter 2 [5] the boundedness of T from $L_{p,1}$ into L_p follows from this inequality. ■

Theorem 3 fails if $1<p<q<2$. More exaetly in this case the imbedding

$$B(L_q,L_p)\cap B(L_2,L_2)\cap\mathcal{U}\subset V_{qp}$$

is not true. A corresponding counterexample is constructed in [1]. Naturally there arises a question of finding additional conditions on the operator T in order that

$T \in B(L_q, L_p) \cap B(L_2, L_2) \cap \mathcal{U}$ should imply $T \in V_{qp}$. Starting to solve this problem we show first of all the validity of

LEMMA 2. Let $2 < p < q < \infty$ and $\frac{1}{p} = \frac{\alpha}{2} + \frac{1-\alpha}{q}$. For each $x \in L_q$ the inequality

$$\|x\|_2^2 + \|x\|_q^q \geq \|x\|_2^{\alpha p} \|x\|_q^{(1-\alpha)p} \tag{4}$$

is true.

PROOF. For any $a, b > 0$, $\theta \in [0, 1]$ we have

$$a + b \geq \max(a, b) \geq a^{1-\theta} b^{\theta}$$

Assuming $a = \|x\|_2^2$, $b = \|x\|_q^q$, $\theta = \frac{\alpha p}{2}$ we obtain (4).

In [3] the following interesting statement was proved. For any $q \in [2, \infty[$ there exists a constant $A = A(q)$ such that for each $Z \in H$

$$\frac{1}{A} \int_0^1 (|Z(t)|^2 + |Z(t)|^q)\, dt \leq \int_0^1 |\mathbb{1} + Z(t)|^q dt - 1 \leq$$

$$\leq A \int_0^1 (|Z(t)|^2 + |Z(t)|^q)\, dt \tag{5}$$

So for any $d \geq 1$ and $q \in [2, \infty[$ there exists a constant $K = K(q, d)$ possessing the following property: if $\|Z\|_q \leq d$, $Z \in H$ then

$$\|\mathbb{1} + Z\|_q \geq 1 + K(q, d)(\|Z\|_2^2 + \|Z\|_q^q) \tag{6}$$

As it was mentioned above the imbedding $V_{qp} \subset \mathcal{U} \cap B(L_2, L_2)$ is valid.

THEOREM 4. Suppose that $T \in \mathcal{U} \cap B(L_q, L_p) \cap B(L_2, L_2)$ and $1 < p < q < 2$ or $2 < p < q < \infty$. Then $T \in V_{qp}$ iff

$$T \in B(L_q, L_{q,\infty}), \quad 1 < p < q < 2$$

$$T \in B(L_{p,1}, L_p), \quad 2 < p < q < \infty$$

PROOF. The first case can be obtained from the second one with the help of standard duality argument. Therefore we shall consider only the case $2 < p < q < \infty$.

Necessity. By assumption we can find $\varepsilon \neq 0$ such that $\|S + \varepsilon T\|_{q,p} = 1$. Suppose that $z \in H \cap L_q$, $\|z\|_q \leq 1$ Applying (5) we obtain

$$\begin{aligned} \|\mathbb{1} + \varepsilon T z\|_p &= \|(S + \varepsilon T)(\mathbb{1} + z)\|_p \leq \|\mathbb{1} + z\|_q \leq \\ &\leq \left(1 + A(q)\left(\|z\|_2^2 + \|z\|_q^q\right)\right)^{\frac{1}{q}} \leq \\ &\leq 1 + \frac{A(q)}{q}\left(\|z\|_2^2 + \|z\|_q^q\right) \end{aligned} \tag{7}$$

As

$$|\varepsilon| \, \|Tz\|_p \leq |\varepsilon| \, \|T\|_{qp}$$

then by (6)

$$\|\mathbb{1} + \varepsilon T z\|_p \geq 1 + K(p, |\varepsilon| \|T\|_{qp}) \|\varepsilon T z\|_p^p \tag{8}$$

Comparing (7) and (8) we have

$$K(p, \varepsilon \|T\|_{q,p}) \|\varepsilon Tz\|_p^p \leqslant \frac{A(q)}{q} \left(\|z\|_2^2 + \|z\|_q^q \right)$$

and

$$\|Tz\|_p^p \leqslant \frac{A(q)}{q K(p, \varepsilon \|T\|_{q,p}) \varepsilon^p} \left(\|z\|_2^2 + \|z\|_q^q \right)$$

Let us choose an arbitrary element $x \in H$ and let us set $z = \lambda \cdot x$ where

$$\lambda = \|x\|_2^{\frac{2(q-p)}{q-2}} \|x\|_q^{\frac{q(p-2)}{q-2}}$$

Note that

$$\|z\|_q = \|\lambda x\|_q = \left(\frac{\|x\|_2}{\|x\|_q} \right)^{\frac{q(p-2)}{q-2}} \leqslant 1$$

Dividing inequality by λ^p we obtain

$$\|Tx\|_p^p \leqslant \frac{A(q)}{q K(p, \varepsilon \|T\|_{qp}) \varepsilon^p} \left(\|x\|_2^2 \lambda^{2-p} + \|x\|_q^q \lambda^{q-p} \right) =$$

$$= \frac{2A(q)}{q K(p, \varepsilon \|T\|_{qp}) \varepsilon^p} \|x\|_2^{\frac{2(q-p)}{q-2}} \|x\|_q^{\frac{q(p-2)}{q-2}}$$

Hence

$$\|Tx\|_p \leqslant \left(\frac{2A(q)}{q K(p, \varepsilon \|T\|_{qp}) \varepsilon^p} \right)^{\frac{1}{p}} \|x\|_2^{\alpha} \|x\|_q^{1-\alpha}$$

where $\frac{1}{p} = \frac{\alpha}{2} + \frac{1-\alpha}{q}$. For any $c \in R^1$, $x \in H$

$$T(x+c) = Tx, \ \|x\|_2 \leq \|x+c\|_2, \ \|x\|_q \leq 2\|x+c\|_q$$

Therefore the inequality

$$\|Tx\|_p \leq 2\left(\frac{2A(q)}{q\,K(p,\varepsilon\|T\|_{qp})\varepsilon^p}\right)^{\frac{1}{p}} \|x\|_2^{\alpha}\,\|x\|_q^{1-\alpha}$$

is true for each $x \in L_q$. From lemma 1 we have the following inclusion $T \in B(L_{p,1}, L_p)$.

Sufficiency. Suppose that $T \in \mathcal{U} \cap B(L_2, L_2) \cap B(L_{p,1}, L_p)$, $|\varepsilon| \leq 1$ and $Z \in H \cap L_q$. By (5)

$$\|(S+\varepsilon T)(\mathbb{1}+Z)\|_p = \|\mathbb{1}+\varepsilon TZ\|_p \leq (1 + A(p)(\varepsilon^2\|TZ\|_2^2 + \varepsilon^p\|TZ\|_p^p))^{\frac{1}{p}}$$

Applying lemmas 1 and 2 we obtain

$$\|TZ\|_p \leq C\|Z\|_2^{\alpha}\,\|Z\|_q^{1-\alpha}$$

here $\frac{1}{p} = \frac{\alpha}{2} + \frac{1-\alpha}{q}$ and

$$(\|Z\|_2^{\alpha}\,\|Z\|_q^{1-\alpha})^p \leq \|Z\|_2^2 + \|Z\|_q^q$$

Hence

$$\|(S+\varepsilon T)(\mathbb{1}+Z)\|_p \leq 1 + \frac{A(p)}{p}\varepsilon^2(\|T\|_{2,2}^2 + C^p)\cdot$$

$$\cdot(\|Z\|_2^2 + \|Z\|_q^q) = 1 + C_1\varepsilon^2(\|Z\|_2^2 + \|Z\|_q^q)$$

where

$$C_1 = \frac{A(P)}{P}\left(\|T\|_{2,2}^{2} + C^{P}\right)$$

Let us introduce a additional assumption $\|Z\|_q \leq 3$. If ε satisfies the inequality $C_1\varepsilon^2 \leq K(q,3)$ then according to (6)

$$1 + C_1\varepsilon^2\left(\|Z\|_2^2 + \|Z\|_q^q\right) \leq \|\mathbb{1} + Z\|_q$$

and

$$\|(S+\varepsilon T)(\mathbb{1}+Z)\|_p \leq \|\mathbb{1}+Z\|_q \tag{9}$$

Now let $\|Z\|_q \geq 3$. Then

$$\|(S+\varepsilon T)(\mathbb{1}+Z)\|_p = \|\mathbb{1}+\varepsilon Tz\|_p \leq 1 + |\varepsilon|\,\|T\|_{q,p}\,\|Z\|_q$$

If $|\varepsilon|\,\|T\|_{q,p} < \frac{1}{3}$ then

$$1 + |\varepsilon|\,\|T\|_{q,p}\,\|Z\|_q \leq 1 + \frac{1}{3}\|Z\|_q \leq \|Z\|_q - 1 \leq \|\mathbb{1}+Z\|_q$$

and we obtain once more the inequality (9). So if

$$|\varepsilon| \leq \min\left(\sqrt{\frac{K(q,3)}{C_1}},\ \frac{1}{3\|T\|_{q,p}}\right)$$

then (9) is true for each $Z \in H \cap L_q$. It means that $\|S+\varepsilon T\|_{q,p} = 1$. ■

REFERENCES

1. E.M.Semenov and I.Ya.Shneiberg: Hipercontractig operators and Khintchine's inequality. Funct. Anal. and its Appl., 22, N 3, 87-88 (1988) (Russian).

2. E.M.Semenov and I.Ya.Shneiberg: Contracting operator and Khintchine's inequality. Sib. Math. J. (to appear) (Russian).

3. C.Fefferman and H.Shapiro: A planar face on the unit sphere of the multiplier space M_p, $1 < p < \infty$. Pros. Amer. Math. Soc., 36, N 2, 435-439 (1972).

4. Y.Benjamini and Pei-Kee Lin: Norm one multipliers on $L^p(G)$ Arkiv for matematik, 24, N 2, 159-173 (1986).

5. S.G.Krein, Ju.I.Petunin and E.M.Semenov: Interpolation of Linear Operators. AMS, Providence, 1982.

6. J.Bergh and J.Lofstrom: Interpolation spaces. Springer, Berlin-Heidelberg-New York, 1976.

7. J.Diestel: Geometry of Banach Spaces. Springer, Berlin-Heidelberg-New York, 1975.

8. B.S.Kashin, A.A.Saakyan: Orthogonal series. Nauka, Moscow, 1984 (Russian).

Dep.of Mathematics Voronezh
State University, 394693, Voronezh, USSR

Voronezh Forest Tecnology Institute,
394043, Voronezh, USSR

Operator Theory:
Advances and Applications, Vol. 41

C^*-ALGEBRAS OF CRYSTAL GROUPS

Keith F. Taylor

A explicit description is given for the C^*-algebra of a group with an abelian subgroup of finite index. An example is given to illustrate the ease of the construction in any particular case. Finally an extension of the main theorem to the case where the subgroup of finite index is not necessarily abelian is given.

Let G be a locally compact group with an abelian normal subgroup A of finite index n in G. The symmetry group of any crystal is of this type. In this paper the C^*-algebra, $C^*(G)$, of G is described in detail.

In general, the structure of group C^*-algebras have been difficult to understand with detailed descriptions given in only a few isolated cases, see [1] or [3], for examples. If G splits as a semidirect product of a finite group D with A, then $C^*(G)$ is isomorphic to the cross-product C^*-algebra of D acting on $C_0(\hat{A})$. In this case Rieffel has shown that D acts as automorphisms of $M_n(\mathbb{C}) \otimes C_0(\hat{A})$ and $C^*(G)$ is isomorphic to the fixed point algebra for this finite group of automorphisms. Actually Rieffel's result, 4.3 of [2] is more general in that $C_0(\hat{A})$ can be any C^*-algebra on which a finite group is acting.

Even when G is not a semidirect product of A by a finite group, let $D = G/A$. The main result here is that there is an injective homomorphism of D into the automorphism group of $M_n(\mathbb{C}) \otimes C_0(\hat{A})$ such that $C^*(G)$ is isomorphic to the fixed point algebra. Furthermore, the method of proof provides a very easy technique for giving an extremely explicit description of the C^*-algebra of any particular group with an abelian subgroup of finite index.

Using a detailed analysis of the Mackey procedure, Raeburn in [1] gave explicit descriptions of the C^*-algebras of two particular groups of the kind considered here. Raeburn's paper was an original motivation for this work and the procedure given below reproduces his descriptions when applied to the two examples he considered.

All the necessary terminology and the main result are given in section 2. An example of a non-symmorphic two dimensional crystal group is given in section 3 and its C^*-algebra is easily constructed using the main results. Section 4 is an addendum essentially due to the referee who noticed that the arguments of the main theorem, when properly viewed, applied to the case where the subgroup of finite index is not abelian.

2. DESCRIPTION OF $C^*(G)$

For a general locally compact group G, $C^*(G)$ is the completion of $L^1(G)$ with respect to the C^*-norm determined by the set of all *-representations of $L^1(G)$. If G is an amenable group then this C^*-norm is already determined by the left regular representation λ^G of $L^1(G)$ on $L^2(G)$. For $f \in L^1(G)$ and $h \in L^2(G)$, $\lambda^G_f h = f * h$, where $f * h(x) = \int_G f(y)h(y^{-1}x)dy$, for almost every x in G. Then λ^G is a *-representation of $L^1(G)$ on $L^2(G)$. The reduced C^*-algebra of G is $C^*_\lambda(G)$, the norm closure of $\lambda^G[L^1(G)]$ in $\mathcal{B}(L^2(G))$. If the group G is amenable, then $C^*(G)$ is isomorphic to $C^*_\lambda(G)$.

The abelian case will be used to establish some notation and illustrate the above definition. Of course, abelian groups are amenable. Let A be an abelian locally compact group with Pontryagin dual $\hat{A}$. Fix a Haar measure on A. For $f \in L^1(A)$, define the Fourier transform of f, $\hat{f}$ on $\hat{A}$, by $\hat{f}(\chi) = \int_A f(a)\chi(a)da$, for all $\chi \in \hat{A}$. Then $\hat{f} \in C_0(\hat{A})$ and the Stone-Weierstrass theorem can be used to show that $\{\hat{f}: f \in L^1(A)\}$ is dense in $(C_0(\hat{A}), \|\cdot\|_\infty)$. Fix a Haar measure on $\hat{A}$ so that, for any $f \in L^1(G) \cap L^2(G)$, $\|\hat{f}\|_2 = \|f\|_2$. Let $\mathcal{P}$ be the unitary map of $L^2(A)$ onto $L^2(\hat{A})$ such that $\mathcal{P}(h) = \hat{h}$ for any $h \in L^1(A) \cap L^2(A)$. For any $f \in L^1(A)$ and $h \in L^2(A)$, $\mathcal{P}(f * h) = \hat{f}\mathcal{P}(h)$.

Consider the representation $\mathcal{M}$ of $C_0(\hat{A})$ into $\mathcal{B}(L^2(\hat{A}))$ defined, for $f \in C_0(\hat{A})$ and $h \in L^2(\hat{A})$, by $\mathcal{M}_f h = fh$. Then $\mathcal{M}$ is a C^*-isomorphism of $C_0(\hat{A})$ with a C^*-subalgebra of $\mathcal{B}(L^2(\hat{A}))$. From the above discussion we have that $\mathcal{P}\lambda^A_f\mathcal{P}^{-1} = \mathcal{M}_{\hat{f}}$ for each $f \in L^1(A)$. Then $C^*(A)$ is isomorphic to $\mathcal{M}(C_0(\hat{A}))$, the closure of $\{\mathcal{P}\lambda^A_f\mathcal{P}^{-1}: f \in L^1(A)\}$ in $\mathcal{B}(L^2(\hat{A}))$ and $f \to \mathcal{M}^{-1}(\mathcal{P}\lambda^A_f\mathcal{P}^{-1})$ extends to a C^*-isomorphism of $C^*(A)$ with $C_0(\hat{A})$.

Now suppose G is a locally compact group with A as an abelian normal subgroup of finite index n in G. All the above notation will be retained for A. Let $D = G/A$. This group G is amenable, so $C^*(G) = C^*_\lambda(G)$. For any coset d of A, the set of elements of $L^2(G)$ supported on d is a subspace isometric with $L^2(A)$ and $L^2(G)$ is the orthogonal direct sum of these subspaces. Thus there is a natural isometry Ψ of $L^2(G)$ with $\mathcal{H} = \sum_{d\in D} \oplus L^2(\hat{A})$. Then $C^*(G)$ is isomorphic with the closure of $\{\Psi\lambda_f^A\Psi^{-1} : f \in L^1(G)\}$ in $\mathcal{B}(\mathcal{H})$. It is very useful to make this explicit.

Let $\gamma: D \to G$ be a fixed cross-section of the cosets of A in G. It is no loss of generality to assume γ takes the identity in D to the identity in G. By definition a cross-section satisfies $\gamma(d) \in d$ for each $d \in D$ and each d acts on A by $d\cdot a = \gamma(d)a\gamma(d)^{-1}$ for all $a \in A$. For $b, c \in D$, $\gamma(b)\gamma(c) \in bc$ and thus there exists $\alpha(b,c) \in A$ such that $\gamma(b)\gamma(c) = \gamma(bc)\alpha(b,c)$. Then α is the 2-cocycle which determines the isomorphism class of G given D, A and the action of D on A. Up to isomorphism $G = \{(d,a): d \in D,\ a \in A\}$, with group product $(b,a)(c,a') = (bc, \alpha(b,c)(c^{-1}\cdot a)a')$. The cocycle identity satisfied by α is vital in making the calculations below: $\alpha(b,cd)\alpha(c,d) = \alpha(bc,d)(d^{-1}\cdot\alpha(b,c))$ for all $b,c,d, \in D$.

To get an expression for the unitary Ψ mentioned above, for $h \in L^2(G)$ and each $d \in D$, let $h_d \in L^2(A)$ be such that $h_d(a) = h(d,a)$, for almost all a in A. The elements $\underline{h}$ of $\mathcal{H} = \sum_{d\in D} \oplus L^2(\hat{A})$ will be considered as n-tuples $\underline{h} = (h_d)_{d\in D}$, where each $h_d \in L^2(\hat{A})$. Then, for $h \in L^2(G)$, $\Psi(h) = \big(\mathcal{P}(h_d)\big)_{d\in D}$.

Let $M_n(C_0(\hat{A}))$ denote the algebra of $n\times n$-matrices over $C_0(\hat{A})$. It is convenient to index the matrix entries by elements of D. Thus $F \in M_n(C_0(\hat{A}))$ can be written $F = (F_{b,c})_{b,c\in D}$, with $F_{b,c} \in C_0(\hat{A})$ for each $b,c \in D$. If $M_n(\mathbb{C})$ is given the operator norm, then it is a C^*-algebra and $M_n(C_0(\hat{A}))$ is naturally identified with $C_0(\hat{A}, M_n(\mathbb{C}))$. Thus, $\|F\| = \sup_{x\in\hat{A}} \|F(x)\|$ defines a norm on $M_n(C_0(\hat{A}))$ with respect to which it is a C^*-algebra. It is also often convenient to realize $M_n(C_0(\hat{A}))$ as $M_n(\mathbb{C}) \otimes C_0(\hat{A})$.

For $F = (F_{b,c})_{b,c\in D}$ in $M_n(C_0(\hat{A}))$ define $\mathcal{M}(F)$ in $\mathcal{B}(\mathcal{H})$ by for $\underline{h} = (h_c)_{c\in D}$, $(\mathcal{M}(F)\underline{h})_b = \sum_{c\in D} F_{b,c}h_c$, for each $b \in D$. Then $\mathcal{M}$ is a C^*-isomorphism of $M_n(C_0(\hat{A}))$ into $\mathcal{B}(\mathcal{H})$.

Proposition 1. *For each $f \in L^1(G)$, $\Psi\lambda_f^G\Psi^{-1}$ is in the range of $\mathcal{M}$. Let $\mathcal{F}(f) = \mathcal{M}^{-1}(\Psi\lambda_f^G\Psi^{-1})$. Then $\mathcal{F}$ extends to a C^*-isomorphism of $C^*(G)$ onto a C^*-*

subalgebra of $M_n(C_0(\hat{A}))$.

Proof. The last statement of the proposition follows from the first and the facts that G is amenable, Ψ is a unitary and $\mathcal{M}$ is a C^*-isomorphism onto its range. The first statement is established by computing the effect of $\Psi\lambda_f^G$ on $h \in L^2(G)$, for $f \in L^1(G)$.

As before for $d \in D$, $f_d \in L^1(A)$ and $h_d \in L^2(G)$ are such that $f_d(a) = f(d,a)$ and $h_d(a) = h(d,a)$ for almost all $a \in A$. Then using the cocycle identity where required,

$$\begin{aligned}(f*h)_d(a') &= f*h(d,a') = \int_G f(y)h(y^{-1}(d,a'))dy \\ &= \sum_{c\in D}\int_A f(c,a)h((c,a)^{-1}(d,a'))da \\ &= \sum_{c\in D}\int_A f_c(a)h(c^{-1}d,\alpha(c,c^{-1}d)^{-1}(d^{-1}c\cdot a^{-1})a')da \\ &= \sum_{c\in D}\int_A f_c(c^{-1}d\cdot a)h_{c^{-1}d}(\alpha(c,c^{-1}d)^{-1}a^{-1}a')da \\ &= \sum_{c\in D}\int_A f_c(c^{-1}d\cdot(\alpha(c,c^{-1}d)^{-1}a))h_{c^{-1}d}(a^{-1}a')da \\ &= \sum_{c\in D} g_{c,d} * h_{c^{-1}d}(a'),\end{aligned}$$

where $g_{c,d} \in L^1(A)$ is such that $g_{c,d}(a) = f_c(c^{-1}d\cdot(\alpha(c,c^{-1}d)^{-1}a))$ for almost all a in A. Therefore

$$\begin{aligned}(\Psi(f*h))_d &= \sum_{c\in D}\hat{g}_{c,d}\mathcal{P}(h_{c^{-1}d}) \\ &= \sum_{c\in D}\hat{g}_{dc^{-1},d}\mathcal{P}(h_c).\end{aligned}$$

Since $\hat{g}_{dc^{-1},d} \in C_0(\hat{A})$, for each $c, d \in D$, it is now clear that $\Psi\lambda_f^G\Psi^{-1}$ is in the range of $\mathcal{M}$.

The action of D on A generates an action on $\hat{A}$. For $d \in D$ and $\chi \in \hat{A}$, let $d\cdot\chi(a) = \chi(d^{-1}\cdot a)$, for all $a \in A$.

Proposition 2. *For any* $f \in L^1(G)$, $\mathcal{F}(f)$ *is the element of* $M_n(C_0(\hat{A}))$ *whose entries are given by*

$$(\mathcal{F}(f))_{b,c}(\chi) = \chi(\alpha(bc^{-1},c))(\hat{f}_{bc^{-1}})(c\cdot\chi),$$

for all $\chi \in \hat{A}$, $b, c \in D$.

Proof. The proof of proposition 1 shows that $(\mathcal{F}(f))_{b,c} = \hat{g}_{bc^{-1},b}$, where

$g_{c,d}(a) = f_c\big(c^{-1}d \cdot (\alpha(c,c^{-1}d)^{-1}a)\big)$. A direct computation shows that $\hat{g}_{c,d}(\chi) = \chi\big(\alpha(c,c^{-1}d)\big)\hat{f}_c(c^{-1}d \cdot \chi)$, for each $\chi \in \hat{A}$. A simple change of indices then establishes the formula in the proposition.

The next step is to characterize the range of $\mathcal{F}$ inside $M_n\big(C_0(\hat{A})\big)$.

Theorem 1. *There is an injective homomorphism of the finite group D into the automorphism group of $M_n\big(C_0(\hat{A})\big)$. The fixed point algebra $M_n\big(C_0(\hat{A})\big)^D$ under the resulting group of automorphisms is a C^*-subalgebra of $M_n\big(C_0(\hat{A})\big)$ and $\mathcal{F}$ extends from $L^1(G)$ to a C^*-isomorphism of $C^*(G)$ with $M_n\big(C_0(\hat{A})\big)^D$.*

Proof. Let ρ denote the right regular representation of G on $L^2(G)$ given by $\rho(y)h(x) = h(xy)$ for $h \in L^2(G)$, $x, y \in G$. Each $\rho(y)$ is a unitary operator on $L^2(G)$. Then, for any $y \in G$ and $f \in L^1(G)$, $\rho(y)\lambda_f^G\rho(y)^* = \lambda_f^G$. In particular, for $d \in D$, $\rho(\gamma(d))$ commutes with $\lambda^G\big(L^1(G)\big)$. Let $U(d) = \Psi\rho\big(\gamma(d)\big)\Psi^{-1}$, a unitary on $\mathcal{H}$ for each $d \in D$. For $\underline{h} = (h_c)_{c\in D}$ in $\mathcal{H}$, direct computations show that, for any $b \in D$ and $\chi \in \hat{A}$, $(U(d)\underline{h})_b(\chi) = \overline{d^{-1} \cdot \chi(\alpha(b,d))}h_{bd}(d^{-1} \cdot \chi)$, and $(U(d)^*\underline{h})_b(\chi) = \chi(\alpha(bd^{-1},d))h_{bd^{-1}}(d \cdot \chi)$. For $F = (F_{b,c})_{b,c\in D}$ in $M_n\big(C_0(\hat{A})\big)$, $\big[U(d)\mathcal{M}(F)U(d)^*\underline{h}\big]_b(\chi) = \sum_{c\in D} d^{-1} \cdot \chi\big(\alpha(b,d)^{-1}\alpha(c,d)\big)F_{bd,cd}(d^{-1} \cdot \chi)h_c(\chi)$. Thus conjugation by $U(d)$ leaves the range of $\mathcal{M}$ invariant and an automorphism $\beta(d)$ of $M_n\big(C_0(\hat{A})\big)$ can be defined by $\beta(d)(F) = \mathcal{M}^{-1}\big(U(d)\mathcal{M}(F)U(d)^*\big)$, for each $F \in M_n\big(C_0(\hat{A})\big)$. Then $\big[\beta(d)(F)\big]_{b,c}(\chi) = d^{-1} \cdot \chi\big(\alpha(b,d)^{-1}\alpha(c,d)\big)F_{bd,cd}(d^{-1} \cdot \chi)$ and further easy computations show that β is a homomorphism of D into the automorphism group of $M_n\big(C_0(\hat{A})\big)$. Since any F with only one entry nonzero has that entry moved by any d in D different then the identity, β is cleary injective.

Let $M_n\big(C_0(\hat{A})\big)^D = \{F \in M_n\big(C_0(\hat{A})\big)\colon \beta(d)(F) = F, \text{ for all } d \in D\}$. Then, for any $f \in L^1(G)$, $\rho(\gamma(d))\lambda_f^G\rho(\gamma(d))^* = \lambda_f^G$, implies that $\mathcal{F}(f) \in M_n\big(C_0(\hat{A})\big)^D$. The proof of theorem 1 will be completed by showing that $\mathcal{F}\big(L^1(G)\big)$ is dense in $M_n\big(C_0(\hat{A})\big)^D$. Let e denote the identity in D. For any $F \in M_n\big(C_0(\hat{A})\big)^D$, the e-column, $(F_{b,e})_{b\in D}$ and D-invariance completely determine F. For $f \in L^1(G)$, $(\mathcal{F}(f))_{b,e} = \hat{f}_b$, for each $b \in D$.

For fixed $F \in M_n\big(C_0(\hat{A})\big)^D$ and $\varepsilon > 0$, choose $\delta > 0$ so that, $F' \in M_n\big(C_0(\hat{A})\big)$ with $\|F'_{b,c} - F_{b,c}\|_\infty < \delta$ for all $b, c \in D$ implies $\|F' - F\| < \varepsilon$. For each $b \in D$, pick $f_b \in L^1(A)$ such that $\|\hat{f}_b - F_{b,e}\|_\infty < \delta$. Define $f \in L^1(G)$ by $f(b,a) = f_b(a)$, for all $b \in D,\ a \in A$. Then $\|\mathcal{F}f - F\| < \varepsilon$. Thus $\mathcal{F}(L^1(G))$ is dense in $M_n(C_0(\hat{A}))^D$.

Remark. $M_n(C_0(\hat{A}))^D$ is the algebra of matrices F whose e-column consists of n arbitrary $C_0(\hat{A})$ elements $(F_{d,e})_{d\in D}$ with every other entry determined by D-invariance. From the formulas in the above proof one easily shows that
$F_{b,c}(\chi) = c^{-1} \cdot \chi\big(\alpha(b,c^{-1})^{-1}\alpha(c,c^{-1})\big)F_{bc^{-1},e}(c^{-1} \cdot \chi)$, for $\chi \in \hat{A},\ b,c \in D$. Thus, $C^*(G)$, being isomorphic to this algebra, is given a very detailed description.

3. AN EXAMPLE

The description of sectin 2 will be illustrated with an example from the two-dimensional crystal groups. In this example $A \approx \mathbb{Z}^2$ and $D = \{1,-1\} \approx \mathbb{Z}_2$. Let $G = \{(1,n,m),\ (-1,n,m): n,m \in \mathbb{Z}\}$ with group products given by the following list, for $k,l,m,n \in \mathbb{Z}$,

$$(1,k,l)(1,n,m) = (1,k+n,l+m)$$

$$(1,k,l)(-1,n,m) = (-1,k+n,m-l)$$

$$(-1,k,l)(1,n,m) = (-1,k+n,l+m)$$

$$(-1,k,l)(-1,n,m) = (1,k+n+1,m-l)$$

with $A = \{(1,n,m): n,m \in \mathbb{Z}\}$, $\hat{A}$ is isomorphic to the torus $\mathbf{T}^2 = \{(z,w): z,w \in \mathbf{C},\ |z| = |w| = 1\}$. For $(z,w) \in \mathbf{T}^2$, let $\chi^{z,w} \in \hat{A}$ be defined by $\chi^{z,w}(1,n,m) = z^n w^m$, for all $(1,n,m) \in A$. The actions of D on A and $\hat{A}$ are given by $(-1)\cdot(1,n,m) = (1,n,-m)$ and $(-1)\cdot\chi^{z,w} = \chi^{z,\overline{w}}$. A cross-section γ from D into G is given by $\gamma(1) = (1,0,0)$ and $\gamma(-1) = (-1,0,0)$. Then the 2-cocycle α has only one nonidentity value, $\alpha(-1,-1) = (1,1,0)$.

For ease of notation, write functions on $\hat{A}$ as functions of $(z,w) \in \mathbf{T}^2$. The remark at the end of section 2 can now be used to see that $C^*(G)$ is isomorphic to the algebra of matrix-valued functions F on $\mathbf{T}^2$ of the form:

$$F(z,w) = \begin{pmatrix} F_{11}(z,w) & zF_{1-1}(z,\overline{w}) \\ F_{1-1}(z,w) & F_{11}(z,\overline{w}) \end{pmatrix},$$

for all $(z,w) \in \mathbf{T}^2$, where F_{11} and F_{1-1} are arbitrary continuous functions.

Let $\Omega = \{(z,w) \in \mathbf{T}^2: Im(w) \geq 0\}$. For each $z \in \mathbf{T}$, let
$\mathcal{R}_z = \{\left(\begin{smallmatrix} a & zb \\ b & a \end{smallmatrix}\right): a,b \in \mathbf{C}\}$, a $*$-subalgebra of $M_2(\mathbf{C})$. Then the map which restricts F in $M_2(C(\mathbf{T}^2)) = C(\mathbf{T}^2, M_2(\mathbf{C}))$ to Ω provides an isomorphism of $C^*(G)$ with

$$\{F \in C(\Omega, M_2(\mathbf{C})): F(z,\pm 1) \in \mathcal{R}_z, \text{ for each } z \in \mathbf{T}\}.$$

4. AN ADDENDUM

The arguments used to prove theorm 1 apply in greater generality if one removes the Fourier transform converting $C^*(A)$ to $C_0(\hat{A})$. This is the observation of the referee, who was kind enough to work out the details. Before stating the theorem some notation is required.

Let G be a locally compact group with a normal subgroup N of finite index, n, in G. Let $D = G/N$ and let $\gamma : D \to G$ be a fixed cross-section of the cosets of N in G. Let $U : L^2(G) \to \sum_{d\in D} \oplus L^2(N)$ denote the isomorphism given by $(Uh)_d(t) = h(\gamma(d)t)$, for $t \in N$, $d \in D$, and $h \in L^2(G)$. Let ρ denote the right regular representation . As before, $C^*_\lambda(N)$ and $C^*_\lambda(G)$ denote the reduced C^*-algebras of N and G respectively. For any $y \in G$ and $f \in L^1(G)$, $\rho(y)\lambda^G(f)\rho(y^{-1}) = \lambda^G(f)$, so $\rho(y)$ commutes with $C^*_\lambda(G)$.

For $F = (f_{b,c})_{b,c\in D} \in M_n(L^1(N))$, define $\tilde{\lambda}(F)$ in $\mathcal{B}(\sum_{d\in D} \oplus L^2(N))$ by

$$(\tilde{\lambda}(F)\underline{h})_d = \sum_{c\in D} \lambda^N(f_{d,c})h_c\ , \qquad \text{for } \underline{h} = (h_c)_{c\in D} \in \sum_{d\in D} \oplus L^2(N)\ .$$

Then $\tilde{\lambda}$ extends to an isomorphism, also denoted $\tilde{\lambda}$ of $M_n(C^*_\lambda(N))$ into $\mathcal{B}(\sum_{d\in D} \oplus L^2(N))$. For ease of notation, identify $M_n(C^*_\lambda(N))$ with its image under $\tilde{\lambda}$.

Theorem 2. (the referee) *There is an injective homomorphism of D into the automorphism group of $M_n(C^*_\lambda(N))$ such that $C^*_\lambda(G)$ is isomorphic with the fixed point algebra $M_n(C^*_\lambda(N))^D$.*

Proof. Most of the computational details will be left out since they are very similar to those in the proof of theorem 1.

It is easy to see that for $f \in L^1(G)$, $U\lambda^G(f)U^* \in M_n(C^*_\lambda(N))$ so $UC^*_\lambda(G)U^* \subseteq M_n(C^*_\lambda(N))$. For $y \in G$, define $\alpha(y)T = U\rho(y)U^*TU\rho(y^{-1})U^*$, for each $T \in \mathcal{B}(\sum_{d\in D} \oplus L^2(N))$. One checks, by computation, that $\alpha(y)$ maps $M_n(C^*_\lambda(N))$ onto itself. Consider α as a map of G into the automorphism group of $M_n(C^*_\lambda(N))$. Clearly α is a homomorphism and one can check that the kernel of α is N (because $\rho(t)$ commutes with $C^*_\lambda(N)$ for each $t \in N$). Thus, α defines an injective homomorphism, also denoted α, of D into $M_n(C^*_\lambda(N))$. Since $\rho(y)$ commutes with $C^*_\lambda(G)$, $UC^*_\lambda(G)U^*$ is in the fixed point algebra of $\alpha(D)$.

If $T = (T_{b,c})_{b,c\in D} \in M_n(C^*_\lambda(N))$ is invariant under $\alpha(D)$ and $\varepsilon > 0$, let $F = \left(\lambda^N(f_{b,c})\right)_{b,c\in D} \in M_n(\lambda^N(L^1(N)))$, be such that $\| T - F \| < \varepsilon$. Replacing F by

$\frac{1}{n}\sum_{d\in D}\alpha(d)F$, which is still in $M_n(\lambda^N(L^1(N)))$ one can assume F is $\alpha(D)$ invariant. Define $f \in L^1(G)$ by $f(\gamma(d)t) = f_{d,e}(t)$ for $d \in D$, $t \in N$, where e is the identity in D. Then computing carefully yields $U\lambda^G(f)U^* = F$. Thus $U\lambda^G(L^1(G))U^*$ is dense in the fixed point algebra which implies that $UC^*_\lambda(G)U^*$ is the fixed point algebra in $M_n(C^*_\lambda(N))$ under $\alpha(D)$.

REFERENCES

[1] Raeburn, Iain. On group C^*-algebras of bounded representation dimension. *Trans. Amer. Math. Soc.*, 262 (1982), 629-644.

[2] Rieffel, Marc A. Actions of finite groups on C^*-algebras. *Math. Scand.* 47 (1980), 157-176.

[3] Rosenberg, Jonathan. The C^*-algebras of some real and p-adic solvable groups. *Pac. J. Math.* 65 (1976), 175-192.

Department of Mathematics
Univeristy of Saskatchewan
Saskatoon, Saskatchewan S7N OWO
CANADA

Operator Theory:
Advances and Applications, Vol. 41

ON WIENER-HOPF DETERMINANTS

Harold Widom*

Dedicated to Israel Gohberg on his sixtieth birthday.

Any extension of the "strong" limit theorem for Toeplitz determinants to continuous convolution operators requires an assumption that allows one to define the determinant in question. In this paper a natural assumption on a matrix-valued symbol is presented which guarantees that the regularized determinant $\det^{(p)}$ is defined, and a limit theorem for it is proved.

1. INTRODUCTION

Before discussing determinants of (finite) Wiener-Hopf operators, we shall describe the situation for the (simpler) Toeplitz case. Given a matrix-valued function belonging to L_1 of the unit circle one defines the Toeplitz matrices

$$T_N(\varphi) = (\hat{\varphi}(i-j))_{i,j=1,\ldots,N}.$$

The "standard" asymptotic formula for the determinants

$$D_N(\varphi) = \det T_N(\varphi)$$

is

$$D_N(\varphi) \sim G(\varphi)^N \det T(\varphi)T(\varphi^{-1}) \qquad (N \to \infty). \tag{1.1}$$

Here $G(\varphi)$ is given by

$$G(\varphi) = \exp\left\{\frac{1}{2\pi}\int_0^{2\pi} \log\det\varphi(e^{i\theta})d\theta\right\},$$

$T(\varphi)$ denotes the Toeplitz operator with symbol φ on $l_2(\mathbf{Z}^+)$, and the "det" on the right side of (1.1) is the determinant defined for an operator of the form $I - A$ with A trace class. (There is a different expression for this last factor in (1.1) in the scalar case, but

* Research sponsored in part by NSF grant DMS-8601605.

this will not concern us here.) This holds under various "suitable" conditions, of which the ones in [11], which we now describe, are quite natural. Set

$$K = \left\{ \varphi : ||\varphi||_\infty + \sum_{-\infty}^{+\infty} |k||\hat{\varphi}(k)|^2 < \infty \right\}.$$

This is a Banach algebra under a suitable norm. If φ is an invertible element of K then the determinant of the "harmonic extension"

$$\varphi_\rho(e^{i\theta}) = \sum_{-\infty}^{+\infty} \rho^{|k|} \hat{\varphi}(k) e^{ik\theta}$$

is bounded away from zero in some annulus $\rho_0 < \rho < 1$ and one defines the index of φ by

$$i(\varphi) = \Big[\arg \det \varphi_\rho(e^{i\theta}) \Big]_0^{2\pi},$$

which is independent of $\rho > \rho_0$. If $i(\varphi) = 0$ one defines the geometric mean $G(\varphi)$ by

$$G(\varphi) = \lim_{\rho \to 1} G(\varphi_\rho);$$

this limit is guaranteed to exist. Theorem 6.1 of [11] asserts that (1.1) holds under just this assumption that φ is an invertible element of K of index zero. (We mention also that if φ is invertible in K, then $I - T(\varphi)T(\varphi^{-1})$ is trace class and $T(\varphi)$ is a Fredholm operator of index $-i(\varphi)$.)

For the Wiener-Hopf analogue, we take a matrix-valued function σ on the real line, define $W_\alpha(1-\sigma)$ to be I minus convolution by

$$\hat{\sigma}(x) = \frac{1}{2\pi} \int_{-\infty}^{+\infty} e^{-i\xi x} \sigma(\xi) d\xi$$

on $L_2(0, \alpha)$ and set

$$D_\alpha(1-\sigma) = \det W_\alpha(1-\sigma). \tag{1.2}$$

The analogue of (1.1) would be

$$D_\alpha(1-\sigma) \sim G(1-\sigma)^\alpha \det W(1-\sigma) W((1-\sigma)^{-1}) \qquad (\alpha \to \infty) \tag{1.3}$$

where $G(1-\sigma)$ is given by

$$G(1-\sigma) = \exp \left\{ \frac{1}{2\pi} \int_{-\infty}^{+\infty} \log \det(1 - \sigma(\xi)) d\xi \right\}$$

and $W(\tau)$ denotes the Wiener-Hopf operator on $L_2(\mathbf{R}^+)$ with symbol τ. But now there is the problem of the meaning of the determinant in (1.2). One solution is to assume

$\sigma \in L_1$. Then each $W_\alpha(\sigma)$ is trace class and one could use the usual operator-theoretic determinant. This was done by Dym [5] who then established (1.3) under certain other assumptions on σ. Or, one could use the classical Fredholm definition of determinant, possibly modified to allow for a discontinuity of the kernel on the diagonal. Mikaelyan [9] used this approach and obtained a strengthening of the result of Dym.

The aim of this paper is to prove (1.3) under the sole assumption that $1-\sigma$ is an invertible element of index zero in the real line analogue of the algebra K defined above. This is impossible, since something else is needed to define $D_\alpha(1-\sigma)$. What we shall do is consider instead the regularized determinants [8, Chapter IV]

$$D_\alpha^{(p)}(1-\sigma) = \overset{(p)}{\det} W_\alpha(1-\sigma)$$

which are defined whenever $W_\alpha(\sigma)$ belongs to the Schatten class S_p $(p = 1, 2, 3, \ldots)$ of compact operators on $L_2(\mathbf{R}^+)$, and require as little as is necessary to guarantee that $W_\alpha(\sigma) \in S_p$. The natural assumption $\sigma \in L_p$ is all that one needs. Thus if $1-\sigma$ is an invertible element of K of index zero and $\sigma \in L_p$ then a modification of (1.3) holds for $D_\alpha^{(p)}(1-\sigma)$. (See formula (2.5) below.)

As examples we mention the functions

$$\sigma(\xi) = a(\xi^2+1)^{-r/2}$$

where $|a| < 1$ and $r > 0$. The Fourier transform of σ is exponentially small for large ξ and for small ξ is asymptotically a constant times $|\xi|^{r-1}$. Thus, $\sigma \in K$. (See section 3.) The assumption on a guarantees that $1-\sigma$ is invertible. And $\sigma \in L_p$ precisely when $p > r^{-1}$.

A word about the relationship between Fredholm determinants and regularized operator determinants. If $K(x,y)$ is continuous and the kernel of a trace class operator A, then the Fredholm and operator determinants of $I-A$ coincide. If K is continuous but A is not trace class, then the Fredholm determinant $\det(I-A)$ and regularized operator determinant $\det^{(2)}(I-A)$ are related by

$$\det(I-A) = \overset{(2)}{\det}(I-A)\exp\left\{-\mathrm{tr}\int K(x,x)\,dx\right\}.$$

If $K(x,y) = \hat{\sigma}(x-y)$ and the interval is $(0,\alpha)$ this becomes

$$\det W_\alpha(1-\sigma) = \overset{(2)}{\det} W_\alpha(1-\sigma)e^{-\alpha\,\mathrm{tr}\,\hat{\sigma}(0)}. \tag{1.4}$$

One can now take the right side as the definition of the left, even if $\hat{\sigma}(0)$ exists only in some generalized sense. This is the way the Fredholm determinant was extended in [9] to certain symbols σ with discontinuous Fourier transforms. Our result on the asymptotics of $D_\alpha^{(2)}(1-\sigma)$ yields a strengthening of [9] in that the assumption $\hat{\sigma} \in L_1$ can be dropped.

In a very pretty paper of Gohberg, Kaashoek and van Schagen [7], the matrix-valued function on the circle [resp. line] was assumed analytic and the asymptotic formula for Toeplitz [resp. Wiener-Hopf] determinants was derived and expressed in terms of a "realization" of the matrix-valued function. For the case of the line, one assumes that $\sigma(\xi)$ is an $m \times m$ matrix-valued function analytic in a neighborhood of $\mathbf{R} \cup \{\infty\}$ in the Riemann sphere and which vanishes at ∞. Then there is a representation

$$\sigma(\xi) = C(A - \xi I)^{-1}B$$

where A is a bounded operator on a Hilbert space X such that $\mathrm{sp}(A)$ is disjoint from $\mathbf{R}$ and

$$C : X \to \mathbf{C}^m, \qquad B : \mathbf{C}^m \to X$$

are also bounded operators [6, §II.3]. This is called a realization of σ. In general, of course, σ belongs to L_2 but not to L_1, and in [7] $D_\alpha(1 - \sigma)$ was defined as a certain modification of $D_\alpha^{(2)}(1 - \sigma)$ similar in spirit to (1.4). Their result implies that under the assumptions

$$\det(1 - \sigma) \neq 0, \qquad [\arg \det(1 - \sigma(\xi))]_{-\infty}^{+\infty} = 0$$

one has

$$D_\alpha(1 - \sigma) \sim \Sigma_1^\alpha \Sigma_2 \tag{1.5}$$

where Σ_1 is a modified geometric mean of $1 - \sigma$ and

$$\Sigma_2 = \det((I - P)(I - P^\times) + PP^\times)$$

where P is the Riesz projection for A corresponding to the part of its spectrum in the upper half-plane and $P^\times$ the analogous projection for the operator $A^\times = A - BC$. Reconciling this with formula (2.5) in the case $p = 2$ amounts to showing

$$\det W(1 - \sigma)W((1 - \sigma)^{-1}) = \det((I - P)(I - P^\times) + PP^\times). \tag{1.6}$$

Of course, (1.5) and (2.5) imply this but it seems a natural exercise to prove (1.6) directly, and we shall present such a proof here. In this connection we acknowledge with pleasure help from M.A. Kaashoek and H. Bart, who explained to us the basic facts about realization of matrix-valued functions.

Here now is an outline of the various sections of this paper.

2. The Asymptotic Formula: A general principle will be invoked to derive it formally. We also indicate a proof of it under the additional assumption that the numerical ranges of all the matrices $\sigma(\xi)$ lie in a fixed compact convex subset of $\mathbf{C}$ which does not contain 1.

3. The Algebra K: General properties of this algebra are derived and the regularized geometric means $G^{(p)}(1-\sigma)$ are defined.

4. First Limit Theorem: Here it is proved, under the additional assumption that $W(1-\sigma)$ and $W((1-\sigma)^{-1})$ are both invertible, that $W_\alpha(1-\sigma)$ is invertible for sufficiently large α and that as $\alpha \to \infty$

$$\lim_{\alpha\to\infty} \frac{d}{d\alpha} \log D_\alpha^{(p)}(1-\sigma) = \log G^{(p)}(1-\sigma).$$

We make use of a generalization to regularized determinants of the fact that if A_α is the integral operator on $L_2(0,\alpha)$ with kernel $K(x,y)$ (which is continuous for $x, y \in \mathbf{R}^+$, say) then under suitable conditions

$$\frac{d}{d\alpha} \log \det(I - A_\alpha) = -\mathrm{tr}\Gamma_\alpha(\alpha,\alpha)$$

where $\Gamma_\alpha(x,y)$ is the resolvent kernel for $K(x,y)$ over the interval $(0,\alpha)$ (See [9, Lemma 1] or [6, Theorem 3.1]). The proof we present is quite simple, even in the case of nonregularized determinants.

5. Proof of the Asymptotic Formula: The overall strategy is that of [11], adapted to the present situation.

6. Analytic Symbols: A direct proof of identity (1.6) is presented. This section of the paper may be read independently of the rest.

2. THE ASYMPTOTIC FORMULA

Given a symbol σ we define $\tilde{\sigma}$ by $\tilde{\sigma}(\xi) = \sigma(-\xi)$. The "general principle" alluded to in the introduction is the relation

$$(2.1)\quad \lim_{\alpha\to\infty} \mathrm{tr}\big[f(W_\alpha(\sigma)) - W_\alpha(f(\sigma))\big] = \mathrm{tr}\big[f(W(\sigma)) - W(f(\sigma)) + f(W(\tilde{\sigma})) - W(f(\tilde{\sigma}))\big].$$

Here the operators $f(W_\alpha(\sigma))$, etc., are defined by some functional calculus. Conditions assuring the validity of this can be found in [12-13]. Of course

$$(2.2)\qquad \mathrm{tr}W_\alpha(f(\sigma)) = \frac{\alpha}{2\pi} \int_{-\infty}^{+\infty} \mathrm{tr} f(\sigma(\xi))\, d\xi$$

so (2.1) gives the second-order asymptotics of $\mathrm{tr} f(W_\alpha(\sigma))$.

If $A \in S_p$ the regularized determinant $\det^{(p)}(I - A)$ is defined to be

$$\prod (1-\lambda_i) \exp\left\{\sum_{n=1}^{p-1} \frac{1}{n} \lambda_i^n\right\}$$

where λ_i are the nonzero eigenvalues of A repeated according to algebraic multiplicity. If $I - A$ is invertible and we take a simple curve joining 1 to ∞ which is disjoint from sp(A) then in the region complementary to this curve there is a unique analytic branch of $\log(1-\lambda)$ satisfying $\log 1 = 0$, and

$$\overset{(p)}{\det}(I-A) = \exp\left\{ \operatorname{tr}\frac{1}{2\pi i}\int_{\Gamma}\left(\log(1-\lambda)+\sum_{n=1}^{p-1}\frac{1}{n}\lambda^n\right)(\lambda I - A)^{-1}d\lambda \right\} \tag{2.3}$$

where Γ is a Jordan curve lying in the region and surrounding sp(A). This is certainly true when $p=1$, follows easily for general p if $A \in S_1$, and then for general $A \in S_p$ by an approximation argument. Note that the integral represents a trace class operator since it equals that of

$$\lambda^{-p}\left(\log(1-\lambda)+\sum_{n=1}^{p-1}\frac{1}{n}\lambda^n\right)A^p(\lambda I - A)^{-1}. \tag{2.4}$$

If we formally apply (2.1) with the functions

$$f(\lambda) = \log(1-\lambda)+\sum_{n=1}^{p-1}\frac{1}{n}\lambda^n, \qquad f(\lambda)=\log(1-\lambda)$$

(all operators being defined by the analytic functional calculus) and subtract the results we obtain, using (2.2) and (2.3),

$$\begin{aligned}\lim_{\alpha\to\infty}&\left\{\log\frac{D_\alpha^{(p)}(1-\sigma)}{D_\alpha(1-\sigma)} - \alpha\operatorname{tr}\frac{1}{2\pi}\int_{-\infty}^{+\infty}\sum_{n=1}^{p-1}\frac{1}{n}\sigma(\xi)^n d\xi\right\}\\ &= \sum_{n=2}^{p-1}\frac{1}{n}\operatorname{tr}\left[W(\sigma)^n - W(\sigma^n)+W(\tilde{\sigma})^n - W(\tilde{\sigma}^n)\right].\end{aligned}$$

(The term on the right corresponding to $n=1$ equals zero and so was dropped.) Thus from (1.3) we obtain for $D_\alpha^{(p)}(1-\sigma)$ the asymptotic formula

$$\begin{aligned}D_\alpha^{(p)}(1-\sigma) &\sim G^{(p)}(1-\sigma)^\alpha\cdot\det W(1-\sigma)W((1-\sigma)^{-1})\\ &\cdot\exp\left\{\sum_{n=2}^{p-1}\operatorname{tr}\left[W(\sigma)^n - W(\sigma^n)+W(\tilde{\sigma})^n - W(\tilde{\sigma}^n)\right]\right\}\end{aligned} \tag{2.5}$$

where

$$G^{(p)}(1-\sigma) = \exp\left\{\frac{1}{2\pi}\int_{-\infty}^{+\infty}\left[\log\det(1-\sigma(\xi))+\operatorname{tr}\sum_{n=1}^{p-1}\frac{1}{n}\sigma(\xi)^n\right]d\xi\right\}. \tag{2.6}$$

Here is how one can prove this under the assumption that $\sigma \in K$ and that the numerical ranges of the matrices lie in a compact convex set not containing 1. Relation (2.1) is true for any $\sigma \in K$ and any function f which is analytic on a neighborhood of the closed convex hull of the numerical ranges of the matrices $\sigma(\xi)$. (See [13, pp. 294 ff.] for a proof in the scalar case; the matrix case requires only trivial modification. The matrix case can be found in [12], even in higher dimensions, but the condition on σ is stronger.) The assumption on σ guarantees the existence of an analytic branch of

$$f(\lambda) = \log(1-\lambda) + \sum_{n=1}^{p-1} \frac{1}{n}\lambda^n$$

in an appropriate region, which we specify by taking $\log 1 = 0$. Then $f(\sigma) \in L_1$ so $W_\alpha(f(\sigma))$ is trace class and (2.1) gives (2.5) but with

$$\det W(1-\sigma)W((1-\sigma)^{-1}) \tag{2.7}$$

replaced by

$$\operatorname{tr}\left[\log W(1-\sigma) - W(\log(1-\sigma)) + \log W(1-\tilde{\sigma}) - W(\log(1-\tilde{\sigma}))\right]. \tag{2.8}$$

The equality of (2.7) and (2.8) is an identity whose validity for general σ would follow from its validity for all σ's from an appropriately dense set. The identity must hold in any of the cases for which (1.3) is already known to hold, e.g., if σ and $\tilde{\sigma}$ both belong to L_1, a collection of symbols which is dense enough. A direct proof of the identity can be found in [11] under this same assumption.

Formulas for the traces appearing on the right side of (2.5) will be given at the end of section 5.

3. THE ALGEBRA K

To distinguish notationally the algebras on the circle and the line we denote the former by $K_{\mathbf{T}}$ and the latter simply by K. The norm on $K_{\mathbf{T}}$ is given by $||\varphi||_\infty + |||\varphi|||$ where

$$|||\varphi|||^2 = \sum_{-\infty}^{+\infty} |k||\hat{\varphi}(k)|^2$$

and $|\ldots|$ for a matrix denotes its Hilbert-Schmidt norm. The easily established identity

$$|||\varphi|||^2 = \frac{1}{4\pi^2}\int\int \left|\frac{\varphi(s)-\varphi(t)}{s-t}\right|^2 |ds||dt| \tag{3.1}$$

(the integrals being taken over the unit circle) gives an alternative description of the elements of $K_{\mathbf{T}}$.

The norm on K is $||\sigma||_\infty + |||\sigma|||$, in analogy with K_{T}, but now $|||\sigma|||$ is defined by the analogue of (3.1),

$$|||\sigma|||^2 = \frac{1}{4\pi^2} \int_{-\infty}^{+\infty}\int_{-\infty}^{+\infty} \left| \frac{\sigma(\xi)-\sigma(\eta)}{\xi-\eta} \right|^2 d\xi d\eta. \tag{3.2}$$

It turns out that if $\sigma \in L_\infty$ then $\sigma \in K$ if and only if its distributional Fourier transform $\hat{\sigma}$ is equal on $R \setminus \{0\}$ to a function for which

$$\int |x| |\hat{\sigma}(x)|^2 dx \tag{3.3}$$

in finite, and then $|||\sigma|||^2$ is equal to this integral. (See [13, Prop. 1] for the scalar case. The matrix case is no different.)

The first lemma says that K is an inverse-closed subalgebra of L_∞. We denote by K^* the set of invertible elements of K.

LEMMA 3.1. *If $\sigma \in K$, $\sigma^{-1} \in L_\infty$ then $\sigma \in K^*$.*

PROOF: We have

$$\sigma(\xi)^{-1} - \sigma(\eta)^{-1} = \sigma(\xi)^{-1}(\sigma(\eta) - \sigma(\xi))\sigma(\eta)^{-1}.$$

The rest is left to the reader. □

The algebras K and K_{T} are not only analogous but also isometric. In fact, a straightforward computation using (3.1) and (3.2) shows that the mapping

$$\varphi(s) \to \sigma(\xi) = \varphi\left(\frac{\xi - i}{\xi + i}\right) \tag{3.4}$$

is an isometry between K and K_{T}. The algebra K_{T} is a subalgebra of $H_{\mathrm{T}}^\infty + C_{\mathrm{T}}$ (since $\varphi \in K_{\mathrm{T}}$ implies the Hankel operator $H(\tilde{\varphi})$ is Hilbert-Schmidt and so compact, and Hartman's theorem implies $\varphi \in H_{\mathrm{T}}^\infty + C_{\mathrm{T}}$) and the statements made in the introduction concerning the harmonic extension φ_ρ apply to all invertible elements of $H_{\mathrm{T}}^\infty + C_{\mathrm{T}}$. (See [3] for the scalar case, [4] for the matrix.) Transferring to K by means of (3.4) gives us much useful information.

For $\sigma \in K$ denote by σ_λ $(\lambda > 0)$ the harmonic extension of σ to the upper half-plane in $\xi + i\lambda$ space,

$$\sigma_\lambda(\xi) = \frac{1}{\pi} \int_{-\infty}^{+\infty} \frac{\lambda}{\lambda^2 + (\xi - \eta)^2} \sigma(\eta) d\eta. \tag{3.5}$$

Then if $\sigma \in K^*$, $\det \sigma_\lambda$ is bounded away from zero outside some compact set. We define

$$i(\sigma) = \frac{1}{2\pi} \Delta_\Gamma \arg \det \sigma_\lambda(\xi)$$

where Γ is a Jordan curve, described counterclockwise, lying in the upper half-plane and surrounding the compact set.

It is an easy exercise to show that if $\sigma \in L_p$ then in each half-plane $\lambda > \lambda_0 > 0$

(3.6) $$\lim_{|\xi+i\lambda|\to\infty} \sigma_\lambda(\xi) = 0.$$

If follows that if, in addition, $1-\sigma$ is invertible, its index is given by

(3.7) $$i(1-\sigma) = \frac{1}{2\pi}\left[\arg\det(1-\sigma_\lambda(\xi))\right]_{\xi=-\infty}^{\infty}$$

for all sufficiently small λ.

The next lemma lists some other properties of the harmonic extension σ_λ.

LEMMA 3.2.

a) *For any $\sigma \in K$ we have*

$$||\sigma_\lambda||_\infty = O(1), \qquad \lim_{\lambda\to 0+} |||\sigma_\lambda - \sigma||| = 0.$$

b) *If $\sigma \in K^*$ then $||\sigma_\lambda^{-1}|| = O(1)$ for sufficiently small λ.*

c) *If $\sigma \in L_p$ then each $\sigma_\lambda \in C_0$ (the continuous functions vanishing at $\pm\infty$) and*

$$\lim_{\lambda\to 0+} ||\sigma_\lambda - \sigma||_p = 0.$$

PROOF: All the assertions follow from what has already been said, or from standard results on convolutions, except for the second assertion of part (a). To prove it observe that

$$|||\sigma_\lambda - \sigma|||^2 = \int |x|\left(e^{-\lambda|x|}-1\right)^2|\hat{\sigma}(x)|^2dx$$

and this has limit zero by dominated convergence. □

Next we state some density theorems. The useful notion of convergence is not with respect to the norm of K (convergence in L_∞ is too restrictive) but rather the sorts that are asserted in Lemma 3.2. We make the following definitions:

i) *A sequence σ_n converges to σ in K if*

$$||\sigma_n||_\infty = O(1), \qquad |||\sigma_n - \sigma||| \to 0.$$

ii) *If B is a Banach space of functions on $\mathbf{R}$ then σ_n converges to σ in $K \cap B$ if σ_n converges to σ in K and in the norm of B.*

LEMMA 3.3.

a) *Given $\sigma \in K$ there exist σ_n converging to σ in K and a.e. such that each $\sigma_n = c_n + \tau_n$ where c_n is a constant and $\tau_n \in K \cap L_2$.*

b) *Given $\sigma \in K \cap L_p$ we can find $\sigma_n \in K \cap C_c^\infty$ converging to σ in $K \cap L_p$.*

c) *The same as (b) but with L_p replaced by $L_p \cap C_0$.*

PROOF: For part (a) see Proposition 2 of [13]. Lemma 3.2 reduces part (b) to part (c). So assume $\sigma \in L_p \cap C_0$ and take any scalar-valued function $\varphi \in C_c^\infty$ such that $\varphi = 1$ on a neighborhood of $\xi = 0$. Then

$$\sigma_n(\xi) = \varphi(\xi/n)\sigma(\xi)$$

belongs to $K \cap C_c$ and is easily seen to converge to σ in the norm of $K \cap L_p \cap C_0$. So we may assume that our given σ belongs to $K \cap C_c$. Now we take any scalar-valued function $\psi \in C_c^\infty$ such that $\int \psi(\xi)d\xi = 1$ and let σ_n be the convolution of σ with $n\psi(n\xi)$. Then $\sigma_n \in C_c^\infty$ and it is again easy to see that $\sigma_n \to \sigma$ in the norm of $K \cap L_p \cap C_0$. □

Next we turn to the regularized geometric mean, which will be defined whenever $\sigma \in L_p$ and $1 - \sigma$ is an invertible element of K of index zero. Because of (3.6) and (3.7) there is, for sufficiently small $\lambda > 0$, a unique continuous $\log\det(1 - \sigma_\lambda(\xi))$ which vanishes at $\pm\infty$. This function satisfies

$$\log\det(1 - \sigma_\lambda(\xi)) + \operatorname{tr}\sum_{n=1}^{p-1}\frac{1}{n}\sigma_\lambda(\xi)^n = O(|\sigma_\lambda(\xi)|^p)$$

for large $|\xi|$ and so the integrand in (2.6) with σ replaced by σ_λ belongs to L_1. Thus $G^{(p)}(1 - \sigma_\lambda)$ is well-defined for small $\lambda > 0$.

LEMMA 3.4. *If $\sigma \in L_p$, $1 - \sigma \in K^*$, and $i(1 - \sigma) = 0$ then the limit*

$$\lim_{\lambda \to 0+} G^{(p)}(1 - \sigma_\lambda)$$

exists.

PROOF: From the formula

$$\frac{d}{d\lambda}\log\det(1 - \sigma_\lambda) = -\operatorname{tr}\sigma_\lambda'(1 - \sigma_\lambda)^{-1}$$

(where the prime denotes $\frac{d}{d\lambda}$) we obtain for sufficiently small λ_0 and λ

$$\log\det(1 - \sigma_\lambda) = \log\det(1 - \sigma_{\lambda_0}) + \int_\lambda^{\lambda_0} \operatorname{tr}\sigma_\mu'(1 - \sigma_\mu)^{-1}d\mu.$$

The identity

$$(1 - \sigma_\mu)^{-1} = \sum_{n=1}^{p}\sigma_\mu^{n-1} + \sigma_\mu^p(1 - \sigma_\mu)^{-1}$$

allows us to rewrite this as

$$\log\det(1-\sigma_\lambda)+\operatorname{tr}\sum_{n=1}^{p-1}\frac{1}{n}\sigma_\lambda^n=\log\det(1-\sigma_{\lambda_0})+\operatorname{tr}\sum_{n=1}^{p-1}\frac{1}{n}\sigma_{\lambda_0}^n$$
$$(3.8)\qquad +\frac{1}{p}\operatorname{tr}\left(\sigma_{\lambda_0}^p-\sigma_\lambda^p\right)+\operatorname{tr}\int_\lambda^{\lambda_0}\sigma_\mu{}'\sigma_\mu^p(1-\sigma_\mu)^{-1}d\mu.$$

Let us estimate the L_2 norm $||\sigma_\lambda{}'||_2$, and assume to begin with that $\sigma\in K\cap L_2$. Parseval's identity gives

$$||\sigma_\lambda{}'(\xi)||_2^2=2\pi||\widehat{\sigma_\lambda{}'}(x)||_2^2=2\pi||\frac{\partial}{\partial\lambda}e^{-\lambda|x|}\hat\sigma(x)||_2^2$$
$$=2\pi\int e^{-2\lambda|x|}|x|^2|\hat\sigma(x)|^2dx\le|||\sigma|||^2\max_x(2\pi|x|e^{-2\lambda|x|}),$$

and we obtain the inequality

$$(3.9)\qquad ||\sigma_\lambda{}'||_2\le(\pi/\lambda e)^{1/2}|||\sigma|||.$$

To extend this to all σ note that it holds also for a constant plus a function in $K\cap L_2$ since adding a constant changes neither $\sigma_\lambda{}'$ nor $|||\sigma|||$. For general $\sigma\in K$ we use Lemma 3.3 (a) and Fatou's lemma.

Returning to (3.8), the sum of the first two terms on the right is a fixed L_1 function. The next-to-last term on the right converges in L_1 as $\lambda\to\infty$ since σ_λ converges in L_p. (The resulting convergence in L_1 is an exercise in Hölder's inequality.) Finally we have for small μ

$$||\sigma_\mu^p||_2\le||\sigma_\mu||_p^{p/2}||\sigma_\mu||_\infty^{p/2}=O(1),$$

$$||(1-\sigma_\mu)^{-1}||_\infty=O(1)$$

and, by (3.9),

$$||\sigma_\mu{}'||_2=O(\mu^{-1/2}).$$

Hence

$$\int_0^{\lambda_0}||\sigma_\mu{}'\sigma_\mu^p(1-\sigma_\mu)^{-1}||_1d\mu<\infty$$

and so the integral on the right side of (3.8) converges in L_1 as $\lambda\to 0+$. Hence the left side of (3.8) converges in L_1 as $\lambda\to 0+$ and this proves the lemma. □

Of course, we define $G^{(p)}(1-\sigma)$ to be the limit in the statement of the lemma. It can be shown that under the same hypotheses there exist a determination of $\log\det(1-\sigma)$, unique up to an additive constant of the form $2k\pi i$, such that

$$|||\log\det(1-\sigma)||| < \infty;$$

that for an appropriate choice of this constant we have

$$\log\det(1-\sigma) + \operatorname{tr}\sum_{n=1}^{p-1}\frac{1}{n}\sigma^n \in L_1;$$

and that with this choice $G^{(p)}(1-\sigma)$ is given by formula (2.6). We shall not need this fact.

Next, some implications for Wiener-Hopf operators. For $\sigma \in K$ we define the Hankel operator $H(\sigma)$ on $L_2(\mathbf{R}^+)$ to be, as usual, the integral operator with kernel $\hat{\sigma}(x+y)$. Here we take $x, y > 0$ and $\hat{\sigma}$ denotes the function on $R \setminus \{0\}$ that the distributional Fourier transform is equal to. The operator is Hilbert-Schmidt, the square of its Hilbert-Schmidt norm being equal to the integral (3.3). This fact is of fundamental importance.

LEMMA 3.5. *If $\sigma_1, \sigma_2 \in K$, then*

$$W(\sigma_1\sigma_2) - W(\sigma_1)W(\sigma_2) = H(\sigma_1)H(\tilde{\sigma}_2). \tag{3.10}$$

PROOF: If $\sigma_1, \sigma_2 \in K \cap L_2$ the two operators are easily seen to have the same kernel. Neither side is changed if constants are added to σ_1 and σ_2. To prove the result in general we invoke Lemma 3.3 (a). □

So the left side of (3.10) is trace class. In particular $\sigma \in K^*$ implies $W(\sigma)$ is Fredholm. The transfer (3.4) leads to a Toeplitz operator $T(\varphi)$ unitarily equivalent to $W(\sigma)$ [2]. Since the index of $T(\varphi)$ equals $-i(\varphi)$ [4, Th. 6], the index of $W(\sigma)$ equals $-i(\sigma)$. So $W(\sigma)$ is Fredhom of index zero precisely when σ is an invertible element of K of index zero.

Now for the final lemma of the section, which tells us that $W_\alpha(\sigma) \in S_p$ whenever $\sigma \in L_p$. Note that convolving σ with the inverse Fourier transform of a scalar C_c^∞ function equal to 1 on $[-\alpha, \alpha]$ does not change $W_\alpha(\sigma)$ and so it is not necessary to assume that $\sigma \in L_\infty$; it can always be achieved by such a convolution.

LEMMA 3.6. *The mapping $\sigma \to W_\alpha(\sigma)$ is continuous from $L_p(R)$ to $S_p(L_2(0,\alpha))$.*

PROOF: Let $\sigma = uh$ be the polar decomposition of σ with u partially isometric and h nonnegative. Define the operator family $T_z (0 \le Re\ z \le p)$ by

$$T_z = W_\alpha(uh^z).$$

Then on the line $Re\ z = 0$ the T_z have operator norms at most 1 and on the line $Re\ z = p$ their norms in S_1 are at most $\alpha||\sigma||_p^p$. (Here we use the fact $||W_\alpha(\tau)||_1 \le \alpha||\tau||_1$.) It follows by interpolation that

$$||W_\alpha(\sigma)||_p = ||T_1||_p \le \alpha^{1/p}||\sigma||_p.$$

(See [8, Chap. III, Th. 13.1]. The fact that the space of the bounded operators is used here in place of S_∞ is of no consequence.)

4. FIRST LIMIT THEOREM

We begin with an extension to regularized determinants of the fact [6, Th. 3.1] that if K_α is the integral operator with matrix kernel $K(x,y)$ on $L_2(0,\alpha)$ then under suitable conditions

$$\frac{d}{d\alpha} \log \det(I - K_\alpha) = -\mathrm{tr}\Gamma_\alpha(\alpha,\alpha)$$

where $\Gamma_\alpha(x,y)$ is the resolvent kernel for K on $(0,\alpha)$. Of course Γ_α is the kernel of the operator $K_\alpha(I - K_\alpha)^{-1}$. For the regularized determinants this will be replaced by

$$\Gamma_\alpha^{(p)}(x,y) = \text{kernel of } K_\alpha^p(I - K_\alpha)^{-1}.$$

LEMMA 4.1. *Suppose $K(x,y)$ is continuous for $x,y \in [0,\alpha_0]$ and is the kernel of an operator in $S_p(L_2(0,\alpha_0))$. Then for any $\alpha \in [0,\alpha_0]$ for which the operator $I - K_\alpha$ on $L_2(0,\alpha)$ is invertible we have*

$$\frac{d}{d\alpha} \log \overset{(p)}{\det}(I - K_\alpha) = -\mathrm{tr}\Gamma_\alpha^{(p)}(\alpha,\alpha). \tag{4.1}$$

PROOF: Suppose $\alpha \to A_\alpha$ is a Fréchet differentiabl e mapping from $[0,\alpha_0]$ to $S_p(H)$ for some Hilbert space H and a particular $I - A_\alpha$ is invertible. Then replacing A by A_α in (2.3) and differentiating gives

$$\begin{aligned} &\frac{d}{d\alpha} \log \overset{(p)}{\det}(I - A_\alpha) \\ &= \mathrm{tr}\frac{1}{2\pi i}\int_\Gamma \left(\log(1-\lambda) + \sum_{n=1}^{p-1} \frac{1}{n}\lambda^n\right)(\lambda I - A_\alpha)^{-1}A'_\alpha(\lambda I - A_\alpha)^{-1}d\lambda. \end{aligned} \tag{4.2}$$

Since it is a question of taking traces, the operator

$$(\lambda I - A_\alpha)^{-1}A'_\alpha(\lambda I - A_\alpha)^{-1}$$

may be replaced by

$$(\lambda I - A_\alpha)^{-2} A'_\alpha = -\frac{d}{d\lambda}(\lambda I - A_\alpha)^{-1} A'_\alpha.$$

An integration by parts then shows that the right side of (4.2) equals

$$-\mathrm{tr}\frac{1}{2\pi i}\int_\Gamma \lambda^{p-1}(1-\lambda)^{-1}(\lambda I - A_\alpha)^{-1} A'_\alpha d\lambda$$

and so we have shown

$$\frac{d}{d\alpha}\log \overset{(p)}{\det}(I - A_\alpha) = -\mathrm{tr} A_\alpha^{p-1}(I - A_\alpha)^{-1} A'_\alpha, \tag{4.3}$$

which generalizes the well-known result for $p = 1$.

We shall assume at first that our kernel $K(x, y)$ belongs to C^1 and instead of $L_2(0, \alpha_0)$ use the space $H = H_1(0, \alpha_0)$ of absolutely continuous functions f such that $f' \in L_2(0, \alpha_0)$. This is a Hilbert space under the norm

$$\{\|f\|_2^2 + \|f'\|_2^2\}^{1/2}.$$

We define A_α $(0 \le \alpha \le \alpha_0)$ to be the operator on H given by

$$A_\alpha f(x) = \int_0^\alpha K(x, y) f(y) dy.$$

The usefulness of H lies in the fact that the evaluation operators

$$E_\alpha : f \to f(\alpha)$$

are continuous. In fact, it is easy to show, under our hypothesis on the kernel, that the rank one operators B_α defined by

$$B_\alpha f = K(\cdot, \alpha) f(\alpha)$$

form a continuous family from $[0, \alpha_0]$ to $S_p(H)$ and that

$$A_\alpha = \int_0^\alpha B_\beta d\beta.$$

If follows that A_α is Fréchet differentiable and $A'_\alpha = B_\alpha$.

It is easy to show also that the nonzero eigenvalues of K_α on $L_2(0, \alpha)$ and A_α on H are the same, with the same algebraic multiplicities. Hence the regularized determinants of $I - A_\alpha$ and $I - K_\alpha$ are the same. The operator on the right side of (4.3) is

$$f \to \left[(I - A_\alpha)^{-1} A_\alpha^{p-1} K(\cdot, \alpha)\right] f(\alpha) = \Gamma_\alpha(\cdot, \alpha) f(\alpha)$$

and it remains to show that the trace of this is $\mathrm{tr}\Gamma_\alpha(\alpha, \alpha)$. Of course this is trivial in the scalar case, but perhaps requires some argument in the matrix case. Since the operator can be written

$$f \to \Gamma_\alpha(\cdot, \alpha) E_\alpha f$$

and E_α is idempotent, its trace is equal to that of

$$f \to E_\alpha \Gamma_\alpha(\cdot, \alpha) E_\alpha f = \Gamma_\alpha(\alpha, \alpha) f(\alpha).$$

The trace of an operator is the same as the trace of the operator restricted to any subspace containing its range, and in this case we take the space of constant vectors. The trace of the resulting operator is clearly $\mathrm{tr}\Gamma_\alpha(\alpha, \alpha)$.

So the lemma is proved under the extra assumption that $K \in C^1$. To remove this restriction extend K continuously a little outside $[0, \alpha_0]^2$ and define

$$K_h(x, y) = \frac{1}{h^2} \int_x^{x+h} \int_y^{y+h} K(s, t)\, dt\, ds$$

with corresponding operators $K_{h\alpha}$ and kernels $\Gamma_{h\alpha}^{(p)}$. As $h \to 0$ the operators $K_{h\alpha}$ converge in S_p to K_α (since they are obtained from K_α by left and right multiplication by self-adjoint operators converging strongly to I). If we consider the operators as acting on $C[0, \alpha]$ then $I - K_\alpha$ is invertible by an easy argument and

$$(I - K_{\alpha h})^{-1} K_{\alpha h}^{p-1} \to (I - K_\alpha)^{-1} K_\alpha^{p-1}$$

in norm so

$$\Gamma_{\alpha h}^{(p)}(\cdot, \alpha) = (I - K_{\alpha h})^{-1} K_{\alpha h}^{p-1} K_{\alpha h}(\cdot, \alpha)$$

converges uniformly to

$$(I - K_\alpha)^{-1} K_\alpha^{p-1} K_\alpha(\cdot, \alpha) = \Gamma_\alpha^{(p)}(\cdot, \alpha).$$

Those remarks enable us to deduce (4.1) for K from the already established identity for K_h. □

If the kernel of $W_\alpha(\sigma)$ were continuous, Lemma 4.1 would give us a formula for the logarithmic derivative of $D_\alpha^{(p)}(1 - \sigma)$. It turns out that this formula holds without any further assumption.

LEMMA 4.2. *If $\sigma \in L_p$ and $W_\alpha(1-\sigma)$ is invertible for a particular α then*

$$\Gamma_\alpha^{(p)}(x,y) = \text{kernel of } W_\alpha(\sigma)^p W_\alpha(1-\sigma)^{-1}$$

is continuous and

$$\frac{d}{d\alpha} \log D_\alpha^{(p)}(1-\sigma) = -\Gamma_\alpha^{(p)}(\alpha,\alpha). \tag{4.4}$$

PROOF: By the remark preceding Lemma 3.6, we may assume also that $\sigma \in L_\infty$. First, the continuity of $\Gamma_\alpha^{(p)}$. For $p = 1, \Gamma_\alpha^{(p)}$ is the resolvent kernel of a continuous kernel and its continuity is familiar and easy. For $p \geq 2$ we use the easily established fact that for any operator A on $L_2(0,\alpha)$ the kernel of $W_\alpha(\sigma)AW_\alpha(\sigma)$ is given by

$$\frac{1}{2\pi}\int e^{-i\xi x}\sigma(\xi)\hat{A}\ P_\alpha\big(\sigma(\xi)e^{i\xi y}\big)d\xi \tag{4.5}$$

where $P_\alpha\tau$ is defined by

$$(P_\alpha\tau)\hat{} = \begin{cases} \hat{\tau} & \text{on } [0,\alpha] \\ 0 & \text{elsewhere} \end{cases}$$

and $\hat{A}$ is the operator on $P_\alpha L_2(\mathbf{R})$ obtained from A by conjugating with the Fourier transform. The projection P_α is bounded on L_p and it follows that (4.5) will be continuous whenever $\hat{A}$ is a bounded operator from L_p to L_q ($p^{-1}+q^{-1}=1$). In particular the case $p=2$ is established since we just take $A = W_\alpha(1-\sigma)^{-1}$. For $p > 2$ we use the fact that $\sigma \in L_r$ for any $r \in [p,\infty]$ to deduce the boundedness of the operators

$$W_\alpha(\sigma) : L_s \to L_t \qquad \text{if } t^{-1} \in [s^{-1}, s^{-1}+p^{-1}], t > 1$$

and so inductively that of

$$W_\alpha(\sigma)^k : L_p \to L_2 \qquad \text{if } t^{-1} \in [s^{-1}, s^{-1}+kp^{-1}], t > 1. \tag{4.6}$$

In particular we have the boundedness of

$$W_\alpha(\sigma)^k : L_p \to L_2 \qquad \text{if } k+1 \geq p/2$$

and so also for such k that of

$$W_\alpha(1-\sigma)^{-1}W_\alpha(\sigma)^k : L_p \to L_2.$$

Again from (4.6) we have boundedness of

$$W_\alpha(\sigma)^j : L_2 \to L_q \qquad \text{if } j \geq \frac{3}{2}p - 1$$

and so for such j and k

$$W_\alpha(\sigma)^{j+k}W_\alpha(1-\sigma)^{-1} : L_p \to L_q$$

is bounded. With this as the operator $\hat{A}$ in (4.5) we deduce that

$$W_\alpha(\sigma)^{n_0}W_\alpha(1-\sigma)^{-1}$$

has continuous kernel for some sufficiently large n_0. On the other hand we also know from (4.6) that we have boundedness of

$$W_\alpha(\sigma)^{n-2} : L_p \to L_q \qquad \text{if } n \geq p$$

and so $W_\alpha(\sigma)^n$ has continuous kernel for such n. Consequently so also does the operator

$$W_\alpha(\sigma)^p W_\alpha(1-\sigma)^{-1} = \sum_{n=p}^{n_0-1} W_\alpha(\sigma) + W_\alpha(\sigma)^{n_0} W_\alpha(1-\sigma)^{-1}.$$

Finally, (4.4) follows from Lemma 4.1 in case $W_\alpha(\sigma)$ has continuous kernel, say if $\sigma \in L_p \cap L_1$. To prove it in general we remark that one can show a little more by the above argument, namely that the mapping $\sigma \to \Gamma_\alpha^{(p)}$ is continuous from L_p to $C([0,\alpha]^2)$ at any σ such that $W_\alpha(1-\sigma)$ is invertible. From Lemma 3.6 it follows that $D_\alpha{}^{(p)}(1-\sigma)$ is also L_p continuous. So since $L_p \cap L_1$ is dense in L_p, relation (4.4) holds general σ. □

REMARK: We mention for later use that (as the proof shows) the first assertion of the lemma also holds for $\alpha = \infty$ as long as we add $\sigma \in L_\infty$ as an assumption, that $\Gamma_\alpha^{(p)}$ converges pointwise as $\alpha \to \infty$ to $\Gamma_\infty^{(p)}$, and that the mapping $\sigma \to \Gamma_\alpha^{(p)}$ is continuous from L_p to $C([0,\infty]^2)$ on any subset of L_p for which the norms $||\sigma||_\infty$ and $||W(1-\sigma)^{-1}||$ are uniformly bounded.

We can now prove the first limit theorem. We define

$$\mathcal{I} = \{\sigma \in K : W(\sigma) \text{ and } W(\sigma^{-1}) \text{ are invertible}\}.$$

It is a fact which we shall not use that in this definition $W(\sigma^{-1})$ may be replaced by $W(\tilde{\sigma})$; another is that if $\sigma \in K$ then $\sigma \in \mathcal{I}$ if and only if $\det W(\sigma) W(\sigma^{-1}) \neq 0$.

LEMMA 4.3. *If $\sigma \in L_p$ and $1 - \sigma \in \mathcal{I}$ then $W_\alpha(1-\sigma)$ is invertible for sufficiently large α and*

$$\lim_{\alpha \to \infty} \frac{d}{d\alpha} \log D_\alpha^{(p)}(1-\sigma) = \log G^{(p)}(1-\sigma). \tag{4.7}$$

PROOF: For the invertibility assertion see [11, Th. 5.1] for the circle analogue, or [13, p. 295] for the scalar case, either of which is easily adapted to the present situation.

For (4.7) observe first that the operators $W_\alpha(\sigma)$ and $W_\alpha(\tilde{\sigma})$ are unitarily equivalent via the unitary $f(x) \to f(\alpha - x)$. It follows that $\Gamma_\alpha^{(p)}(\alpha,\alpha) = \tilde{\Gamma}_\alpha^{(p)}(0,0)$ where $\tilde{\Gamma}$ is associated with the symbol $\tilde{\sigma}$, and by the remark following the proof of Lemma 4.2

$$\lim_{\alpha \to \infty} \tilde{\Gamma}_\alpha^{(p)}(0,0) = \tilde{\Gamma}_\infty^{(p)}(0,0).$$

Consequently, by Lemma 4.2, the limit in (4.7) exists and equals $-\tilde{\Gamma}_\infty^{(p)}(0,0)$. And it follows from this that (4.7) itself is equivalent to the assertion

$$\tilde{\Gamma}^{(p)}_{\infty}(0,0) = -\log G^{(p)}(1-\sigma) \tag{4.8}$$

and also to the assertion

$$\lim_{\alpha\to\infty} \frac{1}{\alpha} \log D^{(p)}_{\alpha}(1-\sigma) = \log G^{(p)}(1-\sigma). \tag{4.9}$$

We shall first establish (4.9) under the assumption that $\sigma, \hat{\sigma} \in L_1$. For the case $p = 1$ we can quote either [6, Th. 4.1] or [9, Th. 4]. To prove (4.9) for general p (under the same assumption) observe that

$$\log D^{(p)}_{\alpha}(1-\sigma) = \log D_{\alpha}(1-\sigma) + \mathrm{tr} \sum_{n=1}^{p-1} \frac{1}{n} W_{\alpha}(\sigma)^n$$

and by (2.1) (which holds for powers for any $\sigma \in K$)

$$\mathrm{tr} W_{\alpha}(\sigma)^n = \mathrm{tr} W_{\alpha}(\sigma^n) + O(1) = \frac{\alpha}{2\pi} \int \sigma(\xi)^n d\xi + O(1).$$

Thus (4.9) holds for general p.

So (4.9), and therefore its equivalent statement (4.8), is established if $\sigma, \hat{\sigma} \in L_1$ and we want to remove this extra condition.

Assume first the less stringent condition $\sigma \in C_0$ is satisfied and let σ_n be as in Lemma 3.3(c). Then each side of (4.8) for σ_n converges as $n \to \infty$ to the corresponding expression for σ; the right obviously and the left by the remark after Lemma 4.2. So the identity holds if $\sigma \in C_0$, in particular for each σ_λ if σ is arbitrary. Let $\lambda \to 0+$. We claim that each side of (4.8) for σ_λ converges to the corresponding expression for σ. For the right side this is true by the definition of $G^{(p)}(1-\sigma)$. This will be shown to be true for the left, again by the remark following Lemma 4.2, once we show that

$$||W(1-\sigma_\lambda)^{-1}|| = O(1) \tag{4.10}$$

for sufficiently small λ.

The assumption $\sigma \in I$ guarantees that

$$W(1-\sigma)W((1-\sigma)^{-1}) = I - H(1-\sigma)H((1-\tilde{\sigma})^{-1}) \tag{4.11}$$

is invertible. Denote the operator on the right side of (4.11) by H_0, and by H_λ the corresponding operator with σ replaced by σ_λ. It follows from Lemma 3.2(a) and (b) (and the general fact $||H(\tau)||_2 = |||\tau|||$) that $||H_\lambda - H_0||_1 \to 0$ as $\lambda \to 0+$. Certainly then, H_λ is invertible for small λ and $||H_\lambda^{-1}|| = O(1)$. Then

$$W(1-\sigma_\lambda)^{-1} = W((1-\sigma_\lambda)^{-1})H_\lambda^{-1}$$

and (4.10) follows. □

LEMMA 4.4. *Under the assumptions of Lemma 4.3 we have*

$$\lim_{\beta\to\infty} \frac{D^{(p)}_{\alpha+\beta}(1-\sigma)}{D^{(p)}_{\beta}(1-\sigma)} = G^{(p)}(1-\sigma)^{\alpha}$$

for each α.

PROOF: Immediate from Lemma 4.3. □

5. PROOF OF THE ASYMPTOTIC FORMULA

For the statement of the first lemma we introduce the notation

$$t_{\alpha,n}(\sigma) = \operatorname{tr}\left[W_\alpha(\sigma)^n - W_\alpha(\sigma^n)\right].$$

That this makes sense is a consequence of the identity, analogous to (3.10),

$$W_\alpha(\sigma_1\sigma_2) - W_\alpha(\sigma_1)W(\sigma_2) = P_\alpha H(\sigma_1)H(\tilde{\sigma}_2)P_\alpha + Q_\alpha H(\tilde{\sigma}_1)H(\sigma_2)Q_\alpha \tag{5.1}$$

where here P_α is the projection from $L_2(0,\infty)$ to $L_2(0,\alpha)$ and Q_α is P_α followed by the flip $f(x) \to f(\alpha - x)$. Using this one can show inductively that $W_\alpha(\sigma^n) - W_\alpha(\sigma)^n \in S_1$ for any $\sigma \in K$.

5.1 *Assume that $\sigma \in K \cap L_p$ and that $W_\alpha(1-\sigma)$ and $W_\beta(1-\sigma)$ are invertible. Then*

$$\begin{aligned} \frac{D_{\alpha+\beta}{}^{(p)}(1-\sigma)}{D^{(p)}_\alpha(1-\sigma)D_\beta{}^{(p)}(1-\sigma)} &= \exp\left\{\sum_{n=2}^{p-1}\frac{1}{n}\left[t_{\alpha+\beta,n}(\sigma) - t_{\alpha,n}(\sigma) - t_{\beta,n}(\sigma)\right]\right\} \\ &\cdot \det\left(I - W_\beta(1-\sigma)^{-1}P_\beta H(\sigma)P_\alpha W_\alpha(1-\tilde{\sigma})^{-1}P_\alpha H(\tilde{\sigma})P_\beta\right). \end{aligned} \tag{5.2}$$

PROOF: We mention first that Q_α (restricted to $L_2(0,\alpha\,)$) is a unitary transforming $W_\alpha(\sigma)$ to $W_\alpha(\tilde{\sigma})$. Thus our hypothesis assures the invertibility of $W_\alpha(1-\tilde{\sigma})$ also.

Let us consider the continuity with respect to convergence in $K \cap L_p$ of the ingredients of the formula. The regularized determinants are continuous by Lemma 3.6 and therefore so is the ratio in question. The continuity of the $t_{\alpha,n}(\sigma)$ (with respect to convergence in K) can be proved inductively using identity (5.1). And the continuity of the determinant on the right is straightforward. The upshot of this is that it suffices to prove the identity for any subset of $K \cap L_p$ which is dense with respect to convergence. Such a set is $K \cap L_1$, by Lemma 3.3(b), and so we may assume $\sigma \in L_1$.

For such σ the identity for general p will follow from that for $p = 1$. Here is why. The left side of (5.2) for p divided by the same expression for $p = 1$ is equal (in this case where $W_\alpha(\sigma)$ is trace class) to

$$\operatorname{tr} \sum_{n=1}^{p} \frac{1}{n} \left[W_{\alpha+\beta}(\sigma)^n - W_\alpha(\sigma)^n - W_\beta(\sigma)^n \right]. \tag{5.3}$$

Since

$$\operatorname{tr} W_\alpha(\sigma^n) = \operatorname{tr} \frac{\alpha}{2\pi} \int \sigma(\xi)^n d\xi$$

we have

$$\operatorname{tr} \left[W_{\alpha+\beta}(\sigma^n) - W_\alpha(\sigma^n) - W_\beta(\sigma^n) \right] = 0.$$

Therefore (5.3) is equal to the sum in the exponential on the right side of (5.2), the term corresponding to $n = 1$ vanishing. So the formula for general p is indeed a consequence of the formula for $p = 1$.

To prove it we identify $L_2(0, \alpha + \beta)$ with the direct sum $L_2(0, \alpha) \oplus L_2(0, \beta)$. The operator corresponding to $W_{\alpha+\beta}(\sigma)$ under this identification has the matrix representation

$$\begin{bmatrix} A & B \\ C & D \end{bmatrix}$$

where

$$A = W_\alpha(1 - \sigma) \qquad B = -Q_\alpha H(\tilde{\sigma}) P_\beta$$
$$C = -P_\beta H(\sigma) Q_\alpha \qquad D = W_\beta(1 - \sigma).$$

This is easy to check. Write

$$\begin{bmatrix} A & B \\ C & D \end{bmatrix} = \begin{bmatrix} A & 0 \\ 0 & D \end{bmatrix} \begin{bmatrix} I & A^{-1}B \\ D^{-1}C & I \end{bmatrix}.$$

The left side has determinant $D_{\alpha+\beta}(1 - \sigma)$ while the first factor on the right has determinant $D_\alpha(1 - \sigma) D_\beta(1 - \sigma)$. The second factor on the right has determinant

$$\det(I - D^{-1} C A^{-1} B).$$

(Note that $D^{-1}C$ and $A^{-1}B$ are trace class.) We have

$$D^{-1} C A^{-1} B = W_\beta(1 - \sigma)^{-1} P_\beta H(\sigma) Q_\alpha W_\alpha(1 - \sigma)^{-1} Q_\alpha H(\tilde{\sigma}) P_\beta.$$

Since

$$Q_\alpha W_\alpha(1 - \sigma)^{-1} Q_\alpha = W_\alpha(1 - \tilde{\sigma})^{-1}$$

the result is established. □

LEMMA 5.2 *Assume* $\sigma \in L_p$ *and* $1-\sigma \in \mathcal{I}$. *Then*

$$\lim_{\alpha,\beta\to\infty} \frac{D_{\alpha+\beta}{}^{(p)}(1-\sigma)}{D_\alpha^{(p)}(1-\sigma)D_\beta{}^{(p)}(1-\sigma)}$$
$$= \exp\left\{\operatorname{tr}\sum_{n=2}^{p-1}\frac{1}{n}\left[W(\sigma^n)-W(\sigma)^n+W(\tilde\sigma^n)-W(\tilde\sigma)^n\right]\right\}$$
$$\cdot \det W(1-\sigma)^{-1}W((1-\sigma)^{-1})^{-1}.$$

PROOF: As $\alpha \to \infty$ the operators $W_\alpha(1-\sigma)^{-1}$ converge strongly to $W(1-\sigma)^{-1}$ (for the circle case see [11, Th. 5.1, Cor. 1]), the operators $H(\sigma), H(\tilde\sigma)$ are Hilbert-Schmidt, and consequently the determinant on the right side of (5.2) converges as $\alpha,\beta\to\infty$ to

(5.4) $$\det\left(I - W(1-\sigma)^{-1}H(\sigma)W(1-\tilde\sigma)^{-1}H(\tilde\sigma)\right).$$

Next, we use an identity similar to (3.10) and proved in an analogous way, namely

$$H(\tau_1)W(\tilde\tau_2) = H(\tau_1\tau_2) - W(\tau_1)H(\tau_2).$$

We apply it with $\tau_1 = 1-\tilde\sigma$, $\tau_2 = (1-\tilde\sigma)^{-1}$ and deduce that

$$W(1-\tilde\sigma)^{-1}H(\tilde\sigma) = H((1-\tilde\sigma)^{-1})W((1-\tilde\sigma)^{-1})^{-1}.$$

It follows that the operator in (5.4) is equal to

$$I - W(1-\sigma)^{-1}H(\sigma)H((1-\tilde\sigma)^{-1})W((1-\sigma)^{-1})^{-1}$$
$$= I + W(1-\sigma)^{-1}\left[I - W(1-\sigma)W((1-\sigma)^{-1})\right]W((1-\sigma)^{-1})^{-1}$$
$$= W(1-\sigma)^{-1}W((1-\sigma)^{-1})^{-1}.$$

For the first factor on the right side of (5.2) we invoke (2.1) for powers to obtain

(5.5) $$\lim_{\alpha\to\infty} t_{\alpha,n}(\sigma) = \operatorname{tr}\left[W(\sigma)^n - W(\sigma^n) + W(\tilde\sigma)^n - W(\tilde\sigma^n)\right].$$

It follows that

$$\lim_{\alpha,\beta\to\infty}\left[t_{\alpha+\beta,n}(\sigma) - t_{\alpha,n}(\sigma) - t_{\beta,n}(\sigma)\right]$$

equals the negative of the right side of (5.5) and this completes the proof. □

THEOREM. *If* $\sigma \in L_p$ *and* $1-\sigma$ *is an invertible element of* K *of index zero then the asymptotic formula (2.5) holds.*

PROOF: As in the proof of Theorem 6.1 of [11] we may assume that $1-\sigma \in \mathcal{I}$. (The argument is briefly as follows. The main theorem of [10] guarantees the existence of a $\tau \in K$ such that $\sigma + \varepsilon\tau \in 1 - \mathcal{I}$ for all sufficiently small nonzero complex numbers ε. If we have the result for $\sigma + \varepsilon\tau$ with ε on a circle $|\varepsilon| = \varepsilon_0$ we integrate over the circle, using the analyticity in ε of all the ingredients in the asymptotic formula, to obtain the result for $\varepsilon = 0$.) For such a σ let δ_α denote $D_\alpha^{(p)}(1-\sigma)$ divided by the right side of (2.5). The content of Lemma 5.2 is that

$$\lim_{\alpha,\beta\to\infty} \frac{\delta_{\alpha+\beta}}{\delta_\alpha \delta_\beta} = 1$$

while Lemma 4.4 implies that

$$\lim_{\beta\to\infty} \frac{\delta_{\alpha+\beta}}{\delta_\beta} = 1$$

for each fixed α. It follows easily that

$$\delta_\alpha = \frac{\delta_{\alpha+\beta}}{\delta_\beta} \Big/ \frac{\delta_{\alpha+\beta}}{\delta_\alpha \delta_\beta}$$

has limit 1 as $\alpha \to \infty$ and this is the assertion of the theorem. □

Traces more general than the ones appearing on the right side of (2.5) were considered in [13]. In our case formula (5.10) of this reference gives

$$\operatorname{tr}\left[W(\sigma^n) - W(\sigma)^n\right] = \sum_{j=1}^{n-1} \int_0^\infty \cdots \int_0^\infty u_j \hat{\sigma}(u_j - u_{j-1}) \ldots \hat{\sigma}(u_2 - u_1)\hat{\sigma}(u_1)$$
$$\cdot\, \hat{\sigma}(-v_1)\hat{\sigma}(v_1 - v_2) \ldots \hat{\sigma}(v_{n-j-1} - v_{n-j})\, du_1 \ldots du_j dv_1 \ldots dv_{n-j-1}$$

where in the integrand v_{n-j} is set equal to u_j.

6. ANALYTIC SYMBOLS

Here we present a direct proof of formula (1.6). First, two lemmas.

LEMMA 6.1. *Let X, Y be Hilbert spaces, $(S, d\mu)$ a measure space, and φ and ψ square integrable functions on S taking values in $\mathcal{L}(X,Y)$ and $\mathcal{L}(Y,X)$ respectively. Denote by $\mathcal{Y}$ the space of square-integrable functions on S taking values in Y. Define the operators*

$$U \in \mathcal{L}(\mathcal{Y}), \qquad V \in \mathcal{L}(X)$$

by

$$Uf = \varphi \int \psi f\, d\mu, \qquad Vv = \int \psi\varphi v\, d\mu.$$

Then $\det(I - U) = \det(I - V)$ if U and V are both trace class.

PROOF: Define $M_\varphi \in \mathcal{L}(X, \mathcal{Y})$ by $M_\varphi v = \varphi v$ and $I_\psi \in \mathcal{L}(\mathcal{Y}, X)$ by $I_\psi f = \int \psi f d\mu$. Then $U = M_\varphi I_\psi$ and $V = I_\psi M_\varphi$. □

LEMMA 6.2. *Let $P, P^\times$ be projection (idempotent) operators on the Hilbert space X such that*

$$X = \ker P \oplus \mathrm{im} P^\times, \qquad P^\times - P \in S_1.$$

Then

$$\det\left[(I-P)(I-P^\times) + P^\times\right] = \det\left[(I-P)(I-P^\times) + PP^\times\right].$$

PROOF: Our operators go from X to X. Let X as initial space be decomposed as

$$\mathrm{im} P^\times \oplus \ker P^\times$$

while as final space decomposed as

$$\mathrm{im} P \oplus \ker P.$$

With respect to these decompositions the operator $(I-P)(I-P^\times) + P^\times$ has matrix representation

$$\begin{bmatrix} P\big|_{\mathrm{im} P^\times} & 0 \\ (I-P)\big|_{\mathrm{im} P^\times} & (I-P)\big|_{\ker P^\times} \end{bmatrix}$$

while the operator $(I-P)(I-P^\times) + PP^\times$ has representation

$$\begin{bmatrix} P\big|_{\mathrm{im} P^\times} & 0 \\ 0 & (I-P)\big|_{\ker P^\times} \end{bmatrix}$$

The operator in both upper left-hand corners (call it A) is an isomorphism between $\mathrm{im} P^\times$ and $\mathrm{im} P$ by our first assumption. The operator in the lower left-hand corner of the first matrix (call it B), is trace class by our second assumption. The first matrix is obtained from the second by multiplying it on the left by

$$\begin{bmatrix} I & 0 \\ BA^{-1} & I \end{bmatrix}$$

(the I's denote the identity operators on $\mathrm{im} P$ and $\ker P$) and this matrix has determinant 1. □

PROOF OF (1.6): By (3.10)

$$W((1-\sigma)^{-1})W(1-\sigma) = I - H((1-\sigma)^{-1})H(1-\tilde{\sigma}).$$

The kernel of $H(1-\tilde{\sigma})$ is $-\hat{\sigma}(-x-y)$ and it follows from the representation

$$\sigma(\xi) = C(A-\xi I)^{-1}B \tag{6.1}$$

and the definition of P (as the Riesz projection for A corresponding to the part of its spectrum in the upper half-plane) that for $x > 0$

$$\hat{\sigma}(-x) = -iCe^{ixA}PB.$$

For $(1-\sigma)^{-1}$ we have the representation

$$(1-\sigma(\xi))^{-1} = 1 + C(A^{\times} - \xi I)^{-1}B$$

(by [1, Sec I.1] or an easy check) and we find that for $x > 0$

$$[(1-\sigma)^{-1} - 1]^{\wedge}(x) = iC(I-P^{\times})e^{-ixA^{\times}}B.$$

Hence the kernel of $H((1-\sigma)^{-1})H(1-\tilde{\sigma})$ is equal to

$$-\int_0^{\infty} C(I-P^{\times})e^{-i(x+z)A^{\times}}BCe^{i(z+y)A}PB\,dz.$$

Writing $BC = A - A^{\times}$ shows that this equals

$$i\int_0^{\infty} \frac{d}{dz}C(I-P^{\times})e^{-i(x+z)A^{\times}}e^{i(z+y)A}PB\,dz \tag{6.2}$$

$$= -iC(I-P^{\times})e^{-ixA^{\times}}e^{iyA}PB.$$

We apply Lemma 6.1 with $(S, d\mu)$ equal to $\mathbf{R}^+$ with Lebesque measure, with $Y = \mathbf{C}^m$, with $X = X$ and with

$$\varphi(x) = -iC(I-P^{\times})e^{-ixA^{\times}}, \qquad \psi(x) = e^{ixA}PB.$$

Then the operator U of the lemma is the operator with kernel (6.1), in other words $H((1-\sigma)^{-1})H(1-\tilde{\sigma})$. The operator V is given by the integral

$$-i\int_0^{\infty} e^{ixA}PBC(I-P^{\times})e^{-ixA^{\times}}dx.$$

Writing $BC = A - A^{\times}$ as before shows that this equals

$$-\int_0^{\infty} \frac{d}{dx}e^{ixA}P(I-P^{\times})e^{-ixA^{\times}}dx = P(I-P^{\times}).$$

Hence applying the lemma shows that the left side of (1.6) equals

$$\det(I - P(I-P^{\times})) = \det((I-P)(I-P^{\times}) + P^{\times})$$

and by Lemma 6.2 this equals $\det((I-P)(I-P^{\times}) + PP^{\times})$. For the satisfaction of the requirements of this lemma see [1, Sec. I.2] and [7, p.29].

REFERENCES

1. H. Bart, I. Gohberg, and M.A. Kaashoek, Minimal factorization of matrix and operator functions, Operator Theory: Advances and Applications, vol. 1, Birkhäuser Verlag, 1979.
2. A. Devinatz, On Wiener-Hopf operators, in Functional Analysis, Proc. Conf. Irvine, CA, Academic Press, 1966.
3. R.G. Douglas, Banach algebra techniques in operator theory, Academic Press, 1972.
4. R.G. Douglas, Banach algebra techniques in the theory of Toeplitz operators, CBMS Lecture Notes, Amer. Math. Soc. 15 (1973).
5. H. Dym, Trace formulas for blocks of Toeplitz-like operators, J. Functional Anal. 31 (1979), pp. 69-100.
6. H. Dym and S. Ta'assan, An Abstract version of a limit theorem of Szegő, J. Functional Anal. 43 (1981), pp. 294–312.
7. I. Gohberg, M.A. Kaashoek and F. van Schagen, Szegő-Kac-Achiezer formulas in terms of realizations of the symbol, J. Functional Anal. 74 (1987) pp. 24–51.
8. I.C. Gohberg, and M.G. Krein, Introduction to the theory of linear nonselfadjoint operators, vol. 18, Transl. Math. Monographs, Amer. Math. Soc., 1969.
9. L.M. Mikaelyan, Continual matrix analogs of Szegő's theorems on Toeplitz determinants J. Contemporary Soviet Analysis 17 (1982) pp. 1–26.
10. H. Widom Perturbing Fredholm operators to obtain invertible operators, J. Functional Anal. 20 (1975) pp. 26–31,
11. H. Widom, Asymptotic behavior of block Toeplitz matrices and determinants II, Adv. in Math. 21 (1976) pp. 1–29,
12. H. Widom, Szegő's limit theorem: The higher-dimensional matrix case, J. Functional Anal. 39 (1980) pp. 182–198.
13. H. Widom, A trace formula for Wiener-Hopf operators, J. Operator Th. 8 (1982) pp. 279–298.
14. H. Widom, Asymptotic expansions for pseudodifferential operators on bounded domains, Lecture notes in Math. No. 1152, Springer-Verlag, 1986.

H. Widom
Department of Mathematics
University of California
Santa Cruz, CA 95064

Table of contents of Volume I

Part B: Papers on Matrix Theory and its Applications

ERRATA

The Editors of the Gohberg Anniversary Collection regret that the paper of M.M. Djrbashian on "Differential Operators of Fractial Order and Boundary Value Problems in the Complex Domain" had to be published without a careful proof-reading.

A list of typographic errors follows:

p.153	line 14	For	$[0, \sigma)$	read	$[0, \sigma]$	
p.154	line 4	For	Z	read	z	(twice)
	eqn.(0.3)	For	$\mid g(z)^2$	read	$\mid g(z) \mid^2$	
	eqn.(0.4)	For	ψk	read	ψ_k	
p.155	line 11	For	$(i - \mu)$	read	$(1 - \mu)$	
p.156	eqn.(0.11)	For	rk	read	η_k	(twice)
	eqn.(0.13)	For	$k)^2$	read	$k) \mid^2$	
p.157	line 4	For	K	read	k	(twice)
	eqn.(0.18)	For	$\pm$	read	$\mp$	(twice)
	eqn.(0.19)	For	$\frac{2}{\sigma}$	read	$\sqrt{\frac{2}{\sigma}}$	(twice)
p.158	lines 11& 13	For	ε	read	Ξ	
p.160	eqn.(1.16)	For	$\delta_{k,r}$	read	$= \delta_{k,r}$	
p.161	line 1	For	Z	read	z	
		and insert "where $\tilde{\mu} = 3 + \nu - \mu$."				
	lines 4 & 9	For	Z	read	z	
	eqn. (1.26)	Insert parenthesis under square root.				
	line 6 up	In the second expression replace μ by $\tilde{\mu}$				(twice)
p.162	line 6 up	For	Z	read	z	
p.163	line 12 up	For	$Y_{(\omega)}$	read	$Y(\omega)$	
p.166	line 9	For	Z	read	z	
	eqn.(2.25)	For	Y	read	ν	
p.167	line 3	Insert parenthesis under square root.				
p.168	line 1 up	Delete period.				
p.169	eqn.(3.3)	Enclose the function of ζ in parentheses and then evaluate at $\zeta = z^2$.				
	line 14	For	$(-1 - \mu)$	read	$- (1 - \mu)$	
p.171	eqn.(3.17)	For	z	read	ζ	